# Vehicle-Bridge Interaction Dynamics

## With Applications to High-Speed Railways

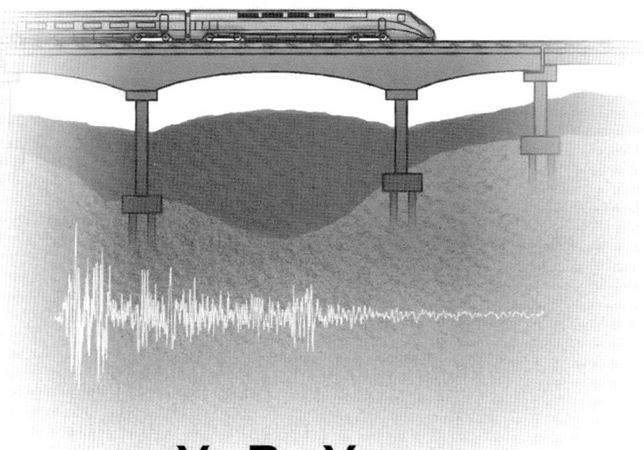


**Y. B. Yang**
National Taiwan University, Taiwan

**J. D. Yau**
Tamkang University, Taiwan

**Y. S. Wu**
Sinotech Engineering Consultants, Ltd., Taiwan



World Scientific

NEW JERSEY • LONDON • SINGAPORE • BEIJING • SHANGHAI • HONG KONG • TAIPEI • CHENNAI

*Published by*

World Scientific Publishing Co. Pte. Ltd.
5 Toh Tuck Link, Singapore 596224
*USA office:* 27 Warren Street, Suite 401-402, Hackensack, NJ 07601
*UK office:* 57 Shelton Street, Covent Garden, London WC2H 9HE

**British Library Cataloguing-in-Publication Data**
A catalogue record for this book is available from the British Library.

**VEHICLE–BRIDGE INTERACTION DYNAMICS**
**With Applications to High-Speed Railways**



ISBN-13 978-981-238 847-6
ISBN-10 981-238-847-8

Printed in Singapore

# Contents

# Preface

The commercial operation of the first high-speed (or bullet) train in 1964 with a speed of 210 km/hr in the Japanese railways connecting Tokyo and Osaka marked the beginning of a new era in railway engineering. Since then, high-speed trains with speeds over 200 km/hr or higher have emerged as a competitive tool for inter-city transportation in several countries including Japan, Germany, France, Italy, Spain, United Kingdom and Sweden. Such a trend continues to spread in different parts of the world. While Japan and many European countries have been working on expanding their high-speed railway networks or improving their existing railway lines, Asian countries, such as Korea, Taiwan and China, have reached the stage of planning, constructing, or field-testing their high-speed railway systems. Undoubtedly, high-speed train will become a key tool for inter-city passenger transportation, at least in the aforementioned countries.

Partly enhanced by the rapid expansion of high-speed railway systems, research on the moving load problems in general, and *vehicle–bridge interactions*[a] in particular, has been booming in the past two decades. Nevertheless, there is an apparent lack of a timely book that can adequately address most of the problems encountered in the design of high-speed railway bridges, which for the reasons stated

[a]In the literature, the term "bridge–vehicle interaction" was also used. It is realized that such a term was used by those who place more emphasis on the bridge than on the moving vehicles. In this text, we prefer to use the term "vehicle–bridge interaction", since we place equal weights on the dynamic behavior of the bridge and moving vehicles.

below, are different from those encountered in traditional railway or highway bridges. This book is intended to fill such a gap. It has been developed as a result of the research works conducted by the authors and their co-workers. In preparing this book, special attention was paid to the problems that may be encountered by engineers in practice, with clear physical meanings given for each of the phenomena involved. It is hoped that the book in the present form can serve as a most updated source of reference for engineers and researchers working in high-speed railways, and possibly to those working in the broad area of railway or bridge engineering.

One problem encountered in high-speed railways is the impact and vibration of bridges caused by the moving trains. This problem is substantially different from that encountered in highway bridges for the following reasons. First, the loads induced by a moving train on the bridge are *repetitive* in nature, as characterized by the sequentially moving wheel loads, implying that certain frequencies of excitation will be imposed on the bridge during the passage of a train. In contrast, the loads implied by a highway traffic are random in nature, when expressed in terms of the wheel loads and wheel distance. Second, high-speed trains can travel at a speed much higher than the vehicles moving on highways, making it possible for the excitation frequencies to coincide with the vibration frequencies of the bridge, resulting in the so-called *resonance* phenomenon. Whenever the condition of resonance is reached, the bridge response will be continuously amplified as there are more wheel loads passing the bridge. Such a phenomenon can hardly be observed in highway bridges. Third, the *mass ratio* of the vehicles to the bridge is generally larger for railways than for highways, due to the fact that a train consists of a number of cars in connection and the railway bridge deck is relatively narrow, it carries no more than two tracks in most cases. In contrast, a highway bridge deck may be so wide that it can afford four or more lanes of running vehicles in each of the two directions. For this reason, the *interaction* between the moving vehicles and bridge appears to be much stronger for railways than for highways. Finally, concerning the maneuverability of the train in motion, the *riding comfort* or *vehicle response* is an issue that should be taken into account in the design of high-speed railways. Moreover, the response of a moving vehicle is more sensitive to the *vehicle–bridge interaction* (VBI) compared with that of the bridge. However, the analysis of the dynamic behavior of a VBI system is not straightforward as there are two subsystems, i.e., the moving vehicles and the bridge, interacting with each other through the *contact forces* existing between the wheels and rails surface, which, in essence, is a nonlinear, coupled and time-dependent problem.

This book intends to give a broad and systematic coverage of the vibration problems encountered in high-speed railway bridges, with particular emphasis placed on the interaction between the moving vehicles and supporting bridge. In general, the book is divided into two parts, with Part I dedicated to the moving load problems and Part II to the interaction dynamics problems. These two parts can also be distinguished by the fact that the moving load problems (i.e., those treated in Part I) can generally be solved by analytical means, for which closed form solutions are possible, while the interaction dynamics problems (i.e., those treated in Part II) can only be solved by numerical means, say, using the vehicle–bridge interaction elements derived.

Starting with a general review of the related previous works in Chapter 1, an analytical formulation was presented for simply-supported beams subjected to a sequence of moving loads in Chapter 2, from which the phenomena of *resonance* and *cancellation* were identified, along with the optimal design criteria established for bridges. The closed-form solution presented for simple beams allows us to identify the key parameters involved. Conventionally, elastic bearings are installed at the supports of bridge girders for isolating the earthquake forces transmitted from the ground to the superstructure. However, such devices may adversely result in amplification of the response of the bridge during the passage of a train. The problem of elastically supported beams subjected to moving loads has received little attention in the literature, which was studied by an analytical approach in Chapter 3. The envelope impact formulas presented in Chapter 3 can be used as a useful aid for preliminary designs. Moreover, the mechanism for thc occurrence of resonance and cancellation was thoroughly investigated in Chapter 4, with which the measured results obtained from the field test for two adjacent bridges was interpreted with clear physical meanings.

The dynamic behavior of a horizontally-curved beam subjected to a series of moving masses was formulated and studied in Chapter 5. This problem was not well-treated before, due to the overlook of the centrifugal forces induced by masses moving over a circular path, which are functions of both the speed of the moving masses and radius of the curved beam. In Chapter 5, a complete theory was presented for the vertical and horizontal vibrations of a horizontally-curved beam under the excitation of the gravitational and centrifugal forces, respectively, that are induced by the moving masses. Particular emphasis was placed on the impact effect caused by the moving masses on the radial response of the curved beam.

One feature of the book is the derivation of a number of efficient VBI elements by condensing the vehicle's degrees of freedom to those of the bridge in contact, based on the concept of dynamic condensation in Chapter 6. These elements can be used to simulate problems with bridges and moving vehicles of various complexities. The VBI element presented in Chapter 6 was extended in Chapter 7 to include the pitching motion of the moving vehicle. Using the VBI elements derived, the dynamic properties of the vehicles and bridge, as well as rail irregularities, can be duly taken into account, while the dynamic response of the moving vehicle can be solved in addition to the bridge response.

Another way to analyze the VBI dynamics is to treat the moving vehicles and bridge as two separate systems, which interact with each other through the contact forces. By solving for the contact forces from the vehicles equations, one can treat them as external forces acting on the bridge, which can then be solved using conventional finite element procedures. Such a concept was utilized in developing the VBI element in Chapter 8, which was then extended in Chapter 9 to include the effect of rails with profile irregularities that form part of a railway track in the two-dimensional sense. Because of its versatility, the VBI element derived, based on the concept of contact forces, can be used in the simulation of various three-dimensional vehicle-rails-bridge systems considering, for instance, the crossing of two trains on a bridge, the risk of derailment of a moving train (Chapter 10), and the stability of trains moving over bridges simultaneously shaken by earthquake (Chapter 11).

The authors wish to express their sincerest gratitude to their great teacher in civil engineering and education, Dr. Chao-Chung Yu, the former dean of the College of Engineering, National Taiwan University (NTU) (1972–1979) and the former President of the NTU (1981–1984), for his strong influence and continuous advice through their careers of development, both as students and teachers. His experience as a teacher, researcher, educationist, and in some sense as an engineer, has always been a source of inspiration to all the young fellows under his instruction or working with him.

A large portion of the research results presented in this book has been sponsored through a series of research projects granted by the National Science Council of the Republic of China on subjects related to vehicle–bridge interactions, as well as on bridge dynamics. The senior author has been the principal investigator of all these projects. Without such a continuous support, it would be difficult to maintain such a large research group at the NTU working on different aspects of the VBI problem, ranging from the vibration of substructure and superstructure of railway bridges to wave propagations in soils and nearby buildings; the latter forms an independent subject that requires further research, which was not covered in this book. Besides, we are also grateful to the China Engineering Consultants, Inc., for their continuous support to our research group, especially through the 1st Structural Department previously led by Senior Vice President Mr. Dyi-Wei Chang. Some research results presented in this book have been made possible through such a support.

This book was prepared as part of the results of the research group led by the senior author at the National Taiwan University. Many of the graduate students have contributed directly or indirectly to the success of this work, including Chia-Hung Chang, Chon-Min Wu, Chin-Lu Lin, Bing-Houng Lin, Lin-Ching Hsu, Shyh-Rong Kuo, Hsiao-Hui Hung, Chern-Hwa Chen, Jiann-Tsair Chang, Cheng-Wei Lin, and Kuo-Wei Chang. The assistance from the administrative staff of the College of Engineering, NTU, especially Ms. Hong-Hua Chang, during the preparation of this book is greatly appreciated. Finally, a book can never be completed without the continuous support, and expectation, from the families of the authors, colleagues, friends, and the society in which they live in.

*Y. B. Yang*
*J. D. Yau*
*Y. S. Wu*

Taipei, Taiwan, Republic of China

# Acknowledgments

Parts of the materials presented in this book have been revised from the papers published by the authors and their co-workers in a number of technical journals. Efforts have been undertaken to update, digest and rewrite the materials acquired from each source, such that a unified and progressive style of presentation can be achieved throughout the book. In particular, the authors like to thank the copyright holders for permission to use the materials contained in the following papers:

Wu, Y. S. and Yang, Y. B. (2003). "Steady-state response and riding comfort of trains moving over a series of simply supported bridges," *Eng. Struct.*, **25**(2), 251–265. Reproduced with permission from Elsevier.

Wu, Y. S., Yang, Y. B., and Yau, J. D. (2001). "Three-dimensional analysis of train-rail-bridge interaction problems," *Vehicle System Dyn.*, **36**(1), 1–35. ©Swets & Zeitlinger.

Yang, Y. B., Chang, C. H., and Yau, J. D. (1999). "An element for analysing vehicle–bridge systems considering vehicle's pitching effect," *Int. J. Numer. Meth. Eng.*, **46**, 1031–1047. ©John Wiley & Sons Limited, reproduced with permission.

Yang, Y. B., Lin, C. L., Yau, J. D., and Chang, D. W. (2004). "Mechanism of resonance and cancellation for train-induced vibrations on bridges with elastic bearings," *J. Sound & Vibr.*, **269**(1–2), 345–360. Reproduced with permission from Elsevier.

Yang, Y. B. and Yau, J. D. (1997). "Vehicle–bridge interaction element for dynamic analysis," *J. Struct. Eng.*, ASCE, **123**(11), 1512–1518 (Errata: **124**(4), 479). Reproduced by permission of ASCE.

Yang, Y. B., Yau, J. D., and Hsu, L. C. (1997b). "Vibration of simple beams due to trains moving at high speeds," *Eng. Struct.*, **19**(11), 936–944. Reproduced with permission from Elsevier.

Yang, Y. B., Wu, C. M., and Yau, J. D. (2001). "Dynamic response of a horizontally curved beam subjected to vertical and horizontal moving loads," *J. Sound & Vibr.*, **242**(3), 519–537. Reproduced with permission from Elsevier.

Yang, Y. B. and Wu, Y. S. (2001). "A versatile element for analysing vehicle–bridge interaction response," *Eng. Struct.*, **23**, 452–469. Reproduced with permission from Elsevier.

Yang, Y. B. and Wu, Y. S. (2002). "Dynamic stability of trains moving over bridges shaken by earthquakes," *J. Sound & Vibr.*, **258**(1), 65–94. Reproduced with permission from Elsevier.

Yau, J. D., Wu, Y. S., and Yang, Y. B. (2001). "Impact response of bridges with elastic bearings to moving loads," *J. Sound & Vibr.*, **248**(1), 9–30. Reproduced with permission from Elsevier.

Yau, J. D., Yang, Y. B., and Kuo, S. R. (1999). "Impact response of high speed rail bridges and riding comfort of rail cars," *Eng. Struct.*, **21**(9), 836–844. Reproduced with permission from Elsevier.

# List of Symbols

The following is a list of symbols used throughout this book. All the symbols are defined at the place where they first appear in the text. Matrices, column and row vectors are enclosed by [ ], { }, and $\langle\ \rangle$, respectively. A quantity occurring at time $t$ and $t + \Delta t$ are denoted with subscript $t$ and $t + \Delta t$, respectively. A dot placed over a quantity is used to denote the derivative of the quantity with respect to time $t$. And a prime attached to a quantity is used to denote the derivative of the quantity with respect to coordinate $x$. Only quantities that are not confined to local use are listed below.

| | |
|---|---|
| $A$ | cross-sectional area of beam |
| $a$ | acceleration of vehicle |
| $a_0 \sim a_7$ | coefficients as defined in Eq. (B.4) |
| $B(f_i)$ | weighting factor, Eq. (9.48) |
| BYQ | bogie-side lateral to vertical force ratio |
| $b_0 \sim b_7$ | coefficients as defined in Eq. (8.8) |
| $[C]$ | damping matrix of structure |
| $[C_b]$ | damping matrix of bridge free of any vehicle actions |
| $C_n$ | $n$th modal damping coefficient |
| $[C_0]$ | damping matrix of railway bridge free of any vehicle actions |
| $[c_b]$ | damping matrix of bridge element |
| $[c_{bi}]$ | damping matrix of $e_i$th element of bridge |
| $[c_c]$ | contact matrix as defined in Eq. (8.14) |

| | |
|---|---|
| $[c^*_{cij}]$ | damping matrix caused by linking action of car body |
| $[c_d]$ | rail damping matrix due to interaction with the bridge |
| $c_e$ | external damping coefficient |
| $c_i$ | internal damping coefficient |
| $[c_{ii}]$, $[c_{jj}]$ | damping matrices for element $i$ and $j$, Eqs. (7.25) and (7.26) |
| $[c_{ij}]$, $[c_{ji}]$ | damping matrices related to pitching actions for element $i$ and $j$, Eqs. (7.25) and (7.26) |
| $[c_r]$ | damping matrix of rail element |
| $[c_{rl}]$ | damping matrix of LSR element |
| $[c_{rr}]$ | damping matrix of RSR element |
| $c_t$ | material damping coefficient for translational motion of track |
| $c^*_t$ | material damping coefficient for torsional motion of track |
| $[c_{uu}]$ | a partitioned damping matrix of vehicle |
| $[c_{uw}]$ | a partitioned damping matrix of vehicle |
| $c_v$ | damping coefficient of suspension unit |
| $[c_v]$ | damping matrix of vehicle |
| $[c_{wu}]$ | a partitioned damping matrix of vehicle |
| $[c_{ww}]$ | a partitioned damping matrix of vehicle |
| $D$ | determinant of matrix |
| $\{D\}$ | structural displacement vector |
| $d$ | length of train car |
| $\{d_b\}$ | displacement vector of bridge |
| $\{d_{bi}\}$ | nodal displacement vector of $e_i$th element of bridge |
| $\{d^n_b\}$ | natural deformations of bridge |
| $\{d^r_b\}$ | rigid displacements of bridge |
| $\{d_c\}$ | displacement vector of contact points of bridge or rail |
| $\{d_e\}$ | displacement vector of car body |
| $\{d_f\}$ | displacement vector of front bogie |

| | |
|---|---|
| $\{d_r\}$ | displacement vector of rail element |
| $\{d_r\}$ | displacement vector of rear bogie, Chapter 10 |
| $\{d_{rl}\}$ | displacement vector of LSR element |
| $\{d_{rr}\}$ | displacement vector of RSR element |
| $\{d_u\}$ | displacement vector of upper part of car body |
| $\{d_v\}$ | displacement vector of vehicle, $\{d_v\} = \langle\{d_u\}\ \{d_w\}\rangle^T$ |
| $\{d_w\}$ | displacement vector of wheel part of car body |
| $\{d_{wi}\}$ | displacement vector for $i$th wheel |
| $d_{wb}$ | wheelbase of each wheelset |
| $E$ | modulus of elasticity |
| $F(t)$ | load function |
| $\{F\}$ | external nodal forces of structure |
| $\{F_b\}$ | external nodal forces of bridge |
| $\{\bar{F}_b\}$ | effective resistant force vector of structure |
| $F_k(v,t)$ | generalized forcing function |
| $\{f_A\}$ | nodal loads of rail element of Track A |
| $\{f_A^t\}$ | total equivalent nodal forces of element under earthquake |
| $f_b$ | frequency of vibration of bridge unit |
| $\{f_b\}$ | external nodal forces of bridge element |
| $\{f_{bci}\}$ | vector of consistent nodal forces for $i$th contact force |
| $\{f_{bi}\}$ | vector of external forces for $e_i$th element of bridge |
| $f_c$ | contact force |
| $\{f_c\}$ | vector of contact forces |
| $\{f_c^*\}$ | general vector of contact forces |
| $f_{c1} \sim f_{c4}$ | contact forces |
| $\{f_e\}$ | external force components excluding contact forces |
| $f_h$ | horizontal moving load |
| $\{f_r\}$ | nodal loads of rail element |
| $\{f_{rci}\}$ | equivalent nodal forces caused by $i$th vertical contact force |
| $\{f_{rl}\}$ | nodal loads of LSR element |

| | |
|---|---|
| $\{f_{rr}\}$ | nodal loads of RSR element |
| $f_{rot}$ | rotational moment experienced by car body |
| $\{f_s\}$ | resistant forces of sprung mass unit |
| $\{f_{ue}\}$ | external forces acting on upper part of vehicle |
| $f_v$ | gravitational load of moving vehicle, $f_v = -m_v g$ |
| $\{f_v\}$ | force vector of vehicle |
| $f_{ver}$ | vertical force experienced by car body |
| $\{f_{we}\}$ | external forces acting on wheel part of vehicle |
| $G$ | shear modulus |
| $g$ | acceleration of gravity |
| $H$ | unit step function |
| $H_i$ | $i$th horizontal contact force |
| $h$ | vertical distance between deck and torsional center of cross section |
| $h_{ci}$ | lateral displacement of $i$th contact point |
| $I$ | impact factor |
| $I$ | moment of inertia (used together with $E$) |
| $I_M$ | impact factor for bending moment |
| $I_p$ | polar moment of inertia of beam |
| $I_u$ | impact factor for deflection |
| $I_V$ | impact factor for end shear force |
| $I_v$ | rotatory inertia of car body |
| $I_y$, $I_z$ | moments of inertia of beam about $y$ and $z$ axes |
| $J$ | torsional constant |
| $K$ | stiffness of elastic bearings |
| $[K]$ | stiffness matrix of structure |
| $[K_b]$ | stiffness matrix of bridge structure |
| $[\bar{K}_b]$ | effective stiffness matrix of structure |
| $[K_{eff}]$ | effective stiffness matrix |
| $[K_0]$ | stiffness matrix of railway bridge free of any vehicle actions |
| $k_B$ | stiffness of ballast |
| $[k_b]$ | stiffness matrix of bridge element |
| $[\hat{k}_b]$ | stiffness matrix of condensed system |
| $[k_{bi}]$ | stiffness matrix of $e_i$th element of bridge |

| | |
|---|---|
| $[k_c]$ | contact matrix as defined in Eq. (8.14) |
| $[k^*_{cij}]$ | stiffness matrix caused by linking action of car body |
| $[k_{ii}]$, $[k_{jj}]$ | stiffness matrices for element $i$ and $j$, Eqs. (7.25) and (7.26) |
| $[k_{ij}]$, $[k_{ji}]$ | stiffness matrices related to pitching actions for element $i$ and $j$, Eqs. (7.25) and (7.26) |
| $[k_r]$ | stiffness matrix of rail element |
| $[k_{rl}]$ | stiffness matrix of LSR element |
| $[k_{rr}]$ | stiffness matrix of RSR element |
| $[k_s]$ | rail stiffness matrix due to interaction with bridge |
| $[k_{uu}]$ | a partitioned stiffness matrix of vehicle |
| $[k_{uw}]$ | a partitioned stiffness matrix of vehicle |
| $k_v$ | stiffness of suspension unit |
| $[k_v]$ | stiffness matrix of vehicle |
| $[k_{wu}]$ | a partitioned stiffness matrix of vehicle |
| $[k_{ww}]$ | a partitioned stiffness matrix of vehicle |
| $L$ | span length or characteristic length of bridge |
| $L_c$ | distance between two wheel assemblies of train car |
| $L_d$ | length equal to car length minus the axle distance |
| $L_r$ | length of irregularity |
| $l$ | length of beam element |
| $[l]$ | transformation matrix, Chapters 8 and 9 |
| $l_a$ | half of axle length of wheelset |
| $M$ | car mass lumped at each load |
| $M(x,t)$ | bending moment |
| $[M]$ | mass matrix of structure |
| $[M_b]$ | mass matrix of bridge structure |
| $M_c$ | mass of car body |
| $M_n$ | $n$th modal mass |
| $M_t$ | mass of bogie |
| $M_v$ | half of the mass of car body |
| $M_w$ | mass of vehicle |
| $[M_0]$ | mass matrix of railway bridge free of any vehicle actions |

| | |
|---|---|
| $m$ | mass per unit length |
| $[m_b]$ | mass matrix of bridge element |
| $[m_{bi}]$ | mass matrix of $e_i$th element of bridge |
| $[m_c]$ | contact matrix as defined in Eq. (8.14) |
| $[m^*_{cij}]$ | mass matrix caused by linking action of car body |
| $[m_{ii}]$, $[m_{jj}]$ | mass matrices for element $i$ and $j$, Eqs. (7.25) and (7.26) |
| $[m_r]$ | mass matrix of rail element |
| $[m_{rl}]$ | mass matrix of LSR element |
| $[m_{rr}]$ | mass matrix of RSR element |
| $[m_{uu}]$ | a partitioned mass matrix of vehicle |
| $[m_{uw}]$ | a partitioned mass matrix of vehicle |
| $m_v$ | mass of moving vehicle |
| $[m_v]$ | mass matrix of vehicle |
| $m_w$ | mass of wheel assembly |
| $[m_{wu}]$ | a partitioned mass matrix of vehicle |
| $[m_{ww}]$ | a partitioned mass matrix of vehicle |
| $N$ | total number of moving loads |
| $\{N_c\}$ | interpolation vector for beam displacement evaluated at contact point $x_c$, $\{N_c\} = \{N(x_c)\}$ |
| $\{N^h_{ci}\}$ | interpolation vector of bridge element evaluated for $i$th horizontal contact force |
| $\{N^v_{ci}\}$ | interpolation vector of bridge element evaluated for $i$th vertical contact force |
| $N_{min}$ | minimum number of bridge units used in analysis |
| $\langle N_u \rangle$ | interpolation vector for axial displacement |
| $\langle N_v \rangle$ | interpolation vector for vertical displacement |
| $n_A$ | number of train cars moving on Track A |
| $n_B$ | number of train cars moving on Track B |
| $n_b$ | number of CFR elements considered within minimal bridge segment |
| $n_v$ | number of vehicles comprising the train |
| $P$ | unified load function as defined in Eq. (5.45) |
| $\{P\}$ | applied loads of structure |
| $\{P_b\}$ | external loads of bridge |

| | |
|---|---|
| $\{P_c^*\}$ | equivalent contact forces of structure |
| PD | axle load decrement ratio |
| $\{P_{eff}\}$ | effective load vector |
| $P_n(x,t)$ | function as defined in Eq. (2.50) |
| PQ | axle decrement ratio |
| $P_1(v,t)$, $P_2(v,t)$ | response functions as defined in Eq. (3.15) |
| $\bar{P}_1(v,t)$ | response function as defined in Eq. (2.54) |
| $p$ | load magnitude, $p = -(M_v + m_w)g$, Chapter 6 |
| $\{p_b\}$ | nodal loads of bridge element |
| $\{p_c\}$ | load vector as defined in Eq. (8.15) |
| $(p_c, p_k)$ | unit axial interaction forces between rail and bridge elements |
| $\{p_{ci}^*\}$ | equivalent loads as defined in Eq. (8.24) |
| $\{p_v\}$ | load vector of sprung mass model, $\{p_v\}^T = \langle p, 0\rangle$ |
| $\{p_w\}$ | loads induced by wheels |
| $p_{x1p}$, $p_{z1p}$ | particular solutions for in-plane vibrations of curved beam |
| $p_{x1h}$, $p_{z1h}$ | homogeneous solutions for in-plane vibrations of curved beam |
| $p_{y1p}$, $p_{\theta 1p}$ | particular solutions for out-of-plane vibrations of curved beam |
| $p_{y1h}$, $p_{\theta 1h}$ | homogeneous solutions for out-of-plane vibrations of curved beam |
| $\{Q_c^*\}_t$ | equivalent contact forces of structure |
| $Q_1(v,t)$, $Q_2(v,t)$ | response functions as defined in Eq. (3.21) |
| $\{q_c\}$ | load vector as defined in Eq. (8.15) |
| $(q_c, q_k)$ | unit vertical interaction forces between rail and bridge elements |
| $\{q_{ci}^*\}$ | equivalent loads as defined in Eq. (8.24) |
| $q_n(t)$ | $n$th generalized coordinate |
| $\{q_s\}$ | internal resistant forces of sprung mass unit |
| $\{q_u\}$ | equivalent nodal loads for upper part of car body |
| $\{q_{uc}\}$ | equivalent contact forces for upper part of car body |

| | |
|---|---|
| $q_{x1}$ | first generalized coordinate for axial displacement $u_x$ |
| $q_{yi}$ | $i$th generalized coordinate for displacement $u_y$ |
| $q_{zi}$ | $i$th generalized coordinate for radial displacement $u_z$ |
| $q_{\theta i}$ | $i$th generalized coordinate for angle of twist $\theta_x$ |
| $R$ | radius of curvature of curved beam |
| $R_d$ | maximum dynamic response |
| $R_s$ | maximum static response |
| $R_{sM}$ | maximum bending moment under static loads |
| $R_{su}$ | maximum static deflection |
| $R_{sV}$ | maximum static shear force |
| $r(x)$ | profile of rail irregularity |
| $\{r\}$ | vector of irregularity profile at contact points |
| $r_c$ | elevation of rail irregularity at contact point $x_c$ |
| $(r_c, r_k)$ | unit vertical interaction forces between rail and bridge elements |
| $r_h(x)$ | deviation in lateral alignment of two rails |
| $r_i$ | track profile evaluated at $i$th contact point |
| $r_{v1}(x)$, $r_{v2}(x)$ | vertical deviations of two rails |
| $r_0$ | amplitude of irregularities |
| $r_0$ | nominal radius of wheel, Chapter 10 |
| $S$ | speed parameter |
| $S_c$ | speed parameter for cancellation |
| $S_{h1}$ | speed parameter for the horizontal vibration of curved beam |
| $S_n$ | $n$th speed parameter |
| $S_r$ | speed parameter for resonance |
| $S_{v1}$ | speed parameter for vertical vibration of curved beam |
| SYQ | single wheel lateral to vertical force ratio |
| $S(\Omega)$ | power spectral density (PSD) function for track irregularity |
| $t$ | time |

| | |
|---|---|
| $t_c$ | time lag between two sets of moving loads $(= L_c/v)$ |
| $t_{end}$ | ending time for analysis |
| $t_j$ | arriving time of $j$th load on beam |
| $t_N$ | arriving time of $N$th load on beam |
| $\{U_b\}$ | total displacements of bridge |
| $U_{N,1}$ | forced vibration caused by $N$th moving load, Eq. (5.51) |
| $U_{N,2}$ | residual vibration caused by $N-1$ moving loads, Eq. (5.52) |
| $U_j(t,v,L)$ | load configuration as defined in Eq. (2.34) |
| $U_j(x,t)$ | residual vibration caused by $j$th moving load, Eq. (5.50) |
| $U(x,t)$ | unified displacement function as defined in Eq. (5.45) |
| $\{U\}$ | displacements of structure |
| $\{U_b\}$ | displacements of bridge structure |
| $u_b$ | axial displacement of bridge element, Chapter 9 |
| $\{u_b\}$ | nodal displacements of beam element |
| $\{u_g\}$ | support displacements of bridge |
| $u(x,t)$ | deflection of beam at section $x$ and time $t$ |
| $u_i$, $u_j$ | vertical displacements of elements $i$ and $j$ |
| $u_r$ | axial displacement of rail element |
| $u_x$, $u_y$, $u_z$ | cross-sectional displacements of curved beam along three axes |
| $V(x,t)$ | shear force |
| $V_i$ | vertical force acting at $i$th contact point of bridge |
| $v$ | vehicle speed |
| $v_b$ | vertical displacement of bridge element, Chapter 9 |
| $v_{ci}$ | vertical displacement of $i$th contact point of bridge |
| $v_r$ | vehicle speed at resonance |
| $v_r$ | vertical displacement of rail element, Chapter 9 |
| $v_u$ | displacement of car body mass |

| | |
|---|---|
| $v_w$ | displacement of wheel mass |
| $v_{wi}$ | displacement of $i$th wheel |
| $v_0$ | initial velocity of vehicle |
| $W$ | half weight of car body, $W = 0.5\ M_v\mathrm{g}$ |
| $W$ | static weight of each wheelset, Chapter 10 |
| $W_z$ | Sperling's ride index |
| $W_{z_i}$ | comfort index |
| $x$ | beam axis |
| $x_{bs0}$ | position for first wheel to start acceleration or braking |
| $x_{bsf}$ | position for first wheel to stop acceleration or braking |
| $x_c$ | position of contact point |
| $x_{ci}$, $x_{cj}$ | position of contact point for elements $i$ and $j$ |
| $x_{end}$ | ending position of first wheel |
| $x_0$ | reference distance in Eq. (6.38) |
| $Y_{lim}$ | limit on lateral track force |
| $y$ | a cross-sectional axis |
| $y_1$, $y_2$ | coordinates of the two masses of sprung mass model, Chapter 6 |
| $\{y\}$ | nodal vector of sprung mass model, $\{y\}^T = \langle y_1, y_2\rangle$, Chapter 6 |
| $\{y\}$ | displacements of car body as rigid beam, $\{y\}^T = \langle y_v\ \theta_v\rangle$, Chapter 7 |
| YQ | wheelset lateral to vertical force ratio |
| $y_v$ | vertical displacement of car body as rigid beam |
| $z$ | a cross-sectional axis |
| $\beta$ | coefficient related to variation of acceleration |
| $[\Gamma]$ | constraint matrix |
| $\gamma$ | coefficient related to numerical damping |
| $\gamma_0$ | wavelength of corrugation |
| $\{\Delta d_u\}$ | upper-part vehicle displacement increments |
| $\Delta_{st}$ | maximum static deflection of beam with hinge supports |
| $\Delta t$ | time increment |

| | |
|---|---|
| $\Delta y_1$ | displacement increment of wheel mass |
| $\{\Delta y\}$ | displacement increments of sprung mass unit |
| $\{\Delta U_b\}$ | displacement increments of bridge |
| $\{\Delta u_b\}$ | displacement increments of bridge element |
| $\delta$ | delta function |
| $\theta_b$ | rotation about $x$ axis of bridge element |
| $\theta_v$ | rotation of car body as rigid beam |
| $\theta_x$, $\theta_y$, $\theta_z$ | rotations of a cross section of curved beam about three axes |
| $\kappa$ | stiffness ratio of beam to elastic springs |
| $[\lambda]$ | transformation matrix, Eq. (10.14) |
| $\lambda_r$ | wavelength of track irregularity |
| $\lambda_u$ | horizontal characteristic number of beam-Winkler foundation |
| $\lambda_v$ | vertical characteristic number of beam-Winkler foundation system |
| $\mu_i$ | coefficient of friction for $i$th wheel |
| $\xi$ | damping coefficient |
| $\xi_n$ | damping coefficient of $n$th mode |
| $\rho$ | density of beam |
| $\varphi$ | subtended angle of curved beam, $\varphi = L/R$ |
| $\varphi_u$ | rotation of car body as rigid bar, Chapter 8 |
| $\phi_n$ | $n$th vibration mode |
| $\{\phi\}_n$ | $n$th vibration mode of structure |
| $\Psi(t)$ | unified amplitude function as defined in Eq. (5.45) |
| $[\Psi_{uu}]$ | equivalent matrix for upper part of car body, Eq. (8.10) |
| $[\Psi_{wu}]$ | matrix as defined in Eq. (8.16a) |
| $\Omega$ | exciting frequency implied by the moving load |
| $\Omega$ | spatial frequency of track irregularity, Chapters 9 and 10 |
| $\omega$ | frequency of vibration |
| $\omega_d$ | damped frequency |
| $\omega_{dn}$ | damped frequency of vibration of $n$th mode |

| | |
|---|---|
| $\omega_{h1}$ | fundamental frequency for horizontal plane of curved beam |
| $\omega_n$ | frequency of vibration of $n$th mode |
| $\omega_v$ | vertical vibration frequency of car body |
| $\omega_{v1}$ | fundament frequency of vertical vibration of curved beam |
| $\omega_0$ | frequency of vibration for beam with hinge supports |
| $\omega_\theta$ | rotational vibration frequency of car body |

# Chapter 1

# Introduction

The interaction between a bridge and the vehicles moving over the bridge is a coupled, nonlinear dynamic problem. Conventionally, most research has been focused on the dynamic or *impact* response of the bridge, but not of the moving vehicles. For the cases where only the bridge response is desired, the moving vehicles have frequently been approximated to the extreme as a number of moving loads. However, whenever the responses of both the bridge and moving vehicles are desired, as encountered in the design of high-speed railways, models that can adequately account for the dynamic properties of the moving vehicles should be adopted. In this chapter, the key factors involved in the dynamic interaction between the bridge and moving vehicles will be discussed, along with procedures for solving the vehicle–bridge interaction problems. The materials presented in this chapter have been revised from the review paper by Yang and Yau (1998) with supplement of the relevant literature published recently.

## 1.1. Major Considerations

The dynamic interaction between a bridge and the moving vehicles represents a special discipline within the broad area of structural dynamics. The vehicles considered may be those constituting the traffic flow of a highway bridge, in general, or those that form a connected line of railroad cars, in particular. From the theoretical point of view, the two subsystems, i.e., the bridge and moving vehicles, can be simulated as two elastic structures, of which each is characterized by some frequencies of vibration. The two subsystems interact with each other through the *contact forces*, i.e., the forces induced at the contact points between the wheels and rails surface (of the railway bridge) or pavement surface (of the highway bridge). A problem such as this is nonlinear and time-dependent due to the fact that the contact forces may move from time to time, while their magnitudes do not remain constant, as a result of the relative movement of the two subsystems. The way by which the two subsystems interact with each other is determined primarily by the inherent frequencies of the two subsystems and the driving frequency of the moving vehicles. In this book, we prefer to use the term *vehicle–bridge interaction* (VBI) to refer to the interaction between the two subsystems. The vehicle considered in this book is a general term, which can be a car, a truck, a tractor-trailer, or a railroad car that forms part of the train. The term bridge is also a general one. It can be a simply-supported beam, a multi-span continuous beam, or a bridge of any types used in highways and railways, with or with no account of the effects of surface pavement (for highways) or rails and ballast (for railways). The consideration of the VBI is necessary if the vehicle response, in addition to the bridge response, is desired. In the design of high-speed railway bridges, for instance, the maximum vertical and/or lateral accelerations of the moving vehicles are used as indicators for evaluating the riding comfort of passengers carried by the train. Besides, the vertical and lateral contact forces of the wheels of railroad cars with the rails represent a kind of information central to assessment of the risk of derailment for moving trains, especially in the presence of earthquake shaking.

In many cases, especially when the vehicle to bridge mass ratio is small, the elastic and inertial effects of the vehicles may be ignored and much simpler models can be adopted for the vehicles. One typical example is the simulation of a moving vehicle over a bridge as a single moving load, which has been conventionally referred to as the *moving load* model (Fig. 1.1). Since the interaction between the two subsystems has been ignored, the moving load model is good only for computing the response of the larger subsystem, i.e., the bridge,

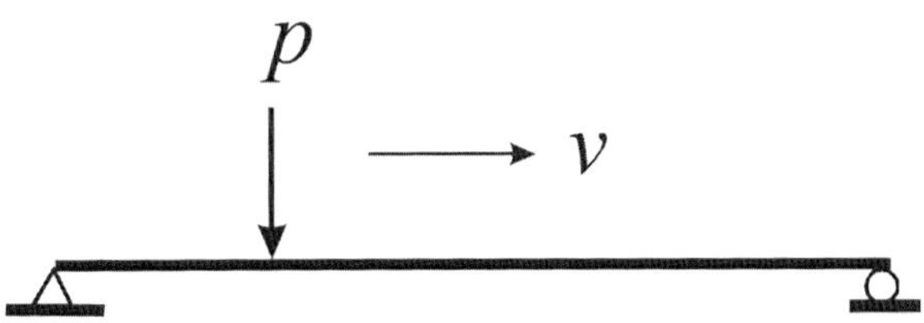

Fig. 1.1. Moving load model.

but not of the smaller subsystem, i.e., the vehicle. In this book, the moving load problem can be regarded as a special case of the more general formulation that considers the various dynamic properties of the moving vehicles.

The objective of this book is to establish some efficient methods within the framework of finite element methods for solving the dynamic response of the VBI systems. The formulation of these methods will be kept as general as possible, so that they can be applied to most conceivable problems. However, in deriving the fundamental theories using the analytical approaches or in conducting the parametric studies to illustrate the various dynamic effects involved, more emphasis will be placed on the problems encountered in the design of high-speed railway bridges, so as to reflect the public concern over the safety and the riding comfort of high-speed trains. It is believed that the methodologies established herein can be applied to solving similar problems encountered in traditional railways and mass rapid transit systems.

From the point of view of structural dynamics, a railway bridge is different from a highway bridge in that the sources of excitation caused by the moving vehicles are different for the two cases. For example, the vehicles moving over a highway bridge are random in nature. The vehicles constituting the highway traffic may vary in terms of the axle weight, axle interval, moving speed, and even the headway. However, a train moving over a railway bridge can generally be regarded as a sequence of identical vehicles in connection, plus one or two locomotives. Conventionally, a train has been simplified as a sequence of moving masses, or in the extreme case as a sequence of concentrated loads, of regular intervals. Because of the repetitive character of the wheel or bogie loads, a moving train usually contains some inherent frequencies, plus an excitation frequency associated with the moving speed. If any of these frequencies coincides with any of the frequencies of vibration of the bridge, the so-called *resonance* phenomenon will be induced on the bridge by the moving train, in the sense that the response will be continuously built up, as there are more railroad cars passing the bridge. Under the condition of resonance, great amplification in the bridge responses, as well as in the vehicles response, can be expected, which is likely to affect the life span of the bridge and the riding quality of running vehicles. It is advisable that the phenomenon of resonance be circumvented from the onset in the design of railway bridges.

Research on the dynamic response of bridges caused by the vehicular movement dates back to the mid-nineteenth century, following primarily the works of Willis (1849) and Stokes (1849) in investigating the collapse of the Chaster Rail Bridge in England in 1847, the first case for the collapse of a railway bridge in history. In these pioneer works, the effect of inertia of the beam was ignored, and the vehicle is modeled as a concentrated moving mass traveling at constant speeds. Although for this particular case, an exact solution can be obtained, its applicability remains rather limited due to the omission of the inertial effect of the beam. Nevertheless, the contribution of Stokes and Willis is considered historical, since they are among the first to bring the problem of vehicle impacts to the design desks of bridge engineers.

In the past two decades, the amount of research conducted on the vibration of bridges under moving vehicles has been increasing at a rate much faster than ever, partly due to the successful operation of high-speed railways in Japan and some European countries. It is difficult, if not impossible, to have a complete count of all the works conducted by previous researchers on this subject. For the days when hand calculations and slide rules play the most important role in design offices, i.e., before the advent of digital computers in the 1940s, investigations on bridge dynamics were concerned mainly with the development of analytical or approximate solutions for some simple, fundamental problems. Researchers of this period who were frequently cited in the literature include Timoshenko (1922), Jeffcott (1929) and Lowan (1935). The work by Inglis (1934) contains an early general treatment on the dynamics of railway bridges, which also lays the foundation for the following development.

The advent of digital computers, later followed by workstations, has enabled researchers to adopt more realistic bridge and vehicle models in analysis. The general texts by Timoshenko and Young (1955) and Biggs (1964) on structural dynamics contain some partial treatment on the moving load problems. Other texts that should be mentioned include the one by Frýba (1972) in analyzing the vibration of structures under moving loads, and those by Garg and Dukkipati (1984) and Frýba (1996) in dealing with the vibration of railway bridges. Starting from 1975, literature reviews were conducted by Ting and co-workers from time to time to update the related researches on vehicle–guideway interactions (Ting *et al.*, 1975; Genin and Ting, 1979; Ting and Genin, 1980; Ting and Yener, 1983; Taheri *et al.*, 1990). Nowadays, very powerful numerical methods, especially those based on the finite element methods, can be employed to analyze the dynamic behavior of bridges and moving vehicles, with virtually no limit placed on the level of complexity of the models used for the two subsystems. It should be noted that most of the works mentioned above were concerned primarily with the vibration of the bridge or supporting structure, but not of the moving vehicles.

## 1.2. Vehicle Models

By neglecting the inertia effect of the vehicle and considering a vehicle as a moving load or pulsating force, Timoshenko (1922) derived an enormous number of approximate solutions to the problem of simple beams under moving loads. Similar models were adopted by Ayre *et al.* (1950) and Ayre and Jacobsen (1950) in studying the dynamic responses of a two-span beam, and later by Vellozzi (1967) in studying the vibration of suspension bridges. The moving load model was also adopted by Chen (1978) in analyzing the dynamic response of continuous beams. Research on the vibration of bridges traveled by moving loads is abundant. It is only possible to cite a few of the most related ones, for instance, the works by Tan and Shore (1968a), Frýba (1972), Fertis (1973), Sridharan and Mallik (1979), Wu and Dai (1987), Weaver *et al.* (1990), Galdos *et al.* (1993), Gbadeyan and Oni (1995), Wang (1997), Zheng *et al.* (1998), Rao (2000), Chen and Li (2000), and Dugush and Eisenberger (2002), among others.

The *moving load* model is the simplest model that can be conceived, which has been frequently adopted by researchers in studying the vehicle-induced bridge vibrations. With this model, the essential dynamic characteristics of the bridge caused by the moving action of the vehicle can be captured with a sufficient degree of accuracy. However, the effect of interaction between the bridge and the moving vehicle was just ignored. For this reason, the moving load model is good only for the case where the mass of the vehicle is small relative to that of the bridge, and only when the vehicle response is not of interest.

For cases where the inertia of the vehicle cannot be regarded as small, a *moving mass* model (Fig. 1.2) should be adopted instead. The inertial effects of both the beam and the moving vehicle were studied as early as in 1929 by Jeffcott (1929) by the method of successive approximations. The investigations along this line were later carried out by a number of researchers. Stanišić and Hardin (1969) determined the response of a simple beam under an arbitrary number of moving masses by employing the Fourier series expansion. By the use of Green's function, algorithms for dealing with the moving mass problem has been studied by Ting *et al.* (1974) and Sadiku and Leipholz (1987). For a simple beam carrying a single moving mass, an exact, closed form solution was derived by Stanišić (1985) by means of expansion of the eigenfunctions in a series. The same

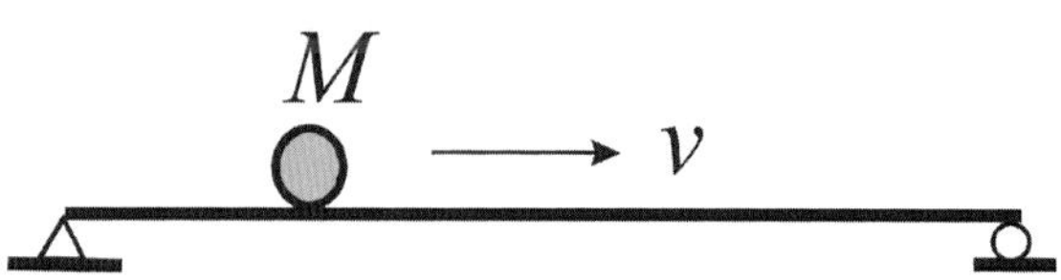


Fig. 1.2. Moving mass model.

moving mass model was adopted by Akin and Mofid (1989) in their study of the dynamic response of beams with various boundary conditions using an analytical–numerical approach.

One drawback with the moving mass model is that it excludes consideration of the bouncing action of the moving mass relative to the bridge. Such an effect is expected to be significant in the presence of rail irregularities or pavement roughness, or for vehicles moving at rather high speeds. Occasionally, it may be necessary to consider the separation and recontact of the moving vehicle with the bridge for some very bad road conditions, in which the bouncing action of the vehicles plays a decisive role in the seperation–recontact process.

The vehicle model can still be enhanced through consideration of the elastic and damping effects of the suspension systems. The simplest model in this case is a moving mass supported by a spring-dashpot unit, the so-called *sprung mass* model (Fig. 1.3). Biggs (1964) presented a semi-analytical solution to the problem of a simple beam traversed by a sprung mass. By using the series expansion technique, Pesterev *et al.* (2001) examined the response of an elastic continuum to multiple moving oscillations. Later, Pesterev *et al.* (2003) studied in depth the asymptotics of the solution of the moving oscillator problem and found that in the limiting case the moving oscillator problem and the moving mass problem for a simply supported beam are equivalent in terms of the beam displacements, but not in terms of the beam stresses. Also, it was shown that for small values of spring stiffness, the moving oscillator problem is equivalent to the moving load problem. In the book by Frýba (1972), a comprehensive treatment was given for the various vehicle models,

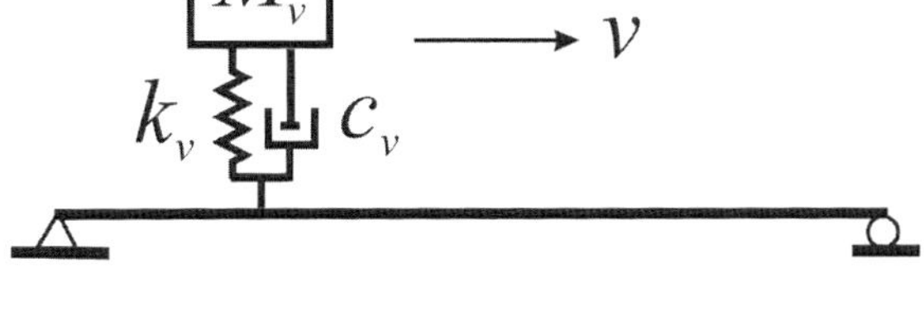


Fig. 1.3. Sprung mass model.

i.e., the moving load, moving mass, and moving sprung mass, concerning primarily the dynamic response of the structure traveled by vehicles. The analytical solutions as well as numerical solutions for some problems were presented in this book.

Because of the emergence of high-performance computers and the advance in computation technologies, it becomes feasible to have a more realistic modeling of the dynamic properties of the various components constituting a moving vehicle. Previously, the elastic effect of the tires and suspension mechanisms has been modeled by springs, the damping effect of the tires, suspension systems, and air-cushion by dashpots (Tan and Shore, 1968b; Genin *et al.*, 1975; Blejwas *et al.*, 1979; Genin and Chung, 1979; Humar and Kashif, 1993; Green and Cebon, 1994), and the energy dissipating effect of the interleaf mechanism by frictional devices (Veletsos and Huang, 1970; Chatterjee *et al.*, 1994; Tan *et al.*, 1998). Using such techniques, a multiple-axle truck or tractor-trailer can be represented as a number of discrete masses each supported by a set of spring and dashpot or frictional device. In the study by Yang *et al.* (1999), a railroad car was simulated as a rigid beam supported by two sets of spring-dashpot unit each resting on a wheel mass. Such a model enables us to consider the pitching effect of the car body.

To represent the various dynamic properties of railway freight cars, vehicle models that contain dozens of degrees of freedom (DOFs) have been devised and used by Chu *et al.* (1986), Wang *et al.* (1991), Xia *et al.* (2000), and Zhang *et al.* (2001a). In order to study the train–rails–bridge interaction, a train composed of a sequence of identical cars was considered by Wu *et al.* (2001), in which each car is assumed to consist of a car body, assumed to be rigid, resting on the front and rear bogies, each of which in turn is supported by two wheelsets. A total of 5 DOFs was assigned to the car body and also to each bogie, to account for the vertical, lateral, rolling, yawing, and pitching motions. In contrast, only three DOFs are assigned to each wheelset, which relate to the vertical, lateral and rolling motions.

Although the use of a more sophisticated vehicle model can make the simulation more realistic, it does create certain computation problems. For instance, in the simulation of bridges subjected to a series of railroad cars or highway vehicles that appear as a random flow (Yang *et al.*, 1996), divergence or slow convergence may occur in the process of iteration searching for a large number of contact forces at the wheels/rails or wheels/girder contact points in a step-by-step time-history analysis. The other concern here is that using simplified models can help identify the key parameters dominating the dynamic response of the bridge, which is beneficial for the developing of rational formulas for use in the design codes (Humar and Kashif, 1993).

## 1.3. Bridge Models

A beam that is simply-supported at both ends is the most popular structure that has ever been adopted in the study of vehicle-induced vibrations. Except for the research works that rely exclusively on analytical approaches, there is basically no restriction on the type of structures considered for the VBI problems, as the structures can always be represented by finite elements of various forms; the only difference being that a simpler bridge model requires less preparation and computation efforts.

In the past, various types of bridges have been considered in study of the vehicle-induced vibrations, which include the truss bridges (Chu *et al.*, 1979; Wiriyachai *et al.*, 1982), multispan uniform or nonuniform bridges (Wu and Dai, 1987; Yang *et al.*, 1995; Kou and DeWolf, 1997; Cheung *et al.*, 1999; Marchesiello *et al.*, 1999), girder or multigirder bridges (Chu *et al.*, 1986; Hwang and Nowak, 1991; Huang *et al.*, 1993; Cai *et al.*, 1994), continuous beams (Wu and Dai, 1987; Yang *et al.*, 1995), curved girder bridges (Tan and Shore, 1968a,b; Galdos *et al.*, 1993; Chang, 1997; Yang *et al.*, 2001), guideways (Genin *et al.*, 1975), steel plate girder bridges (Kawatani and Kim, 2001), and arch bridges (Chatterjee and Datta, 1995; Ju and Lin, 2003). The impact factor of horizontally-curved box bridges was studied by Galdos *et al.* (1993) and Senthilvasan *et al.* (2002). The dynamic response of a flat plate under the moving load was studied by Wu *et al.* (1987).

The dynamic response of cable-stayed bridges to moving vehicles has been studied by a number of researchers. By simulating the cable-stayed bridge as a beam resting on an elastic foundation, Meisenholder and Weidlinger (1974) proposed an approach for modeling the dynamic effects of cable-stayed guideways subjected to track levitated vehicles moving at high speeds. The effect of road surface roughness was considered by Wang and Huang (1992) in studying the cable-stayed bridge vibrations. By using an approximate bridge model, taking into account the nonlinear effect of cables, the dynamic response of cable-stayed bridges under moving loads was analyzed by Yang and Fonder (1998). In the review paper by Diana *et al.* (2000) for the railway runability of long-span cable supported bridges, it was noted that the impact effect of cable-stayed bridges is more sensitive than that of suspension bridges. Recently, Au *et al.* (2001a,b) investigated the impact effects of cable-stayed bridges under railway traffic using various vehicle models, and concluded that the moving force and moving mass models significantly underestimate the impact effects and the effects of random road surface roughness on the impact response of the bridge deck are more significant at sections close to the bridge towers. Guo and Xu (2001) studied the interaction between a cable-stayed bridge and a tractor-trailer moving over the bridge by a fully computerized approach. Recently, a hybrid tuned mass damper system composed of several subsystems was proposed by Yau and Yang (2004) for suppressing the multiple resonant peaks of cable-stayed bridges that may be excited by high-speed trains.

The vibration of suspension bridges under the vehicular movement was investigated by Chatterjee *et al.* (1994) with the torsional vibration taken into account. The dynamic interaction between a long suspension bridge, which has a main span length of 1377 m, and the running train was shown to be insignificant by Xia *et al.* (2000). The same suspension bridge was later studied by Xu *et al.* (2003), considering that there are high winds acting on the bridge, but not directly on the running train; the latter being protected from exposure to the high wind. Their results indicated that the wind-induced vibration on the bridge is detrimental to the running safety of the train and also to the riding comfort of passengers.

Another concern in simulation of the bridge response has been the inclusion of road *surface roughness* or *rail irregularities*. It has been reported that road surface or pavement roughness can significantly affect the impact response of bridges (Paultre *et al.*, 1992). However, the elevation of roughness or surface profile depends primarily on the workmanship involved in the construction of pavement or rail tracks and on how they are maintained, which, though random in nature, may contain some inherent frequencies. In most cases, the surface roughness or rail irregularities, which is three-dimensional in nature, is often approximated by a two-dimensional profile. As for the railways, it is realized that the profiles of irregularities on the two rails of a track may be different.

The road surface roughness was considered by Gupta (1980) by representing the elevation of road surface by a sine function. To account for its random nature, the road profile can be modeled as a stationary Gaussian random process and generated using certain power spectral density functions. Methods similar to this have been widely adopted by researchers in studying the vehicle-induced bridge vibrations (Inbanathan and Wieland, 1987; Coussy *et al.*, 1989, Hwang and Nowak, 1991; Chatterjee *et al.*, 1994; Chang and Lee, 1994; Henchi *et al.*, 1998; Pan and Li, 2002). The power spectral density functions developed by Dodds and Robson (1973) have been modified and used by Wang and Huang (1992) and Huang *et al.* (1993) in their analyses. The work by Marcondes *et al.* (1991) is of interest in that the power spectral density functions used to compute the road elevation have been determined by using the data collected from a field measurement, with distinction made for three different categories of pavement. Such an approach was adopted by Yang and Lin (1995) in the study of simple and continuous beams traveled by vehicles moving at different speeds.

As far as railway bridges are concerned, track irregularities may occur as a result of initial installation errors, degradation of support materials, and dislocation of track joints. Four geometric parameters can be used to quantitatively describe the rail irregularities, i.e., the vertical profile, cross level, alignment, and gauge (Wiriyachai *et al.*, 1982; Chu *et al.*, 1986; Wang *et al.*, 1991). From the point of structural dynamics, it is the wavelengths or frequencies implied by the rail irregularities that are crucial to the dynamic behavior of the VBI system. The frequencies implied by the surface roughness of a bridge plays a role similar to that of the bridge frequencies, in that resonance may occur on the bridge and traversing vehicles, if any of the excitation or vehicle frequencies coincides with, or are close to, any of the frequencies implied by the surface roughness.

## 1.4. Railway Bridges and Vehicles

Most of the research works cited above consider only a single or very small number of vehicular loads. In contrast, comparatively few works have been conducted on the dynamic response of bridge structures under the action of a sequence of moving loads with regular intervals, to simulate the effect of a connected line of train loads (Fig. 1.4). Bolotin (1964) studied a beam subjected to an infinite sequence of equal loads with uniform interval $d$ and constant speed $v$. In his study, the period $d/v$ of the moving loads has been identified as a key parameter. For the same problem, Frýba (1972) concluded that the response of the forced steady-state vibration will attain its maximum when the time intervals between two successive moving loads are equal to some periods of vibration of the beam in free vibration or to an integer multiple thereof. Kurihara and Shimogo (1978a,b) investigated the vibration and stability problems of a simple beam subjected to a series of discrete moving loads. The dynamic response of a girder or truss bridge during the passage of a series of railway vehicles was studied by Chu *et al.* (1979). By the transfer matrix method, Wu and Dai (1987) studied the response of multispan nonuniform beams subjected to two sets of identical loads moving in the same or opposite directions. Savin (2001) derived an analytical expression of the dynamic amplification factor and response spectrum for beams with various boundary conditions under successive moving loads.

Partly enhanced by the successful operation of high-speed railways worldwide, the dynamic response of railway bridges is receiving much more attention from researchers than ever. Matsuura (1976)

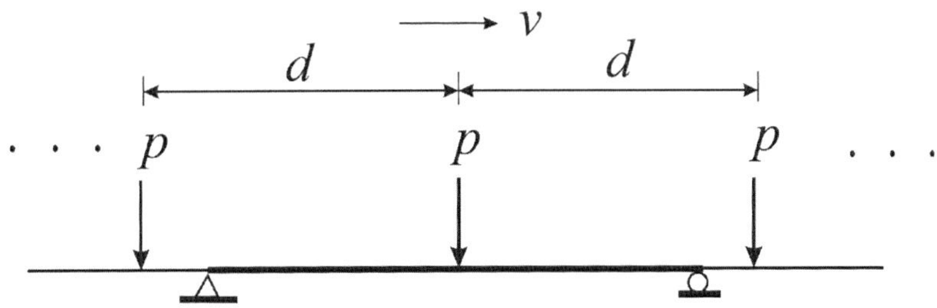


Fig. 1.4. Railway vehicles in series.

studied the dynamic behavior of various bridge girders used in the Shinkansen system. With the rail irregularities represented by power spectral densities, the impact responses of various railway bridges were investigated by Wiriyachai *et al.* (1982) and Chu *et al.* (1986). Following a brief review of the state-of-the-art methodologies for simulating the train–bridge interactions, Diana and Cheli (1989) studied the dynamic behavior of a train running over a long span bridge. By modeling a vehicle as a moving force or sprung mass, Cai *et al.* (1994) investigated the dynamic characteristics of single- and two-span beams subjected to vehicles moving at high speeds.

With the advancement in locomotive and control technologies, railway trains that have a design speed of 350 km/h or higher are not uncommon nowadays. In the literature, a maximum speed of 515.3 km/h has been reported during a field test (Delfosse, 1991). As far as high-speed trains are concerned, one needs to consider not only the vibration amplitudes of the bridge, but also the riding comfort of passengers carried by the trains, which can be assessed from the vertical or lateral accelerations of the moving vehicles (Diana and Cheli, 1989; Yau *et al.*, 1999). Due to the relatively stringent requirements imposed on the allowable deflection of the bridge and on the riding comfort of moving vehicles, the design of high-speed railway bridges is generally governed by the conditions of serviceability, rather than by strength and yielding, as learned from the design practices in Taiwan.

Recently, the dynamic response of a typical bridge under the passage of various commercial high speed trains was studied by Hsu

(1996), Yau (1996), Chang (1997), and Wu (2000). In particular, the effect of column stiffness on the dynamic response of bridges traveled by high-speed trains was studied by Hsu (1996), and the effects of ballast and elastic bearings by Yau (1996). It was demonstrated by Yau *et al.* (2001) that the insertion of elastic bearings at the supports of bridge girders for the purpose of isolating the earthquake forces may adversely amplify the dynamic response of the beam to moving train loads. Museros *et al.* (2002) investigated the influence of sleepers and ballast layers, as well as train–bridge interactions, on the response of short high-speed railway bridges. They concluded that inclusion of these factors can result in smaller maximum displacements and accelerations on the bridge, compared with those obtained using barely the moving loads model. The mechanism involved in the phenomena of resonance and cancellation for elastically supported beams was further explored in the study by Yang *et al.* (2004).

Other related effects that have been investigated include the torsional vibration of bridges caused by vehicles moving along one of the two tracks on a bridge, the crossing of two vehicles moving in opposite directions, and the mass ratio of the railway vehicles to the bridge (Hsu, 1996). A parametric study was carried out by Shen (1996) on a number of factors affecting the dynamic response of the bridge, in which both the modal superposition method and finite element method were employed. In the study by Wu *et al.* (2001), a bridge containing two railway tracks was considered, with which two trains are allowed to move over the bridge in opposite directions. Such a vehicle–rails–bridge interaction model was adopted by Wu (2000) and Yang and Wu (2002) in evaluating the risk of derailment for trains traveling over a bridge and simultaneously subjected to an earthquake excitation, and further by Wu and Yang (2003) in assessing the steady-state response and riding comfort of trains moving over a series of simply supported beams.

Based on an analytical approach, closed form solution has been obtained by Yang *et al.* (1997b) for the response of simple beams subjected to the passage of a high-speed train modeled as a sequence of moving loads with regular nonuniform intervals (Fig. 1.5), in which

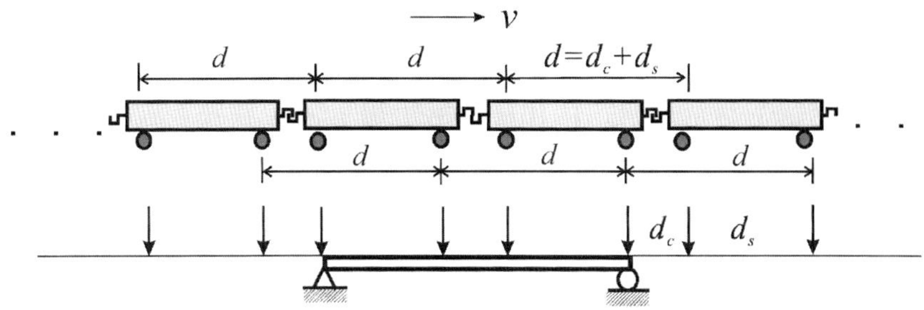


Fig. 1.5. Train load model.

the conditions for the phenomena of *resonance* and *cancellation* to occur have been identified. Based on these conditions, optimal design criteria that are effective for suppressing the resonant response of the VBI systems have been proposed. The same problem was later examined by Li and Su (1999) by using an alternative analytical approach, with similar findings obtained.

## 1.5. Methods of Solution

In studying the dynamic response of a VBI system, two sets of equations of motion can be written, one for the bridge and the other for the vehicles. It is the interaction or *contact forces* existing at the contact points of the two subsystems that make the two sets of equations coupled. As the contact points move from time to time, the system matrices are generally time-dependent, which must be updated and factorized at each time step. To solve these two sets of equations, procedures of an iterative nature have often been used (Hwang and Nowak, 1991; Green and Cebon, 1994; Yang and Fonder, 1996; Delgado and dos Santos, 1997). One way to do this is to start by assuming some values of displacements for the contact points, with which the contact forces can be solved from the vehicle equations. Next, by substituting the contact forces into the bridge equations, improved values of displacements for the contact points can be solved. The advantage of the iterative procedures is that the responses of both the vehicles and the bridge at any instant can be simultaneously made available. However, the convergence rate of iteration is likely to be low, when dealing with the more realistic case of bridges traveled by a large number of vehicles, whether in a random traffic flow, as encountered in highways, or in a connected line, as encountered in railways, for there exists twice the number of contact points if each vehicle is assumed to consist of two sets of wheel assemblies.

In the literature, Lagrange's equation with multipliers and constraint equations has also been applied to the analysis of VBI systems (Blejwas *et al.*, 1979). As it is well known, the use of Lagrange multipliers will increase the total number of unknowns and, therefore, the effort of computation. Theoretically speaking, a third approach is still possible, namely, by eliminating the interaction or contact forces from the original two sets of equations, one can form a new set of coupled equations for the entire VBI system. However, if the condensation procedure is performed on the structure level, the symmetry and other advantageous properties of the dynamic matrices associated with each subsystem will be destroyed (Yang and Lin, 1995).

Perhaps, one of the most efficient approaches for solving the VBI equations is to perform the condensation technique on the element level. Garg and Dukkipati (1984) used the Guyan (1965) reduction scheme to condense the DOFs of the vehicles to those of the bridge. Recently, Yang and Lin (1995) used the *dynamic condensation method* to eliminate all the DOFs associated with each vehicle on the element level. Such approaches are good if only the response of the bridge (the larger subsystem) is desired. They may not yield accurate solutions for the response of the moving vehicles (the smaller subsystem), due to the approximations adopted in relating the vehicle (slave) DOFs to the bridge (master) DOFs.

Other methods that have been employed in solving the second-order differential equations of motion of the VBI problems include: (1) the direct integration methods, such as Newmark's $\beta$ method (1959) (Inbanathan and Wieland, 1987; Yang and Lin, 1995), Wilson's $\theta$ method (Sridharan and Mallik, 1979), and fourth-order Runge–Kutta method (Chu *et al.*, 1986); (2) the modal superposition method (Blejwas *et al.*, 1979; Wu and Dai, 1987; Galdos *et al.*, 1993; Cai *et al.*, 1994), along with various integration schemes; and (3) the

Fourier transformation method (Green and Cebon, 1994; Chang and Lee, 1994).

One essential feature with the VBI problem is that the number of vehicles acting on the bridge is time-dependent. The more the number of vehicles simultaneously acting on the bridge, the higher is the level for the vehicles to interact with the bridge. To overcome the dependency of the system matrices on the wheel load positions, i.e., the contact point positions, one feasible approach is to eliminate the DOFs of the vehicles not in direct contact with the bridge on the element level by the method of *dynamic condensation* (Yang and Lin, 1995). This will result in a VBI element that takes into account all the coupling effects. The following is a summary of the procedure presented by Yang and Yau (1997) for deriving the VBI element.

Consider a beam simulated by a number of elements traversed by a train, of which each railroad car is idealized as two lumped masses, each supported by a spring-dashpot unit, as shown in Fig. 1.6. For

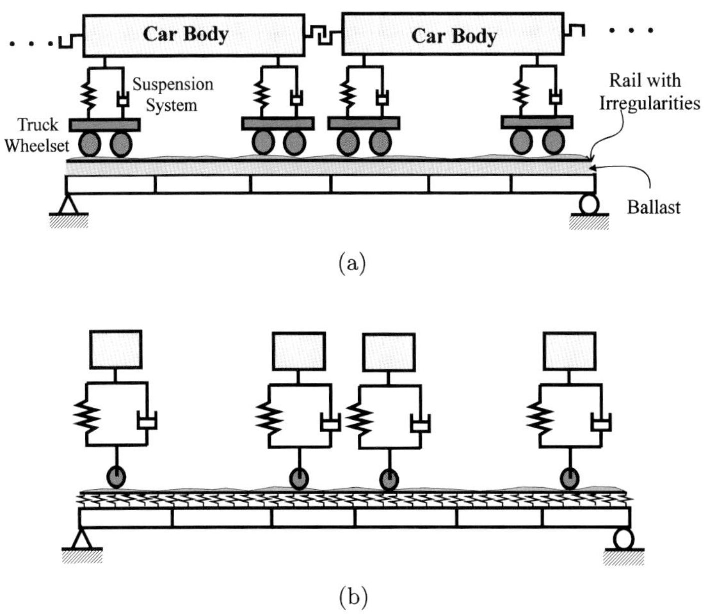


Fig. 1.6. Train–bridge system: (a) general model and (b) sprung mass model.

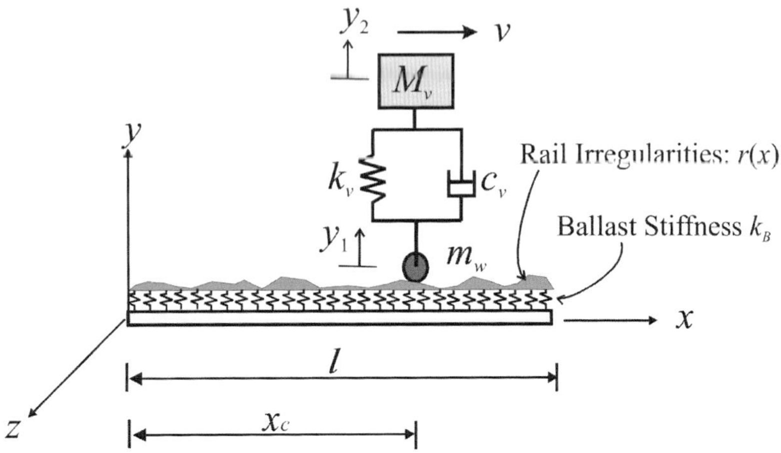


Fig. 1.7. Vehicle–bridge interaction element.

the present purposes, an *interaction element* is defined such that it consists of a beam element and a car-body mass and wheel mass connected by a suspension (spring-dashpot) unit directly acting over the beam element (Fig. 1.7). For the parts of the beam that are not directly under the action of the railroad cars, they can be modeled by conventional beam elements. However, for the remaining parts that are in direct contact with the wheel loads, the interaction elements have to be used instead.

With reference to the interaction element shown in Fig. 1.7, two sets of equations of motion can be written, one for the beam element and the other for the sprung mass unit. By Newmark's single-step finite difference formulas, the sprung mass equation can be discretized in time domain, from which the vehicle DOFs can be solved. Further, by the method of dynamic condensation, the sprung mass DOFs can be condensed to the associated DOFs of the beam element in contact. This will result in a VBI element with the effect of interaction fully taken into account. Since the VBI element possesses the same number of DOFs as the parent element, while retaining the properties of symmetry and bandedness in element matrices, it can be directly assembled with the conventional beam elements to form the structure equations. Such an element is particularly suitable for modeling bridges under a series of moving sprung masses with constant or varying intervals. It has the advantage that the response of the sprung mass can be recovered at any time step of the time-history analysis, which serves as a measure of the passengers' riding comfort. Using the VBI element, various dynamic properties of the beam and vehicles can be considered in the formulation, including the rail irregularities, ballast stiffness, damping of the beam, and stiffness and damping of the suspension units.

Recently, a more versatile approach for dealing with the VBI problems was proposed by Yang and Wu (2001). This method hinges on computation of the *contact forces* from the vehicles equations, in terms of the contact displacements. Before this can be done, the vehicle equations should first be discretized in time domain, say, using finite-difference equations of the Newmark type. The contact forces solved from the vehicles can then be treated as external loads and transformed as *consistent nodal loads* onto the bridge structure. With the bridge equations discretized in time domain, the bridge displacements can be solved as well. Such a procedure has been demonstrated to be quite flexible for treating vehicles of various complexities that appear in a sequence, in which both the vertical and horizontal contact forces are involved.

## 1.6. Impact Factor and Speed Parameter

In design practice, the dynamic response of a bridge has been considered indirectly by increasing the forces and stresses caused by the static live loads by an *impact factor*, defined as the ratio of the maximum dynamic to the maximum static response of the bridge under the same load minus one. One typical definition for the impact factor $I$ is (Yang *et al.*, 1995):

$$I = \frac{R_d(x) - R_s(x)}{R_s(x)}, \tag{1.1}$$

where $R_d(x)$ and $R_s(x)$ denote respectively the maximum dynamic and static responses of the bridge calculated at the cross section $x$ of the bridge of interest. The responses that may be considered for a beam include the deflection, bending moment, and shear force. The definition given in Eq. (1.1) is more rational and computationally more convenient than the *dynamic increment factor* (DIF) suggested by the AASHTO (*Guide*, 1980; Galdos *et al.*, 1993), or the *dynamic amplification factor* (DAF) discussed in Paultre *et al.* (1992) and Zhang *et al.* (2001b), since both the maximum dynamic and static responses are calculated at the same cross section of the bridge. Such an advantage will become obvious when dealing with moving loads that appear as a sequence, such as those of a train, or as a random flow, such as those encountered in highways, or in treating problems involving the resonance response, in the sense that the response of the bridge at the same cross section will be continuously amplified, as there are more loads passing the bridge. Care must be taken to distinguish the maximum impact factor from the maximum total response calculated for a beam. Occasionally, unreasonably large impact factors may be computed for a beam at some points due to the fact that the static responses, i.e., the denominator of Eq. (1.1), are very small. For this reason, the impact factor computed or measured for a bridge should not be regarded as the only criterion in the design of bridges.

It is well known that a number of factors may affect the impact factor of a bridge under the excitation of moving vehicular loads, for instance, the dynamic properties of the vehicle, the dynamic properties of the bridge, the vehicle speed, and the pavement roughness. Many *bridge codes*, including the American Association of State Highway and Transportation Officials (AASHTO) Specifications (*Standard*, 1989) and the Ontario Code (*Ontario*, 1983), have related the impact factor to a single parameter of the bridge, such as the span length or frequency of vibration, and have applied the same impact factor to all responses of the bridge including the deflection, shear force, and bending moment. According to the AASHTO Specifications, the impact factor $I$ is related to the span length $L$ of the bridge as

$$I = \frac{50}{L + 125} \leq 0.3 \tag{1.2}$$

when the bridge span length $L$ is expressed in ft, and as

$$I = \frac{15.24}{L + 38.1} \leq 0.3 \tag{1.3}$$

when $L$ is expressed in m. Formulas such as the preceding ones have been established several decades ago, based on limited field measurements, which are valid for the particular types of vehicles and bridges available in those days, if one realizes that modern trucks used are much heavier than those used half a century ago. The preceding formulas may be convenient for practical design, but are not theoretically sound at least for two reasons. First, the formulas are inconsistent in physical units, if one notes that the impact factor $I$ itself is a nondimensional quantity, while the span length has some physical units. Second, the use of span length as the control parameter is not representative of the physical property of the bridge concerning the vehicle–bridge interactions. This is especially true for continuous beams, for which there exist more than one span lengths, none of which can be directly related to the modal vibration shape. Based on the evidence of more extensive theoretical analyses and field measurements, it was reported that impact factors calculated according to current design codes may significantly underestimate the bridge response in many cases (O'Connor and Pritchard, 1984; Inbanathan and Wieland, 1987; Galdos *et al.*, 1993; Huang *et al.*, 1993; Chang and Lee, 1994).

By denoting the velocity of the moving vehicle as $v$ and the *characteristic length* of the beam as $L$, the exciting frequency of the moving vehicle can be expressed as $\pi v/L$. The *speed parameter* $S$ that is particularly useful for expressing the dynamic response of the VBI system is defined as the ratio of the exciting frequency of the moving vehicle to the fundamental frequency $\omega$ of the beam, that is,

$$S = \frac{\pi v}{\omega L} \tag{1.4}$$

which is dimensionless. In the study by Yang *et al.* (1995), it was demonstrated that for VBI systems with a speed parameter $S$ less than 0.5, the impact factor can be related to the speed parameter for the deflection, shear force, and bending moment of simple beams by linear formulas. Moreover, these formulas can be multiplied by some reduction coefficients to yield formulas for continuous beams. One particular point here is that the *characteristic length*, rather than the span length, should be used for the beam in defining the speed parameter. The former relates to the span length of a simple beam or the distance between two adjacent inflection points of the first mode of vibration of a continuous beam. It is with the use of the characteristic length in Eq. (1.4) that simple impact formulas can be established for both the simple and continuous beams (Yang *et al.*, 1997a). In the study by Pan and Li (2002), it was shown that the maximum displacement, velocity, and acceleration responses of the supporting structure appear to be almost linear to the speed parameter, which is consistent with the findings of Yang *et al.* (1995).

The speed parameter is an important parameter in the study of moving load problems, see, for instance, Tan and Shore (1968a,b), Veletsos and Huang (1970), Warburton (1976), Kurihara and Shimogo (1978b), Wang *et al.* (1992), Humar and Kashif (1993), Cai *et al.* (1994), and Chatterjee *et al.* (1994), among others. It was demonstrated that by plotting the response of the bridge against the speed parameter, rather than the span length or frequency of vibration that make up the parameter, generally more compact results can be obtained (Yang *et al.*, 1995). It was also noted by Paultre *et al.* (1992) that the DAF generally increases with the speed parameter. Noteworthy is the fact that the nondimensional character of the speed parameter enables us to extend the range of application of the computed or measured results to bridges beyond those covered in the study. Nevertheless, such a property was not fully recognized by a substantial portion of researchers working on vehicle-induced vibrations.

## 1.7. Concluding Remarks

A brief review of previous researches on the dynamic interaction of the bridge and moving vehicles was presented in this chapter. Vehicle models of increasing complexities, including the moving load, moving mass, sprung mass models, and more sophisticated ones, have been discussed. The factors that need to be considered in analyzing the response of the VBI systems include the dynamic properties and driving frequencies of the moving vehicles, and the dynamic properties and surface roughness of the bridge. Even though vehicle models of higher complexities, e.g., those consisting of dozens of DOFs, can be employed in studying the VBI problems nowadays, the use of simplified vehicle and bridge models is helpful, since it allows us to identify the key parameters dominating the dynamics of the VBI systems.

The impact factor adopted herein is computed based on quantities related to the same cross section of the beam that is of interest. It can be conveniently applied to cases involving a series of moving loads. The impact formulas provided by most current design codes are not consistent in physical units and lack a solid theoretical basis, of which the application should not be extended to bridges traveled by vehicles at high speeds. A more rational approach is to relate the impact factor, which is nondimensional, to the speed parameter, which is also nondimensional, defined as the ratio of the driving frequency of the moving vehicles to the vibration frequency of the bridge.

The VBI problem is a complicated one in that the contact points of the vehicles with the bridge move from time to time. Various methods exist for solving this problem. However, the most effective one appears to be the one based on condensation of the noncontact DOFs of the vehicle to the beam element in contact. The VBI element so derived can be applied to solving a great variety of VBI problems, by which the dynamic response of the moving vehicles, in addition to that for the bridge, can be obtained. Other factors that require further studies for high-speed railway bridges include the braking and acceleration of railroad cars, the torsional vibration of bridges caused by vehicles not moving along the centerline of the bridge girders, the crossing of two vehicles moving in opposite directions, the mass ratio of the vehicles to the bridge, the stability of rails, the risk of derailment of railroad cars under earthquake motions, and the stiffness effects of the ballast, elastic bearings, and supporting columns, among others.

# Part I

# Moving Load Problems

## Chapter 2

# Impact Response of Simply-Supported Beams

The most fundamental problem that should be considered in the study of vehicle-induced vibrations on bridges is the dynamic response of a simply-supported beam subjected to a single moving load. This problem is important in that the solution can be given in closed form. In this chapter, impact formulas will be derived for the deflection, bending moment and shear force of a simple beam under a single moving load. By the principle of superposition, the solution obtained for a single moving load will be expanded to deal with a series of identical, equi-distant moving loads, by which the key parameters dominating the dynamic response of the beam can be identified. Furthermore, based on the conditions of resonance or cancellation for the waves generated by a series of moving loads, optimal design criteria for suppressing the resonant response of beam structures will be presented. In designing a high-speed railway bridge, such criteria are useful for the selection of the span length and cross section of a girder bridge, if the car length, axle distance and operating speed of the train are already decided.

## 2.1. Introduction

Two effects are associated with the motion of a vehicle over a bridge, i.e., the gravitational effect and the inertial effect, both related to the mass of the vehicle. For the cases where the mass of the vehicle is small compared with that of the bridge, the vehicle can be represented as a concentrated load, with the inertial effect neglected.

This is the so-called *moving load* model, the simplest case that can be conceived of a moving vehicle. One advantage of using such a model is that for some special structures, e.g., for simply-supported beams, closed-form solutions can be obtained, from which the key parameters dominating the dynamic response of the supporting structure can be identified. It is from this consideration that the moving load problem for simple beams deserves a special and in-depth treatment.

The dynamic response of bridges subjected to the passage of moving vehicles continues to be a subject of great interest to structural engineers. In early studies, a bridge has been modeled as a beam-like structure and a vehicle as a moving load or moving mass (Timoshenko, 1922; Jeffcot, 1929; Lowan, 1935; Biggs, 1964; Frýba, 1972). Such a model was adopted in later studies including those of Warburton (1976), Stanišič (1985), Sadiku and Leipholz (1987), and Akin and Mofid (1989). In the meanwhile, more delicate vehicle and bridge models that consider the effects of multi-axle loadings, multi-lane loadings, vehicle suspension, surface roughness, etc. have been developed for the analysis of bridge response (Chu *et al.*, 1986; Inbanathan and Wieland, 1987; Galdos *et al.*, 1993; Huang *et al.*, 1993; Humar and Kashif, 1993; Chang and Lee, 1994; Yang and Lin, 1995; Yang *et al.*, 1995). In later chapters, it will be demonstrated that the response of a vehicle–bridge interaction (VBI) system obtained by using more sophisticated models for the moving vehicles remains in essence dominated by the key parameters identified from the analysis based on the moving load model.

A review of the research works cited above indicates that most of them consider only the case of a single or very small number of moving vehicular loads. In comparison, rather few research has been conducted on the dynamic response of bridges under the action of a series of moving loads, to represent the continuous action of the wheels of a moving train on the bridge. Bolotin (1964) studied a beam subjected to an infinite sequence of equal loads with uniform intervals $d$ and constant speed $v$. In his study, the period $d/v$ of the moving loads has been identified as a key parameter. For the same problem, Frýba (1972) concluded that the response of the forced steady-state vibration will attain its maximum when the time intervals between two successive moving loads are equal to some periods of the beam in free vibration or to an integer multiple thereof. Kurihara and Shimogo (1978a,b) investigated the vibration and stability problems of a simple beam subjected to a series of discrete loads with random intervals.

Several features have to be considered in the design of bridges for high-speed railways. First, the moving loads acting on the bridge are not random in nature, as encountered in highway bridges, but are of regular nonuniform intervals in general. Second, compared with the car length of a train, which ranges normally from 18 to 26 m, the span length of the elevated bridges constructed as part of the railway lines in most metropolitan areas is usually not long, which may vary from 10 to 40 m. Finally, because of the rather high operating speed of the train, e.g., with a maximum speed ranging from 250 to 350 km/hr, and because of the repetitive nature of wheel loads, it is likely that the condition of *resonance* be excited on high-speed railway bridges.

In this chapter, only bridges that can be modeled as a simple beam will be considered, which is the most common type of bridges used in railways due to its relative ease in construction and other considerations. The problem of a simple beam subjected to a single moving load will first be investigated in Sec. 2.2. Based on the analytical solution given in Sec. 2.2, impact factor formulas for the midpoint deflection and bending moment of the simple beam will be presented in Secs. 2.3 and 2.4, respectively, followed by the impact formulas for the end shear force in Sec. 2.5. The solution presented in Sec. 2.2 will be expanded to deal with the case of a simple beam under the passage of a train in Sec. 2.6, by modeling a train as the composition of two subsystems of wheel loads of constant intervals, with one subsystem consisting of all the front wheel assemblies and the other the rear assemblies. Of interest herein is the identification of the conditions for the phenomena of *resonance* and *cancellation* to occur on the beam, along with the *optimal design criteria* for the VBI system. Some illustrative examples will be presented in Sec. 2.7, followed by the concluding remarks in Sec. 2.8. The materials presented in Secs. 2.2–2.5 were rewritten from the paper by Lin *et al.*

(2000) and those in Secs. 2.6 and 2.7 were modified from the paper by Yang *et al.* (1997b).

## 2.2. Simple Beam Subjected to a Single Moving Load

A simply-supported beam traversed by a vehicle that is modeled as a concentrated load $p$ of speed $v$ is shown in Fig. 2.1. The following assumptions will be adopted in this study: (1) The beam is homogeneous and of constant cross sections, for which the Bernoulli–Euler hypothesis of plane cross sections remain plane after deformation applies; (2) only a single moving vehicle is allowed to travel on the beam at a time; (3) only the gravitational effect of the vehicle is considered, while the inertia effect of the vehicle is neglected, assumed to be small compared with that of the bridge; (4) the vehicle moves at a constant speed $v$; (5) the damping of the beam is of the Rayleigh type; (6) the beam is initially at rest before the vehicle moves in; and (7) no consideration is made of the road surface roughness of the bridge.

As shown in Fig. 2.1, a simple beam is subjected to a load of magnitude $p$ moving at speed $v$. Here, we shall use $u(x,t)$ to denote the deflection of the beam along the $y$ axis at position $x$ and time $t$, $L$ the length of the beam, $m$ the mass per unit length, $c_e$ the external damping coefficient, $c_i$ the internal damping coefficient, $E$ the modulus of elasticity, and $I$ the moment of inertia of the beam. Based on the aforementioned assumptions, the equation of motion of

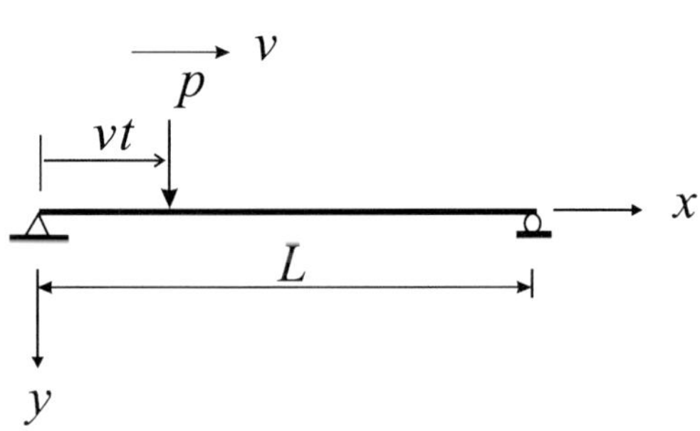


Fig. 2.1. A simply-supported beam subjected to a moving load.

the beam can be written as

$$m\ddot{u} + c_e\dot{u} + c_i I\dot{u}'''' + EIu'''' = p\delta(x - vt)\,, \quad 0 \le vt \le L\,, \tag{2.1}$$

where primes ($'$) and dots ($\dot{}$) denote differentiation with respect to coordinate $x$ and time $t$, respectively, and $\delta$ is the Dirac delta function. For the beam with simple supports, the boundary conditions are

$$\begin{aligned} u(0,t) &= 0 \\ u(L,t) &= 0 \\ EIu''(0,t) &= 0 \\ EIu''(L,t) &= 0 \end{aligned} \tag{2.2}$$

and the initial conditions are

$$\begin{aligned} u(x,0) &= 0 \\ \dot{u}(x,0) &= 0 \end{aligned} \tag{2.3}$$

as the beam is assumed to be at rest prior to the arrival of the moving vehicle.

Let $\phi_n$ denote the $n$th vibration mode of the beam that satisfies the boundary conditions. The deflection of the beam $u(x,t)$ due to only the $n$th mode of vibration is

$$u(x,t) = \phi_n(x)q_n(t)\,, \tag{2.4}$$

where $q_n(t)$ is the generalized coordinate corresponding to the $n$th mode. Substituting Eq. (2.4) into Eq. (2.1), multiplying both sides of the equation by $\phi_n$, and integrating with respect to $x$ over the length $L$ of the beam, one obtains

$$\begin{aligned} &m\ddot{q}_n(t)\int_0^L [\phi_n(x)]^2 dx + \dot{q}_n(t) \\ &\quad\times \left\{ c_e \int_0^L [\phi_n(x)]^2 dx + c_i I \int_0^L \phi_n^{\langle 4 \rangle}(x)\phi_n(x)dx \right\} \\ &\quad + EIq_n(t)\int_0^L \phi_n^{\langle 4 \rangle}(x)\phi_n(x)dx = p\phi_n(vt)\,, \end{aligned} \tag{2.5}$$

where it is realized that

$$\int_0^L \delta(x-a)\phi_n(x)dx = \phi_n(a)\,. \tag{2.6}$$

Let us denote the vibration frequency $\omega_n$ of the $n$th mode of the beam as

$$\omega_n^2 = \frac{EI}{m}\frac{\int_0^L \phi_n^{\langle 4\rangle}(x)\phi_n(x)dx}{\int_0^L[\phi_n(x)]^2dx}\,. \tag{2.7}$$

We shall also let $c_e = \alpha_e m, c_i = \alpha_i E$, and define the damping coefficient $\xi_n$ of the $n$th mode of vibration as

$$\xi_n = \frac{1}{2}\left(\frac{\alpha_e}{\omega_n} + \alpha_i\omega_n\right)\,. \tag{2.8}$$

Consequently, Eq. (2.5) reduces to

$$\ddot{q}_n + 2\xi_n\omega_n\dot{q}_n + \omega_n^2 q_n = \frac{p\phi_n(vt)}{\int_0^L m[\phi_n(x)]^2dx}\,. \tag{2.9}$$

This is exactly the equation of motion for the $n$th mode of vibration, in terms of the generalized coordinate $q_n$, which is valid only when the acting position $vt$ of the moving load is located within the range of the beam, i.e., $0 \le vt \le L$. Once the moving load leaves the beam, only free oscillation remains.

For a simply-supported beam, the $n$th modal shape of vibration is

$$\phi_n(x) = \sin\frac{n\pi x}{L} \tag{2.10}$$

and the frequency of vibration $\omega_n$ obtained from Eq. (2.7) is

$$\omega_n = \frac{n^2\pi^2}{L^2}\sqrt{\frac{EI}{m}}\,. \tag{2.11}$$

Substituting the preceding expression into Eq. (2.9) yields the equation of motion for the $n$th mode of the simply-supported beam as

$$\ddot{q}_n + 2\xi_n\omega_n\dot{q}_n + \omega_n^2 q_n = \frac{2p}{mL}\sin\frac{n\pi vt}{L}\,, \tag{2.12}$$

which is uncoupled from the other modes of vibration. From this equation, the generalized coordinate $q_n$ for the $n$th mode can be solved as

$$q_n = \frac{2pL^3/(EIn^4\pi^4)}{(1-S_n^2)^2 + 4(\xi_n S_n)^2} \times \left\{ (1-S_n^2)\sin\Omega_n t - 2\xi_n S_n \cos\Omega_n t \right.$$
$$\left. + e^{-\xi_n\omega_n t}\left[ 2\xi_n S_n \cos\omega_{dn} t + \frac{S_n}{\sqrt{1-\xi_n^2}}(2\xi_n^2 + S_n^2 - 1)\sin\omega_{dn} t \right] \right\}, \tag{2.13}$$

where $\omega_{dn}$ is the damped frequency of vibration of the beam,

$$\omega_{dn} = \omega_n \sqrt{1-\xi_n^2}\,. \tag{2.14}$$

$\Omega_n$ is the exciting frequency implied by the moving load,

$$\Omega_n = \frac{n\pi v}{L} \tag{2.15}$$

and $S_n$ is a nondimensional speed parameter defined as the ratio of the frequency of excitation of the moving load to the $n$th frequency of vibration of the beam, i.e.,

$$S_n = \frac{\Omega_n}{\omega_n} = \frac{n\pi v}{\omega_n L}\,. \tag{2.16}$$

Consequently, the total displacement $u(x,t)$ of the beam caused by all the vibration modes can be summed as follows:

$$u(x,t) = \sum_{n=1}^{\infty} \frac{2pL^3/(EIn^4\pi^4)}{(1-S_n^2)^2 + (2\xi_n S_n)^2}$$
$$\times \left\{ (1-S_n^2)\sin\Omega_n t - 2\xi_n S_n\cos\Omega_n t + e^{-\xi_n\omega_n t} \right.$$
$$\left. \times \left[ 2\xi_n S_n\cos\omega_{dn} t + \frac{S_n}{\sqrt{1-\xi_n^2}}(2\xi_n^2 + S_n^2 - 1)\sin\omega_{dn} t \right] \right\}$$
$$\times \sin\frac{n\pi x}{L}\,. \tag{2.17}$$

This is exactly the displacement of the beam caused by a single moving load with taking into account the effect of damping. In Eq. (2.17), the terms with $\Omega_n t$ represent the forced vibration of the bridge induced by the moving load, and the terms with $\omega_{dn}t$ are the free vibration, which will eventually be damped out. Again, this equation applies only when the acting position $vt$ of the moving load is located within the range of the beam.

For a wide class of moving load problems encountered in practice, the effect of damping on the bridge is so small, due to the rather short acting time of the moving loads that it can be ignored completely. This is especially true, if one is interested in the response of the bridge of the first few cycles. By neglecting the effect of damping, the total displacement $u(x,t)$ of the beam as given in Eq. (2.17) reduces to

$$u(x,t) = \frac{2pL^3}{EI\pi^4} \sum_{n=1}^{\infty} \frac{1}{n^4} \sin \frac{n\pi x}{L} \left( \frac{\sin \Omega_n t - S_n \sin \omega_n t}{1 - S_n^2} \right). \tag{2.18}$$

This is exactly the deflection of the simple beam at section $x$ caused by the moving load $p$ acting at position $vt$, with the effect of damping neglected. Correspondingly, the bending moment $M(x,t)$ caused by the moving load $p$ on the beam can be computed as $M(x,t) = -EIu''(x,t)$, or

$$M(x,t) = \frac{2pL}{\pi^2} \sum_{n=1}^{\infty} \frac{1}{n^2} \sin \frac{n\pi x}{L} \left( \frac{\sin \Omega_n t - S_n \sin \omega_n t}{1 - S_n^2} \right), \tag{2.19}$$

and the shear force is $V(x,t) = EIu'''(x,t)$ or

$$V(x,t) = \frac{2p}{\pi} \sum_{n=1}^{\infty} \frac{1}{n} \cos \left( \frac{n\pi x}{L} \right) \left( \frac{\sin \Omega_n t - S_n \sin \omega_n t}{1 - S_n^2} \right). \tag{2.20}$$

Here, it should be noted that the solutions obtained above for the problem considered are not new in the literature, see, for instance, Biggs (1964) and Warburton (1976). However, most of the previous researchers have not proceeded further to derive impact formulas from these solutions, which are more useful to practicing engineers.

For the present purposes, let us consider a simple beam of length $L = 20$ m, per unit mass $m = 3000$ kg/m, flexural rigidity $EI = 10^6$

N-m$^2$ and a damping coefficient $\xi$ of 2.5% for all vibration modes, subjected to a moving load $p = 6$ kN of speed $v = 27.8$ m/s (100 km/hr). The displacements for the midpoint of the beam obtained from Eqs. (2.17) and (2.18) for the damped and undamped cases have been compared in Fig. 2.2. As can be seen, the effect of damping on the response of the beam during the action of the moving load is rather small. For this reason, the effect of damping has often been neglected in research concerning the vehicle-induced vibrations on bridges.

To illustrate the effect of multi-modes of vibration, for the three speeds of $S_1 = 0.05$, 0.01 and 0.25, the deflections of the midpoint of the beam obtained from Eq. (2.18) for the undamped case considering various numbers of vibration modes have been plotted in Fig. 2.3. As can be seen, the result obtained for the midpoint deflection by considering only the first mode is good enough, partly due to the fact that all the anti-symmetric modes of vibration contribute nothing to the midpoint deflection. This gives us the impression that using only the first mode can yield generally good approximate solutions for vehicle-induced response, especially when the midpoint deflection

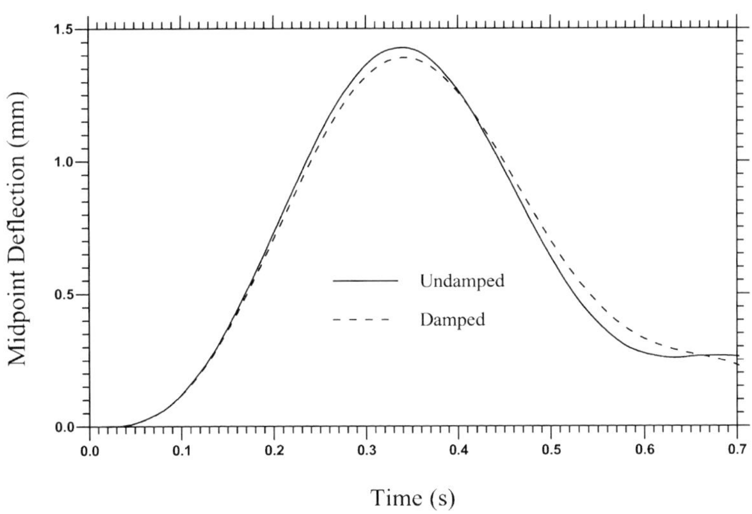


Fig. 2.2. The effect of damping of the beam.

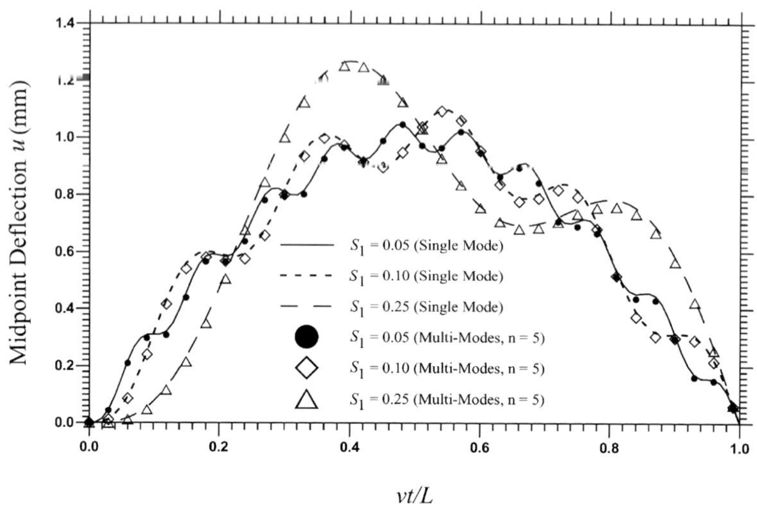


Fig. 2.3. The effect of multi-modes on midpoint deflection of the beam.

of the beam is desired. Such an approximation has been previously adopted by a number of researchers in their analytical studies.

## 2.3. Impact Factor for Midpoint Displacement

In this section, the responses derived in Sec. 2.2 for a single moving load will be used to derive the impact formula for the midpoint displacement of the simply-supported beam. The results presented in this section cover a wide range of applications, as they are all expressed in terms of the nondimensional speed parameter. They also serve as a useful reference for comparison with other results. It is realized that the impact response induced by a single moving load on the beam is generally larger than that induced by multi or continuously moving loads due to the suppression effect of the simultaneous acting loads. Thus, the impact formulas presented in this chapter for a single moving load should be regarded as reasonable upper bounds for the responses considered. The other message to convey here is that the impact factors for the deflection, bending moment and shear force can be quite different, and that the use of identical impact formulas for all physical quantities, as implied by most design codes, is not theoretically sound.

Here, we shall adopt the definition given in Eq. (1.1) for the impact factor $I$,

$$I = \frac{R_d(x) - R_s(x)}{R_s(x)}, \tag{2.21}$$

where $R_d(x)$ and $R_s(x)$ respectively denote the maximum dynamic and static response of the bridge at section $x$ due to passage of the moving load. For a simple beam, both the maximum dynamic and static deflections occur at the midpoint. The following is the maximum static deflection of the beam under the static load $p$:

$$R_{su}\left(\frac{L}{2}\right) = \frac{pL^3}{48EI}. \tag{2.22}$$

In contrast, the dynamic response for the midpoint deflection of the beam can be obtained from Eq. (2.18) by setting $x = L/2$, that is,

$$u\left(\frac{L}{2}, t\right) = \frac{2pL^3}{EI\pi^4} \sum_{n=1}^{\infty} \frac{1}{n^4} \sin\frac{n\pi}{2} \left(\frac{\sin\Omega_n t - S_n \sin\omega_n t}{1 - S_n^2}\right), \tag{2.23}$$

where it is realized that $\Omega_n = n\pi v/L$ and $\omega_n = n\pi v/(S_n L)$. The preceding equation is valid only when the acting position $vt$ of the moving load $p$ is located within the span of the beam. For $n = 2, 4, 6, \ldots$, the shape function $\sin(n\pi/2)$ vanishes at the midpoint, as it turns out to be asymmetrical. Thus, only the modes with $n = 1, 3, 5, \ldots$, i.e., the symmetrical modes, contribute to deflection of the midpoint.

Correspondingly, the impact factor for the midpoint deflection of the simple beam caused by the moving load $p$ acting at position $vt$ is

$$I_u = \frac{96}{\pi^4}\left[\sum_{n=1,3,5\ldots}^{\infty} \frac{1}{n^4} \sin\frac{n\pi}{2} \left(\frac{\sin\Omega_n t - S_n \sin\omega_n t}{1 - S_n^2}\right)\right] - 1, \tag{2.24}$$

which is independent of the magnitude $p$ of the moving load. As can be seen, the contribution of higher order terms decreases by a factor $n^{-4}$. It follows that the effect of higher order terms in Eq. (2.24) can be neglected without losing accuracy. By the relation $96/\pi^4 \cong 1$, Eq. (2.24) reduces to

$$I_u \cong \left( \frac{\sin \Omega_1 t - S_1 \sin \omega_1 t}{1 - S_1^2} \right) - 1 \,. \tag{2.25}$$

The impact factors calculated for the midpoint displacement using Eqs. (2.24) and (2.25), considering only the contribution of either multi-modes or the first mode, have been plotted in Fig. 2.4. As can be seen, the midpoint displacement impact response of the simple beam is dominated by the first mode.

To illustrate the effect of the acting position $vt$ of the moving load, different values have been assumed for $vt$, i.e., $vt/L = 1/8$, $1/4$, $3/8$, $1/2$, $5/8$, $3/4$, and $7/8$, in calculation of the impact factor $I_u$ for the midpoint displacement of the simple beam in Fig. 2.5. As can be seen, there exists an upper-bound envelope for the displacement impact factor $I_u$. The maximum impact factor $I_u$ can be regarded as proportional to the speed parameter $S_1$ for vehicle speeds in the range $S_1 < 0.5$, and as constant for $S_1 \geq 0.5$. Based on such an observation, the following formulas can be proposed for the midpoint deflection:

$$I_u = \begin{cases} 1.54 S_1 & \text{for} \quad S_1 < 0.5 \,, \\ 0.77 & \text{for} \quad S_1 \geq 0.5 \,. \end{cases} \tag{2.26}$$

In arriving at Eq. (2.26), the *effect of damping* of the beam has been neglected. If the effect of damping of the beam has been considered, slightly smaller impact factors can be derived.

With regard to the impact formula presented above, several comments can be made here. First, the present impact formula has been expressed as a function of the speed parameter $S_1$ of the moving vehicle, which is physically more meaningful, compared with formulas using the frequency of vibration $\omega$ or the span length $L$ of the beam as the key parameter, e.g., the one recommended by AASHTO in Eqs. (1.2) or (1.3). Second, since the speed parameter $S_1$, as defined in Eq. (2.16), is nondimensional, the impact formula derived herein based on this parameter remains valid for a wide range of simple beams subjected to vehicles moving at various speeds. Third, for

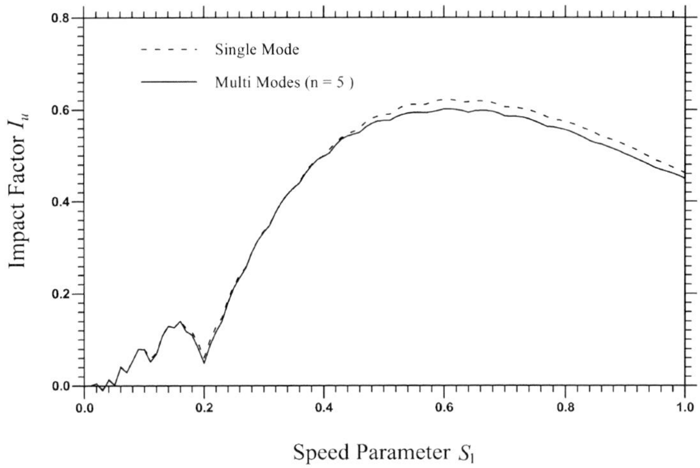


Fig. 2.4. The effect of multi-modes on impact factor — midpoint deflection.

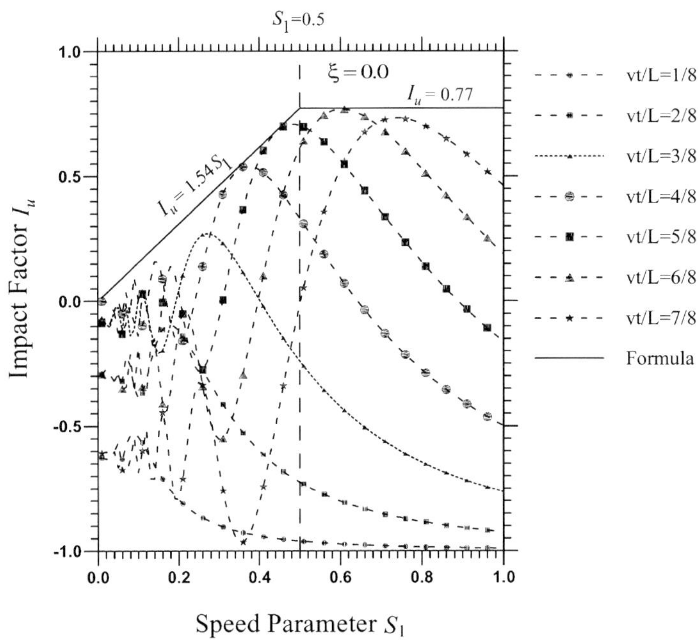


Fig. 2.5. The effect of loading positions — midpoint deflection.

most girder bridges with a span length no shorter than 30 m and traversed by vehicles moving at speeds no greater than 140 km/hr, the speed parameter $S_1$ computed is generally less than 0.5 (Yang *et al.*, 1995). In other words, the upper bound presented for the range with $S_1 \geq 0.5$ should find its application for bridges with rather short span lengths and/or subjected to vehicles moving at extremely high speeds. Finally, the impact factors calculated using the present formula show a trend similar to, but numerically larger than, those based on the finite element analysis (Yang *et al.*, 1995). Such a difference can be attributed to the use of more complicated vehicle and bridge models in the latter. For instance, in the study by Yang *et al.* (1995), a five-axle tractor-trailer has been selected as the target vehicle, and modeled as three lumped masses each resting on a spring-dashpot unit. Moreover, the bridge is represented by a number of finite elements, allowing the effects of multi-vibration modes to be considered. Compared with the finite element analysis results, it is believed that the present formula is generally on the conservative side.

## 2.4. Impact Factor for Midpoint Bending Moment

Conventionally, the same impact factor formula has been used for the deflection, bending moment and shear force of a bridge under the moving loads. Such an approach has the advantage of being convenient, but may not be accurate enough. In this section, an independent impact factor formula will be derived for the midpoint bending moment of the simple beam. The bending moment $M(x,t)$ acting at section $x$ of the simple beam with zero damping due to a single moving load $p$ of velocity $v$ at time $t$ has been given in Eq. (2.19). The bending moment at the midpoint of the beam can be evaluated as

$$M\left(\frac{L}{2},t\right) = \frac{2pL}{\pi^2}\sum_{n=1}^{\infty}\frac{1}{n^2}\sin\frac{n\pi}{2}\left(\frac{\sin\Omega_n t - S_n\sin\omega_n t}{1-S_n^2}\right), \tag{2.27}$$

where it is realized that $\Omega_n = n\pi v/L$ and $\omega_n = n\pi v/(S_n L)$. Correspondingly, the maximum bending moment of the simple beam subjected to a static load $p$ acting at $x = L/2$, i.e.,

$$R_{sM}\left(\frac{L}{2}\right) = \frac{pL}{4}\,. \tag{2.28}$$

It follows that the impact factor $I_M$ for the midpoint bending moment of the beam subjected to a moving load $p$ acting at position $vt$ is

$$I_M = \frac{8}{\pi^2}\left[\sum_{n=1,3,5\ldots}^{\infty} \frac{1}{n^2}\sin\frac{n\pi}{2}\left(\frac{\sin\Omega_n t - S_n \sin\omega_n t}{1 - S_n^2}\right)\right] - 1\,. \tag{2.29}$$

By comparing Eq. (2.29) with Eq. (2.24), one observes that the convergence rate for the bending moment impact factor $I_M$, as represented by the factor $n^{-2}$, is slower than that for the displacement impact factor $I_u$. Thus, the effect of higher modes of vibration on the bending moment may not be as small as that for the midpoint displacement. To investigate such an effect, the impact factors $I_M$ computed for the midpoint bending moment from Eq. (2.29) using either a single mode or multi-modes have been plotted in Fig. 2.6, along with the finite element results. As can be seen, the effect of higher modes appears to be quite significant. For this reason, a total of seven modes, i.e., with $n = 7$, will be included in this study in computing the impact factor for bending moment.

To illustrate the effect of the acting position $vt$ of the moving load, different values have been assumed for $vt$, i.e., $vt/L = 1/8$, $1/4$, $3/8$, $1/2$, $5/8$, $3/4$, and $7/8$, in computing the impact factor $I_M$ for the midpoint bending moment using Eq. (2.29), where it is noted that $\Omega_n = n\pi v/L$ and $\omega_n = n\pi v/(S_n L)$. As can be seen from Fig. 2.7, an upper-bound envelope exists for the impact factor $I_M$ through the range of speed parameter $S_1$ considered, that is, for $S_1 < 0.36$, the maximum impact factor $I_M$ is proportional to the speed parameter $S_1$, and for $S_1 \geq 0.36$, it is a constant,

$$I_M = \begin{cases} 1.24 S_1 & \text{for} \quad S_1 < 0.36\,, \\ 0.45 & \text{for} \quad S_1 \geq 0.36\,. \end{cases} \tag{2.30}$$

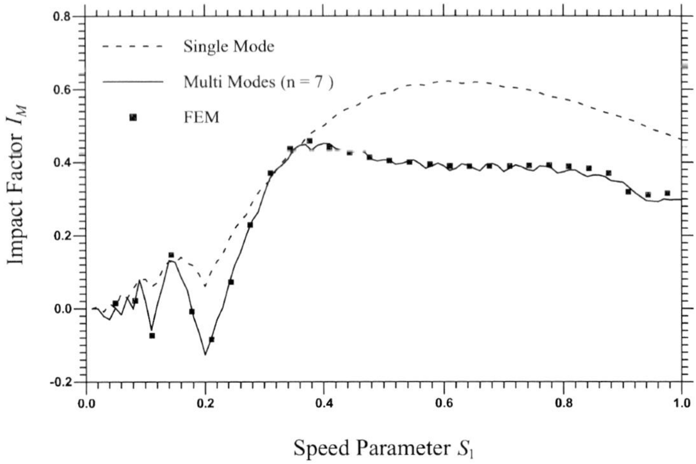


Fig. 2.6. The effect of multi-modes on impact factor — midpoint moment.

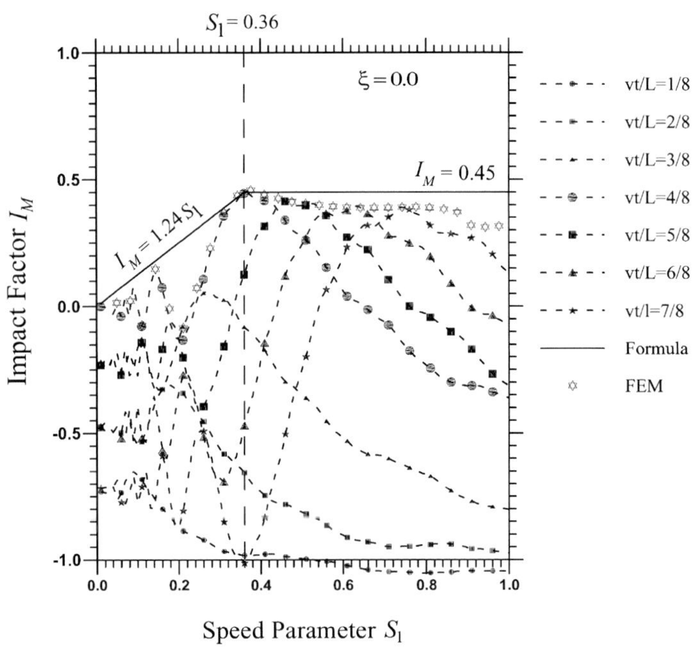


Fig. 2.7. The effect of loading positions — midpoint moment.

In Fig. 2.7, the solutions obtained by the independent finite element analyses were also plotted, which appear to be well represented by the formulas proposed. It should be noted that in deriving the preceding impact factor formula, the *effect of damping* of the beam has been ignored. Moreover, a comparison of Eq. (2.30) with Eq. (2.26) indicates that for simple beams, the impact effect caused by a moving load on the bending moment is generally smaller than that on the displacement.

## 2.5. Impact Factor for End Shear Force

From the shear force $V$ expression in Eq. (2.20), one observes that the shear force $V$ vanishes at the center point, at $x = L/2$, of the simple beam, as indicated by the term $\cos(n\pi x/L)$. Thus, it is meaningless to compute the impact factor $I_V$ for the shear force at the midpoint of the beam. In this section, the shear impact factor $I_V$ will be computed for a point very close to, but not at, the right end of the beam. Note that similar result can be derived from the left end of Eq. (2.20). For a simple beam subjected to a moving load $p$, the maximum static shear force is $R_{sV} = p$, which occurs when the load acts on one end of the beam. Consequently, the shear impact factor $I_V$ at a point near the right end is

$$I_V = \frac{2}{\pi}\left|\sum_{n=1}^{\infty}\frac{1}{n}\cos n\pi\left(\frac{\sin\Omega_n t - S_n\sin\omega_n}{1-S_n^2}\right)\right| - 1\,, \tag{2.31}$$

where it is noted that $\Omega_n = n\pi v/L$ and $\omega_n = n\pi v/(S_n L)$. Compared with the impact factors $I_u$ and $I_M$ given in Eqs. (2.24) and (2.29), the preceding series expression for $I_V$ converges at a rate much slower than those for $I_u$ and $I_M$, as represented by the term $n^{-1}$, which indicates that the influence of higher-order terms can hardly be neglected in computing the impact factor for the end shear force. For this reason, a total of 300 terms, i.e., with $n = 300$, will be included in computation of the end shear force in this section.

To illustrate the effect of the acting position $vt$ of the moving load, different values have been assumed for $vt$, i.e., $vt/L = 1/8, 1/4, 3/8,$

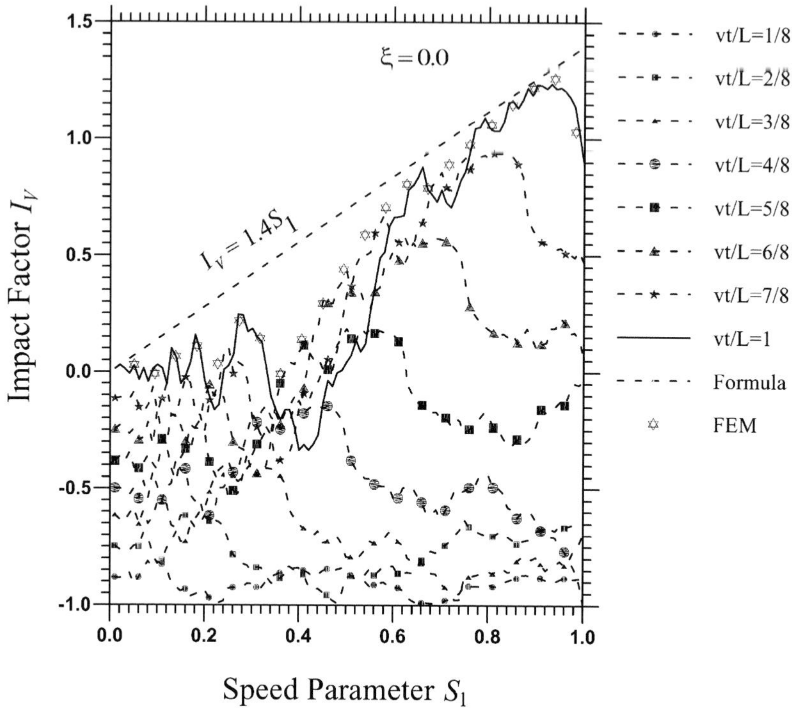


Fig. 2.8. The effect of loading positions — end shear force.

1/2, 5/8, 3/4, and 7/8, in calculation of the impact factor $I_V$ for the near-support shear force of the simple beam using Eq. (2.31). From the results plotted for $I_V$ in Fig. 2.8, which have been obtained with 300 terms for each of the curves shown, it is obvious that the upper-bound envelope for the impact factor $I_V$ can be very well represented by a straight line as

$$I_V = 1.4S_1 \tag{2.32}$$

based on the assumption of zero damping for the beam. It is interesting to note that the result obtained by an independent finite element analysis using a more sophisticated vehicle model, also plotted in Fig. 2.8, is in good agreement with the present ones.

## 2.6. Simple Beam Subjected to a Series of Moving Loads

In this section, a train is modeled as the composition of two sets of moving loads of constant intervals, with the first set representing the wheel loads of all the front wheel assemblies and the second set the rear ones. To simplify the derivation that leads to closed solutions, no consideration will be made of the variations in rail elevations, also known as track irregularities, or any other initial conditions of the beam.

### 2.6.1. *Modeling of Wheel Loads of a Train*

Figure 2.9 shows a simply-supported beam of length $L$ traveled by a train at speed $v$, which consists of a number of identical cars of length $d$. Only the flexural vibration of the beam will be considered herein. This means that the train is assumed to travel along the centerline of the beam, with no consideration made for the torsional action. As a first approximation, the train will be simulated as a series of lumped loads $p$ of constant intervals $d$ moving at speed $v$, as shown in Fig. 2.10(a). The corresponding load function $F(t)$ can be given as

$$F(t) = \sum_{j=1}^{N} p \cdot U_j(t, v, L)\,, \tag{2.33}$$

where

$$U_j(t, v, L) = \delta[x - v(t - t_j)] \cdot \left[H(t - t_j) - H\left(t - t_j - \frac{L}{v}\right)\right]. \tag{2.34}$$

Here, $\delta$ denotes the Dirac delta function, $x$ the coordinate of the beam, $H(\bullet)$ a unit step function, $t_j$ the arriving time of the $j$th load at the beam, $t_j = (j-1)d/v$, and $N$ the total number of moving loads considered. Obviously, the action of the $j$th moving load is turned on by the term $H(t - t_j)$ when it enters the beam, and turned off by the term $H(t - t_j - L/v)$ when it leaves the beam.

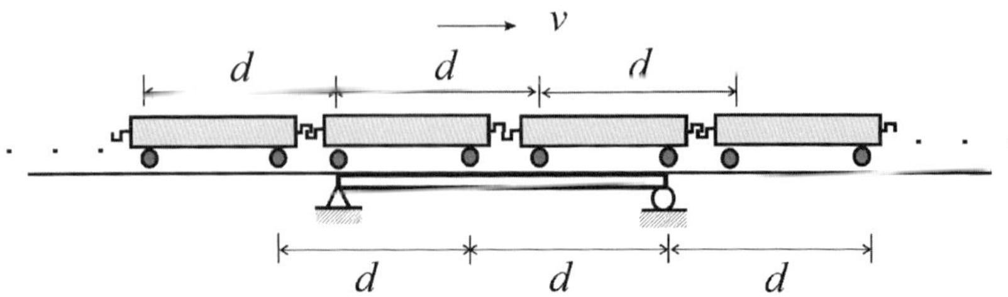


Fig. 2.9. The simply-supported beam under a moving train.

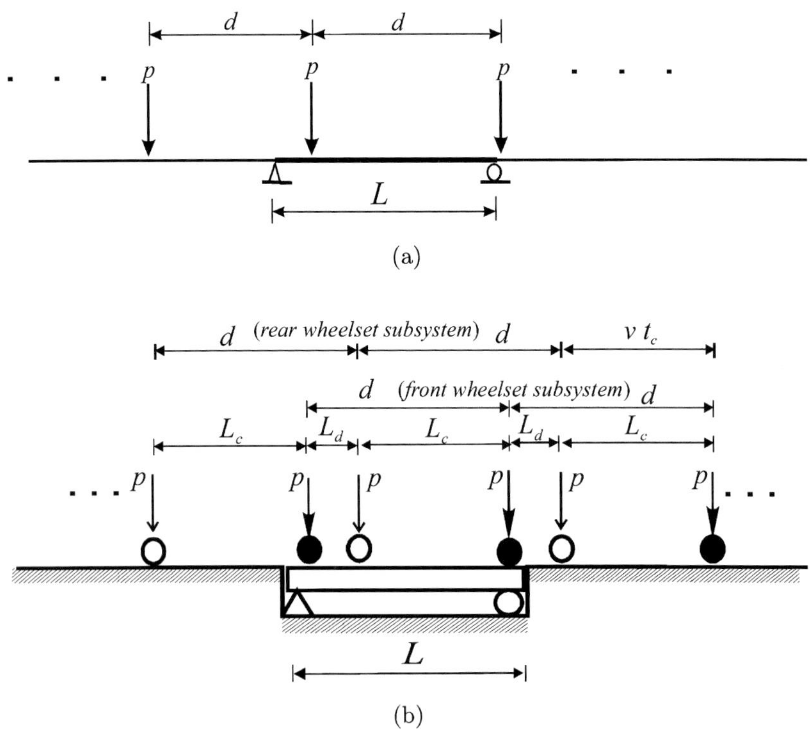


Fig. 2.10. The simple beam subjected to: (a) uniform loads and (b) train loads.

The train is assumed to have $N$ identical cars, and each car is supported by two *bogies*, each of which in turn is supported by two *wheelsets*. In this chapter, each bogie and the associated wheelsets will be grossly referred to as a *wheel assembly*. Other kinds of arrangement for the wheel assemblies that are different from the one considered herein are possible. But in this section, we shall consider the support mechanism of a train car only up to the level of wheel assembly, and we shall treat the load associated with each wheel assembly as a concentrated wheel load. Let $L_c$ denote the distance between the two wheel assemblies of a car, and $L_d$ the distance between the rear wheel assembly of a car and the front wheel assembly of the following car. It follows that the car length $d$ is equal to the sum of $L_c$ and $L_d$, i.e., $d = L_c + L_d$, and that a train can be represented as a sequence of wheel loads $p$ (assumed to be constant) with alternative intervals $L_c$ and $L_d$.

For most commercial trains, the distance $L_c$ between two wheel assemblies of a car is larger than the distance $L_d$ between the rear wheel assembly of a car and the front wheel assembly of the following car, i.e., $L_c > L_d$. For a better representation of the load configuration, we shall conceive each train as the composition of two wheel load sets, with the first set representing the wheel loads of all the front bogies and the second set the rear ones. By doing so, the distance between any two consecutive wheel loads in each set is simply $d$ [see Fig. 2.10(b)]. It is realized, however, that a time lag of $t_c = L_c/v$ exists between the two sets of moving loads. Based on the above considerations, the wheel load function $F(t)$ for the train can be modified from Eq. (2.33) as:

$$F(t) = \sum_{j=1}^{N} p \cdot \left[U_j(t, v, L) + U_j(t - t_c, v, L)\right], \tag{2.35}$$

where $p$ denotes the lumped load of each wheel assembly.

Either the expression given in Eq. (2.33) or Eq. (2.35) considers only the effect of *moving loads*, but neglects the effect of *inertia* of the moving masses and the *interaction* between the train cars and supporting beam. In order to consider these two effects, the term $p$ in Eq. (2.35) should be replaced by the following function $F(p, M, v)$ (Bolotin, 1964):

$$F(p, M, v) = p - M(\ddot{u} + 2v\dot{u}' + v^2 u''), \tag{2.36}$$

where $M$ denotes the mass lumped at each load or wheel position, $u$ is the vertical deflection of the beam, and dots and primes represent differentiation with respect to time $t$ and coordinate $x$, respectively. The physical meanings for the newly added terms in Eq. (2.36) can be given as follows: The term $M\ddot{u}$ represents the inertial force acting along the direction of deflection $u$ of the beam; the term $2M\dot{u}'$ is the *Coriolis force* relating to the rate of inclination of the beam; and the term $Mv^2u''$ is the *centrifugal force* associated with the curvature of the beam induced by the mass of speed $v$ at the position of action.

### 2.6.2. *Method of Solution*

Using the wheel load function $F(p, M, v)$ in Eq. (2.36), the equation of motion for the beam under a moving train with the load configuration described above can be written as

$$m\ddot{u}+c_e\dot{u}+c_iI\dot{u}''''+EIu'''' = \sum_{j=1}^{N} F(p, M, v)\cdot[U_j(t, v, L)+U_j(t-t_c, v, L)] \tag{2.37}$$

subjected to the following boundary conditions:

$$\begin{aligned} u(0, t) &= 0\,, \\ u(L, t) &= 0\,, \\ EIu''(0, t) &= 0\,, \\ EIu''(L, t) &= 0\,, \end{aligned} \tag{2.38}$$

and the initial conditions:

$$\begin{aligned} u(x, 0) &= 0\,, \\ \dot{u}(x, 0) &= 0\,. \end{aligned} \tag{2.39}$$

In Eq. (2.37), $m$ denotes the mass per unit length, $c_e$ the external damping coefficient, $c_i$ the internal damping coefficient, $E$ the elastic modulus, $I$ the moment of inertia of the beam, and $t_c$ is the time lag between the front and rear wheel loads of each car.

By taking into account the boundary conditions, one may express the vertical deflection $u(x, t)$ for the simply-supported beam in a series form as

$$u(x,t) = \sum_{n=1}^{\infty} q_n(t) \sin \frac{n\pi x}{L}\,, \tag{2.40}$$

where $q_n(t)$ denotes the *generalized coordinate* of the $n$th mode. By substituting the preceding expression for $u$ into Eq. (2.37), multiplying both sides of the equation by $\sin(n\pi x/L)$, and integrating with respect to length $L$ of the beam, one can obtain the equation of motion for the simply-supported beam in terms of the generalized coordinates $q_n$ as

$$(1+\varepsilon_{m,n})\ddot{q}_n + (2\xi_n\omega_n + \varepsilon_{c,n})\dot{q}_n + (\omega_n^2 + \varepsilon_{s,n})q_n = F_n(t)\,, \tag{2.41}$$

where the damping property of the beam is assumed to be of the *Rayleigh type*, $\omega_n$ is the frequency of vibration of the $n$th mode, i.e., $\omega_n = n^2\pi^2\sqrt{EI/mL^4}$, and $\xi_n$ the damping ratio of the $n$th mode. It should be noted that the equation of motion given in Eq. (2.41) for the generalized coordinates is as general as the original one in Eq. (2.37), except that the conditions of hinge and roller supports are taken into account. In Eq. (2.41), the forcing function $F_n(t)$ for the $n$th generalized coordinate is

$$F_n(t) = \frac{2p}{mL}\sum_{j=1}^{N}[f_n(t,v,L) + f_n(t-t_c,v,L)] \tag{2.42}$$

and the other coefficients are

$$\varepsilon_{m,n} = \frac{2M}{mL}\sum_{j=1}^{N}[g_n(t,v,L) + g_n(t-t_c,v,L)]\,, \tag{2.43}$$

$$\varepsilon_{c,n} = \frac{2M}{mL}\left(\frac{n\pi v}{L}\right)\sum_{j=1}^{N}[h_n(t,v,L) + h_n(t-t_c,v,L)]\,, \tag{2.44}$$

$$\varepsilon_{s,n} = -\frac{2M}{mL}\left(\frac{n\pi v}{L}\right)^2\sum_{j=1}^{N}[g_n(t,v,L) + g_n(t-t_c,v,L)]\,. \tag{2.45}$$

In Eqs. (2.42)–(2.45), the three functions $f_n$, $g_n$, $f_h$ are defined as

$$f_n(t,v,L) = \sin\frac{n\pi v(t-t_j)}{L}H(t-t_j) + (-1)^{n+1}\sin\frac{n\pi v(t-t_j-L/v)}{L}H\left(t-t_j-\frac{L}{v}\right), \tag{2.46}$$

$$g_n(t,v,L) = \left(\sin\frac{n\pi v(t-t_j)}{L}\right)^2 H(t-t_j) - \left[\sin\frac{n\pi v(t-t_j-L/v)}{L}\right]^2 H\left(t-t_j-\frac{L}{v}\right), \tag{2.47}$$

$$h_n(t,v,L) = \sin\frac{2n\pi v(t-t_j)}{L}H(t-t_j) - \sin\frac{2n\pi v(t-t_j-L/v)}{L}H\left(t-t_j-\frac{L}{v}\right). \tag{2.48}$$

Here, some physical meanings can be given for the terms involved in Eq. (2.41): $\varepsilon_{m,n}$ represents the *effect of added masses* due to the moving wheel loads. Such an effect tends to elongate the period of vibration of the beam. $\varepsilon_{c,n}$ represents the effect of the *Coriolis force* due to the dynamic coupling between the speed of moving masses and the angular velocity of the beam. $\varepsilon_{s,n}$ represents the effect of the *centrifugal force*, which is a function of the speed of masses and the curvature of deflection of the beam. Such an effect tends to reduce the dynamic stiffness of the beam, especially when trains of high speeds are concerned. Theoretically, it is possible that a beam may become unstable because of this effect. In practice, however, the beams used in high-speed railways are constructed to be so stiff that the effects of both $\varepsilon_{c,n}$ and $\varepsilon_{s,n}$ on the dynamic response of the beam are much smaller than that of the added masses, as represented by the term $\varepsilon_{m,n}$. For this reason, the effects of both the Coriolis force and centrifugal force induced by the moving masses can generally be excluded without affecting the accuracy of solution.

If only the effects of moving loads are taken into account, i.e., by neglecting all the three effects mentioned above by setting

$\varepsilon_{m,n} = \varepsilon_{c,n} = \varepsilon_{s,n} = 0$, the generalized coordinate $q_n(t)$ of the beam can be evaluated from Eq. (2.41) as

$$q_n(t) = \frac{1}{m\omega_{dn}} \int_0^t F_n(\tau) e^{-\xi_n \omega_n (t-\tau)} \sin \omega_{dn}(t-\tau) d\tau$$
$$= \frac{2pL^3}{EI\pi^4} [P_n(v,t) + P_n(v, t-t_c)] , \tag{2.49}$$

where $\omega_{dn}$ is the damped frequency of vibration of the beam, i.e., $\omega_{dn} = \omega_n \sqrt{1-\xi_n^2}$, and the function $P_n(v,t)$ can be expressed as

$$P_n(v,t) = \frac{1}{n^4} \sum_{j=1}^{N} \frac{1}{(1-S_n^2)^2 + 4(\xi_n S_n)^2}$$
$$\times \left[ A \cdot H(t-t_j) + (-1)^{n+1} B \cdot H\left(t - t_j - \frac{L}{v}\right)\right] . \tag{2.50}$$

Here,

$$A = (1-S_n^2) \sin \Omega_n (t-t_j) - 2\xi_n S_n \cos \Omega_n (t-t_j) + e^{-\xi_n \omega_n (t-t_j)}$$
$$\cdot \left[ 2\xi_n S_n \cos \omega_{dn}(t-t_j) + \frac{S_n}{\sqrt{1-\xi_n^2}} (2\xi_n^2 + S_n^2 - 1) \sin \omega_{dn}(t-t_j) \right] , \tag{2.51}$$

$$B = (1-S_n^2) \sin \Omega_n \left(t - t_j - \frac{L}{v}\right) - 2\xi_n S_n \cos \Omega_n \left(t - t_j - \frac{L}{v}\right)$$
$$+ e^{-\xi_n \omega_n (t - t_j - L/v)}$$
$$\cdot \left[ 2\xi_n S_n \cos \omega_{dn} \left(t - t_j - \frac{L}{v}\right)\right.$$
$$\left. + \frac{S_n}{\sqrt{1-\xi_n^2}} (2\xi_n^2 + S_n^2 - 1) \sin \omega_{dn} \left(t - t_j - \frac{L}{v}\right)\right] , \tag{2.52}$$

where $\Omega_n$ denotes the exciting frequencies of the moving loads, $\Omega_n = n\pi v/L$, and the speed parameter $S_n$ has been defined in Eq. (2.16).

As was stated previously, a train traveling over a beam can be represented as the composition of two identical wheel load sets, with the first set representing the wheel loads of all the front bogies and the second set the rear ones. The terms $P_n(v,t)$ and $P_n(v,t-t_c)$ in Eq. (2.49) represent the dynamic responses excited by the two sets of wheel loads moving over the beam, in which the second set has a time lag $t_c$ behind the front set.

For a simple beam subjected to moving loads, which is basically a transient problem with very short acting time, only the first mode will be important for determination of the deflection of the beam (Biggs, 1964). By neglecting the effect of damping and considering only the first mode of vibration, the dynamic response of the beam can be derived from Eqs. (2.40) and (2.49) as

$$u(x,t) = \frac{2pL^3}{EI\pi^4}\frac{1}{1-S_1^2}\sin\frac{\pi x}{L}\left[\bar{P}_1(v,t)+\bar{P}_1(v,t-t_c)\right], \tag{2.53}$$

where the response function $\bar{P}_1(v,t)$ for the first set of wheel loads is

$$\begin{aligned}\bar{P}_1(v,t) = \sum_{k=1}^{N}\Bigg\{&[\sin\Omega_1(t-t_j)-S_1\sin\omega_1(t-t_j)]H(t-t_j)\\&+\left[\sin\Omega_1\left(t-t_j-\frac{L}{v}\right)-S_1\sin\omega_1\left(t-t_j-\frac{L}{v}\right)\right]\\&\times H\left(t-t_j-\frac{L}{v}\right)\Bigg\}.\end{aligned} \tag{2.54}$$

In Eq. (2.53), the terms $\bar{P}_1(v,t)$ and $\bar{P}_1(v,t-t_c)$ denote the contribution of the front and rear wheel loads, respectively.

In this section, it is assumed that the beam has a span length $L$ not greater than twice the car length $d$, i.e., $L \leq 2d$. Depending on the bridge/car length ratio $L/d$, there may be two, three, or no wheel loads acting on the beam during the passage of the train, as shown in Fig. 2.11. The most severe case occurs when the front wheel load of the $(N-1)$th car has left the beam, and the front wheel load of the $N$th car has entered the beam, namely, when the rear wheel load of the $(N-1)$th car and the front wheel load of the $N$th car are simultaneously acting on the beam, as shown in Fig. 2.11(a). There are two reasons for this. First, the two wheel loads, of distance $L_d$,

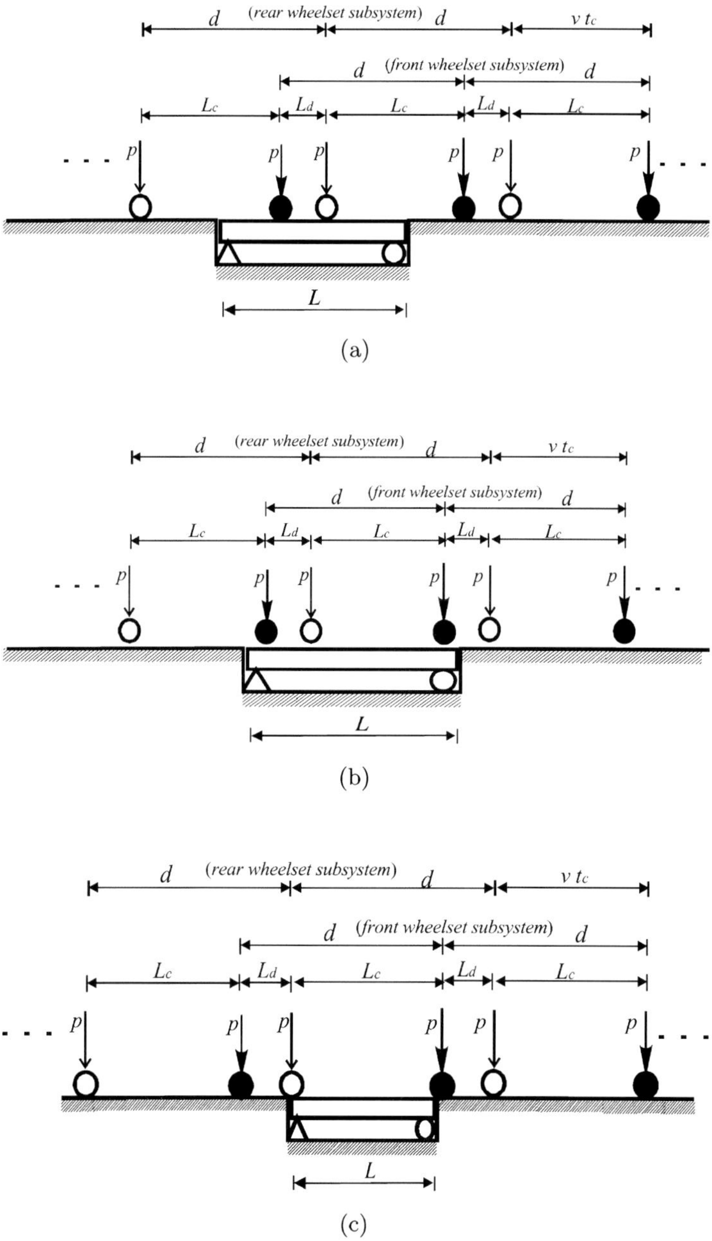


Fig. 2.11. The loading cases: (a) two wheelsets, (b) three wheelsets, and (c) no wheelset.

may induce the largest static response when passing the midpoint of the beam. Second, the dynamic response of the beam has been excited to the utmost by the former $N-1$ cars that have passed the beam. For this case, it is known that $t_N < t < t_N + L/v$ and the preceding equation reduces to

$$\begin{aligned}\bar{P}_1(v,t)&\\ &= [\sin\Omega_1(t-t_N) - S_1\sin\omega_1(t-t_N)]H(t-t_N) - 2S_1\cos\frac{\omega_1 L}{2v}\\ &\quad\cdot\left[\sin\omega_1\left(t-\frac{L}{2v}\right) + \sin\omega_1\left(t-\frac{L}{2v}-\frac{t_N}{2}\right)\frac{\sin\omega_1\left(\frac{t_N}{2}-\frac{d}{2v}\right)}{\sin\frac{\omega_1 d}{2v}}\right]\\ &\quad\times H\left(t-t_{N-1}-\frac{L}{v}\right) \qquad (2.55)\end{aligned}$$

of which the derivation has been given in Appendix A. Again, it can be appreciated that the term containing $H(t-t_N)$ represents the dynamic response of the beam induced by the motion of the $N$th front wheel load of the train, and the term containing $H(t-t_{N-1}-L/v)$ is the free vibration caused by the former $N-1$ front wheel loads that have passed the beam. Since the two response functions $\bar{P}_1(v,t)$ and $\bar{P}_1(v,t-t_c)$ are similar in nature, only the function $\bar{P}_1(v,t)$ for the first set of wheel loads will be considered in the following discussion.

As a side note, whenever the car length $d$ is greater than the span length $L$ of the beam, there exist certain intervals during which the beam is not in direct contact with the wheel loads, as was illustrated in Fig. 2.11(c). The dynamic response for the beam during these intervals, which represents only free vibration, can be obtained from Eq. (2.55) by dropping the term containing $H(t-t_N)$.

### 2.6.3. *Phenomenon of Resonance*

With the present moving load model for the train, it can be seen from Eq. (2.55) that the response of the beam reaches a maximum when the denominator of the second term within the brackets vanishes, i.e., when $\sin(\omega_1 d/2v) = 0$ or $\omega_1 d/2v = i\pi$, with $i = 1, 2, 3, \ldots$.

This is exactly the condition for *resonance* of the beam to occur under repetitive loads. For the case with $\sin(\omega_1 d/2v) = 0$, the response function $\bar{P}_1(v,t)$ in Eq. (2.55) becomes indeterminate. By the relation $t_N = (N-1)d/v$ and L'Hospital's rule, it can be shown that

$$\sin\omega_1\left(t - \frac{L}{2v} - \frac{t_N}{2}\right)\frac{\sin\omega_1\left(\frac{t_N}{2} - \frac{d}{2v}\right)}{\sin\frac{\omega_1 d}{2v}} = (N-2)\sin\omega_1\left(t - \frac{L}{2v}\right). \tag{2.56}$$

Consequently, Eq. (2.55) reduces to

$$\begin{aligned}\bar{P}_1(v,t) &= [\sin\Omega_1(t-t_N) - S_1\sin\omega_1(t-t_N)]H(t-t_N)\\ &\quad - 2(N-1)S_1\cos\frac{\omega_1 L}{2v}\sin\omega_1\left(t-\frac{L}{2v}\right)\\ &\quad \times H\left(t - t_{N-1} - \frac{L}{v}\right),\end{aligned} \tag{2.57}$$

which is the response function for the case when the former $N-1$ wheel loads have passed the beam and the last, i.e., the $N$th wheel load is acting on the beam. In Eq. (2.57), the term containing $H(t - t_{N-1} - L/v)$ indicates that under the condition of resonance, the response of the beam will be continuously built up, as there are more loads passing the beam.

It should be noted that the kind of resonance produced by moving loads on the beam is different from that occurring on a structure due to excitation of a harmonic load that has a driving frequency equal to the fundamental frequency of the structure, for which the response of the structure may theoretically become unbounded in the absence of damping. However, once the resonance condition is met by the moving loads, the response of the beam keeps increasing as there are more wheel loads passing through the beam and reaches a maximum when the last wheel load enters the beam, if no consideration is made of the damping effect. After all the wheel loads have passed the beam, only free vibration will remain on the beam, which eventually will be damped out in reality.

With the condition of resonance, i.e., by letting $\sin(\omega_1 d/2v) = 0$, the *critical car length* $d$ of the train traveling over the beam can be solved as

$$d = 2i\frac{\pi v}{\omega_1} = 2iS_1L\,, \quad i = 1, 2, 3, \ldots\,. \tag{2.58}$$

On the other hand, given the car length $d$ and span length $L$, the speed parameter can be found from the *resonance condition* as

$$S_1 = \frac{d}{2iL}\,, \quad i = 1, 2, 3, \ldots\,. \tag{2.59}$$

which implies that the longer the beam, the lower is the speed for resonance to occur. By letting $i = 1, 2, 3, \ldots$, the preceding equation indicates that resonance may occur at the following speeds: $S_1 = 0.50d/L$, $0.25d/L$, $0.167d/L$, $0.125d/L$, $\ldots$, with diminishing values. Here, we shall call the speed $0.50d/L$ the *primary resonant speed*, and all the remaining the *secondary resonant speeds*. Since a train is accelerated from zero to its full speed, it is obvious that certain resonant speeds will always be encountered by trains in their passage over a bridge. However, if one notes that all the $S_1$ values computed above are small compared with unity for high-speed railway bridges and that the factor $1/(1 - S_1^2)$ in Eq. (2.53) decreases as the speed parameter $S_1$ decreases, the resonant responses induced by the train moving at the secondary speeds are generally small and can be neglected in practice. As will be demonstrated in the numerical study, a good design of railway bridges is to have the primary resonance, as implied by the condition $S_1 = 0.50d/L$, suppressed at all times, say, through adjustment of the span length or cross section of the beam, once the car length has been fixed.

### 2.6.4. *Phenomenon of Cancellation*

As can be seen from Eq. (2.55) or Eq. (2.57), whenever the condition $\cos(\omega_1 L/2v) = 0$ is met, the excitation effects of all the former $N-1$ wheel loads that have passed the beam sum to zero, that is, no *residual response* will be induced by the loads that have passed the beam. Such a condition has been referred to as the condition of *cancellation*, under which the response function $\bar{P}_1(v,t)$ reduces to

$$\bar{P}_1(v,t) = [\sin\Omega_1(t-t_N) - S_1\sin\omega_1(t-t_N)]H(t-t_N)\,. \quad (2.60)$$

This equation indicates that if the condition of cancellation is met, the response of the beam is determined solely by the last or $N$th wheel load acting on the beam, as the free vibrations caused by all the former wheel loads passing the beam have been suppressed. Moreover, whenever the condition of cancellation is met, no residual response will be induced on the beam after the last wheel load leaves the beam. Such a property remains valid even in the absence of damping.

By definition, $S_1 = \pi v/(\omega_1 L)$, the speed parameter can be determined for the *condition of cancellation* as follows:

$$S_1 = \frac{1}{2i-1}\,, \quad i = 1, 2, 3, \ldots \quad (2.61)$$

from which the cancellation points can be computed as $S_1 = 1$, $1/3$, $1/5$, $1/7$, .... As can be seen from Eq. (2.55), the condition for cancellation to occur, i.e., $\cos(\omega_1 L/2v) = 0$, is a condition more decisive than that for resonance. In other words, if the condition of cancellation is met, then the bracketed term in Eq. (2.55) just disappears, meaning that the phenomenon of resonance is entirely suppressed. Theoretically speaking, it is possible to select an optimal speed for the train, aimed at providing better riding quality, such that the condition of cancellation is satisfied during its passage over a bridge.

### 2.6.5. *Optimal Design Criteria*

As was stated in Sec. 2.6.3, the resonance response induced by a high-speed train on the bridge will reach the maximum, when the speed parameter $S_1$ equals the primary resonant speed, i.e., $0.5d/L$, as given in Eq. (2.59), rather than the secondary resonant speeds, due to the fact that the factor $1/(1-S_1^2)$ in Eq. (2.53) is larger for the primary resonance than for the secondary resonance. In practice, the primary resonance can be circumvented through the selection of proper car length $d$ or span length $L$ for the beam such that the condition of cancellation is always enforced. For instance, by letting the first resonance speed, $S_1 = 0.5d/L$, equal to any of the speeds that satisfy the cancellation condition in Eq. (2.61), the span/car length ratio $L/d$ can be solved as follows:

$$\frac{L}{d} = i - 0.5\,, \quad i = 1, 2, 3, \ldots\,. \tag{2.62}$$

This represents exactly the *optimal criterion* for suppressing the resonance response caused by repetitive, regular moving loads.

The meaning of Eq. (2.62) and its applicability to practical design should not be underestimated. First, it is the span to car length ratio $L/d$ that determines whether resonance will be induced by moving loads on the bridge. Second, while the integer $i$ may take any integer value, the value $i = 2$ can be easily met in practice, which implies that $L/d = 1.5$ is an optimal span/car length ratio. Thus, given the car length $d$ for a specific model of train, the optimal span length $L$ can be found for the bridge, and vice versa. Third, the formula given in Eq. (2.62) has the important feature of being independent of the speed parameter $S_1$ or speed $v$ of the train. For this reason, this formula can find a wide range of applications. In the following section and in many other studies conducted by the authors and co-workers, it has been confirmed that whenever the condition $L/d = 1.5$ is met, the first resonance response can be virtually suppressed.

## 2.7. Illustrative Examples

The phenomena of resonance and cancellation presented in Sec. 2.6 are based primarily on the closed form solution presented in Eq. (2.53), which considers only the first mode of vibration of the beam. In this section, the accuracy of the solution presented in Sec. 2.6, as well as the phenomena of resonance and cancellation, will be numerically investigated. In particular, finite element solutions obtained by a computer analysis program based on the vehicle–bridge interaction element derived in Chapter 3 will be used to generate some reference solutions, by which the effects of all modes of vibration of the beam are automatically taken into account.

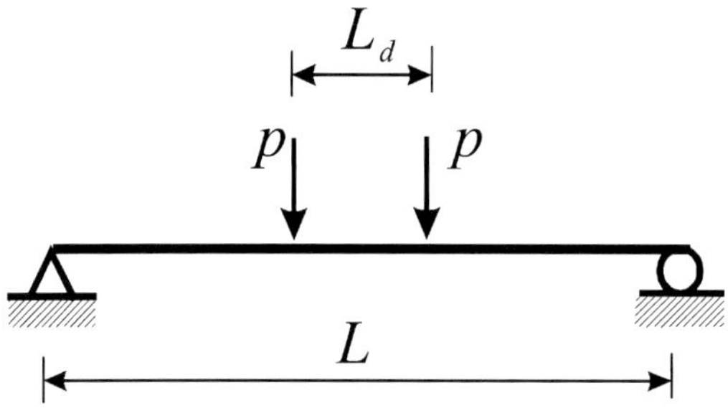


Fig. 2.12. The most severe loading case.

### 2.7.1. *Comparison with Finite Element Solutions*

Consider a simple beam made of prestressed concrete with $L = 20$ m, $I = 3.81$ m$^4$, $E = 29.43$ GPa, $m = 34,088$ kg/m, for which the first frequency of vibration solved is $\omega_1 = 44.75$ rad/s. The train is assumed to have $N = 5$ cars of identical length $d = 24$ m. The two wheel assemblies (or bogies) of the car is separated by 18 m, i.e., $L_c = 18$ m and $L_d = 6$ m. The mass of each wheel assembly is $M = 22\,000$ kg, corresponding to $p = 215.6$ kN. For the present case, the maximum static deflection $R_s$ of the simple beam occurs when two wheel loads $p$ of interval $L_d$ are located symmetrically on the beam as shown in Fig. 2.12, which can be computed as (Gere and Timoshenko, 1990):

$$R_s = \frac{pa(3L^2 - 4a^2)}{24EI}, \tag{2.63}$$

where $a = (L - L_d)/2$ for the present case. The maximum static deflection $R_s$ is required in computation of the impact factor $I$ in this section based on the definition of Eq. (2.21).

For the purpose of verification, two cases are considered, particularly to demonstrate the phenomena of resonance and cancellation. In the first case, the speed parameter $S_1$ is selected so that the resonance condition given in Eq. (2.59) is met. By setting $i = 5$, $S_1 = d/(2iL) = 0.12$, the resonance speed is found to be $v = 34$ m/s $= 122.4$ km/h. In the second case, the speed parameter $S_1$ is selected to meet the condition of cancellation given in Eq. (2.61).

By setting $i = 6, S_1 = 1/(2i - 1) = 1/11$, the speed of cancellation is found as $v = 26$ m/s $= 93.6$ km/h. By simulating the train as a set of concentrated wheel loads, the midpoint responses of the beam (undamped) subjected to the action of the wheel loads moving at the above two speeds have been plotted in Fig. 2.13, along with the finite element solutions, which have been obtained by discretizing the beam into 16 elements. As can be seen, for the two cases considered, the present solution, which was obtained by considering only the first mode of vibration of the beam, agrees very well with the finite element result. Moreover, for a train with $v = 34$ m/s, the midpoint response of the beam tends to increases steadily as there are more loads passing the beam, which is indicative of the *resonance phenomenon.* For a train with $v = 26$ m/s, the response of the beam appears merely as a periodic function, with no amplification effect observed during the passage of wheel loads on the bridge. Of interest is the fact that as long as all the wheel loads depart from the beam, no *residual response* remains on the beam, which is a typical cancellation phenomenon.

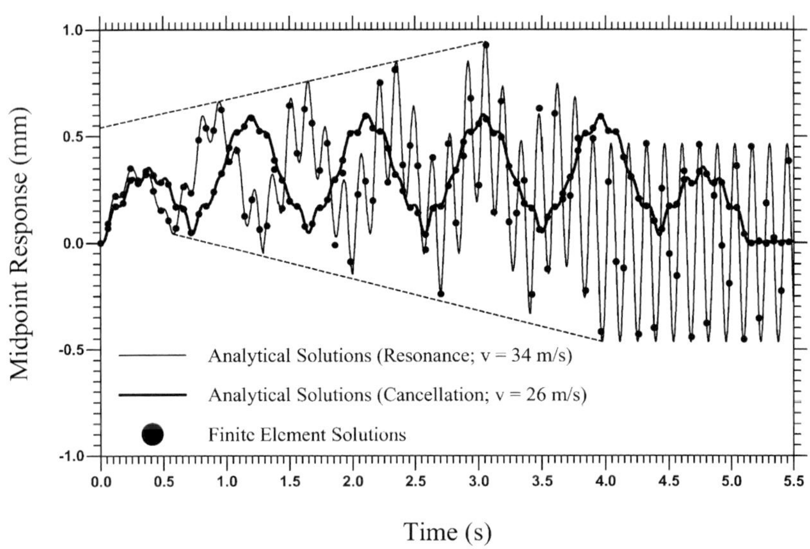


Fig. 2.13. The cases of verification.

For the present case of $d/L = 1.2$, the points of resonance can be found from Eq. (2.59) as follows: $S_1 = 0.60, 0.30, 0.20, 0.15, 0.12, 0.10, \ldots$, and the points of cancellation from Eq. (2.61) as follows: $S_1 = 1.00, 0.33, 0.20, 0.143, 0.111, 0.091, \ldots$. Correspondingly, the speeds of resonance for the train are $v = 615.4$, 307.7, 205.1, 153.9, 123.1, 102.6 km/h, ..., and the speeds of cancellation are $v = 1025.6$, 338.4, 205.1, 146.5, 114.0, 93.0 km/h, .... The impact factor $I_u$ and displacement $u$ computed for the midpoint of the beam have been plotted against the speed parameter $S_1$ in Figs. 2.14 and 2.15, respectively. From these figures, the following observations can be made: First, the present result agrees very well with the finite element solution that considers the contribution of all modes of vibration, indicating that the *effect of higher modes* is rather minor for the present problem. Second, the resonance points with $S_1 = 0.60$, 0.15, 0.12 can generally be observed, while the other resonance points are merely suppressed as they are coincident with or close to the points of cancellation. Third, the first resonance point ($S_1 = 0.6$) should

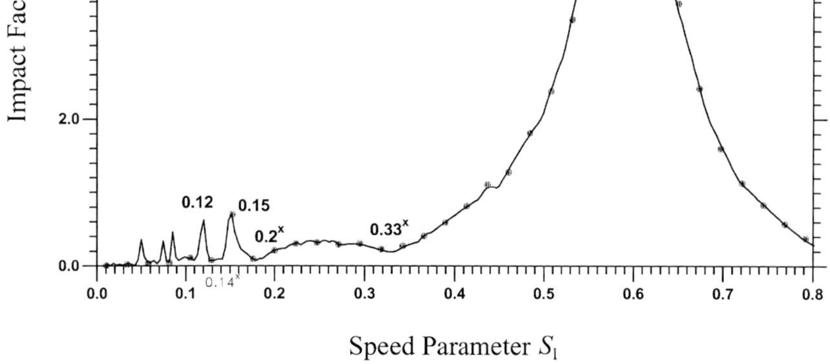


Fig. 2.14. Impact factor versus speed parameter $S_1$.

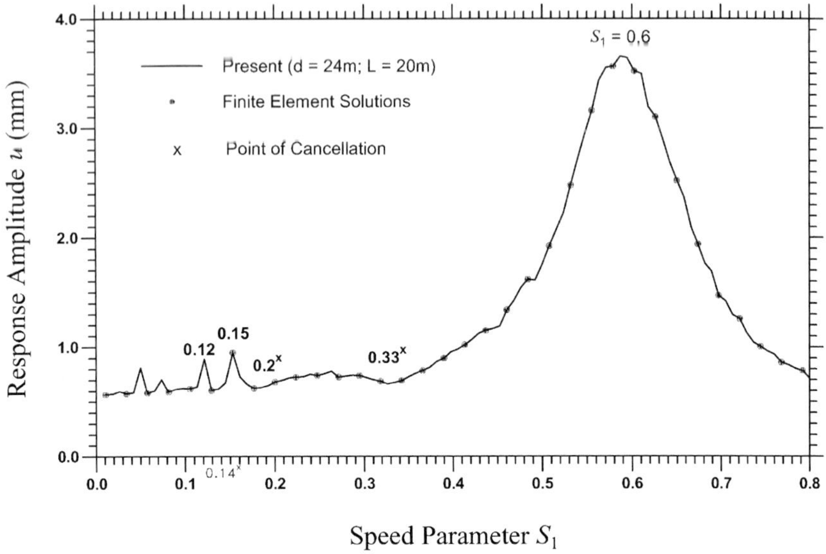


Fig. 2.15. Response amplitude $u$ versus speed parameter $S_1$.

be avoided in practical design for its relatively large magnitude of response. Fourth, if the operation speed of the train can be controlled such that it falls in the range with $0.18 \leq S_1 < 0.33$, or with $184.6\ \text{km/h} \leq v < 338.4\ \text{km/h}$, minimal resonant responses will be induced on the beam. Note that the point with $S_1 = 0.18$ is implied by another cancellation condition $\cos(\omega_1 L_c/2v) = 0$ not mentioned in the text. Finally, the results presented in Figs. 2.14 and 2.15 are nondimensional, which can be applied to simple beams of other cross sections, as long as the car length $d$ and span length $L$ remain unchanged. For instance, for a beam with a smaller cross section, say, with $I = 1.75\ \text{m}^4$, $m = 33,144\ \text{kg/m}$, and $\omega_1 = 30.76\ \text{rad/s}$, the first resonant speed becomes $v = S_1\omega_1 L/\pi = 117.5\ \text{m/s} = 423.0\ \text{km/h}$.

### 2.7.2. *Effects of Moving Masses and Damping*

All the data adopted in this example are identical to those of the previous one in Sec. 2.7.1, except that a damping ratio of $\xi = 2.5\%$

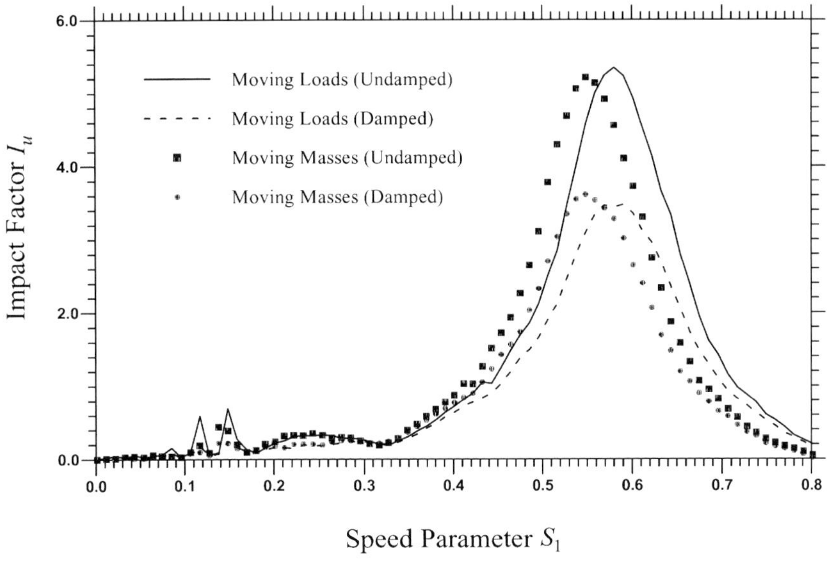


Fig. 2.16. The effects of damping and moving masses.

is considered for the beam. All the results presented in Fig. 2.16 have been obtained by directly integrating the generalized equation of motion in Eq. (2.41) using Newmark's $\beta$ method (see Appendix B) with or without considering the effect of moving masses. As can be seen, the *added mass effect* of the moving masses tends to elongate the period of vibration of the beam, which is the main cause for shifting of the resonance speed to smaller $S_1$ values. On the other hand, the inclusion of 2.5% of damping has resulted in significant reduction of the resonant responses. Nevertheless, the general impact characteristics of the beam, including the speeds for occurrence of resonance and cancellation, can still be observed using the moving load model.

### 2.7.3. *Effect of Span to Car Length Ratio*

In order to demonstrate the effect of span to car length ratio on the impact response, a number of simple beams made of prestressed concrete (with $E = 29.43$ GPa) and with different span lengths $L$ are considered herein, of which the unit mass and first frequency of vibration of each beam are specified as follows: $m = 30 + 0.2L$ t/m and $\omega_1 = 900/L$ rad/s, which represent the interpolation formula for the seven simple beams of span lengths 20, 25, 30, 35, 40, 45, and 50 m studied by T. Y. Lin Taiwan (1993). Zero damping is assumed for all the beams.

Let each of the beams be traveled by the train model considered previously in Sec. 2.7.1. The impact factor $I_u$ computed for the mid-point displacement of the beam has been plotted with respect to the speed parameter $S_1$ and span/car length ratio $L/d$ in Fig. 2.17(a), along with the contour lines in Fig. 2.17(b). An important trend revealed by these figures is that the shorter the span length of a beam, the larger the impact factor for the displacement of the beam. Meanwhile, the resonance speed $S_1$ shifts to smaller values in response to the increase in span length, meaning that a long beam can be more easily excited to resonance by moving loads than a short beam. However, from the response amplitude given in Fig. 2.18(a) and contour lines in Fig. 2.18(b), it can be observed that the maximum responses for long beams are not necessarily smaller than those of short beams. As such, the effect of resonance on long beams has to be considered in design as well. Another observation from the figures is that only the first resonance responses are of engineering significance, while the other resonance responses can generally be neglected in practice.

Noteworthy is the fact that when the span/car length ratio $L/d$ equals 1.5 or 0.5, virtually no resonance response will be induced on the beams under the action of a sequence of continuously moving loads, as can be seen from the contour lines in Figs. 2.17(b) and 2.18(b). In fact, these conditions have been identified to be the *optimal conditions* presented in Eq. (2.62), with $i$ set equal to 2 or 1, by which the first resonance has been suppressed through enforcement of the condition of cancellation. Another merit with the three-phase plots in Figs. 2.17 and 2.18 is that they have all been presented in a nondimensional form, which can be applied virtually to all other simple beams subject to the action of the particular train model considered.

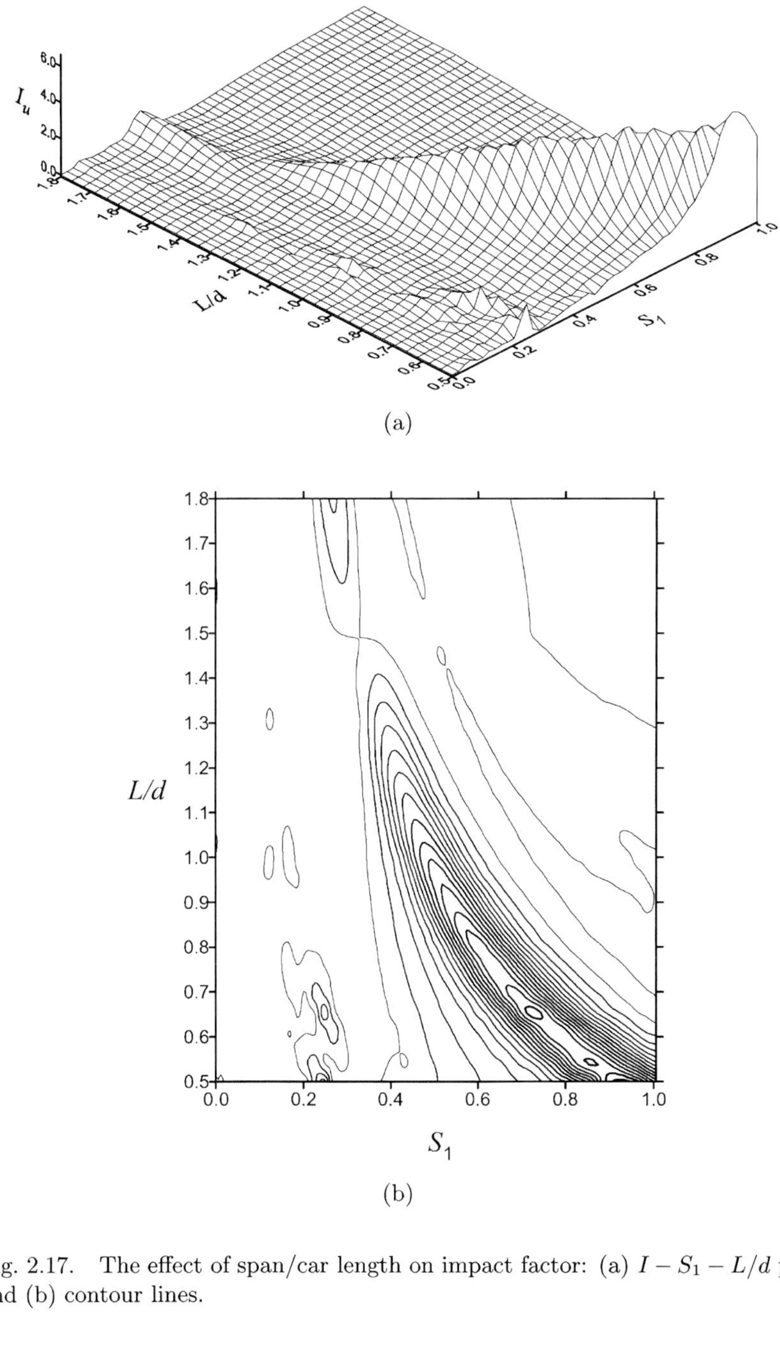


(a)

(b)

Fig. 2.17. The effect of span/car length on impact factor: (a) $I - S_1 - L/d$ plot and (b) contour lines.

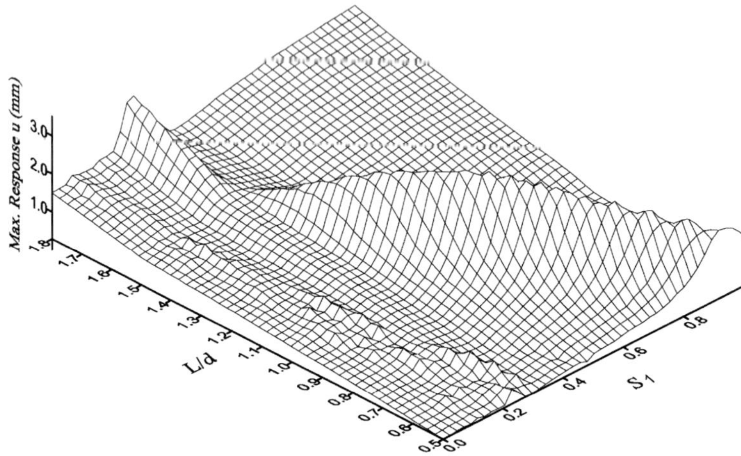


(a)

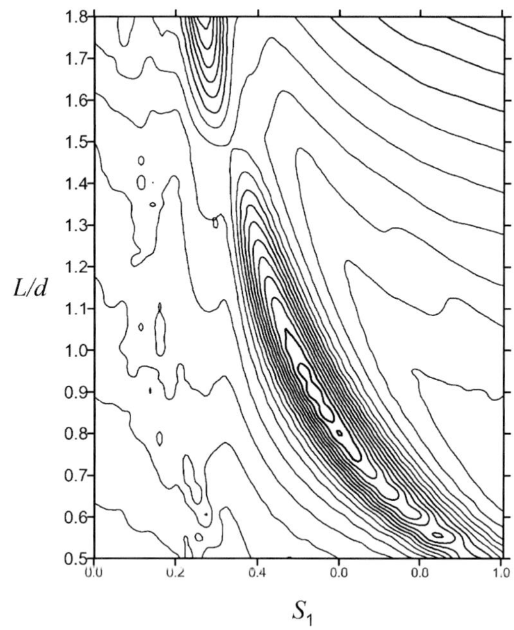


(b)

Fig. 2.18. The effect of span/car length on displacement: (a) $u - S_1 - L/d$ plot and (b) contour lines.

## 2.8. Concluding Remarks

This chapter consists primarily of two parts. In the first part, i.e., in Secs. 2.2–2.5, the impact response of simple beams subjected to a single moving load was analytically studied. The results indicate that there exist upper-bound envelopes for the midpoint deflections, bending moments, and near-support shear forces, which can all be related to the nondimensional speed parameter $S_1$, defined as the ratio of the exciting frequency of the moving load to the fundamental frequency of the beam. For the case with $S_1 < 0.5$, which is normally the case encountered in practice, all the envelopes for the impact factors are linearly proportional to the speed parameter $S_1$, and for the case with $S_1 \geq 0.5$, all the envelopes can be taken as constant. Although the impact factor formulas established herein for a single moving load may be larger than those for multiple moving loads, they serve as conservative and useful upper bounds for the latter cases.

In the second part, i.e., in Secs. 2.6 and 2.7, the moving load problem is extended to deal with the case of train loads, in the sense that a train is modeled as the composition of two sets of wheel loads of constant intervals, each to account for the front and rear wheel assemblies. The following are the conclusions: (1) The first resonance, as indicated by $S_1 = 0.5d/L$, represents the most critical condition and should always be avoided in practical design. (2) Once the condition of cancellation is met, the resonance peak can be effectively suppressed, which forms the basis for the optimal design criteria. (3) The inertia effect of the moving vehicles tends to elongate the period of vibration of the beam, resulting in smaller resonance speeds. (4) Inclusion of damping of the beam can significantly reduce the peak responses. (5) The shorter the span length of a beam, the larger the impact factor for the midpoint deflection of the beam. (6) When the span to car length ratio $L/d$ equals 1.5, virtually no resonance response will be induced on the beam, as the first resonance has been suppressed.

Chapter 3

# Impact Response of Railway Bridges with Elastic Bearings

In earthquake-prone regions, elastic bearings are often inserted at the supports of bridge girders to reduce the earthquake forces transmitted upward from the ground. While they are effective for such a purpose, they may equally prevent the bridge vibrations induced by moving vehicles from transmitting downward to the ground, thereby resulting in accumulated or amplified response on the bridge. The latter problem seems not well addressed in the literature. In this chapter, the dynamic response of an elastically-supported beam to moving train loads will be studied using an analytical approach. The present results indicate that the dynamic response of the beam at resonance remains generally constant, if the effect of damping is taken into account, and the installation of elastic bearings at the supports of the beam to reduce the earthquake forces may adversely amplify the dynamic response of the beam to moving train loads. Envelope impact formulas are derived for the deflection of the beam with light damping, which serve as a useful, preliminary design aid to railway engineers.

## 3.1. Introduction

To prevent the damage of bridges from severe earthquakes, elastic bearings are often used as base isolators in bridge engineering. Conventionally, they are installed at the supports of bridge girders to serve as filters for isolating the vibration energy transmitted from the ground. However, as they are effective for shielding the upward-transmitting forces, they may be equally effective in preventing the vehicle-induced vibrations of the bridge girder from dissipating to the ground, thereby resulting in accumulated, amplified response on the bridge when subjected to a connected line of moving loads, as is the case encountered in railways. In this chapter, an analytical approach is presented for investigating the dynamic response of bridges with elastic bearings to high-speed trains. Particular emphasis is placed on the resonant response that may be induced by high-speed trains.

The dynamic response of bridge structures to moving loads at high speeds is a problem of great concern in the design of high-speed railway bridges. As can be seen from the review presented in Chapter 1, a large number of analytical investigations have been carried out by assuming the bridges to be simply-supported, and by modeling the vehicles as moving loads or moving masses. In Chapter 2, based on the paper by Yang *et al.* (1997b), we have presented a closed-form solution for the dynamic response of simply-supported beams to a series of moving loads at high speeds, in which both the phenomena of resonance and cancellation have been investigated. By considering the effect of damping, Li and Su (1999) investigated the fundamental characteristics and dominant factors for the resonant vibration of a girder bridge under high-speed trains. Using the dynamic stiffness approach and damped Timoshenko beam theory, Chen and Li (2000) investigated the dynamic response of elevated high-speed railways considered by the Bureau of Taiwan High Speed Rail in the preliminary stage of design.

On the other hand, based on the finite element methods, more sophisticated models have been devised to study the dynamic behavior of vehicle–bridge interaction problems by researchers. To resolve the coupling effect between the bridge and moving vehicles, methods that are of iterative nature have been employed (Hwang and Nowak, 1991; Green and Cebon, 1994; Yang and Fonder, 1996). By the concept of dynamic condensation, Yang and Yau (1997) and Yang *et al.* (1999) developed the vehicle–bridge interaction (VBI) elements for the dynamic analysis of railway bridges subjected to moving trains, which will be described in Chapters 6 and 7. Cheung *et al.* (1999)

used the modified beam vibration functions to investigate the response of multi-span nonuniform bridges under moving vehicles and trains. Ichikawa *et al.* (2000) used the modal analysis method to study the dynamic behavior of the continuous beam subjected to a moving mass. They found that the inertial effect of the moving mass has greater influence on the second and successive spans than on the first span. By using the ultimate vertical accelerations of the railway bridge deck to evaluate the operability of high-speed trains, and by comparing the theoretical results with the experiments carried out on high-speed railway lines in Europe, Frýba (2001) proposed some interoperability formulas for quick assessment of the suitability of railway bridges for high-speed trains. For a given train configuration, Savin (2001) derived analytically a pseudo-acceleration spectrum for predicting the maximum accelerations of weakly-damped beams with various boundary conditions.

To the knowledge of the authors, very few studies have been conducted on the dynamic response of elastically-supported beams to a series of moving loads. The objective of this chapter is to analytically investigate the dynamic behavior of elastically-supported beams subjected to moving loads in the high-speed range. Based on the analytical results, envelope impact formulas that take into account the effect of damping will be proposed for the deflection of the beam. The accuracy of such a formula will be demonstrated in the numerical examples through comparison with the finite element solutions. The materials presented in this chapter are based largely on the work by Yau *et al.* (2001).

## 3.2. Equation of Motion

As shown in Fig. 3.1, a beam supported by two elastic bearings of stiffness $K$ at the two ends is considered. The beam is assumed to be of length $L$ and uniform cross sections. The train moving over the beam at speed $v$ is modeled as a sequence of equidistant moving loads. The interval between two adjacent moving loads is $d$ and the weight of each moving load is $p$. The equation of motion for the

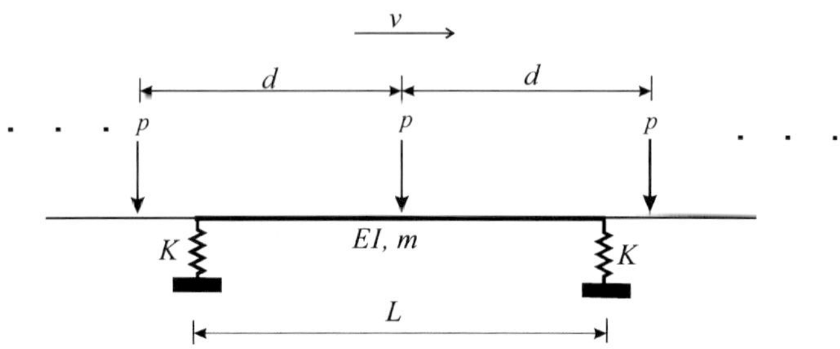


Fig. 3.1. Elastically-supported beam subjected to uniform moving loads.

beam traveled by the moving loads can be written as

$$m\ddot{u} + c_e\dot{u} + c_i\dot{u}'''' + EIu''''$$
$$= p\sum_{k=1}^{N} \delta[x - v(t - t_k)] \times \left[H(t - t_k) - H\left(t - t_k - \frac{L}{v}\right)\right], \quad (3.1)$$

where a prime denotes derivative with respect to coordinate $x$, an overdot denotes derivative with respect to time $t$, $m$ = the mass per unit length of the beam, $u(x,t)$ = vertical displacement, $c_e$ = external damping coefficient, $c_i$ = internal damping coefficient, $E$ = elastic modulus, $I$ = moment of inertia of the beam, $\delta(x)$ = Dirac's delta function, $H(t)$ = unit step function, $N$ = total number of moving loads, and $t_k = (k-1)d/v$ = arriving time of the $k$th load at the beam. Correspondingly, the boundary conditions of the beam are

$$\begin{aligned} EIu''(0,t) &= 0\,, \\ EIu''(L,t) &= 0\,, \\ EIu'''(0,t) &= -Ku(0,t)\,, \\ EIu'''(L,t) &= Ku(L,t)\,, \end{aligned} \quad (3.2)$$

and the initial conditions are

$$\begin{aligned} u(x,0) &= 0\,, \\ \dot{u}(x,0) &= 0\,, \end{aligned} \quad (3.3)$$

assuming that the beam is initially at rest.

## 3.3. Fundamental Frequency of the Beam

To analyze the dynamic response of the elastically-supported beam to a sequence of moving loads, the vibration shape of the beam will be approximated by the combination of a flexural sine mode and a rigid displacement mode, as shown in Fig. 3.2. Thus, the displacement $u(x,t)$ of the elastically-supported beam can be expressed as

$$u(x,t) = q(t)\phi(x) \cong q(t) \times \frac{\sin(\pi x/L) + \kappa}{1+\kappa}, \tag{3.4}$$

where $q(t)$ denotes the *generalized coordinate* of the vibration shape, $\phi(x)$ the assumed shape function, and $\kappa(= EI\pi^3/KL^3)$ is the *stiffness ratio* of the beam to the elastic bearings. In particular, the term $\kappa$ in the numerator of Eq. (3.4) denotes the *rigid displacement mode.* For the special case with $\kappa = 0$, which implies hinge supports, the

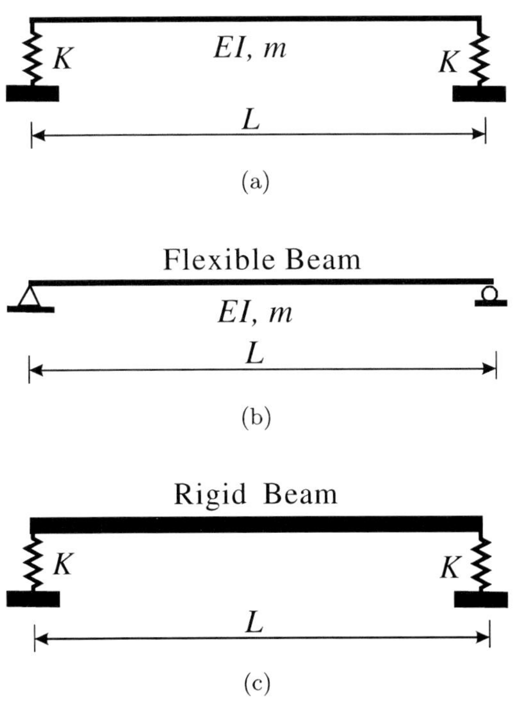


Fig. 3.2. The concept for modal vibration shape of elastically-supported beam.

assumed mode shape of vibration, $\phi(x)$, reduces to the first flexural mode. As can be seen, higher modes other than the first flexural and rigid modes have been excluded from Eq. (3.4), which is justified for the moving load problems, because of the transient nature of the beam in response to the moving loads. This is particularly true if only the midpoint displacement of the beam is desired.

By *Rayleigh's method*, the fundamental frequency $\omega$ of the elastically-supported beam can be computed as

$$\omega^2 = \frac{\int_0^L EI[\phi''(x)]^2 dx + K\{[\phi(0)]^2 + [\phi(L)]^2\}}{\int_0^L m[\phi(x)]^2 dx}$$

$$\cong \frac{\omega_0^2(1 + 4\kappa/\pi)}{1 + 8\kappa/\pi + 2\kappa^2}, \tag{3.5}$$

where $\omega_0 = (\pi/L)^2\sqrt{EI/m}$ = the fundamental frequency of the corresponding beam with hinge supports. When $\kappa$ equals 0, the fundamental frequency $\omega$ of the elastically-supported beam equals that of the corresponding simple beam. On the other hand, if $\kappa$ approaches infinity, the fundamental frequency $\omega$ reduces to zero, meaning that the beam is unsupported. To verify the accuracy of the approximate fundamental frequency given by Eq. (3.5), the exact solution will be computed from the following *frequency equation* for the elastically-supported beam (Gorman, 1975):

$$b^2\left(\cos kL - \frac{1}{\cosh kL}\right) + 2b(\sin kL - \tanh kL \cos kL)$$

$$- 2\sin kL \tanh kL = 0, \tag{3.6}$$

where $kL = \pi\sqrt{\omega/\omega_0}$ and $b = EIk^2$. The fundamental frequencies computed from Eq. (3.5), which are approximate, have been compared with the exact ones obtained from Eq. (3.6) for different stiffness ratios in Fig. 3.3. As can be seen, the approximate frequencies agree excellently with the exact ones, implying that the use of the approximate mode shape for the elastically-supported beam, as given in Eq. (3.4), is acceptable.

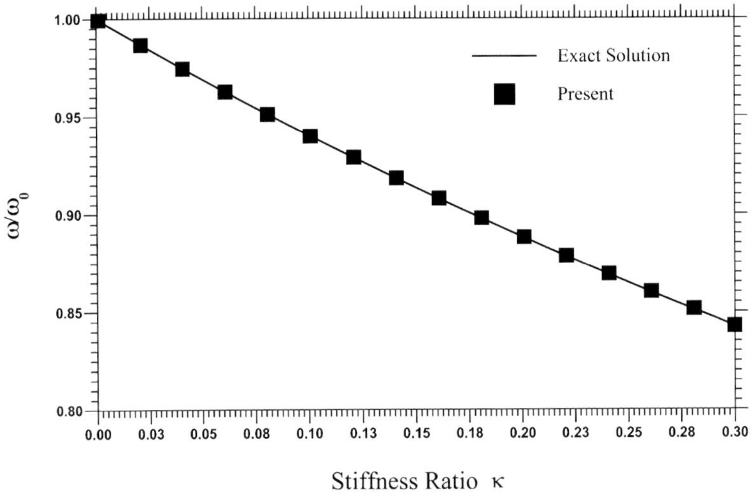


Fig. 3.3. The comparison of fundamental frequencies.

## 3.4. Dynamic Response Analysis

By substituting the expression for the displacement $u(x,t)$ in Eq. (3.4) into Eq. (3.1), multiplying both sides of the equation by the shape function $\phi(x)$, and then integrating with respect to the beam axis $x$ over the length $L$, one obtains the equation of motion in terms of the generalized coordinate $q(t)$ as

$$\ddot{q}(t)+2\xi\omega\dot{q}(t)+\omega^2 q(t)=\frac{2p(1+\kappa)}{mL(1+8\kappa/\pi+2\kappa^2)}\sum_{k=1}^{N}F_k(v,t)\,, \tag{3.7}$$

where $\xi$ is the *modal damping ratio* and $F_k(v,t)$ is the *generalized forcing function*,

$$\begin{aligned}F_k(v,t)&=[\kappa+\sin\Omega(t-t_k)]H(t-t_k)\\&\quad+\left[-\kappa+\sin\Omega\left(t-t_k-\frac{L}{v}\right)\right]H\left(t-t_k-\frac{L}{v}\right).\end{aligned} \tag{3.8}$$

Here, $\Omega(=\pi v/L)$ is the *driving frequency*, as implied by the moving loads.

First, consider the case when only a single moving load is crossing the bridge. The equation of motion, Eq. (3.7), becomes

$$\ddot{q}(t) + 2\xi\omega\dot{q}(t) + \omega^2 q(t) = \frac{2p(1+\kappa)}{mL(1+8\kappa/\pi+2\kappa^2)}\left(\sin\frac{\pi vt}{L}+\kappa\right). \tag{3.9}$$

By *Duhamel's integral*, the generalized coordinate $q(t)$ can be solved from Eq. (3.9) as

$$\begin{aligned} q(t) = {} & \frac{[\Delta_{st}(1+\kappa)]/[1+4\kappa/\pi]}{(1-S^2)^2+(2\xi S)^2}\left\{(1-S^2)\sin\Omega t - 2\xi S\cos\Omega t \right. \\ & \left. + e^{-\xi\omega t}\left[2\xi S\cos\omega_d t - \frac{S(1-S^2-2\xi^2)}{\sqrt{1-\xi^2}}\sin\omega_d t\right]\right\} \\ & + \frac{\Delta_{st}(1+\kappa)}{1+4\kappa/\pi}\times\kappa\{1-e^{-\xi\omega t}[\cos\omega_d t + \xi\sqrt{1-\xi^2}\sin\omega_d t]\}, \end{aligned} \tag{3.10}$$

where $\Delta_{st} = 2pL^3/\pi^4 EI \approx pL^3/48EI$ = the *maximum static deflection* of the corresponding simple beam, $S = \Omega/\omega$ = the *speed parameter*, and $\omega_d = \omega\sqrt{1-\xi^2}$ = the damped frequency. It should be noted that the speed parameter $S$ represents the ratio of the driving frequency to the frequency of the beam. For most of the vehicle–bridge problems encountered in practice, the speed parameter $S$ is less than 0.3.

In this chapter, only elastically-supported beams with *light damping* ($\xi < 0.05$) are considered, which implies that terms involving $\xi^2$, $\xi S$, and $\xi\kappa$ are so small that they can be neglected. As a result, the response in Eq. (3.10) reduces to

$$q(t) \cong \frac{\Delta_{st}(1+\kappa)}{1+4\kappa/\pi}\left[\frac{1}{1-S^2}G_1(v,t)+\kappa G_2(t)\right], \quad 0 \le vt \le L, \tag{3.11}$$

where the functions $G_1(v,t)$ and $G_2(t)$ are

$$\begin{aligned} G_1(v,t) &= \sin \Omega t - S e^{-\xi\omega t} \sin \omega t\,, \\ G_2(t) &= 1 - e^{-\xi\omega t} \cos \omega t\,. \end{aligned} \tag{3.12}$$

Consider the case when a series of moving loads of constant intervals $d$ are crossing the bridge. The dynamic response can be extended from Eq. (3.11) as follows:

$$\begin{aligned} q(t) \cong \frac{\Delta_{st}(1+\kappa)}{1+4\kappa/\pi} \Bigg\{ & \frac{1}{1-S^2} \sum_{k=1}^{N} \Bigg[ G_1(v, t-t_k) H(t-t_k) \\ & + G_1\left(v, t-t_k-\frac{L}{v}\right) H\left(t-t_k-\frac{L}{v}\right)\Bigg] \\ & + \kappa \sum_{k=1}^{N} [G_2(t-t_k)H(t-t_k) \\ & - G_2(t-t_k-L/v)H(t-t_k-L/v)] \Bigg\}\,, \end{aligned} \tag{3.13}$$

where $t_k = (k-1)d/v$ denotes the arriving time of the $k$th load on the bridge, the unit step function $H(t-t_k)$ is used to represent the direct action of the $k$th moving load on the beam, while the function $H(t-t_k-L/v)$ is the residual action of the $k$th moving load. It is easy to see that for the undamped case, $\xi = 0$, as $\kappa = 0$, the preceding expression for the generalized coordinate reduces to that for the simply-supported beam, as the one implied by Eq. (2.53).

## 3.5. Phenomena of Resonance and Cancellation

In this chapter, the span length $L$ of the beam is assumed to be no greater than twice the interval $d$ between two consecutive moving loads, i.e., $L \leqq 2d$ which is the case implied by the most high-speed railway constructions. In order to derive the conditions for the phenomena of resonance and cancellation to occur under the action of a series of moving loads of constant intervals, we shall neglect the effect of damping on the beam in this section. For this special case, $\xi = 0$, the generalized coordinate of the response of the beam, as given in Eq. (3.13), reduces to

$$q(t) = \frac{\Delta_{st}(1+\kappa)}{1+4\kappa/\pi}\left[\frac{P_1(v,t)}{1-S^2} + \kappa P_2(v,t)\right], \tag{3.14}$$

where

$$\begin{aligned} P_1(v,t) = \sum_{k=1}^{N} \Bigg\{ & [\sin\Omega(t-t_k) - S\sin\omega(t-t_k)]H(t-t_k) \\ & + \left[\sin\Omega\left(t-t_k-\frac{L}{v}\right) - S\sin\omega\left(t-t_k-\frac{L}{v}\right)\right] \\ & \times H\left(t-t_k-\frac{L}{v}\right)\Bigg\}, \end{aligned} \tag{3.15a}$$

$$\begin{aligned} P_2(v,t) = \sum_{k=1}^{N} \Bigg\{ & [1-\cos\omega(t-t_k)]H(t-t_k) \\ & - \left[1-\cos\omega\left(t-t_k-\frac{L}{v}\right)\right] H\left(t-t_k-\frac{L}{v}\right)\Bigg\}. \end{aligned} \tag{3.15b}$$

In Eqs. (3.14) and (3.15), the function $P_1(v,t)$ indicates the contribution of the flexural vibration mode of the simple beam, which is identical to the function $\bar{P}_1(v,t)$ given in Eq. (2.54) for the simply-supported beam, and $P_2(v,t)$ the rigid displacement mode of the elastic bearings. The beam will be excited to its utmost, in the sense that the response will reach the maximum, when the last (i.e., the $N$th) moving load enters the beam. For this case, $t_N < t < t_N + L/v$, the function $P_1(v,t)$ in Eq. (3.15a) can be simplified as

$$\begin{aligned} P_1(v,t) &= [\sin \Omega(t-t_N) - S\sin\omega(t-t_N)]H(t-t_N) \\ &\quad - 2S\cos\frac{\omega L}{2v} \times \left[\sin\omega\left(t-\frac{L}{2v}\right) + \sin\omega\left(t-\frac{t_N+L/v}{2}\right)\right. \\ &\quad \left. \times \sin\frac{\omega}{2}\left(t_N-\frac{d}{v}\right) \Big/ \sin\left(\frac{\omega d}{2v}\right)\right] H\left(t-t_{N-1}-\frac{L}{v}\right) \end{aligned} \tag{3.16}$$

by following the procedure presented in Sec. 2.6.2 leading to Eq. (2.55). In the meantime, the function $P_2(v,t)$ in Eq. (3.15b) reduces to

$$\begin{aligned} P_2(v,t) &= [1-\cos\omega(t-t_N)]H(t-t_N) \\ &\quad + \left\{\sum_{k=1}^{N-1}\left[\cos\omega\left(t-t_k-\frac{L}{v}\right) - \cos\omega(t-t_k)\right]\right\} \\ &\quad \times H\left(t-t_{N-1}-\frac{L}{v}\right). \end{aligned} \tag{3.17}$$

Through introduction of the following relations:

$$\begin{aligned} &\cos\omega\left(t-t_k-\frac{L}{v}\right) - \cos(t-t_k) \\ &\quad = 2\sin\left(\frac{\omega L}{2v}\right)\sin\left[\omega\left(t-t_k-\frac{L}{2v}\right)\right], \\ &\sum_{k=1}^{N-1}\sin\omega\left(t-t_k-\frac{L}{2v}\right) \\ &\quad = \sin\omega\left(t-\frac{L}{2v}\right) + \sin\omega\left(t-\frac{t_N+L/v}{2}\right)\frac{\sin\frac{\omega}{2}(t_N-\frac{d}{v})}{\sin(\frac{\omega d}{2v})}, \end{aligned} \tag{3.18}$$

the function $P_2(v,t)$ in Eq. (3.17) can be rearranged as

$$\begin{aligned}P_2(v,t) &= [1-\cos\omega(t-t_N)]H(t-t_N)\\ &\quad + 2\sin\frac{\omega L}{2v} \times \left[\sin\omega\left(t-\frac{L}{2v}\right) + \sin\omega\left(t-\frac{t_N+L/v}{2}\right)\right.\\ &\quad \left.\times \sin\frac{\omega}{2}\left(t_N-\frac{d}{v}\right)\Big/\sin\left(\frac{\omega d}{2v}\right)\right] H\left(t-t_{N-1}-\frac{L}{v}\right).\end{aligned} \tag{3.19}$$

By using Eqs. (3.14), (3.16), and (3.19), the dynamic response for the midpoint of the elastically-supported beam can be obtained from Eq. (3.4) as

$$\begin{aligned}u\left(\frac{L}{2},t\right) &= \frac{\Delta_{st}(1+\kappa)}{1+4\kappa/\pi}\\ &\quad \times \left[Q_1(v,t)H(t-t_N) + Q_2(v,t)H\left(t-t_{N-1}-\frac{L}{v}\right)\right],\end{aligned} \tag{3.20}$$

where the dynamic response factors $Q_1(v,t)$ and $Q_2(v,t)$ are

$$\begin{aligned}Q_1(v,t) &= \frac{\sin\Omega(t-t_N) - S\sin\omega(t-t_N)}{1-S^2}\\ &\quad + \kappa[1-\cos\omega(t-t_N)]\,,\end{aligned} \tag{3.21a}$$

$$\begin{aligned}Q_2(v,t) &= 2\left[\kappa\sin\left(\frac{\pi}{2S}\right) - S\frac{\cos(\pi/2S)}{1-S^2}\right]\\ &\quad \times \left[\sin\omega\left(t-\frac{L}{2v}\right) + \sin\omega\left(t-\frac{t_N+L/v}{2}\right)\right.\\ &\quad \left.\times \frac{\sin((\omega/2))(t_N-(d/v))}{\sin(\omega d/2v)}\right].\end{aligned} \tag{3.21b}$$

It should be noted that the relation $\omega L/2v = \pi/2S$ has been utilized in arriving at Eq. (3.21). From Eq. (3.20), it can be seen that the term $Q_1(v,t)$ is associated with $H(t-t_N)$, which represents the *forced response* induced by the $N$th moving load directly acting on the beam, and the term $Q_2(v,t)$ is associated with $H(t-t_{N-1}-L/v)$, which represents the *residual response* induced by the $N-1$ moving loads that have passed the beam.

From Eq. (3.21b), it can be seen that the response reaches a maximum when $\sin(\omega d/2v) = 0$. This is exactly the condition for *resonance* to occur. Correspondingly, the resonant speed is denoted as $v_r$. Note that the resonance speed $v_r$ as implied by the condition $\sin(\omega d/2v) = 0$ is independent of the stiffness ratio $\kappa$. By the L'Hospital rule, the dynamic response factor $Q_2(v_r,t)$ for resonance in Eq. (3.21b) can be manipulated to yield

$$\begin{aligned} Q_2(v_r,t) = 2(N-1)\left[\kappa \sin\left(\frac{\pi}{2S_r}\right) - S_r \frac{\cos(\pi/2S_r)}{1-S_r^2}\right] \\ \times \sin\omega\left(t - \frac{L}{2v_r}\right), \end{aligned} \tag{3.22}$$

where $S_r$ is the speed parameter for resonance to occur. The preceding equation indicates that under the condition of resonance, larger response will be induced on the beam if there are more loads passing the beam, as implied by $(N-1)$. The other observation from Eq. (3.22) is that when the signs of $\sin(\pi/2S_r)$ and $\cos(\pi/2S_r)$ are different, i.e., when the vibration phases of the elastic bearings and the beam are the same, the response factor $Q_2(v_r,t)$ attains its maximum, meaning that the beam response will be amplified. In contrast, when the elastic bearings and beam are out of phase, less severe response will be induced on the beam.

On the other hand, as can be observed from Eq. (3.21b), whenever the following condition is met, that is,

$$\kappa \sin\left(\frac{\pi}{2S}\right) - S\frac{\cos(\pi/2S)}{1-S^2} = 0, \tag{3.23}$$

the residual response caused by the previous $N-1$ moving loads that have passed the beam disappears. Because of this, the condition in Eq. (3.23) has been referred to as the condition of *cancellation*. Under this condition, the midpoint dynamic response of the beam in

Eq. (3.20) becomes

$$u\left(\frac{L}{2},t\right) = \frac{\Delta_{st}(1+\kappa)}{1+4\kappa/\pi} \times \left\{ \frac{\sin\Omega(t-t_N) - S_c \sin\omega(t-t_N)}{1-S_c^2} \right.$$
$$\left. + \kappa[1-\cos\omega(t-t_N)] \right\}, \tag{3.24}$$

where $S_c$ denotes the speed parameter for cancellation to occur. As can be seen, whenever the condition of cancellation is met, the response of the beam is determined solely by the $N$th moving load, while the effect of the moving loads that have passed the beam is fully suppressed.

## 3.6. Effect of Structural Damping

Consider the case when the resonance condition implied by Eq. (3.21b) is met, i.e., $\sin(\omega d/2v_r) = 0$, or $\omega d/v_r = 2n\pi$, with $n = 1, 2, 3\ldots$, and when the $N$th moving load is acting on the beam at time $t = t_E + t_N$, that is, $t = t_E + (N-1)d/v_r$, where $t_E$ denotes the time for the load to reach the position shown in Fig. 3.4. Then $\omega(t-t_k) = \omega(t_E+t_N-t_k) = \omega t_E+(N-k)\omega d/v_r = \omega t_E+2n\pi(N-k)$ and the following relations can be derived:

$$\begin{aligned}
&\sin\Omega(t-t_k) + \sin\Omega\left(t-t_k-\frac{L}{v_r}\right) = 0\,,\\
&\sin\omega(t-t_k) = \sin\omega t_E\,,\\
&\sin\omega\left(t-t_k-\frac{L}{v_r}\right) = \sin\omega\left(t_E-\frac{L}{v_r}\right),\\
&\cos\omega(t-t_k) = \cos\omega t_E\,,\\
&\cos\omega\left(t-t_k-\frac{L}{v_r}\right) = \cos\omega\left(t_E-\frac{L}{v_r}\right),
\end{aligned} \tag{3.25}$$

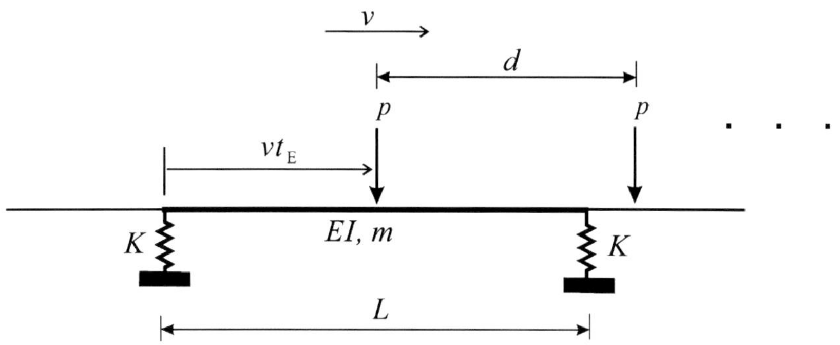


Fig. 3.4. The $N$th moving load is acting at position $vt_E$.

for $0 < t < t_N$. By introducing the preceding relations into Eq. (3.13), where the effect of damping of the beam has been considered, the resonance response for the elastically-supported beam under the action of the last (i.e., the $N$th) moving load can be expressed as

$$q_r(t_E) \cong \frac{\Delta_{st}(1+\kappa)}{1+4\kappa/\pi} \times \left\{ \left[ \frac{\sin \Omega t_E - S_r e^{-\xi\omega t_E} \sin \omega t_E}{1-S_r^2} \right.\right.$$
$$\left. + \kappa(1 - e^{-\xi\omega t_E} \cos \omega t_E) \right] H(t_E) + e^{-\xi\omega(t_E+t_N)} \sum_{k=1}^{N-1} e^{\xi\omega t_k}$$
$$\times \left[ \frac{-S_r}{1-S_r^2} \left( \sin \omega t_E + e^{\xi\pi/S_r} \sin \omega \left( t_E - \frac{L}{v_r} \right) \right) \right.$$
$$\left. + \kappa \left( e^{\xi\pi/S_r} \cos \omega \left( t_E - \frac{L}{v_r} \right) - \cos \omega t_E \right) \right]$$
$$\left. \times H\left( t_E - \frac{L-d}{v_r} \right) \right\}. \tag{3.26}$$

Furthermore, by using the approximation in expansion for the exponential function, i.e., $\exp(\xi\pi/S_r) \cong 1 + \xi\pi/S_r$ for $\xi\pi/S_r \leq 0.3$, and the following relations for the series sum:

$$\begin{aligned} \sum_{k=1}^{N-1} e^{\xi\omega t_k} &= \sum_{k=1}^{N} e^{\xi\omega d(k-1)/v_r} = \frac{e^{\xi\omega d(N-1)/v_r} - 1}{e^{\xi\omega d/v_r} - 1}, \\ e^{-\xi\omega t_N} \sum_{k=1}^{N-1} e^{\xi\omega t_k} &= \frac{1 - e^{-\xi\omega d(N-1)/v_r}}{e^{\xi\omega d/v_r} - 1}, \end{aligned} \tag{3.27}$$

the dynamic response in Eq. (3.26) can be further expressed as

$$\begin{aligned} q_r(t_E) \cong \frac{\Delta_{st}(1+\kappa)}{1+4\kappa/\pi} \times \Bigg\{ \Bigg[ & \frac{\sin\Omega t_E - S_r e^{-\xi\omega t_E} \sin\omega t_E}{1-S_r^2} \\ & + \kappa(1-e^{-\xi\omega t_E}\cos\omega t_E) \Bigg] H(t_E) \\ & + \Bigg[ 2\left(\kappa\sin\frac{\pi}{2S_r} - \frac{S_r}{1-S_r^2}\cos\frac{\pi}{2S_r}\right)\sin\omega\left(t_E - \frac{L}{2v_r}\right) \\ & + \frac{\xi\pi}{S_r}\left(\kappa\cos\omega\left(t_E - \frac{L}{v_r}\right) - \frac{S_r}{1-S_r^2}\sin\omega\left(t_E - \frac{L}{v_r}\right)\right)\Bigg] e^{-\xi\omega t_E} \\ & \times \frac{1-e^{-\xi(N-1)\omega d/v_r}}{e^{\xi\omega d/v_r}-1} H\left(t_E - \frac{L-d}{v_r}\right)\Bigg\}. \end{aligned} \tag{3.28}$$

For the special case of zero damping, i.e., by letting $\xi = 0$, the dynamic response in Eq. (3.28) reduces to

$$\begin{aligned} q_r(t_E) = \frac{\Delta_{st}(1+\kappa)}{1+4\kappa/\pi} \times \Bigg\{ \Bigg[ & \frac{\sin\Omega t_E - S_r \sin\omega t_E}{1-S_r^2} \\ & + \kappa(1-\cos\omega t_E) \Bigg] H(t_E) \\ & + 2(N-1)\left(\kappa\sin\frac{\pi}{2S_r} - \frac{S_r}{1-S_r^2}\cos\frac{\pi}{2S_r}\right)\sin\omega\left(t_E - \frac{L}{2v_r}\right) \end{aligned}$$

$$\times H\left(t_E - \frac{L-d}{v_r}\right)\Bigg\}$$

$$= \frac{\Delta_{st}(1+\kappa)}{1+4\kappa/\pi}\Bigg[Q_1(v_r, t_E)H(t-t_N)$$

$$+ Q_2(v_r, t)H\left(t - t_{N-1} - \frac{d}{v_r}\right)\Bigg], \tag{3.29}$$

which is identical to the one given in Eq. (3.20) for the condition of resonance.

For the purpose of obtaining closed-form solutions, let us assume that there is an infinite number of moving loads crossing the beam. By letting the number of vehicles approach infinity, $N \to \infty$, and using the relation $\exp(\xi\omega d/v_r) - 1 \cong \xi\omega d/(S_r L)$ for light damping, Eq. (3.28) becomes

$$q_r(t_E) \cong \frac{\Delta_{st}(1+\kappa)}{1+4\kappa/\pi} \times \Bigg\{\Bigg[\frac{\sin\Omega t_E - S_r e^{-\xi\omega t_E}\sin\omega t_E}{1-S_r^2}$$

$$+ \kappa(1 - e^{-\xi\omega t_E}\cos\omega t_E)\Bigg]H(t_E) + e^{-\xi\omega t_E}$$

$$\times \frac{S_r L}{\xi\pi d}\Bigg[2\left(\kappa\sin\frac{\pi}{2S_r} - \frac{S_r}{1-S_r^2}\cos\frac{\pi}{2S_r}\right)\sin\left(\omega t_E - \frac{\pi}{2S_r}\right)$$

$$+ \frac{\xi\pi}{S_r}\left(\kappa\cos\left(\omega t_E - \frac{\pi}{S_r}\right) - \frac{S_r}{1-S_r^2}\sin\left(\omega t_E - \frac{\pi}{S_r}\right)\right)\Bigg]$$

$$\times H\left(t_E - \frac{L-d}{v_r}\right)\Bigg\}. \tag{3.30}$$

As can be verified from Eq. (3.30) and in the examples to follow, if the *effect of damping* is considered, the resonant response of the beam subjected to an infinite number of moving loads remains more or less constant. Furthermore, by the use of the following relations:

$$\begin{aligned}\cos\left(\omega t_E - \frac{\pi}{S_r}\right) &= \cos\frac{\pi}{2S_r}\cos\left(\omega t_E - \frac{\pi}{2S_r}\right)\\ &\quad + \sin\frac{\pi}{2S_r}\sin\left(\omega t_E - \frac{\pi}{2S_r}\right),\\ \sin\left(\omega t_E - \frac{\pi}{S_r}\right) &= \cos\frac{\pi}{2S_r}\sin\left(\omega t_E - \frac{\pi}{2S_r}\right)\\ &\quad - \sin\frac{\pi}{2S_r}\cos\left(\omega t_E - \frac{\pi}{2S_r}\right),\end{aligned} \tag{3.31}$$

the resonance response in Eq. (3.30) becomes

$$\begin{aligned}q_{r,\max}(t_E) \approx \frac{\Delta_{st}(1+\kappa)}{1+4\kappa/\pi} \times &\left\{\left[\frac{\sin\Omega t_E - S_r e^{-\xi\omega t_E}\sin\omega t_E}{1-S_r^2}\right.\right.\\ &\left.+\kappa(1-e^{-\xi\omega t_E}\cos\omega t_E)\right]H(t_E)\\ &+\frac{L}{d}e^{-\xi\omega t_E}\times\left[\left(\frac{2S_r}{\xi\pi}+1\right)\right.\\ &\times\left(\kappa\sin\frac{\pi}{2S_r} - \frac{S_r}{1-S_r^2}\cos\frac{\pi}{2S_r}\right)\sin\left(\omega t_E - \frac{\pi}{2S_r}\right)\\ &\left.+\left(\kappa\cos\frac{\pi}{2S_r} + \frac{S_r}{1-S_r^2}\sin\frac{\pi}{2S_r}\right)\cos\left(\omega t_E - \frac{\pi}{2S_r}\right)\right]\\ &\left.\times H\left(t_E - \frac{L-d}{v_r}\right)\right\}.\end{aligned} \tag{3.32}$$

Since $\sin(\omega t_E - \pi/2S_r)$ and $\cos(\omega t_E - \pi/2S_r)$ are out of phase, when the function $\sin(\omega t_E - \pi/2S_r)$ reaches the maximum, the function $\cos(\omega t_E - \pi/2S_r)$ is at its minimum. As far as the maximum response is concerned, the preceding expression can be approximated by dropping the term containing $\cos(\omega t_E - \pi/2S_r)$ as follows:

$$q_{r,\max}(t_E) \approx \frac{\Delta_{st}(1+\kappa)}{1+4\kappa/\pi} \times \left\{ \left[ \frac{\sin \Omega t_E - S_r e^{-\xi\omega t_E} \sin \omega t_E}{1 - S_r^2} \right.\right.$$
$$\left. + \kappa(1 - e^{-\xi\omega t_E} \cos \omega t_E) \right] H(t_E) + \frac{L}{d} e^{-\xi\omega t_E} \left( \frac{2S_r}{\xi\pi} + 1 \right)$$
$$\times \left( \kappa \sin \frac{\pi}{2S_r} - \frac{S_r}{1 - S_r^2} \cos \frac{\pi}{2S_r} \right) \sin \left( \omega t_E - \frac{\pi}{2S_r} \right)$$
$$\left. \times H\left( t_E - \frac{L-d}{v_r} \right) \right\}, \tag{3.33}$$

where it is recognized that $2S_r/\xi\pi + 1 > 1$. At this point, we have derived the maximum response for the elastically-supported beam in Eq. (3.33) considering the effect of damping.

## 3.7. Envelope Formula for Resonance Response

In this section, envelope formulas will be derived for the elastically-supported beam based on the maximum response presented in Eq. (3.33) assuming that the beam is lightly-damped and is subjected to an infinite number of moving loads. When the condition of resonance is met, while the condition of cancellation as given in Eq. (3.23) is not, the dynamic response of the beam is dominated by the term containing $H(t_E - (L-d)/v_r)$ in Eq. (3.33). Moreover, the function $\sin(\omega t_E - \pi/2S_r)$ reaches its maximum when

$$\omega t_E = \frac{\pi}{2} + \frac{\pi}{2S_r} = \frac{\pi(1+S_r)}{2S_r}. \tag{3.34}$$

Substituting Eq. (3.34) into Eq. (3.33) and noting that $\exp(-\xi\omega t_E) \cong 1$ for light damping, one obtains

$$
\begin{aligned}
q_{r,\max}(t_E) \approx{}& \frac{\Delta_{st}(1+\kappa)}{1+4\kappa/\pi} \times \left\{ \left[ \kappa + \frac{\cos(S_r\pi/2)}{1-S_r^2} \right.\right. \\
&+ \left.\left( \kappa \sin\frac{\pi}{2S_r} - \frac{S_r}{1-S_r^2}\cos\frac{\pi}{2S_r} \right) e^{-\xi\pi(1+S_r)/2S_r} \right] H(t_E) \\
&+ \frac{L}{d}\left( 2\frac{S_r}{\xi\pi} + 1 \right)\left( \kappa \sin\frac{\pi}{2S_r} - \frac{S_r}{1-S_r^2}\cos\frac{\pi}{2S_r} \right) \\
&\times \left. e^{-\xi\pi(1+S_r)/2S_r} H\left( t_E - \frac{L-d}{v_r} \right) \right\}. \qquad (3.35)
\end{aligned}
$$

Accordingly, the absolute maximum response for the beam is

$$
\begin{aligned}
q_{r,\max} \approx{}& \frac{\Delta_{st}(1+\kappa)}{1+4\kappa/\pi} \times \left\{ \left[ \kappa + \frac{|\cos(S_r\pi/2)|}{1-S_r^2} \right] \right. \\
&+ \left| \kappa \sin\frac{\pi}{2S_r} - \frac{S_r}{1-S_r^2}\cos\frac{\pi}{2S_r} \right| \\
&\times \left. \left[ \frac{L}{d}\left( \frac{2S_r}{\xi\pi} + 1 \right) + 1 \right] e^{-\xi\pi(1+S_r)/2S_r} \right\}. \qquad (3.36)
\end{aligned}
$$

For the case of light damping considered in this study, $\xi\pi/S_r < 0.3$, the following relations may be adopted:

$$
\begin{aligned}
&\frac{\cos(S_r\pi/2)}{1-S_r^2} \cong 1\,, \\
&1 - S_r^2 \cong 1\,, \qquad (3.37) \\
&\left[ \frac{L}{d}\left( \frac{2S_r}{\xi\pi} + 1 \right) + 1 \right] e^{-\xi\pi(1+S_r)/2S_r} \cong \frac{L}{d} \times \frac{2S_r}{\xi\pi}\,.
\end{aligned}
$$

It follows that the maximum response in Eq. (3.36) reduces to

$$q_{r,\max} \approx \frac{\Delta_{st}(1+\kappa)}{1+4\kappa/\pi} \times \left(1+\kappa+\frac{L}{d}\frac{2S_r}{\xi\pi}\left|\kappa\sin\frac{\pi}{2S_r}-S_r\cos\frac{\pi}{2S_r}\right|\right), \tag{3.38}$$

which is applicable for the case where the resonance condition is met, but the cancellation condition is not.

On the other hand, when both the resonance condition, as implied by $\sin(\omega d/2v)=0$, and the cancellation condition, as given in Eq. (3.23), are satisfied, i.e., when $S_r = S_c$, the response in Eq. (3.33) becomes

$$q_r(t_E) \approx \frac{\Delta_{st}(1+\kappa)}{1+4\kappa/\pi} \times \left[\frac{\sin\Omega t_E - S_c e^{-\xi\omega t_E}\sin\omega t_E}{1-S_c^2} + \kappa(1-e^{-\xi\omega t_E}\cos\omega t_E)\right] H(t_E)\,. \tag{3.39}$$

Here, due to the damping effect and consideration of operating speeds in the range $0 < S_r < 0.3$, the forced vibration term $\sin\Omega t_E$ and the constant $\kappa$ will dominate the response. By letting $\sin\Omega t_E = 1$, or $\omega t_E = \pi/(2S_r)$, Eq. (3.39) becomes

$$q_r(t_E) \approx \frac{\Delta_{st}(1+\kappa)}{1+4\kappa/\pi} \times \left[\frac{1}{1-S_c^2} + \kappa - e^{-\xi\pi/2S_c} \times \left(\frac{S_c\sin(\pi/2S_c)}{1-S_c^2} - \kappa\cos\frac{\pi}{2S_c}\right)\right] H(t_E)\,, \tag{3.40}$$

which can further be expressed as follows:

$$q_{r,\max} \approx \frac{\Delta_{st}(1+\kappa)}{1+4\kappa/\pi} \times \left[\kappa + \frac{1}{1-S_c^2} + \left(1-\frac{\xi\pi}{2S_c}\right)\sqrt{S_c^2+\kappa^2}\right], \tag{3.41}$$

if the relations for approximations are used: $\exp(-\xi\pi/2S_c) \cong 1 - \xi\pi/2S_c$ and $S_c/(1-S_c^2) \cong S_c$. These formulae are valid for the case when both the condition of resonance and the condition of cancellation are satisfied.

## 3.8. Impact Factor and Envelope Impact Formulas

The impact factor defined in Eq. (1.1) is adopted for the deflection of an elastically-supported beam subjected to the moving loads, i.e.,

$$I = \frac{R_d(x) - R_s(x)}{R_s(x)}, \tag{3.42}$$

where $R_d(x)$ and $R_s(x)$, respectively, denote the maximum dynamic and static responses of the beam at position $x$ due to the action of the moving loads. The following is the maximum static deflection for an elastically-supported beam subjected to a lumped load $p$ at the midpoint:

$$R_s(x) = \frac{\Delta_{st}(1+\kappa)^2}{1+4\kappa/\pi}, \tag{3.43}$$

where $\Delta_{st} = 2pL^3/\pi^4 EI \cong pL^3/48EI$ = the maximum midpoint static deflection of the corresponding beam with hinge supports. By using Eqs. (3.38), (3.41) and (3.42), the deflection impact formula for the elastically-supported beam subjected to a sequence of moving loads can be expressed as:

$$\begin{aligned} I &\approx \frac{1}{(1+\kappa)}\left[1+\kappa+\frac{L}{d}\frac{2S_r}{\xi\pi}\left|\kappa\sin\left(\frac{\pi}{2S_r}\right) - S_r\cos\left(\frac{\pi}{2S_r}\right)\right|\right] - 1 \\ &\approx \frac{L}{(1+\kappa)d}\frac{2S_r}{\xi\pi}\left|\kappa\sin\left(\frac{\pi}{2S_r}\right) - S_r\cos\left(\frac{\pi}{2S_r}\right)\right|, \end{aligned} \tag{3.44}$$

for the case of resonance, and

$$\begin{aligned} I &\approx \frac{1}{(1+\kappa)}\left[\kappa+\frac{1}{1-S_c^2}+\left(S_c-\frac{\xi\pi}{2}\right)\sqrt{1+\left(\frac{\kappa}{S_c}\right)^2}\right] - 1 \\ &\approx \frac{1}{(1+\kappa)}\left[\frac{S_c^2}{1-S_c^2}+\left(S_c-\frac{\xi\pi}{2}\right)\sqrt{1+\left(\frac{\kappa}{S_c}\right)^2}\right], \end{aligned} \tag{3.45}$$

for the case when both the conditions of resonance and cancellation are met. Because of the impact factors for $S$ other than $S_r$ or $S_c$ are smaller than those under the resonance or cancellation conditions, it is possible to construct an envelope formula by connecting all the peak values computed from Eqs. (3.44) or (3.45) for all the resonance points in a piecewise manner, which serves as an upper bound for the impact factor of elastically-supported beams subjected to a sequence of moving loads at various speeds.

## 3.9. Numerical Examples

### 3.9.1. *Phenomenon of Resonance*

As shown in Table 3.1, two bridges supported by elastic bearings are considered in this example. The train moving over the bridge is assumed to have $N = 8$ cars, each of length $d = 25$ m. The weight of each car is represented by a moving load of $p = 300$ kN. Different values of resonant speeds computed for the two beams have been listed in Table 3.2, together with those for the corresponding beams with hinge supports. Evidently, most of the resonance speeds can be encountered by modern high-speed trains. The responses of the beam traversed by the moving loads have been plotted in Figs. 3.5–3.7, along with those obtained by the finite element method to be

Table 3.1. Properties of bridges.

| $L$ (m) | $m$ (t/m) | $EI$ (kN-m$^2$) | $\kappa$ | $\omega_0$ (rad/s) | $\omega$ (rad/s) |
|---|---|---|---|---|---|
| 23 | 30 | $1.4 \times 10^8$ | 0.24 | 40.3 | 35.1 |
| 27 | 32 | $2.0 \times 10^8$ | 0.2 | 33.9 | 29.9 |

Table 3.2. Resonant speeds.

| Resonant speed $v_r = (\omega d/2\pi)/n$ | $L = 23$ m ($n = 2$) | $L = 27$ m ($n = 2$) | $L = 27$ m ($n = 4$) |
|---|---|---|---|
| Simple beam $v_0$ [$S_r = d/2nL$] | 80 m/s (= 288 km/h) [$S_r = 0.272$] | 67 m/s (= 242 km/h) [$S_r = 0.231$] | 34 m/s (= 122 km/h) ($S_r = 0.116$) |
| Elast. supp. beam $v$ [$S_r = d/2nL$] | 70 m/s (= 252 km/h) [$S_r = 0.272$] | 59 m/s (= 214 km/h) [$S_r = 0.231$] | 30 m/s (= 108 km/h) ($S_r = 0.116$) |

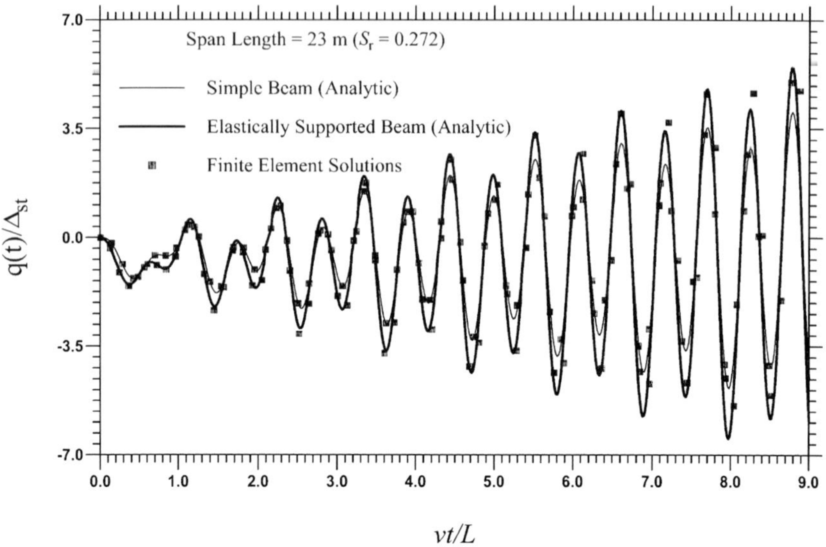


Fig. 3.5. The time history responses at resonant speed $S = d/4L$ for 23 m beam.

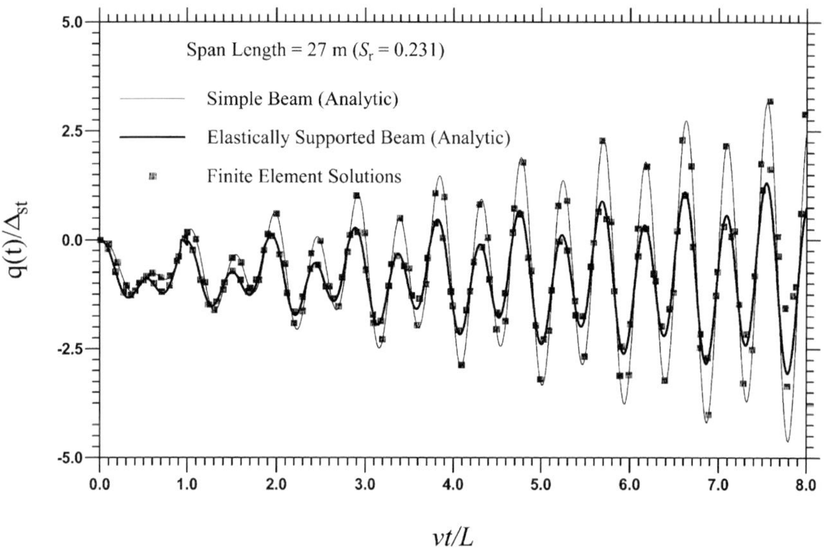


Fig. 3.6. The time history responses at resonant speed $S = d/4L$ for 27 m beam.

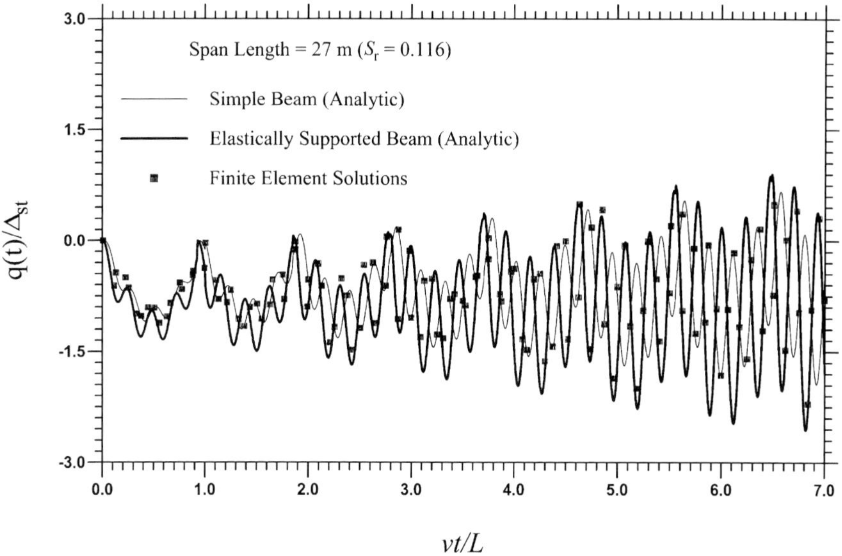


Fig. 3.7. The time history responses at resonant speed $S = d/8L$ for 27 m beam.

presented in Chapter 6. As can be seen, all the solutions obtained by the present approach agree excellently with the finite element solutions. For the case when the motions of the simple beam and elastic bearings are in phase, the elastic bearings inserted at the supports can significantly increase the dynamic response of the bridge to the moving loads, as can be seen from Figs. 3.5 and 3.7. Such a phenomenon is harmful to the riding comfort or maneuverability of the train. However, for the case when the motions of the simple beam and elastic bearings are out of phase, as indicated in Fig. 3.6, the dynamic response of the beam is reduced through the introduction of the elastic bearings.

### 3.9.2. *Effect of Structural Damping*

To investigate the effect of damping on the resonant response of an elastically-supported beam due to an infinite series of moving loads, 30 moving loads are considered in this example. A damping ratio of 2% is assumed for the beam. As can be seen from Figs. 3.8–3.11,

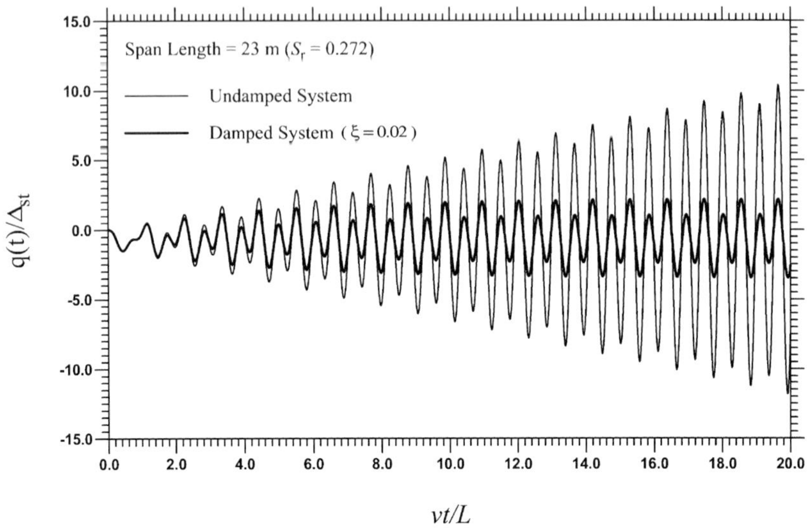


Fig. 3.8. The comparison of time history responses at resonant speed $S = d/4L$ for 23 m simple beam.

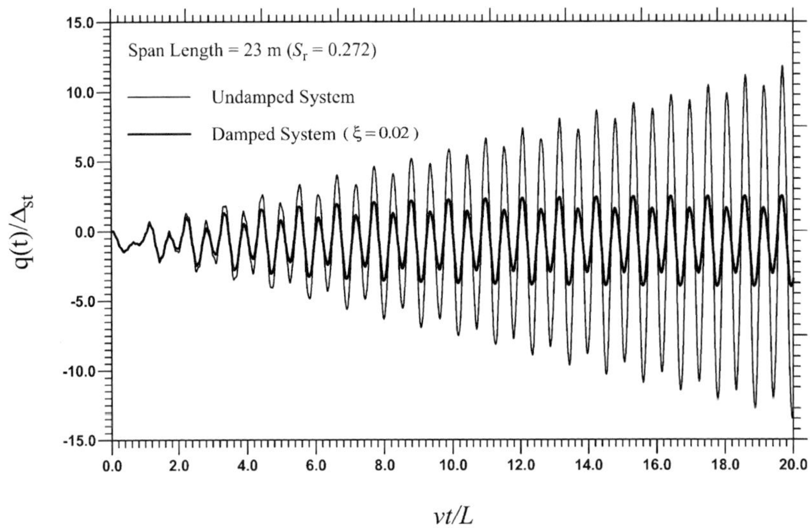


Fig. 3.9. The comparison of time history responses at resonant speed $S = d/4L$ for 23 m elastically-supported beam.

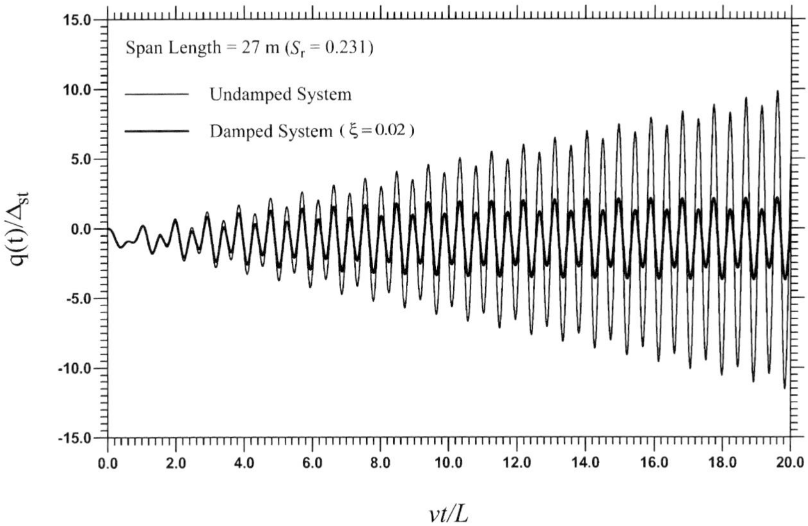


Fig. 3.10. The comparison of time history responses at resonant speed $S = d/4L$ for 27 m simple beam.

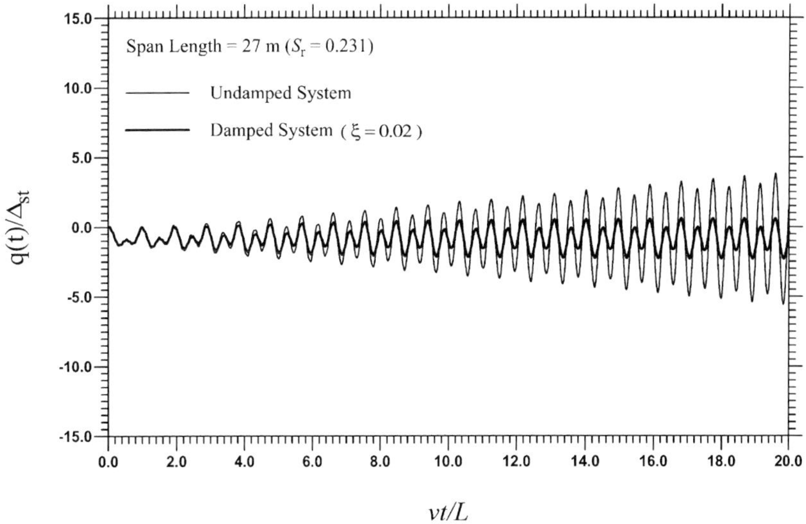


Fig. 3.11. The comparison of time history responses at resonant speed $S = d/4L$ for 27 m elastically-supported beam.

due to the presence of damping, the vibration of the beam remains rather bounded, behaving in a steady state manner, even when the resonance condition is met. This is very different from the undamped case, in which the response amplitude tends to grow continuously, as there are more loads passing the beam under the resonance condition.

### 3.9.3. *Envelope Impact Formula*

From the condition of resonance, $\sin(\omega d/2v_r) = 0$, the resonance speed can be solved: $v_r = \omega d/2n\pi$ or $S_r = d/2nL$, with $n = 1, 2, 3 \ldots$. It is possible to connect all the peak values computed from the envelope impact formulas in Eqs. (3.44) and (3.45) for all the resonance points in a piecewise manner, as shown in Figs. 3.12–3.15. In these figures, both the elastically-supported beams and simple beams are considered, assuming two different values of damping ratios, $\xi = 0.02$ and 0.04. The impact factors $I$ computed using the more accurate Eq. (3.33) in Sec. 3.6 are also plotted in the figures. The figures shown here are known as the $I - S$ *plots*, i.e., impact factor versus speed parameter plots. As can be seen, the envelope impact formulas show a trend in good consistency with the more accurate $I - S$ plots for the two values of damping ratios throughout the entire range of speed parameters considered.

A comparison of Fig. 3.12 with Fig. 3.13 for the 23 m beams is that installation of elastic bearings at the supports tends to drastically increase the impact response of the beam for the entire speed range considered. On the other hand, by comparing Fig. 3.14 with Fig. 3.15 for the 27 m beams, one observes that the installation of elastic bearings may amplify the impact response only over the low-speed range ($S < 0.125$), but may suppress the response for the high-speed range ($S > 0.125$). To give the readers an overview of the effect of elastic bearings on the bridge response, two three-dimensional $I - S - L/d$ plots for the impact factor of the midpoint deflection of the beam were given in Figs. 3.16 and 3.17 for two values of stiffness ratios, $\kappa = 0.15$ and 0.25. A comparison of the two figures indicates that the use of less rigid elastic bearings, as implied by a larger stiffness ratio $\kappa$, tends to increase the impact response of the elastically-supported beam.

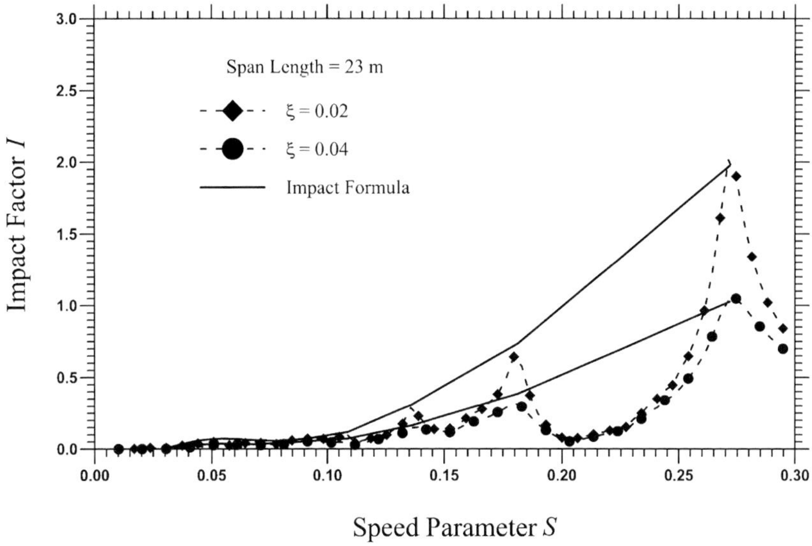


Fig. 3.12. The $I - S$ plot for 23 m simple beam.

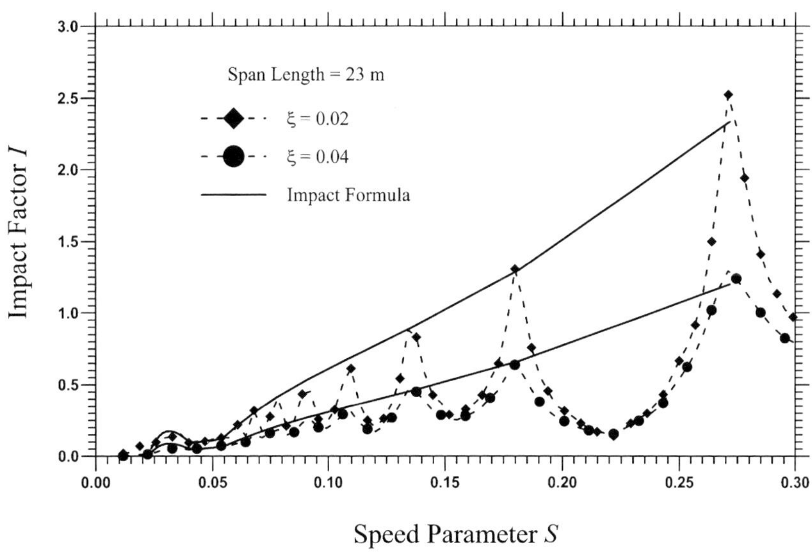


Fig. 3.13. The $I - S$ plot for 23 m elastically-supported beam.

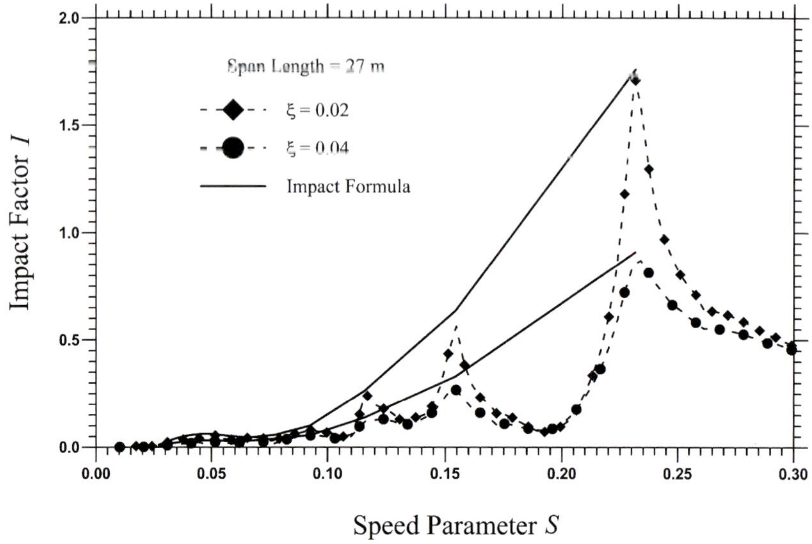


Fig. 3.14. The $I - S$ plot for 27 m simple beam.

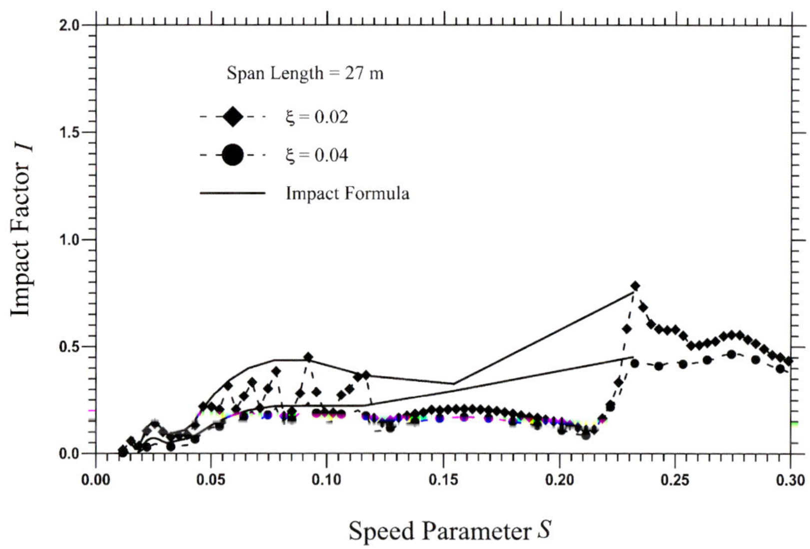


Fig. 3.15. The $I - S$ plot for 27 m elastically-supported beam.

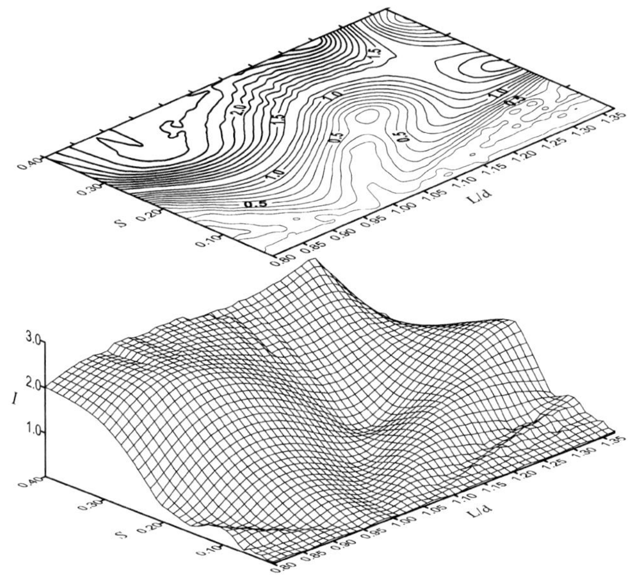


Fig. 3.16. The $I - S - L/d$ diagram for $\xi = 0.02$, $\kappa = 0.15$.

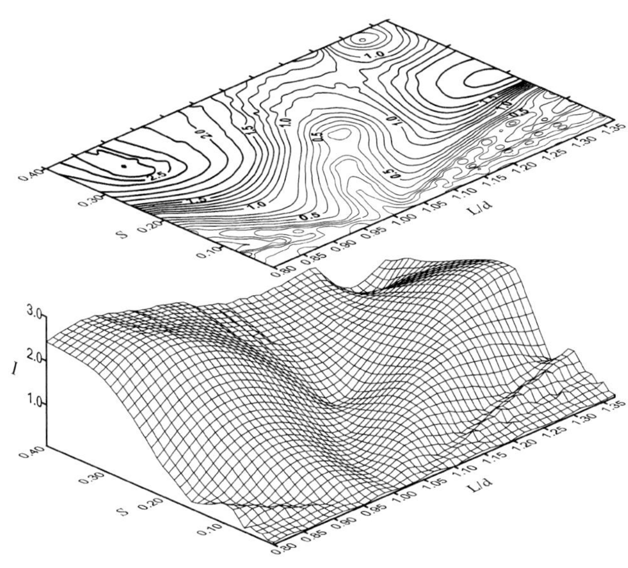


Fig. 3.17. The $I - S - L/d$ diagram for $\xi = 0.02$, $\kappa = 0.25$.

## 3.10. Concluding Remarks

In this chapter, the impact response of elastically-supported beams subjected to a sequence of moving loads has been investigated by an analytical approach. Light damping is assumed for the beams considered. Both the conditions for resonance and cancellation to occur have been identified. Unlike the case for a beam with zero damping, whose resonance response tends to grow continuously as there are more moving loads passing the beam, the resonance response for a damped beam remains more or less constant, regardless of the number of moving loads that have moved over the beam. For the case of an infinite number of moving loads, envelope impact formulas have been derived for the elastically-supported beam with light damping. It is concluded that the installation of elastic bearings may generally increase the response of the beam under most resonance conditions. The more flexible the elastic bearings, the larger is the peak response of the beam.

In the following chapter, the mechanisms for occurrence of the resonance and cancellation phenomena on an elastically-supported beam will be unveiled. In particular, it will be shown that the cancellation condition is a condition that is more decisive than the resonance condition. In case there occurs the cancellation condition, all the resonance response will just be suppressed.

Chapter 4

# Mechanism of Resonance and Cancellation for Elastically-Supported Beams

This chapter can be regarded as a supplement to the preceding chapter on elastically-supported beams. It is aimed at unveiling the mechanism underlying the phenomena of resonance and cancellation on elastically-supported beams induced by passing trains, which is very useful for the mitigation of train-induced vibrations. Basically the same approach as in the preceding chapter was followed. The train is modeled as a sequence of equi-distant moving loads. The vibration shape of the elastically-supported beam is approximated as the composition of a flexural sine mode and a rigid body mode. The analysis indicates that *resonances* of *much higher peaks* can be excited on beams with elastic supports by trains moving at *much lower speeds*, compared with those for beams with rigid supports. On the other hand, the speed for *cancellation* to occur is generally independent of the support stiffness. Of particular interest is that *cancellation* is a phenomenon more decisive than resonance. Whenever the condition of cancellation is met, the dynamic response of the beam will be suppressed, whether the beam be under resonance or not. The analytical results presented herein were verified by a field test on two elastically-supported bridges located at existing railway lines.

## 4.1. Introduction

Taiwan is located at one of the most active seismic zones in the Pacific Rim. In order to prevent the bridge structures from damages or collapse under severe earthquakes, various protective measures have been adopted by structural engineers. Elastic bearings represent a kind of devices commonly installed at the supports of bridge girders for isolating the earthquake forces transmitted from the ground. While they are effective for isolating the ground-borne seismic forces, they can equally prevent the vehicle-induced vibrations from transmission or dissipation to the supports and then to the ground or soils. This is certainly one disadvantage with the use of elastic bearings. For high-speed railway bridges installed with elastic bearings, it is likely that the huge amount of vibration energy brought by the high-speed train be accumulated in the bridge, which may result in amplified or resonant vibrations on the bridge at some critical speeds. The repetitive nature of resonant vibrations may cause early fatigue problems on the associated railway tracks, as well as deterioration in the riding quality of passengers, while increasing the cost of maintenance for the railway lines.

Due to the regular, repetitive nature of the wheel loads that constitute a train, both the phenomena of resonance and cancellation may be induced on the bridge by the train moving at high speeds. The *resonance* phenomenon relates to the continuous *built-up* of the free-vibration response on the bridge as there are more loads passing by. In contrast, the *cancellation* phenomenon implies that the waves associated with the free-vibration responses of the bridge generated by the sequential moving loads cancel out each other. If the resonance condition can be reached by a train within its range of speed of operation, then some detrimental effects can be expected on the track system, as well as on the moving train itself. Such a problem will be aggravated when the factor of elastic bearings is taken into account.

Previously, rather few research works have been conducted on the dynamic response of elastically-supported beams to moving loads. Based on the work of Yau *et al.* (2001), envelope formulas were presented in the preceding chapter for elastically-supported beams with light damping subjected to moving loads. Recently, Lin (2001) investigated the vibrations of railway bridges installed with elastic bearings, together with measures for vibration reduction. In this chapter, focus is placed on the *physical interpretation* of the mechanism involved in the phenomena of resonance and cancellation of the dynamic response of elastically-supported beams to moving loads. The key factors affecting the dynamic response of bridges will be investigated, with comments made concerning the suitability of using elastic bearings as seismic isolation devices for railway bridges. The results obtained from a field test that serve to verify the theory are also presented. This chapter has been rewritten mainly from the paper by Yang *et al.* (2004).

## 4.2. Formulation of the Theory

The bridge model adopted is the one shown in Fig. 4.1, in which a beam supported by two identical elastic bearings is considered. As was done in Chapter 3, the following assumptions will be adopted in the derivation of a closed-form solution for the elastically-supported beam under the moving loads: (1) The vertical stiffness of each elastic bearing is $K$. The mass of the spring is negligible compared with that of the bridge. (2) The train is modeled as a sequence of equally-spaced moving loads, with the inertial effect neglected. (3) For beams subjected to moving loads, which is basically a transient problem with very short acting time, only the fundamental mode of vibration of the bridge needs to be considered, while the higher modes can be neglected without losing accuracy (Biggs, 1964). (4) The damping of the beam can be neglected, also due to the transient nature of the moving loads over the beam.

### 4.2.1. *Assumed Modal Shape of Vibration*

By the concept of modal superposition, the deflection $u(x,t)$ of the elastically-supported beam can be expressed as

$$u(x,t) = \sum \phi_n(x) q_n(t)\,, \tag{4.1}$$

where $q_n(t)$ denotes the generalized coordinate and $\phi_n(x)$ the shape function of the $n$th vibration mode. As was stated above, only the fundamental mode of vibration will be considered in analyzing the vibrational response of the beam, as it is essentially a transient problem

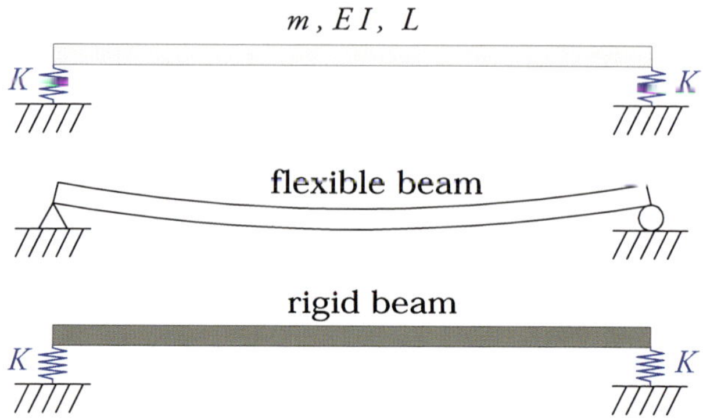


Fig. 4.1. The model beam as a superposition of simple and rigid beams.

with very short acting time. If a mathematically exact approach is employed to find the first modal shape of vibration for the elastically-supported beam, the final form of the modal shape will be rather complex, rendering it impossible to obtain simple closed-form solutions. As the first priority herein is to derive a closed-form solution, by which the mechanism behind the key phenomena can be interpreted, the vibration shape of the elastically-supported beam will be approximated as the *superposition* of the first modal shape of the *flexural deflection* of the beam *with simple supports* and the first modal shape of a *rigid beam supported by two elastic springs*, as indicated in Fig. 4.1, that is,

$$\phi(x) = \sin\frac{\pi x}{L} + \kappa\,, \tag{4.2}$$

where $\kappa = EI\pi^3/KL^3$ denotes the ratio of the flexural rigidity $EI$ of the beam to the stiffness $K$ of the elastic bearing, and $L$ the length of the beam. As can be seen, a higher stiffness ratio $\kappa$ means a softer elastic spring and a zero stiffness ratio means the special case of simple supports. The shape function $\phi(x)$ in Eq. (4.2) differs from that used in Chapter 3, i.e., the one implied by Eq. (3.4) by a factor $1/(1+\kappa)$, which is acceptable, since it is known that shape functions do not have absolute magnitudes.

### 4.2.2. *Single Moving Load*

For an elastically-supported beam subjected to a single moving load $p$ of speed $v$, the equation of motion for the deflection $u(x,t)$ of the beam is

$$m\ddot{u} + EIu'''' = p\delta(x - vt) \qquad \text{for} \qquad 0 \le vt \le L\,, \tag{4.3}$$

where $m$ is the mass per unit length and $EI$ the flexural rigidity of the beam. The boundary conditions are

$$\begin{aligned} EIu''(0,t) &= 0\,, \\ EIu''(L,t) &= 0\,, \\ EIu'''(0,t) &= -Ku(0,t)\,, \\ EIu'''(L,t) &= Ku(L,t)\,. \end{aligned} \tag{4.4}$$

By multiplying both sides of Eq. (4.3) by the shape function $\phi(x)$ in Eq. (4.2) and integrating with respect to the length $L$ of the beam, one obtains

$$\ddot{q} + \omega^2 q = \frac{2p}{mL}\left(1 + \frac{8\kappa}{\pi} + 2\kappa^2\right)^{-1}\left(\sin\frac{\pi vt}{L} + \kappa\right), \tag{4.5}$$

where the frequency of vibration $\omega$ is

$$\omega = \omega_0\sqrt{\frac{\pi + 4\kappa}{\pi + 8\kappa + 2\pi\kappa^2}}\,, \tag{4.6}$$

which is a function of the stiffness ratio $\kappa$. Here, $\omega_0$ indicates the frequency of vibration of the associated beam with simple supports,

$$\omega_0 = \left(\frac{\pi}{L}\right)^2\sqrt{\frac{EI}{m}}\,. \tag{4.7}$$

It has been demonstrated that the fundamental frequency $\omega$ of vibration solved using the present approximate shape function $\phi(x)$, as given in Eq. (4.6), is in excellent agreement with the exact one, as indicated in Chapter 3.

The generalized coordinate $q(t)$ can be solved from Eq. (4.5), together with zero initial conditions, as

$$q(t) = \frac{2pL^3}{EI\pi^4}\left(1+\frac{4\kappa}{\pi}\right)^{-1}\left[\left(\frac{\sin\Omega t - S\sin\omega t}{1-S^2}\right) + \kappa(1-\cos\omega t)\right], \tag{4.8}$$

where $\Omega$ denotes the driving frequency implied by the moving load,

$$\Omega = \frac{\pi v}{L}, \tag{4.9}$$

and $S$ is a *speed parameter* defined as the ratio of the driving frequency $\Omega$ to the frequency $\omega$ of vibration of the bridge, i.e.,

$$S = \frac{\Omega}{\omega} = \frac{\pi v}{\omega L}. \tag{4.10}$$

In Eq. (4.8), the term containing the stiffness ratio $\kappa$ represents the effect of the elastic supports. For the special case of simple supports, i.e., with $\kappa = 0$, the present solution reduces to that given in Chapter 2 for simply-supported beams.

### 4.2.3. *A Series of Moving Loads*

As shown in Fig. 4.2, consider now that the elastically-supported beam is subjected to a series of concentrated loads $p$ of equal intervals $d$ moving at speed $v$, as a representation of the loading action of a train consisting of $N$ cars of length $d$. The equation of motion for the beam should now be modified as

$$m\ddot{u} + EIu'''' = p\sum_{k=1}^{N}\delta[x - v(t-t_k)] \cdot \left[H(t-t_k) - H\left(t - t_k - \frac{L}{v}\right)\right], \tag{4.11}$$

where $\delta(x)$ denotes the Dirac delta function, $H(x)$ the unit step function, $t_k = (k-1)d/v$ the arriving time of the $k$th load at the beam, and $N$ is the total number of moving loads. The term $H(t - t_k)$ indicates the arrival of the $k$th load at the beam and the term

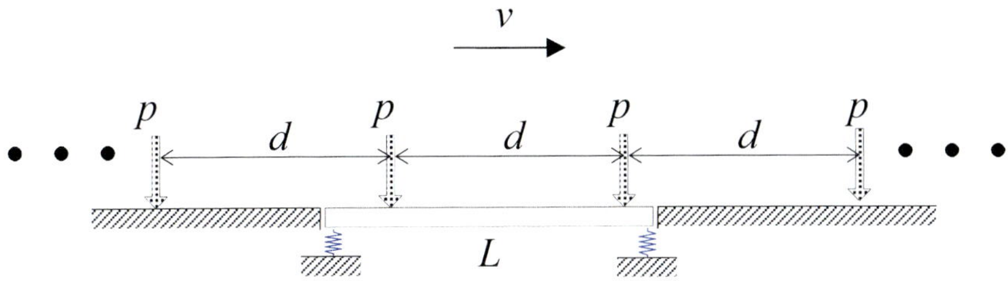


Fig. 4.2. The elastically-supported beam subjected to equi-distant moving loads.

$H(t - t_k - L/v)$ the departure of the same load from the beam. The boundary conditions given in Eq. (4.4) remain valid.

Based on the hypothesis of linear theories, the deflection of the beam induced by a sequence of moving loads can be obtained as the superposition of the deflection induced by each of the moving loads, if the time lag of each moving load is taken into account. Consequently, the generalized deflection $q(t)$ of the beam for the present case can be obtained as a generalization of Eq. (4.8) as

$$q(t) = \frac{2pL^3}{EI\pi^4}\left(1 + \frac{4\kappa}{\pi}\right)^{-1}\left[Q_1(t) + Q_2(t)\right], \tag{4.12}$$

where $Q_1(t)$ represents the contribution caused by the *flexural vibration* of the beam with simple supports, and $Q_2(t)$ the *rigid displacement* of the elastic bearings, namely,

$$\begin{aligned} Q_1(t) = \frac{1}{1-S^2}\sum_{k=1}^{N}\Bigg\{&[\sin\Omega(t-t_k) - S\sin\omega(t-t_k)]H(t-t_k) \\ &+ \left[\sin\Omega\left(t - t_k - \frac{L}{v}\right) - S\sin\omega\left(t - t_k - \frac{L}{v}\right)\right] \\ &\times H\left(t - t_k - \frac{L}{v}\right)\Bigg\}, \end{aligned} \tag{4.13a}$$

$$Q_2(t) = \kappa \sum_{k=1}^{N} \left\{ [1 - \cos\omega(t - t_k)]H(t - t_k) - \left[1 - \cos\omega\left(t - t_k - \frac{L}{v}\right)\right] H\left(t - t_k - \frac{L}{v}\right) \right\}, \quad (4.13b)$$

where $\Omega$ denotes the driving frequency implied by the moving loads and $\omega$ the frequency of vibration of the elastically-supported beam. For the special case of $\kappa = 0$, the solution given in Eq. (4.12) reduces to that given in Chapter 2 for a beam with simple supports.

## 4.3. Conditions of Resonance and Cancellation

The generalized deflection of the elastically-supported beam given in Eq. (4.12) consists of two parts, that is, the *forced vibration* caused by the moving loads that are directly acting on the beam, as indicated by the terms containing the driving frequency $\Omega$, and the *residual* or *free vibration* caused by the moving loads that have passed the beam, as indicated by the terms containing the bridge frequency $\omega$. When all the moving loads have passed the beam, the forced vibration part terminates immediately. However, the free vibration part, which is of sinusoidal form, continues to be functional until the vibration is eventually damped out.

Both the phenomena of resonance and cancellation relate to the free vibrations induced by the moving loads. When a moving load has passed the beam, waves of the sinusoidal form will be induced on the beam. If the time lag of the wave components induced by each moving load equals a multiple of the period $2\pi/\omega$, then superposition of all the wave components will result in amplified responses. This is the so-called phenomenon of *resonance.* On the contrary, if the time lag equals an odd multiple of half of the period, the wave components induced by the sequentially moving loads will just cancel out, indicating that the phenomenon of *cancellation* has occurred.

Whether the phenomena of resonance or cancellation will occur or not depends only on the free vibration part of the motion. In order to interpret the two phenomena using the analytical solution, let us consider the *critical* case when the $(N-1)$th moving load has left the beam and the $N$th load has entered the beam, that is, when $t_N < t \leq t_N + \Delta t$. Such a case is considered critical, since the beam is excited to the utmost. From Eq. (4.12), one can obtain the following for such a case:

$$q(t) = \frac{2pL^3}{EI\pi^4}\left(1+\frac{4\kappa}{\pi}\right)^{-1}\left[A(t)H(t-t_N)+B(t)H(t-t_N-\Delta t)\right], \tag{4.14}$$

where the dynamic response factors $A(t)$ and $B(t)$ are

$$A(t) = \frac{1}{1-S^2}[\sin\Omega(t-t_N) - S\sin\omega(t-t_N)] + \kappa[1-\cos\omega(t-t_N)], \tag{4.15a}$$

$$B(t) = \left(\frac{-2S}{1-S^2}\cos\frac{\omega L}{2v} + 2\kappa\sin\frac{\omega L}{2v}\right) \times \left[\sin\omega\left(t-\frac{L}{2v}\right) + \frac{\sin\omega(t-(L/2v)-(t_N/2))\cdot\sin\omega((t_N/2)-(d/2v))}{\sin(\omega d/2v)}\right]. \tag{4.15b}$$

In Eq. (4.14), the term $A(t)H(t-t_N)$ indicates the forced vibration of the beam caused by the $N$th moving load, and the term $B(t)H(t-t_N-\Delta t)$ the sum of all the free vibrations caused by the previous $N-1$ moving loads that have already passed the beam.

Some physical interpretations can be given using Eq. (4.15b). First of all, if the denominator within the brackets vanishes, i.e., when $\sin(\omega d/2v) = 0$, the response of the beam reaches a peak. By L'Hospital's rule, the second term within the brackets of the dynamic response factor $B(t)$ under the resonance condition becomes

$$\frac{\sin\omega(t-(L/2v)-(t_N/2))\cdot\sin\omega((t_N/2)-(d/2v))}{\sin(\omega d/2v)} = (N-2)\sin\omega\left(t-\frac{L}{2v}\right). \tag{4.16}$$

Accordingly, the dynamic response factor $B(t)$ can be written as

$$B(t) = \left( \frac{-S}{1-S^2} \cos\frac{\omega L}{2v} + \kappa \sin\frac{\omega L}{2v} \right) \times 2(N-1) \sin\omega\left(t - \frac{L}{2v}\right). \tag{4.17}$$

As can be seen, *larger* amplitude for the response can be expected as there are *more* loads passing the beam, indicated by the term $2(N-1)$. Such a phenomenon is similar to that observed for beams with simple supports, as discussed in Chapter 2. Corresponding to $\sin(\omega d/2v) = 0$, the *condition of resonance* is $\omega d/2v = i\pi$ with $i = 1, 2, 3, \ldots$, or $v = \omega d/(2i\pi)$, which can also be written in terms of the *speed parameter* $S$ based on the definition of Eq. (4.10) as

$$S_r = \frac{\pi v}{\omega L} = \frac{1}{2i} \cdot \frac{d}{L} \qquad \text{with} \qquad i = 1, 2, 3, \ldots\,, \tag{4.18}$$

which is the same as that for simply-supported beams. As can be seen either from Eq. (4.18) or Fig. 4.3, the resonant speed parameter $S_r$ is a function of the ratio $d/L$ of the car length to the bridge length. For trains of the commercially available models of which the car length $d$ is known, the resonant speed parameter $S_r$, which is dimensionless, is generally small for bridges of practical length $L$. An observation from Eq. (4.18) is that the *longer* the span length $L$ of a beam, the *easier* is for the resonance phenomenon to occur. It is true that the resonant speed parameter $S_r$ computed from Eq. (4.18) is the same for both the elastically-supported and simply-supported beams. However, because the fundamental frequency of the former is much lower than that of the latter, the resonant speed, i.e., $v_r = \omega L S_r/\pi$, for vehicles moving over an elastically-supported beam will be *much lower* than that over a simply-supported beam. Besides, it should be noted that according to Eq. (4.14), the amplitude of the resonance response also depends on the stiffness ratio $\kappa$ of the elastic supports.

On the other hand, by setting the parenthesized term in Eq. (4.15b) equal to zero, all the residual free vibrations caused by the previous $N-1$ moving loads just cancel out. This is exactly the

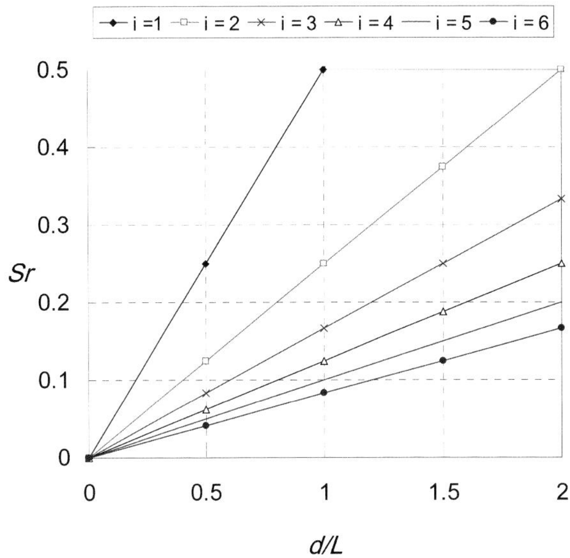


Fig. 4.3. The $S_r - d/L$ plot for resonance.

condition of *cancellation* for the elastically-supported beam:

$$Y(S) = \frac{-S}{1-S^2}\cos\frac{\omega L}{2v} + \kappa\sin\frac{\omega L}{2v} = 0\,. \tag{4.19}$$

For simply-supported beams, $\kappa = 0$, the preceding condition reduces to $\cos(\omega L/2v) = 0$, or $\omega L/2v = \pi(2i-1)/2$ with $i = 1, 2, 3, \ldots$, from which the speed parameter $S$ can be determined as

$$S_c = \frac{1}{2i-1} \qquad \text{with} \qquad i = 1, 2, 3, \ldots\,, \tag{4.20}$$

which is valid only for the simply-supported beam, i.e., for the case with $\kappa = 0$. The cancellation speed for the elastically-supported beam cannot be presented explicitly, since Eq. (4.19) is an implicit function. The solution computed from $Y(S) = 0$ in Eq. (4.19) for the cancellation speed parameter $S_c$ has been plotted with respect to the stiffness ratio $\kappa$ in Fig. 4.4. As can be seen, the cancellation speed parameter $S_c$ increases slightly as the stiffness ratio $\kappa$ increases. For example, for the case with $i = 2$, we have $S_c/S_{c(\kappa=0)} \approx 1 + 0.5\kappa$. However, since the fundamental frequency

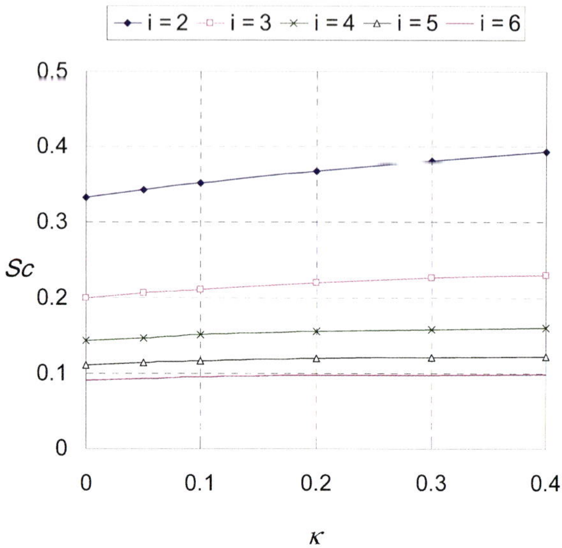


Fig. 4.4. The $S_c - \kappa$ plot for cancellation.

$\omega$ of the elastically-supported beam also decreases as the stiffness ratio $\kappa$ increases, i.e., $\omega/\omega_0 \approx 1 - 0.5\kappa$. It turns out that the difference between the real cancellation speed $v_c$, computed as $v_c = S_c\omega L/\pi$, for vehicles moving over an elastically-supported beam and that over a beam with simple supports is generally small.

## 4.4. Mechanism of Resonance and Cancellation

As was stated before, the impact factor $I$ as defined in Eq. (1.1) has often been used to account for the dynamic amplification effect on the bridge due to the passage of moving vehicles through increase of the design forces and stresses. In order to unveil the mechanism underlying the phenomena of resonance and cancellation of the bridge responses in relation to the effect of elastic supports, two bridges, B1 and B2, will be considered, of which the key properties have been listed in Table 4.1. The train is simulated as eight moving loads of equal weight $p = 220$ kN spaced at an interval $d = 25$ m. By changing the vertical stiffness of the elastic bearings, say, allowing it to

Table 4.1. The properties of the bridges used in analysis.

| | $L$ (m) | $m$ (t/m) | $EI$ (kN-m$^2$) | $\omega_0$ (rad/s) |
|---|---|---|---|---|
| B1 bridge | 23 | 30 | $1.4 \times 10^8$ | 40.3 |
| B2 bridge | 27 | 32 | $2.0 \times 10^8$ | 33.9 |

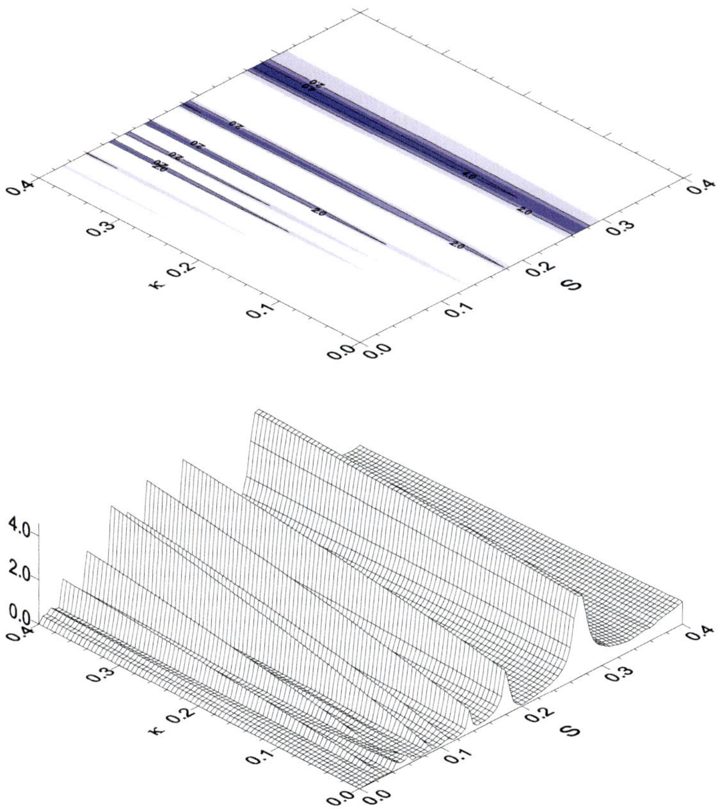


Fig. 4.5. The $I - \kappa - S$ plot and contour map (bridge B1).

vary in terms of the stiffness ratio $\kappa$ from 0 to 0.4, and computing the impact factor $I$ for the midpoint deflection of the beam due to the loads moving at different speeds, one can establish the $I - \kappa - S$ plots as in Figs. 4.5 and 4.6 for the two bridges.

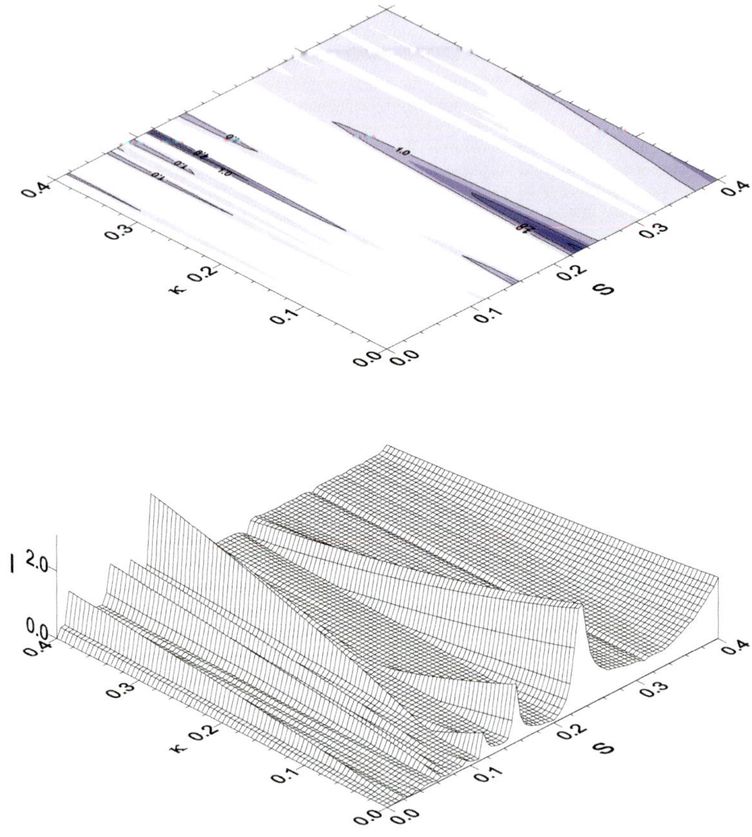


Fig. 4.6. The $I - \kappa - S$ plot and contour map (bridge B2).

As can be seen from Fig. 4.5, for bridge B1, higher stiffness ratios $\kappa$ generally result in higher resonant peaks, indicating that the elastic bearings inserted at the bridge supports tend to amplify the bridge response. Such a phenomenon can be clearly explained using Fig. 4.7, where two impact curves were plotted each for $\kappa = 0$ (i.e., for the beam with simple supports) and $\kappa = 0.2$. As can be seen from Fig. 4.7(a), the resonance phenomenon appears to be *not so visible* for the simply-supported beam in the range $S < 0.1$, which can be practically ignored. However, it is amplified drastically and becomes *rather significant* and *non-negligible* due to the installation

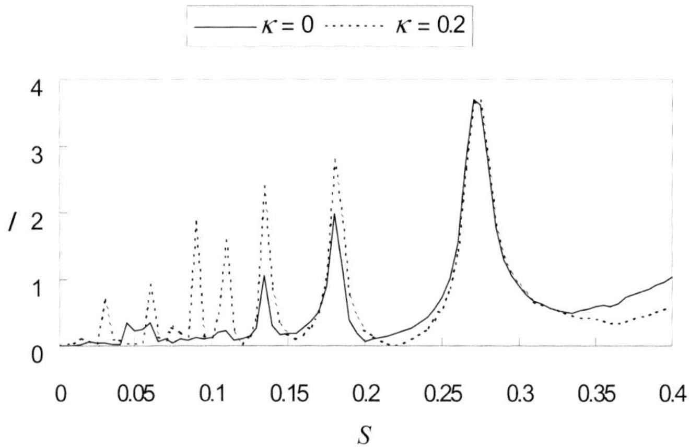


(a)

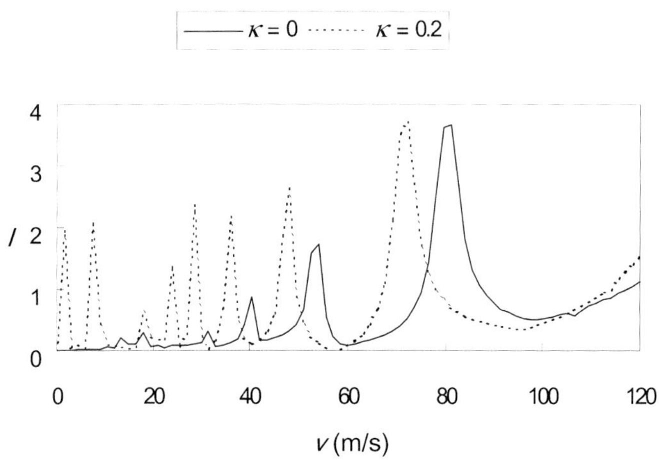


(b)

Fig. 4.7. The impact response of bridge B1: (a) $I - S$ plot and (b) $I - v$ plot.

of the elastic bearings on the bridge. In the design of high-speed railway bridges, the detrimental effect of elastic bearings in amplifying the low-speed resonant responses should be seriously taken into consideration.

On the other hand, a comparison of Fig. 4.7(a) with Fig. 4.7(b) reveals two interesting facts. First, the real resonant speed $v_r$ for an elastically-supported bridge is *much smaller* than that for the bridge with simple supports (see Fig. 4.7(b)), although the resonant speed parameter $S_r$ remains the same for both cases (see Fig. 4.7(a)). Second, the real cancellation speed $v_c$ seems to be *close* for both the elastically- and simply-supported bridges (see Fig. 4.7(b)), although the corresponding speed parameter $S_c$ is slightly larger for the elastically-supported bridge (see Fig. 4.7(a)). Such findings are consistent with the statements made following Eq. (4.20). In studying the impact response on bridges, the results are often presented as an $I - S$ plot for its elegance in expression and convenience in extension to more general cases, as both $I$ and $S$ are nondimensional. However, we should not misinterpret the real physical meanings implied by these nondimensional parameters.

As can be seen from Fig. 4.6, for the speed parameter in the range $S < 0.11$, larger resonant peaks can be expected for bridge B2 for larger stiffness ratio $\kappa$. However, the same is not true for the speed parameter in the range $S > 0.11$. The reason can be given as follows. First, for bridge B2, the car/bridge length ratio is $d/L = 25/27 = 0.926$. By drawing a vertical line at $d/L = 0.926$, one can obtain from the resonance plot (i.e., Fig. 4.3) several intersections of which the ordinates (for $S_r$) represent the points of resonance, as indicated in Fig. 4.8(a). Because the resonance condition in terms of speed parameter is independent of the stiffness ratio $\kappa$, one can draw a *resonance plot* as shown in Fig. 4.8(b) for bridge B2, in which each horizontal line represents one of the ordinates for resonance (i.e., passes through one of the intersections) shown in Fig. 4.8(a). Finally, we can superimpose the resonance plot of Fig. 4.8(b) with the *cancellation plot* of Fig. 4.4 to obtain the *resonance/cancellation plot* as shown in Fig. 4.8(c), which contains all the information we need for explaining the dynamic response of bridge B2.

Take the resonance line $S = 0.23$ in the resonance/cancellation plot, i.e., Fig. 4.8(c), for example. The cancellation line below becomes *closer* to this line as the stiffness ratio $\kappa$ increases, which means that the *resonance peaks will be suppressed* as the stiffness ratio $\kappa$ increases. The fact that cancellation is a condition more decisive than resonance is attributed to the fact that the dynamic

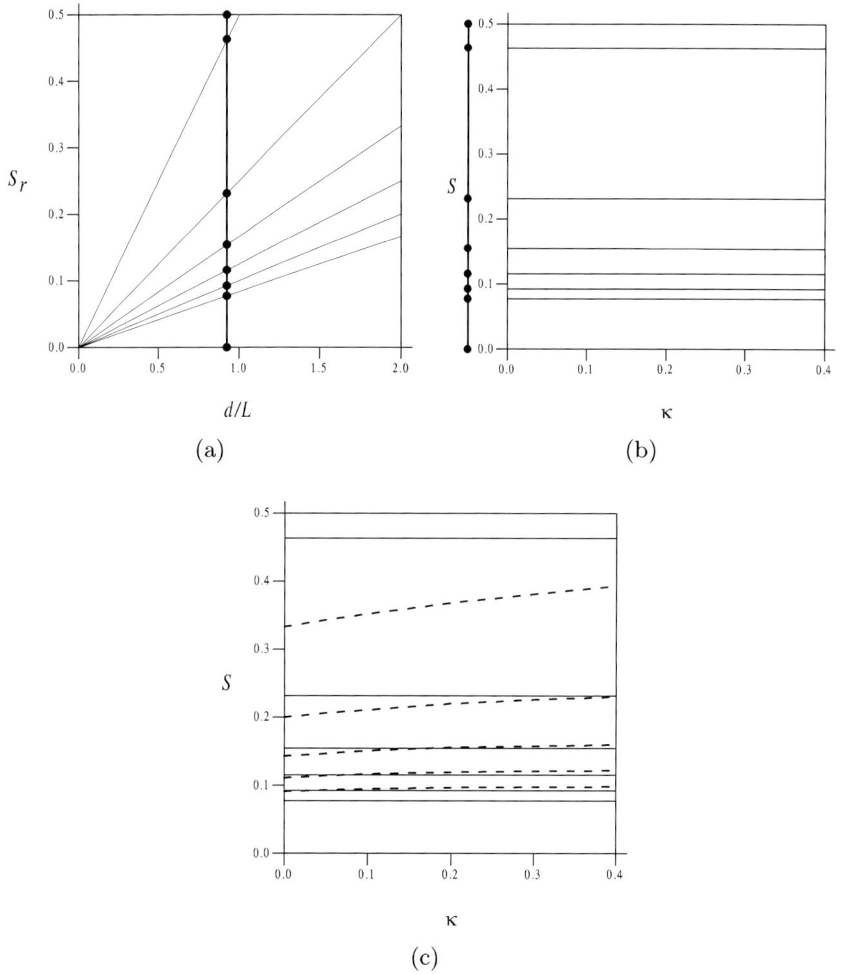


Fig. 4.8. The mechanism of resonance and cancellation: (a) points of resonance, (b) resonance plot, and (c) resonance/cancellation plot.

response factor $B(t)$ simply vanishes under the cancellation condition, regardless of the presence of resonance, as can be observed from Eq. (4.15b). For the same reason, we can explain why the resonant peak at $S = 0.11$ first deminishes and then grows as the stiffness ratio $\kappa$ increases. This is due to the fact that the resonance and cancellation lines are close in the beginning, but are getting apart for increasing $\kappa$. After explaining the mechanism of cancellation and resonance for the train-induced response of elastically-supported bridges, we shall proceed to verify the theory presented herein by some field tests.

## 4.5. Field Measurement of Vibration of Railway Bridges

There have been reports by train drivers on the unusually high levels of oscillations when maneuvering the trains to traverse the bridges over the Fongshan Creek on the Western Railway Lines in northern Taiwan. In order to seek for the reasons behind the problem, a field test was carried out for two adjacent bridges, A1 and A2, located at the Fongshan Creek, which were not known to be elastically-supported at the time of testing. Two locomtives of type E300 (see Fig. 4.9) linked back-to-back were used to generate the action of moving loads at different speeds. Some typical dimensions of the bridges and locomotives were shown in Fig. 4.10. Each locomotive has a bogie-to-bogie distance of 9.6 m and a gross weight of 96 t, much heavier than the normal passenger cars used. The railway gauge is 1067 mm, which is typical in Taiwan. The fundamental frequencies of the two bridges measured from an ambient vibration test are: $f = 5.17$ Hz for bridge A1 and $f = 5.13$ Hz for bridge A2, which represent the combined dynamic effect of all the components constituting the railway bridge, including the continuous rails, sleepers, ballast, elastic bearings, the girder and two side flanges that form a cross section of the U shape. By design, the two bridges are identically the same, but due to degradation in material properties they turn out to be slightly different in the material properties.

During the testing, the two locomotives connected back-to-back are allowed to travel on one side, i.e., the test side, of the two-track railway running through the tested bridges at the following speeds: 15, 30, 45, 60, 75, 85, and 110 km/h. The maximum accelerations measured at the midpoint of the A1 and A2 bridges using seismometers during the passage of the two locomotives at different speeds have been plotted in Fig. 4.11. The maximum accelerations

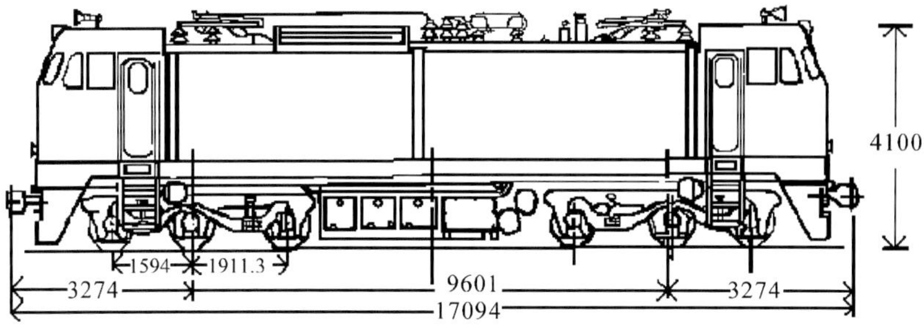


Fig. 4.9. The E300 locomtive (unit: mm).

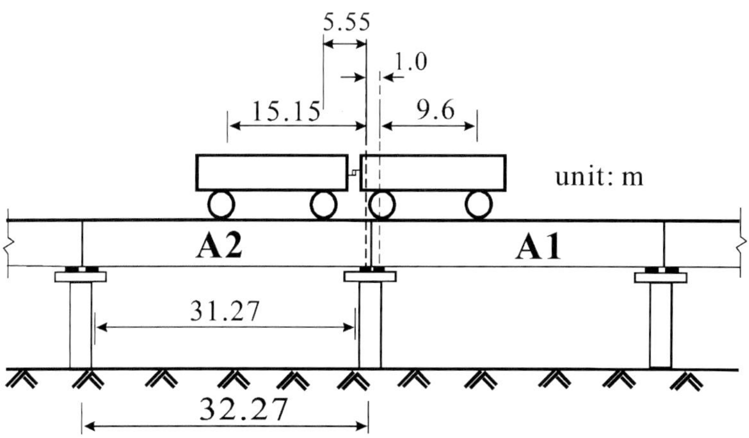


Fig. 4.10. The schematic of tested bridges.

*computed* for the midpoint of the two bridges by simulating the two locomotives as four equal-weight moving loads, but with *simple support* conditions, using the finite element analysis program developed in Chapter 6 is also plotted.

One observation from the measured and computed results shown in Fig. 4.11 is that they both show the occurrence of a peak response at the speed arround 60 km/h. Such a speed can be recognized as one of the resonance speeds, as can be verified from Eq. (4.18),

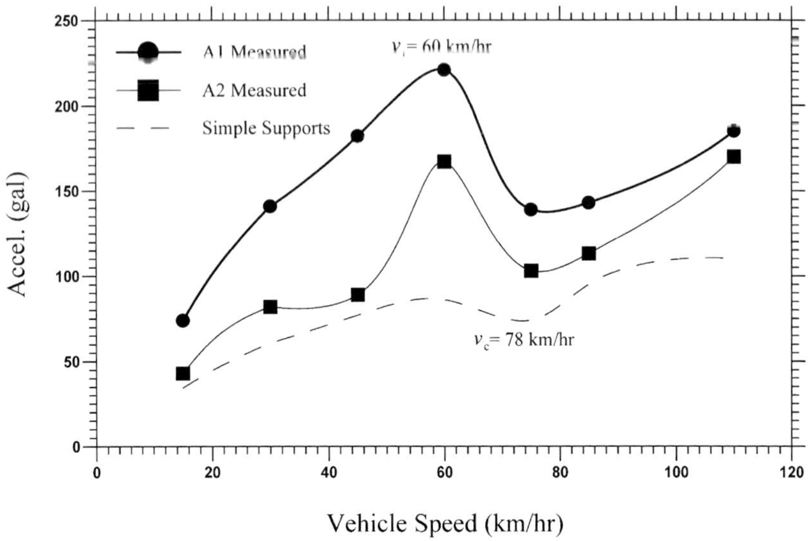


Fig. 4.11. The maximum midpoint accelerations.

that is,

$$\text{Bridge A1: } v_r = \frac{\omega d}{2n\pi} = \left.\frac{fd}{n}\right|_{\substack{f=5.17\\d=9.6\\n=3}} = 16.55\ \text{m/s} = 59.6\ \text{km/h}\,.$$

$$\text{Bridge A2: } v_r = \frac{\omega d}{2n\pi} = \left.\frac{fd}{n}\right|_{\substack{f=5.13\\d=9.6\\n=3}} = 16.42\ \text{m/s} = 59.1\ \text{km/h}\,.$$

However, the computed response in Fig. 4.11 appears to be much smaller than the measured ones for the two bridges considered, while the peak response around the speed of 60 km/h is not quite visible. This is primarily due to the adoption of simple supports for the bridges in the finite element analysis. We assumed the bridges to be supported by hinges simply because we were not aware of the existence of elastic bearings at the bridge supports. Nevertheless, the relatively high amplitudes of the measured responses for the two bridges, compared with the computed one, suggested that the substantial *magnification* in the bridge response may be caused by the existence of elastic bearings, which was latter known to the authors. Such a problem seems to have received little attention from researchers working on railway bridges in the past.

Based on the static deflection tests under the locomotive loads, which were conducted as part of the preliminary tests, the spring constants measured for the A1 and A2 bridges are $K_1 = 3.5 \times 10^6$ kN/m and $K_2 = 8 \times 10^6$ kN/m, respectively, assuming that the elastic bearings installed at the two ends of a bridge are the same. All the key properties identified for the tested bridges, including the flexural rigidity $EI$ and stiffness ratio $\kappa$, have been listed in Table 4.2. With the data given in Table 4.2, the midpoint responses computed by the finite element program for the A1 and A2 bridges using the moving loads assumption were plotted in Fig. 4.12. As can be seen, because of the inclusion of elastic bearings, the computed responses turn out to be generally consistent with the measured ones for the two bridges. Moreover, larger response exists for the A1 bridge simply because it has softer support bearings.

Let us now turn to the phenomenon of cancellation. From Eq. (4.20), for the case with simple supports, the speed parameter $S$ for cancellation to occur is:

$$S_c = \left.\frac{1}{2i-1}\right|_{i=8} = 0.067\,.$$

Correspondingly, the cancellation speeds for the two bridges are

$$\begin{aligned}\text{Bridge A1: } v_c &= \frac{\omega_0 L S_c}{\pi} = 2fLS_c = 2 \times 5.17 \times 31.3 \times 0.067 \\ &= 21.68 \text{ m/s} = 78 \text{ km/h}\,.\end{aligned}$$

$$\begin{aligned}\text{Bridge A2: } v_c &= \frac{\omega_0 L S_c}{\pi} = 2fLS_c = 2 \times 5.13 \times 31.3 \times 0.067 \\ &= 21.51 \text{ m/s} = 77.5 \text{ km/h} \simeq 78 \text{ km/h}\,.\end{aligned}$$

This is exactly the cancellation speed for the bridge with simple supports, as can be seen from the numerical solution shown in Fig. 4.11.

Table 4.2. The properties of tested bridges.

| | $L$ (m) | $f$ (Hz) | $K$ (kN-m) | $EI$ (kN-m$^2$) | $\kappa = EI\pi^3/KL^3$ |
|---|---|---|---|---|---|
| A1 bridge | 31.3 | 5.17 | $3.5 \times 10^6$ | $2.40 \times 10^8$ | 0.069 |
| A2 bridge | 31.3 | 5.13 | $8 \times 10^6$ | $2.44 \times 10^8$ | 0.031 |

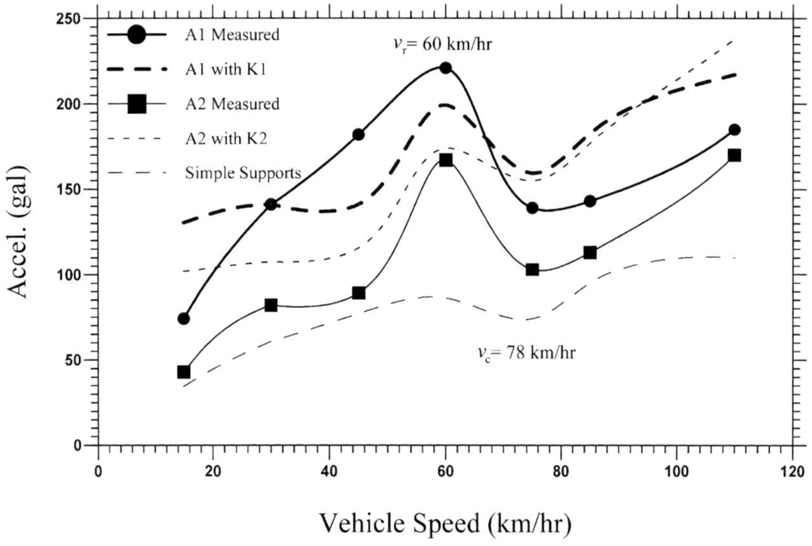


Fig. 4.12. The computed solutions for elastic supports.

To consider the effect of elastic bearings, the speed parameter $S$ for the cancellation to occur can be solved from Eq. (4.19) or

$$Y(S) = \frac{-S_c}{1 - S_c^2} \cos \frac{\pi}{2S_c} + \kappa \sin \frac{\pi}{2S_c} = 0\,. \tag{4.21}$$

By substituting the stiffness ratios $\kappa$ of 0.063 and 0.031, as given in Table 4.2, into the preceding equation, the speed parameter $S$ solved for the A1 and A2 bridges respectively are 0.069 and 0.068. Correspondingly, the cancellation speeds $v$ for the two bridges are

$$\text{Bridge A1: } v_c = 0.069 \times 2fL = 0.069 \times 2 \times 5.17 \times 31.3$$

$$= 22.3 \text{ m/s} = 80.4 \text{ km/h}\,.$$

$$\text{Bridge A2: } v_c = 0.068 \times 2fL = 0.068 \times 2 \times 5.13 \times 31.3$$
$$= 21.8 \text{ m/s} = 78.5 \text{ km/h}\,.$$

Clearly, the above (computed) speeds are consistent with the (measured) speeds for the occurrence of minimal responses for the two bridges, as shown in Fig. 4.12, which is an indication of the reliability of the present theory concerning the occurrence of resonance and cancellation.

## 4.6. Concluding Remarks

In this chapter, the mechansims underlying the resonance and cancellation phenomenona of elastically-supported bridges caused by a sequence of equi-distant moving loads have been analytically studied. A field test on two adjacent bridges traveled by two back-to-back connected locomotives was also conducted to confirm the phenomena of resonance and cancellation identified. The conclusions drawn from this chapter are: (1) The resonance condition in terms of the speed parameter $S$ is the same for the beam with both the elastic and simple supports. Since an elastically-supported beam has a lower frequency of vibration, it therefore has a lower resonant speed $v$, meaning that it can be more easily excited than a beam with simple supports. (2) The speed parameter for the cancellation condition to occur increases slightly as the stiffness ratio increases. However, since the frequency of vibration is slightly smaller for an elastically-supported beam, it turns out that the real cancellation speed for an elastically-supported beams remains close to that for the simply-supported beam. (3) Whenever the cancellation speed comes close to or coincides with the resonance speed, the phenomenon of resonance will be suppressed, meaning that the cancellation condition is more decisive than the resonance condition. (4) Once a resonance condition is reached for an elastically-supported beam, much larger peak responses will be induced on the beam, compared with those of the simply-supported beam.

There is no doubt that elastic bearings are effective devices for isolating the earthquake forces transmitted from the ground to the superstructure. However, the installation of these devices can also prevent the transmission or dissipation of vehicle-induced forces from the superstructure to the ground. Thus, the huge amount of vibration energy brought by a train may be accumulated and amplified on the bridge during its passage. Such a fact should not be overlooked in the design of railway bridges, especially of those to be traveled by high-speed trains, since it is harmful not only for the riding comfort of the passing trains, but also for the maintenance of track structures, as the repetitive occurrence of high-amplitude resonant peaks may cause fatigue problems on related components.

## Chapter 5

# Curved Beams Subjected to Vertical and Horizontal Moving Loads

Analytic solutions are derived in this chapter for a horizontal curved beam subjected to vertical and horizontal moving loads. The horizontal moving loads may be regarded as the centrifugal forces generated by vehicles moving along a curved path, which were rarely studied by previous researchers. Unlike the vertical moving loads, a horizontal moving load is not constant, but is proportional to the square of the speed of the moving vehicle. The case of a single moving load will be studied first, which will then be extended to the case of a series of equidistant moving loads. By superposition of the waves generated by consecutively moving loads on the curved beam, the conditions for the resonance and cancellation phenomena to occur will be derived. As compared with the approaches that are based fully on numerical simulation, the present analytical approach has the advantage of providing clear physical insight into the various phenomena induced by moving vehicles, thereby enabling us to grasp the key parameters involved in the curved-beam dynamics.

## 5.1. Introduction

The vehicle-induced vibration on bridges has been a subject of interest for more than one and half centuries. The pioneer works on this subject include those of Willis (1849) and Stokes (1849) following the collapse of the Chaster Rail Bridge in England in 1847. The problem of simple beams under moving vehicular loads was studied by Timoshenko (1922) neglecting the inertia effect of the vehicle.

In addition to the inertia of the beam, the inertia of the vehicle was included by Jeffcott (1929) in his study, followed by Stanišić and Hardin (1969) and Ting *et al.* (1974). By taking into account the suspension properties of the vehicle, the sprung mass model was adopted in the studies by Tan and Shore (1968a,b), Veletsos and Huang (1970), Blejwas *et al.* (1979), etc. The book by Frýba (1972) represents an early treatise on this subject, in which vehicle models of various complexities were studied. Partly enhanced by the operation of high-speed railways worldwide, research on the vehicle-induced vibrations of bridges continued to grow in the past two decades. The problems that have been studied include the implementation of more realistic vehicle models (Chu *et al.*, 1986; Hwang and Nowak, 1991) and bridge models (Galdos *et al.*, 1993; Chatterjee *et al.*, 1994), the inclusion of surface roughness (Hwang and Nowak, 1991; Inbanathan and Wieland, 1987), or the derivation of various solution schemes, including the vehicle–bridge interaction (VBI) elements that will be presented in later chapters. Some partial reviews of the research on vehicle-induced vibrations of the bridge are available in Diana and Cheli (1989), Paultre *et al.* (1992), and Yang and Yau (1998). In a recent book by Frýba (1996) on the dynamics of railway bridges, 231 references have been cited.

Previously, a majority of research conducted on vehicle-induced vibrations has been aimed at straight beams. While some research has been conducted for horizontally-curved beams under moving loads (Tan and Shore, 1986a,b; Genin *et al.*, 1982), concern was generally placed on the vertical or *out-of-plane vibration* of the curved beam. To the knowledge of the authors, the radial or *in-plane vibration* of curved beams under the action of *centrifugal forces*, generated by masses moving along circular paths, has rarely been studied. Just as a vertical moving load will cause some impact effect on the vertical vibration of a straight beam, a centrifugal force generated by a mass moving over a horizontally-curved beam will also induce certain impact effect on the radial response of the beam. The objective of this chapter is to establish a general theory for treating the vibration of a horizontally-curved beam subjected to a series of moving masses, of which will be simulated as a pair of *gravitational force* and centrifugal force. The conditions for superposition of the waves generated by consecutively moving loads to result in the phenomena of resonance and cancellation on the curved beam will be identified. The reliability of the present theory will be demonstrated in the exemplary studies. The materials presented in this chapter have been revised mainly from the paper by Yang *et al.* (2001).

## 5.2. Governing Differential Equations

Consider the horizontally-curved beam shown in Fig. 5.1, in which $\varphi$ denotes the subtended angle, $R$ the radius of curvature, and $L$ the length of the beam. For the present purposes, a right-handed coordinate system is chosen, of which the $y$ and $z$ axes coincide with the principal axes of the cross section, and the $x$ axis is tangent to the centroidal axis of the beam. Let $u_x$, $u_y$ and $u_z$ denote the displacements of the centroid of each cross section of the curved beam along the three axes, and $\theta_x$, $\theta_y$ and $\theta_z$ the rotations about the three

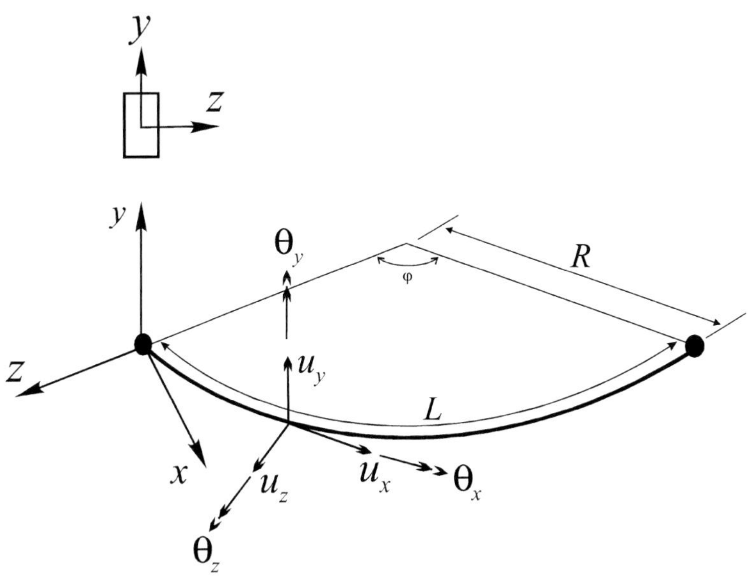


Fig. 5.1. The coordinates of curved beam.

axes. All the deformations are assumed to be small so that the linear theory applies. The curved beam is assumed to be made of constant, bi-symmetric cross sections with negligible warping resistance.

The following are the linear differential equations for a horizontally-curved beam of solid cross sections (Yang and Kuo, 1987, 1994):

- Axial displacement:

$$EA\left(u_x'' + \frac{u_z'}{R}\right) = 0\,. \tag{5.1}$$

- Radial displacement:

$$EI_y\left(u_z'''' + 2\frac{u_z''}{R^2} + \frac{u_z}{R^4}\right) + \frac{EA}{R}\left(u_x' + \frac{u_z}{R}\right) = 0\,. \tag{5.2}$$

- Vertical displacement:

$$EI_z\left(u_y'''' - \frac{\theta_x''}{R}\right) - \frac{GJ}{R}\left(\theta_x'' + \frac{u_y''}{R}\right) = 0\,. \tag{5.3}$$

- Torsional rotation:

$$\frac{EI_z}{R}\left(u_y'' - \frac{\theta_x}{R}\right) + GJ\left(\theta_x'' + \frac{u_y''}{R}\right) = 0\,, \tag{5.4}$$

where a prime denotes differentiation with respect to the axis $x$, $E$ and $G$ denote the moduli of elasticity and rigidity, respectively, of the beam, $A$ the cross-sectional area, $I_y$ and $I_z$ respectively the moments of inertia about the $y$ and $z$ axes, and $J$ the torsional constant. From Eqs. (5.1)–(5.4), one can observe that the differential equations for the in-plane displacements, i.e., $u_x$ and $u_z$, are independent of those for the out-of-plane displacement, i.e., $u_y$ and $\theta_x$. Furthermore, the differential equation for the axial displacement $u_x$ and that for the radial displacement $u_z$ are coupled, and the same is true for the vertical displacement $u_y$ and torsional rotation $\theta_x$.

## 5.3. Curved Beam Subjected to a Single Moving Load

The two ends of the curved beam are assumed to be simply-supported, in the sense that the flexural displacements and twisting rotation of the beam are restrained at the supports, but their first derivatives are not zero. In general, the action of the moving vehicle can be replaced by a *vertical* moving load, to simulate the gravitational effect, and a *horizontal* moving load, to simulate the centrifugal effect, as shown in Fig. 5.2(a).

### 5.3.1. *Vertical Moving Load*

As the in-plane and out-of-plane behaviors of the curved beam are uncoupled, we shall consider first the vertical vibration of the curved beam under the action of a vertical moving load (Fig. 5.2(c)). Let $m_v$ denote the mass of the vehicle moving at speed $v$. The load of the vehicle is $f_v = -m_v g$, where $g$ is the acceleration of gravitation. By taking into account the effect of inertia, the equations of motion for the vertical vibration of the curved beam can be modified from Eqs. (5.3) and (5.4) as follows:

$$m\ddot{u}_y + EI_z\left(u_y'''' - \frac{\theta_x''}{R}\right) - \frac{GJ}{R}\left(\theta_x'' + \frac{u_y''}{R}\right) = f_v\delta(x - vt)\,, \quad (5.5a)$$

$$\rho J\ddot{\theta}_x + \frac{EI_z}{R}\left(u_y'' - \frac{\theta_x}{R}\right) + GJ\left(\theta_x'' + \frac{u_y''}{R}\right) = 0\,, \quad (5.5b)$$

where $m$ denotes the mass per unit length, $\rho$ the density of the curved beam and $\delta(x)$ is Dirac's delta function. The term on the right-hand side of Eq. (5.5a) represents the effect of the vertical moving load $f_v$, where $f_v = -m_v g$. For the present problem, the vertical displacement $u_y$ can be expressed as the summation of a series of sine functions that satisfy the boundary conditions:

$$u_y(x,t) = \sum_{i=1}^{\infty} q_{yi}(t)\sin\frac{i\pi x}{L}\,, \quad (5.6)$$

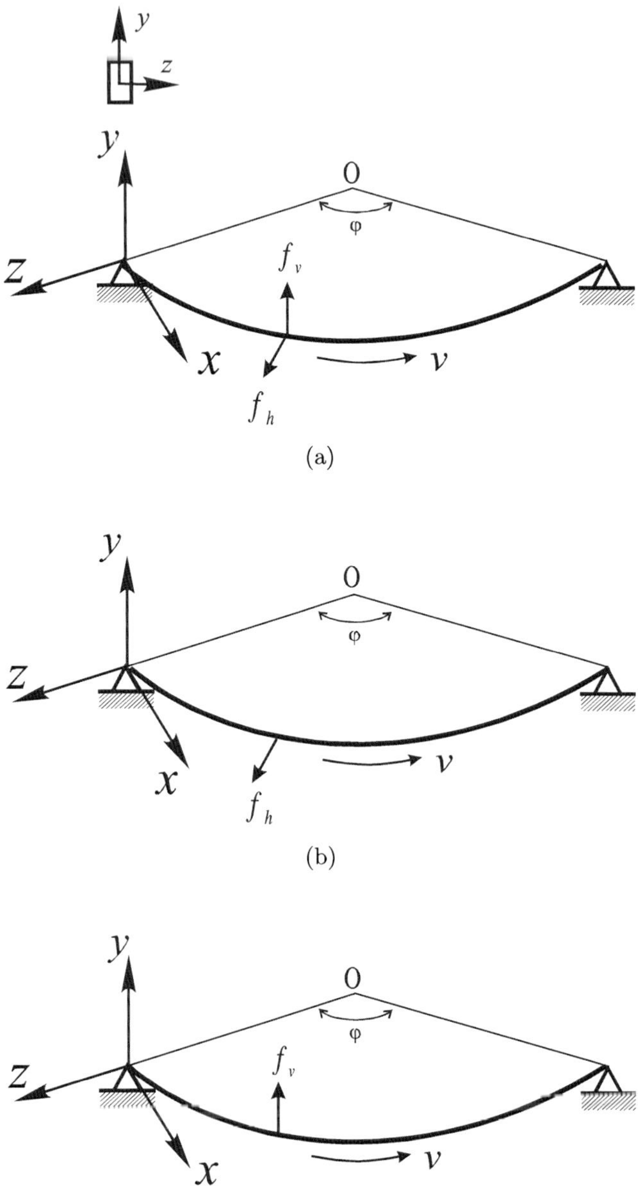


Fig. 5.2. The curved beam subjected to moving loads: (a) general, (b) horizontal, and (c) vertical.

where $q_{yi}$ denotes the generalized coordinate for the $i$th mode. The expression for the angle of twist $\theta_x$ is not arbitrary, but must be determined from Eq. (5.4). Substituting Eq. (5.6) into Eq. (5.4) and making use of boundary conditions for twisting, one obtains

$$\theta_x(x,t) = \sum_{i=1}^{\infty} \gamma_i q_{yi}(t) \sin \frac{i\pi x}{L}, \tag{5.7}$$

where $\gamma_i = -R(i\pi/L)^2(GJ + EI_z)/[(i\pi/L)^2 GJR^2 + EI_z]$, or equivalently,

$$\theta_x(x,t) = \sum_{i=1}^{\infty} q_{\theta i}(t) \sin \frac{i\pi x}{L}, \tag{5.8}$$

where $q_{\theta i}$ denotes the $i$th generalized coordinate for the angle of twist $\theta_x$. For the present purposes, let us consider only the contribution of the first modes, i.e.,

$$u_y(x,t) = q_{y1}(t) \sin \frac{\pi x}{L}, \tag{5.9a}$$

$$\theta_x(x,t) = q_{\theta 1}(t) \sin \frac{\pi x}{L}, \tag{5.9b}$$

where $q_{y1}$ and $q_{\theta 1}$ denote the first generalized coordinates for $u_y$ and $\theta_x$, respectively.

To solve the two differential equations in Eq. (5.5), *Galerkin's method* will be used. First, one multiplies both sides of Eq. (5.5a) by the variation $\delta u_y$ and those of Eq. (5.5b) by $\delta\theta_x$. Use the first-mode approximations for $u_y$ and $\theta_x$ in Eq. (5.9). Then integrate the two differential equations each with respect to $x$ from 0 to $L$. The results are as follows:

$$\begin{aligned} (\ddot{q}_{y1} + a_1 q_{y1} + a_2 q_{\theta 1})\delta q_{y1} &= \frac{2f_v}{mL} \sin \frac{\pi v t}{L} \delta q_{y1}, \\ (\ddot{q}_{\theta 1} + b_1 q_{\theta 1} + b_2 q_{y1})\delta q_{\theta 1} &= 0, \end{aligned} \tag{5.10}$$

where

$$
\begin{aligned}
a_1 &= \frac{1}{m}\left(\frac{\pi}{L}\right)^2 \left[EI_z\left(\frac{\pi}{L}\right)^2 + \frac{GJ}{R^2}\right], \\
a_2 &= \frac{1}{mR}\left(\frac{\pi}{L}\right)^2 (EI_z + GJ), \\
b_1 &= -\frac{1}{\rho J}\left[\frac{EI_z}{R^2} + GJ\left(\frac{\pi}{L}\right)^2\right], \\
b_2 &= -\frac{1}{\rho J}\frac{1}{R}\left(\frac{\pi}{L}\right)^2 (EI_z + GJ).
\end{aligned}
\tag{5.11}
$$

Since the variations $\delta q_{y1}$ and $\delta q_{\theta 1}$ are arbitrary, the two equations in Eq. (5.10) reduce to

$$
\begin{aligned}
\ddot{q}_{y1} + a_1 q_{y1} + a_2 q_{\theta 1} &= \frac{2f_v}{mL}\sin\frac{\pi v t}{L}, \\
\ddot{q}_{\theta 1} + b_1 q_{\theta 1} + b_2 q_{y1} &= 0.
\end{aligned}
\tag{5.12}
$$

The general solutions to the two differential equations in Eq. (5.12) are composed of two parts, i.e., the *homogenous solution* and *particular solution*,

$$
\begin{aligned}
q_{y1} &= q_{y1h} + q_{y1p}, \\
q_{\theta 1} &= q_{\theta 1h} + q_{\theta 1p},
\end{aligned}
\tag{5.13}
$$

where the subscripts $h$ and $p$ respectively denote the homogeneous and particular solutions.

The homogeneous solutions can be given as follows:

$$
q_{y1h} = h_1 \sin\omega_{v1}t + h_2\cos\omega_{v1}t, \tag{5.14a}
$$

$$
q_{\theta 1h} = k_1 \sin\omega_{v1}t + k_2\cos\omega_{v1}t, \tag{5.14b}
$$

where $\omega_{v1}$ denotes the fundamental frequency of vibration for the vertical direction of the curved beam and $h_1$, $h_2$, $k_1$, $k_2$ are the coefficients to be determined from the initial conditions. By substituting the equations in Eq. (5.14) into Eq. (5.12) and dropping the term containing $f_v$, one obtains

$$\begin{bmatrix} a_1 - \omega_{v1}^2 & a_2 \\ b_2 & b_1 - \omega_{v1}^2 \end{bmatrix} \begin{Bmatrix} q_{y1h} \\ q_{\theta 1h} \end{Bmatrix} = \begin{Bmatrix} 0 \\ 0 \end{Bmatrix}. \tag{5.15}$$

By letting the determinant in Eq. (5.15) equal to zero, the fundamental frequency $\omega_{v1}$ can be solved as (see Appendix C)

$$\omega_{v1} = \sqrt{\frac{a_1 + b_1 + \sqrt{(a_1 - b_1)^2 + 4a_2 b_2}}{2}}. \tag{5.16}$$

As for the particular solutions, the following may be assumed:

$$q_{y1p} = p_{y1} \sin \frac{\pi v t}{L}, \tag{5.17a}$$

$$q_{\theta 1p} = p_{\theta 1} \sin \frac{\pi v t}{L}, \tag{5.17b}$$

where $p_{y1}$ and $p_{\theta 1}$ denote the amplitudes of vibration. Substituting the equations in Eq. (5.17) into Eq. (5.10) yields

$$\begin{bmatrix} a_1 - \left(\frac{\pi v}{L}\right)^2 & a_2 \\ b_2 & b_1 - \left(\frac{\pi v}{L}\right)^2 \end{bmatrix} \begin{Bmatrix} p_{y1} \\ p_{\theta 1} \end{Bmatrix} = \begin{Bmatrix} \frac{2f_v}{mL} \\ 0 \end{Bmatrix}. \tag{5.18}$$

From the preceding equation, $p_{y1}$ can be solved,

$$p_{y1} = \frac{2f_v}{mL} \frac{1}{\omega_{v1}^2} \frac{1}{1 - S_{v1}^2} \beta, \tag{5.19}$$

where

$$S_{v1} = \frac{\pi v}{L \omega_{v1}}, \tag{5.20}$$

and

$$\beta = \frac{b_1 - (\pi v/L)^2}{b_1 + a_1 - \omega_{v1}^2 - (\pi v/L)^2}. \tag{5.21}$$

Here, $S_{v1}$ denotes the *speed parameter* for the vertical vibration of the curved beam, which represents the ratio of the driving frequency, $\pi v/L$, implied by the moving load to the fundamental frequency of the beam, $\omega_{v1}$.

The initial displacement and velocity of the beam are zero before it is subjected to the moving load. Such conditions should be obeyed by the general solution $q_{y1}$ given in Eq. (5.13), or the sum of $q_{y1h}$ and $q_{y1p}$ given in Eqs. (5.14a) and (5.17a), respectively. By these conditions, the coefficients of the homogeneous solution can be determined as: $h_1 = -p_{y1}S_{v1}$ and $h_2 = 0$. It follows that the general solution $q_{y1}$ for the vertical vibration of the curved beam is

$$q_{y1}(t) = p_{y1}\left(\sin\frac{\pi vt}{L} - S_{v1}\sin\omega_{v1}t\right). \tag{5.22}$$

By the expression in Eq. (5.19) and the relation $f_v = -m_v g$, the preceding equation may be rewritten as

$$q_{y1}(t) = -\frac{2m_v g}{mL}\frac{1}{\omega_{v1}^2}\frac{1}{1-S_{v1}^2}\beta\Psi_{v1}(t)\,, \tag{5.23}$$

where the amplitude function $\Psi_{v1}(t)$ is

$$\Psi_{v1}(t) = \sin\frac{\pi vt}{L} - S_{v1}\sin\omega_{v1}t\,. \tag{5.24}$$

Consequently, the vertical displacement of the curved beam is

$$u_y(x,t) = -\frac{2m_v g}{mL}\frac{1}{\omega_{v1}^2}\frac{1}{1-S_{v1}^2}\beta\Psi_{v1}(t)\sin\frac{\pi x}{L}\,. \tag{5.25}$$

For the midpoint of the curved beam, $x = L/2$, the vertical displacement is

$$u_y\left(\frac{L}{2},t\right) = \frac{2m_v g}{mL}\frac{1}{\omega_{v1}^2}\frac{1}{1-S_{v1}^2}\beta\Psi_{v1}(t)\,. \tag{5.26}$$

The above solution has been obtained by considering only the first mode of vibration of the curved beam. More accurate solutions can be obtained through consideration of more vibration modes. In practice, however, the acting time of the moving vehicles on the curved beam is so short that the moving load problem is by nature a transient vibration problem. As a result, only the first mode of the beam will be significantly excited. This is especially true when only the midpoint response of the beam is desired, and when the beam is subjected to a series of moving loads, as encountered in high-speed railways. The accuracy of the present solutions will be demonstrated in the numerical example through comparison with the finite element results.

### 5.3.2. *Horizontal Moving Load*

The horizontal moving load $f_h$ considered herein is the centrifugal force generated by a vehicle of mass $m_v$ moving at speed $v$ along a horizontally-curved beam of radius $R$. By taking into account the effect of inertia, the equations of motion for the radial vibration of a curved beam can be modified from Eqs. (5.1) and (5.2) as

$$m\ddot{u}_x + EA\left(u_x'' + \frac{u_z'}{R}\right) = 0\,, \tag{5.27a}$$

$$m\ddot{u}_z + EI_y\left(u_z'''' + 2\frac{u_z''}{R^2} + \frac{u_z}{R^4}\right) = f_h\delta(x - vt)\,, \tag{5.27b}$$

where the horizontal moving load $f_h$ can be related to the speed $v$ as $f_h = m_v v^2/R$. Similarly, the radial displacement $u_z$ can be expressed as the summation of the sine functions that satisfy the boundary conditions of the curved beam as

$$u_z(x,t) = \sum_{i=1}^{\infty} q_{zi}(t)\sin\frac{i\pi x}{L}\,, \tag{5.28}$$

in which $q_{zi}$ denotes the $i$th generalized coordinate for the radial displacement. The relation between the radial displacement $u_z$ and axial displacement $u_x$ is not arbitrary. By substituting Eq. (5.28) into Eq. (5.1), the axial displacement $u_x$ can be solved as

$$u_x(x,t) = \sum_{i=1}^{\infty} -\frac{q_{zi}(t)}{R}\left(\frac{L}{i\pi}\right)\left\{1 - \cos\frac{i\pi x}{L} - [1 - (-1)^i]\frac{x}{L}\right\}. \tag{5.29}$$

If only the first modes are considered for the axial and radial vibrations, the following may be written:

$$\begin{aligned} u_x(x,t) &= q_{x1}(t)\left(1 - \cos\frac{\pi x}{L} - \frac{2x}{L}\right), \\ u_z(x,t) &= q_{z1}(t)\sin\frac{\pi x}{L}\,, \end{aligned} \tag{5.30}$$

where $q_{x1}$ and $q_{z1}$ respectively denote the first generalized coordinate for the axial and radial displacements.

Again, Galerkin's method will be employed to solve the two second-order differential equations in Eq. (5.27). Namely, by multiplying both sides of Eq. (5.27a) by $\delta u_x$ and those of Eq. (5.27b) by $\delta u_z$, making use of the expressions for the displacement functions $u_x$ and $u_z$ in Eq. (5.30), integrating from 0 to $L$, and taking the arbitrary nature of virtual displacements $\delta q_{x1}$ and $\delta q_{z1}$, one obtains

$$\begin{aligned} &\ddot{q}_{x1} + \bar{a}_1 q_{x1} + \bar{a}_2 q_{z1} = 0\,, \\ &\ddot{q}_{z1} + \bar{b}_1 q_{z1} + \bar{b}_2 q_{x1} \delta q_{z1} = \frac{2f_h}{mL} \sin\frac{\pi v t}{L}\,, \end{aligned} \tag{5.31}$$

where

$$\begin{aligned} \bar{a}_1 &= \frac{EA\pi^2}{L}\left(\frac{4}{\pi^2} - \frac{1}{2}\right) \Big/ mL\left(\frac{8}{\pi^2} - \frac{5}{6}\right), \\ \bar{a}_2 &= \frac{EA\pi}{R}\left(\frac{4}{\pi^2} - \frac{1}{2}\right) \Big/ mL\left(\frac{8}{\pi^2} - \frac{5}{6}\right), \\ \bar{b}_1 &= \frac{EI_y}{m}\left[\left(\frac{\pi}{L}\right)^2 - \frac{1}{R^2}\right]^2 + \frac{EA}{mR^2}\,, \\ \bar{b}_2 &= \frac{EA}{mRL}\left(\pi - \frac{8}{\pi}\right). \end{aligned} \tag{5.32}$$

As was stated, the general solutions to the two differential equations in Eq. (5.31) consist of two parts,

$$q_{x1} = q_{x1h} + q_{x1p}\,, \tag{5.33a}$$

$$q_{z1} = q_{z1h} + q_{z1p}\,. \tag{5.33b}$$

The *homogenous parts* can be given as

$$\begin{aligned} q_{x1h} &= \bar{h}_1 \sin\omega_{h1} t + \bar{h}_2 \cos\omega_{h1} t\,, \\ q_{z1h} &= \bar{k}_1 \sin\omega_{h1} t + \bar{k}_2 \cos\omega_{h1} t\,, \end{aligned} \tag{5.34}$$

where $\omega_{h1}$ denotes the frequency of vibration for the horizontal plane of the curved beam and the coefficients $\bar{h}_1$, $\bar{h}_2$, $\bar{k}_1$, $\bar{k}_2$ are to be determined from the initial conditions. Substituting Eq. (5.34) back into the differential equations in Eq. (5.31), and dropping the term containing $f_h$ yields

$$\begin{bmatrix} \bar{a}_1 - \omega_{h1}^2 & \bar{a}_2 \\ \bar{b}_2 & \bar{b}_1 - \omega_{h1}^2 \end{bmatrix} \begin{Bmatrix} q_{x1h} \\ q_{z1h} \end{Bmatrix} = \begin{Bmatrix} 0 \\ 0 \end{Bmatrix}. \tag{5.35}$$

By letting the determinant equal to zero, the vibration frequency $\omega_{h1}$ can be solved as (Appendix D)

$$\omega_{h1} = \sqrt{\frac{\bar{a}_1 + \bar{b}_1 - \sqrt{(\bar{a}_1 - \bar{b}_1)^2 + 4\bar{a}_2\bar{b}_2}}{2}}. \tag{5.36}$$

The *particular solutions* for the present problem are

$$\begin{aligned} q_{x1p} &= p_{x1} \sin \frac{\pi v t}{L}, \\ q_{z1p} &= p_{z1} \sin \frac{\pi v t}{L}. \end{aligned} \tag{5.37}$$

Substituting the preceding expressions for $q_{x1p}$ and $q_{z1p}$ into the differential equations in Eq. (5.31) yields

$$\begin{bmatrix} \bar{a}_1 - \left(\frac{\pi v}{L}\right)^2 & \bar{a}_2 \\ \bar{b}_2 & \bar{b}_1 - \left(\frac{\pi v}{L}\right)^2 \end{bmatrix} \begin{Bmatrix} p_{x1} \\ p_{z1} \end{Bmatrix} = \begin{Bmatrix} 0 \\ \dfrac{2f_h}{mL} \end{Bmatrix}, \tag{5.38}$$

from which the generalized coordinate $p_{z1}$ can be solved as

$$p_{z1} = \frac{2f_h}{mL} \frac{1}{\omega_{h1}^2} \frac{1}{1 - S_{h1}^2} \alpha, \tag{5.39}$$

where

$$S_{h1} = \frac{\pi v}{L\omega_{h1}}, \tag{5.40}$$

and

$$\alpha = \frac{\bar{a}_1 - (\pi v/L)^2}{\bar{a}_1 + \bar{b}_1 - \omega_{h1}^2 - (\pi v/L)^2}. \tag{5.41}$$

Here, $S_{h1}$ denotes the *speed parameter* for the horizontal vibration of the curved beam, which again is defined as the ratio of the driving frequency, $\pi v/L$, to the fundamental frequency $\omega_{h1}$ of the beam.

By assumption, the zero initial conditions must be obeyed by the general solution $q_{z1}$, or the one given in Eq. (5.32b). From these conditions, the coefficients of the homogeneous solution $q_{z1h}$ can be determined as: $\bar{k}_1 = -p_{z1}S_{h1}$ and $\bar{k}_2 = 0$. By the relation $f_h = m_v v^2/R$ and using Eq. (5.39) for $p_{z1}$, the radial displacement $q_{z1}$ can be derived as

$$q_{z1}(t) = \frac{2m_v\varphi}{m\pi^2}\frac{S_{h1}^2}{1-S_{h1}^2}\alpha\Psi_{h1}(t)\,, \tag{5.42}$$

where $\varphi = L/R$ for the curved beam and the amplitude function $\Psi_{h1}(t)$ is

$$\Psi_{h1}(t) = \sin\frac{\pi v t}{L} - S_{h1}\sin\omega_{h1}t\,. \tag{5.43}$$

As a result, the radial displacement of the curved beam is

$$u_z(x,t) = \frac{2m_v\varphi}{m\pi^2}\frac{S_{h1}^2}{1-S_{h1}^2}\alpha\Psi_{h1}(t)\sin\frac{\pi x}{L}\,. \tag{5.44}$$

Of interest is the fact that when the vehicle speed $v$ approaches zero, $S_{h1} \to 0$, the radial displacement $u_z(x,t)$ of the curved beam also approaches zero. In contrast, the vertical displacement $u_y(x,t)$ as given in Eq. (5.25) approaches a constant under the same condition. This can be realized if one notes that as the vehicle speed $v$ reduces to zero, so does the horizontal moving load $f_h$, as there is no centrifugal force. However, the vertical moving load $f_v$ remains constant, regardless of the vehicle speed.

## 5.4. Unified Expressions for Vertical and Radial Vibrations

The solutions for the vertical and radial displacements, $u_y(x,t)$ and $u_z(x,t)$, as given in Eqs. (5.25) and (5.44) are similar in form. During the travel time $L/v$ of the vehicle on the beam, the two equations can be expressed in the same form as

$$U(x,t) = P\Psi(t)H(t)\sin\frac{\pi x}{L} \qquad \text{for} \qquad 0 \le t \le \frac{L}{v}\,, \tag{5.45}$$

where $H(t)$ is a unit step function, indicating that the function $\Psi(t)$ is turned on at $t = 0$. For the vertical vibration case, the displacement $U(x,t)$ should be interpreted as $u_y(x,t)$, the load $P$ as $-(2m_v g/mL)(1/\omega_{v1})[1/(1-S_{v1}^2)]\beta$, and the amplitude function $\Psi(t)$ as $\Psi_{v1}(t)$. For the horizontal vibration case, the displacement $U(x,t)$ should be interpreted as $u_z(x,t)$, the load $P$ as $(2m_v\varphi/m\pi^2)[S_{h1}^2/(1-S_{h1}^2)]\alpha$, and the amplitude function $\Psi(t)$ as $\Psi_{h1}(t)$. For $t > L/v$, the beam reaches a state of *free vibration*, as the moving load already leaves the beam. By the fact that $L/v$ denotes half of the period of the moving load over the beam, the free vibration response of the beam at this stage is

$$U(x,t) = P\left[\Psi(t)H(t) + \Psi\left(t - \frac{L}{v}\right)H\left(t - \frac{L}{v}\right)\right] \times \sin\frac{\pi x}{L} \qquad \text{for} \qquad t > \frac{L}{v}\,. \tag{5.46}$$

Here, the amplitude function $\Psi(t)$ is

$$\Psi(t) = \sin\frac{\pi v t}{L} - S_1 \sin\omega_1 t\,, \tag{5.47}$$

where the frequency $\omega_1$ and speed parameter $S_1$ should be interpreted as $\omega_{v1}$ and $S_{v1}$ for the vertical vibration, and as $\omega_{h1}$ and $S_{h1}$ for the horizontal vibration. By substitution of Eq. (5.47) for $\Psi(t)$ and noting that $H(t) = H(t - L/v)$ for $t > L/v$, $\sin(\pi v t/L) + \sin[\pi v(t - L/v)/L] = 0$ and $\sin a + \sin b = 2\sin[(a+b)/2]\cos[(a-b)/2]$, one can rearrange the free vibration response in Eq. (5.46) as

$$U(x,t) = -2PS_1\cos\frac{\omega_1 L}{2v}\sin\omega_1\left(t - \frac{L}{2v}\right)H\left(t - \frac{L}{v}\right) \times \sin\frac{\pi x}{L} \qquad \text{for} \qquad t \ge \frac{L}{v}\,. \tag{5.48}$$

This is exactly the *residual response* of the beam after the moving load has left the beam, based on the first mode approximation.

By letting $\cos(\omega_1 L/2v) = 0$ or $\omega_1 L/2v = (2i-1)\pi/2$, with $i = 1, 2, 3, \ldots$, one observes that the residual response of the beam reduces to zero. Such a condition is equivalent to

$$S_c = \frac{1}{2i-1} \quad \text{for} \quad i = 1, 2, 3, \ldots . \tag{5.49}$$

Theoretically speaking, if the condition in Eq. (5.49) is met, the residual response of the beam simply vanishes. For this reason, the condition in Eq. (5.49) is referred to as the *condition of cancellation.*

## 5.5. Solutions for Multi Moving Loads

Based on the assumption of small deformations, the vibration of a beam caused by a series of moving loads can be obtained as the superposition of the vibration induced by each of the moving loads. In such a process, care must be taken in calculating the entering time and the departing time for each moving load from the beam and the time lag between any two consecutive loads.

Consider $N$ identical masses of interval $d$ moving at constant speed $v$. As shown in Fig. 5.3, each of the masses will induce a vertical load $f_v$ and a centrifugal force $f_h$. Assuming that the first moving load enters the beam at $t = 0$, the time lag for the $j$th moving load is $t_j = (j-1)d/v$. By setting $x = L/2$, the *residual vibration* response of the midpoint of the beam caused by the $j$th moving load can be modified from Eq. (5.46) through consideration of the time lag $t_j$ as

$$U_j\left(\frac{L}{2}, t\right) = P\left[\Psi(t-t_j)H(t-t_j) + \Psi\left(t - t_j - \frac{L}{v}\right) \right.$$
$$\left. \times H\left(t - t_j - \frac{L}{v}\right)\right] \quad \text{for} \quad t - t_j \geq \frac{L}{v}. \tag{5.50}$$

The most critical condition for the beam occurs when the first $N-1$ masses have left and only the $N$th mass is acting on the beam. The time interval for such a case is $\max(t_N, t_{N-1} + L/v) < t < t_N + L/v$, as shown in Fig. 5.4. For this case, the midpoint response of the

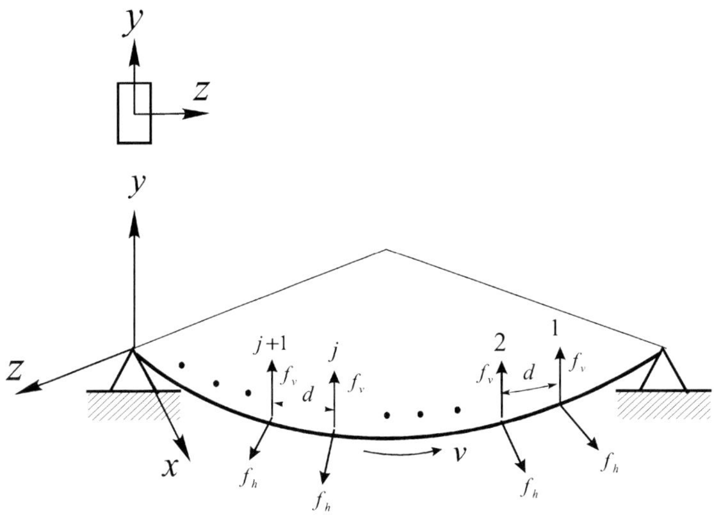


Fig. 5.3. The curved beam subjected to equi-spaced moving loads.

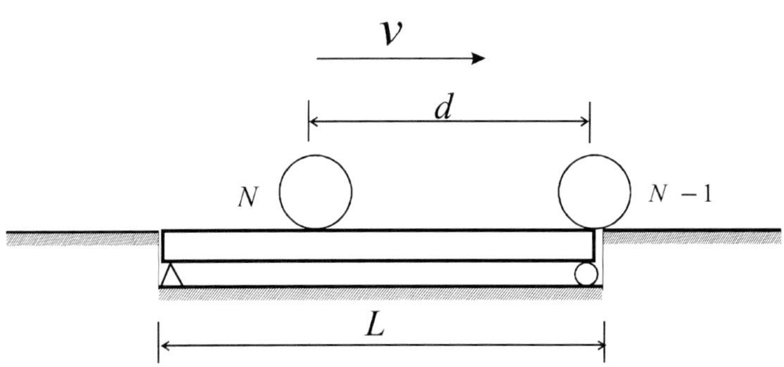


Fig. 5.4. The critical loading case.

beam is composed of two parts. The first part relates to the *forced vibration* caused by the $N$th moving load, which can be obtained by letting $x = L/2$ and replacing $t$ by $t - t_N$ in Eq. (5.45), i.e.,

$$U_{N,1}\left(\frac{L}{2}, t\right) = P\Psi(t - t_N)H(t - t_N) \qquad \text{for} \qquad 0 < t - t_N < \frac{L}{v}\,. \tag{5.51}$$

The second part is simply the *residual vibration* caused by the $N-1$ moving loads that passed the beam. This can be obtained by letting $x = L/2$ and replacing $t$ by $t - t_j$ in Eq. (5.48), and then summing the responses induced by the moving loads from $j = 1$ to $j = N - 1$.[a] Noting that when $H(t - t_{N-1} - L/v) = 1$, it can be ascertained $H(t - t_j - L/v) = 1$ for $j = 1, 2, \ldots, N-1$. Thus, the step function $H(t - t_{j-} - L/v)$ for $j = 1, 2, \ldots, N-1$ can be replaced by $H(t - t_{N-1} - L/v)$. The following is the result for part two or the residual vibration response:

$$U_{N,2}\left(\frac{L}{2}, t\right) = \sum_{j=1}^{N-1}\left\{-2PS_1 \cos\frac{\omega_1 L}{2v} \sin\omega_1\left[(t - t_j) - \frac{L}{2v}\right]\right\} \times H\left(t - t_{N-1} - \frac{L}{v}\right) \quad \text{for} \quad t - t_{N-1} \geq \frac{L}{v}. \tag{5.52}$$

Following the procedure presented in Appendix E, one can remove the summation sign in the preceding equation and obtain the residual vibration response of the beam as

$$U_{N,2}\left(\frac{L}{2}, t\right) = -2PS_1 \cos\frac{\omega_1 L}{2v}\left[\sin\omega_1\left(t - \frac{L}{2v}\right) + \sin\omega_1\left(\frac{N-2}{2}\frac{d}{v}\right) \times \sin\omega_1\left(t - \frac{L}{2v} - \frac{N-1}{2}\frac{d}{v}\right)\sin^{-1}\frac{\omega_1 d}{2v}\right] \times H\left(t - t_{N-1} - \frac{L}{v}\right) \quad \text{for} \quad t - t_{N-1} > \frac{L}{v}. \tag{5.53}$$

Consequently, the midpoint response of the beam under the action of the $N$th moving load can be computed as the summation of $U_{N,1}$ and $U_{N,2}$ given in Eqs. (5.51) and (5.53), respectively.

[a]This is equivalent to summing the expression in Eq. (5.50) for each load that passed the beam.

## 5.6. Conditions of Resonance and Cancellation

The phenomena of resonance and cancellation relate to the free vibration response $U_2$ caused by the $N-1$ moving loads that have passed the beam. From Eq. (5.53), one observes that the response of the beam reaches a maximum when the denominator of some terms vanishes, i.e., when $\sin(\omega_1 d/2v) = 0$ or when $\omega_1 d/2v = i\pi$, with $i = 1, 2, 3, \ldots$, or equivalently,

$$S_r = \frac{1}{2i}\frac{d}{L} \qquad \text{for} \qquad i = 1, 2, 3, \ldots . \tag{5.54}$$

This is exactly the condition for *resonance* of the beam to occur. For the case with $\sin(\omega_1 d/2v) = 0$, the expression in Eq. (5.53) becomes indeterminate. By the condition $\sin(\omega_1 d/2v) = 0$ and L'Hospital's rule, Eq. (5.53) can be manipulated to yield

$$U_{N,2}\left(\frac{L}{2}, t\right) = -2P(N-1)\cos\frac{\omega_1 L}{2v}\sin\omega_1\left(t - \frac{L}{2v}\right) \times H\left(t - t_{N-1} - \frac{L}{v}\right). \tag{5.55}$$

Here, it is interesting to note that the midpoint response of the beam increases as there are more loads passing the beam.

On the other hand, one observes from Eq. (5.53) that whenever $\cos(\omega_1 L/2v) = 0$, or $\omega_1 L/2v = (2i-1)\pi/2$, with $i = 1, 2, 3, \ldots$, or equivalently the one given in Eq. (5.49), the free vibration response $U_2(L/2, t)$ reduces to zero, which means that no residual response will be generated by the moving loads that have passed the beam. Such a condition has been referred to as the condition of *cancellation*. The conditions of resonance and cancellation, as given in Eqs. (5.54) and (5.49), are identical in form for both the vertical and horizontal oscillations in terms of the speed parameter $S_r$. However, because the vertical and horizontal vibration frequencies, $\omega_{v1}$ and $\omega_{h1}$, are generally different for the beam, the resonance or cancellation phenomena for the two directions do not occur at the same speed $v$.

## 5.7. Numerical Examples

The curved beam considered herein is simply-supported, of length $L = 24$ m and subtended angle $\varphi = 30°$. The following properties are assumed for the beam: cross-sectional area $A = 9\ \text{m}^2$, moments of inertia $I_y = 18.75\ \text{m}^4$, $I_z = 2.43\ \text{m}^4$, torsional constant $J = 21.18\ \text{m}^4$, elastic modulus $E = 33.2$ GPa, Poisson's ratio $\nu = 0.2$, and density $\rho = 2.4\ \text{t/m}^3$. Each vehicle has a mass of $m_v = 29.9$ t and the distance between two adjacent vehicles is $d = 25$ m. Unless noted otherwise, all the data assumed here will be used throughout this section. For comparison, finite element solutions obtained by approximating the curved beam by 10 piecewise straight beam elements will also be presented. As the straight beam element was derived from the straight beam theory (Yang and Kuo, 1994), the results obtained from the finite element analysis are totally independent of the present analytical results based on the curved beam theory (Yang and Kuo, 1987), as represented by Eqs. (5.1)–(5.4).

### 5.7.1. *Comparison of Analytic with Finite Element Solutions*

Consider a single moving mass with speed $v = 40$ m/s, which will generate a gravitational force and a centrifugal force. The frequencies of vibration computed for the two directions of the beam using Eqs. (5.16) and (5.36) are: $\omega_{v1} = 32.10$ rad/s and $\omega_{h1} = 115.4$ rad/s, compared with the finite element results of $\omega_{v1} = 32.24$ rad/s and $\omega_{h1} = 116.61$ rad/s.

The static vertical response, dynamic vertical response and dynamic horizontal response of the midpoint of the curved beam have been plotted in Figs. 5.5–5.7, along with the finite element solutions. As can be seen, the present solutions agree very well with the numerical ones, which is a demonstration of the accuracy of the present solutions considering only the first mode of vibration. Noting that the acting time of the mass is $L/v = 0.6$ s, one can easily distinguish between the forced vibration of the beam from the residual response in Figs. 5.5–5.7; the latter does not decay since the damping effect

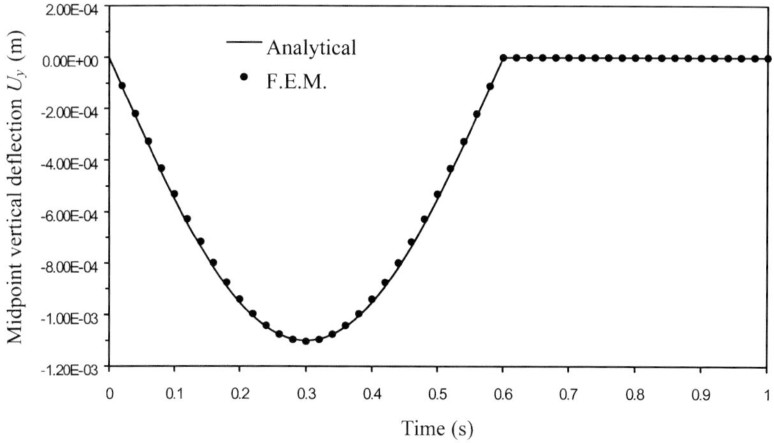


Fig. 5.5. The midpoint static vertical response of curved beam (one moving mass).

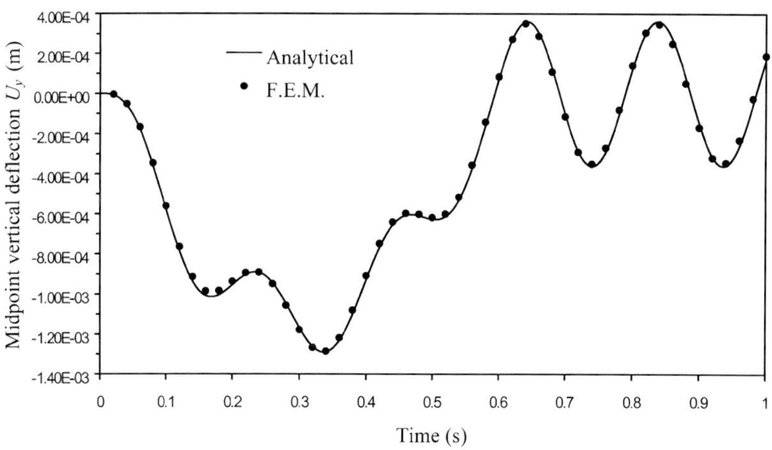


Fig. 5.6. The midpoint dynamic vertical response of curved beam (one moving mass).

was ignored. On the other hand, it is observed that the vertical response is much larger than that of the horizontal one, indicating that the vertical response can be more easily excited by the load moving at the speed $v = 40$ m/s.

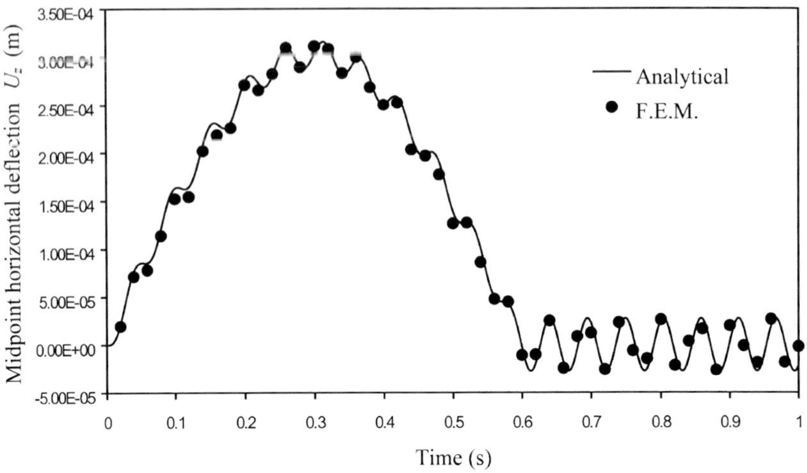


Fig. 5.7. The midpoint dynamic horizontal response of curved beam (one moving mass).

### 5.7.2. *Phenomenon of Cancellation Under Single or Multi Moving Masses*

As an illustration, we shall let $i = 2$ in Eq. (5.49) and compute the speed of cancellation as $S_c = 0.333$. By the definition for $S_1$ in Eqs. (5.20) and (5.40), along with $\omega_{v1} = 32.24$ rad/s and $\omega_{h1} = 116.61$ rad/s, the speed of cancellation can be computed for the two directions as: $v_v = 82$ m/s and $v_h = 297$ m/s. Considering a single mass moving at these speeds, each for one direction, the time-history midpoint response computed for the two directions were plotted in Figs. 5.8 and 5.9. As can be seen, the residual response induced by the moving mass after it leaves the beam, i.e., after 0.29 s and 0.08 s respectively for the vertical and horizontal directions, deviates slightly from the theoretical value of zero response, due to the neglect of higher modes. The close agreement of the present solutions with the finite element ones indicates that the effect of higher modes is generally small.

Consider next the case of eight moving masses and use the same speeds of cancellation, i.e., $v_v = 82$ m/s and $v_h = 297$ m/s, for the

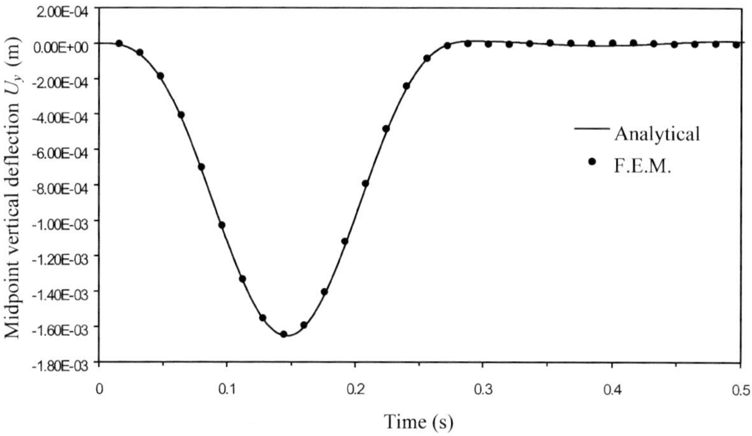


Fig. 5.8. The phenomenon of cancellation for vertical direction (one moving mass).

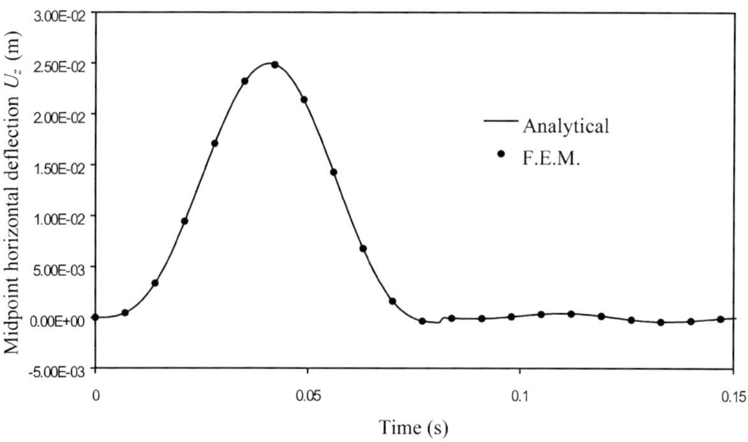


Fig. 5.9. The phenomenon of cancellation for horizontal direction (one moving mass).

two directions. The time-history response computed for the mid-point response for the two directions have been plotted in Figs. 5.10 and 5.11. The time for all the moving loads to depart from the beam is 2.43 s and 0.67 s for the vertical and horizontal direction,

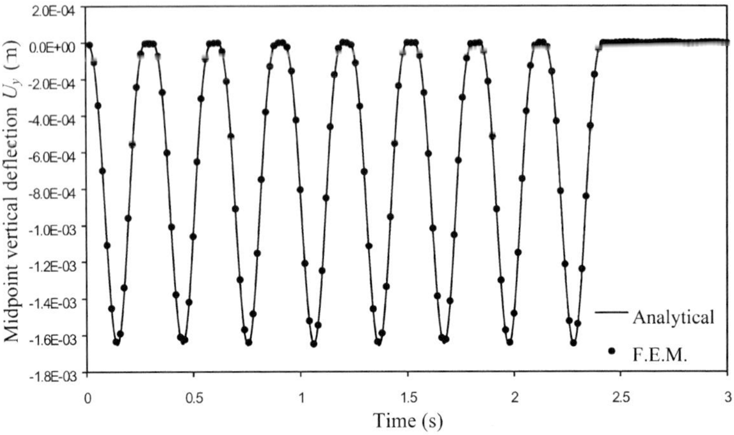


Fig. 5.10. The phenomenon of cancellation for vertical direction (eight moving masses).

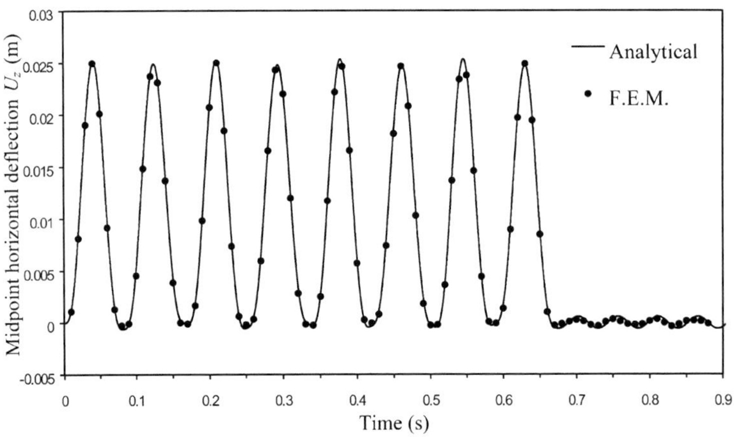


Fig. 5.11. The phenomenon of cancellation for horizontal direction (eight moving masses).

respectively. In the two figures, it is observed that there is a total of eight peaks, each of which corresponds to one moving load. Moreover, the residual responses for both directions remain negligibly small, as was expected. Finally, the effect of higher modes of vibration appears to be generally small, as the present solutions agree very well with the finite element ones.

### 5.7.3. *Phenomenon of Resonance Under Multi Moving Masses*

Consider also the case of eight moving masses. Using $d = 25$ m and $L = 24$ m, the resonance speed computed from Eq. (5.54) for $i = 1$ is $S_1 = 0.521$. By using Eqs. (5.20) and (5.40), along with $\omega_{v1} = 32.24$ rad/s and $\omega_{h1} = 116.61$ rad/s, the corresponding speeds computed of the moving masses for two directions are: $v_v = 128$ m/s and $v_h = 464$ m/s. In Figs. 5.12 and 5.13, the time-history response obtained for the vehicles traveling at the resonance speed for each direction has been plotted. The fact that the response increases for the two directions as there are more masses passing the beam is a typical resonance phenomenon. After all the masses pass the beam, i.e., after 1.55 s and 0.43 s for the vertical and horizontal directions, respectively, the beam tends to oscillate with the largest amplitude that has been excited, as no damping is assumed. Again, the present solutions match very well the finite element solutions.

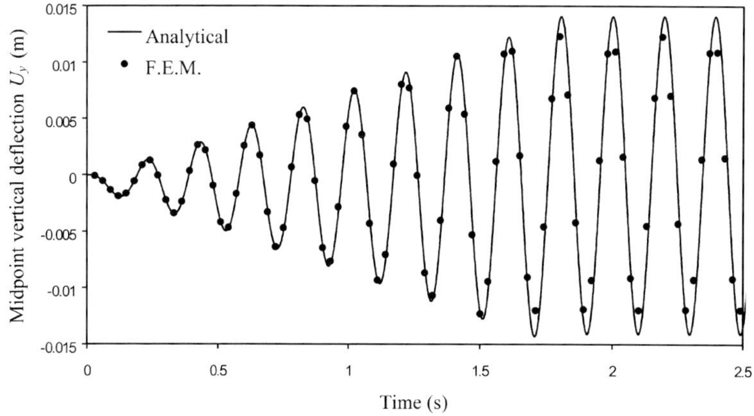


Fig. 5.12. The phenomenon of resonance for vertical direction (eight moving masses).

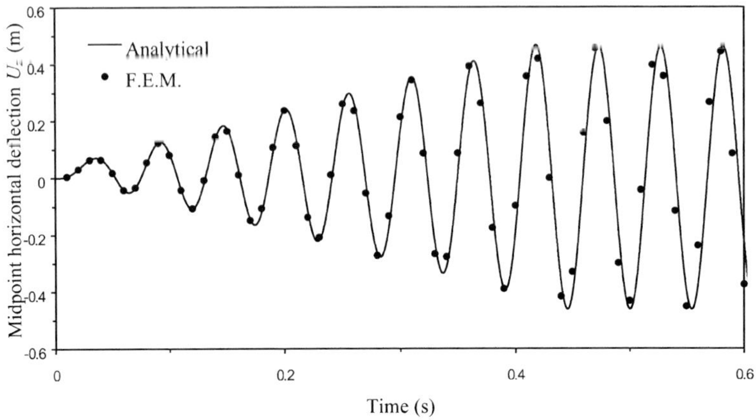


Fig. 5.13. The phenomenon of resonance for horizontal direction (eight moving masses).

### 5.7.4. *I–S Plot — Impact Effect Caused by Moving Loads*

By letting $R_d(x)$ and $R_s(x)$ respectively denote the maximum dynamic and static deflection of the beam at position $x$ due to the action of the moving loads, the impact factor $I$ for the deflection of a beam subjected to the moving loads was defined in Eq. (1.1), i.e.,

$$I = \frac{R_d(x) - R_s(x)}{R_s(x)} . \tag{5.56}$$

For simple beams, both the maximum static and dynamic deflections will occur at the midpoint. Consider the case of eight equally-spaced moving masses. The impact response of the vertical deflection of the beam versus the speed parameter of the moving masses has been plotted in Fig. 5.14, along with the finite element solution. The impact response of the horizontal deflection can hardly be plotted, because of the lack of a static centrifugal force. However, if the centrifugal force $f_h$ computed as $m_v v^2/R$ can be treated as if it was a static force, the impact response of the horizontal deflection can be computed as well (not shown), which appears to be quite similar to that presented in Fig. 5.14 for the vertical vibration (Wu, 1998).

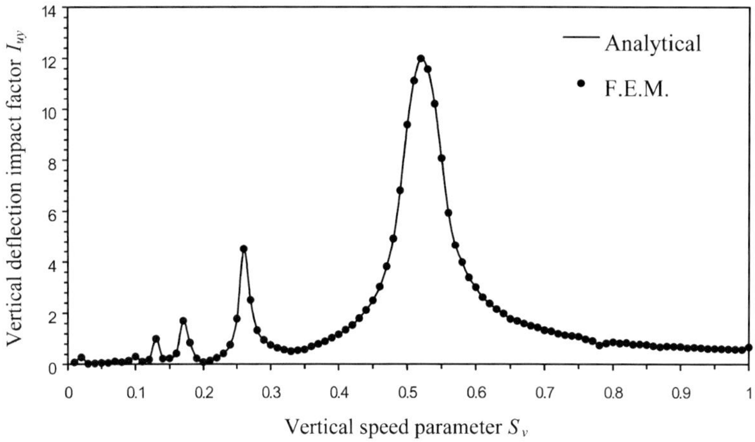


Fig. 5.14. The impact factor for midpoint vertical deflection (eight moving masses).

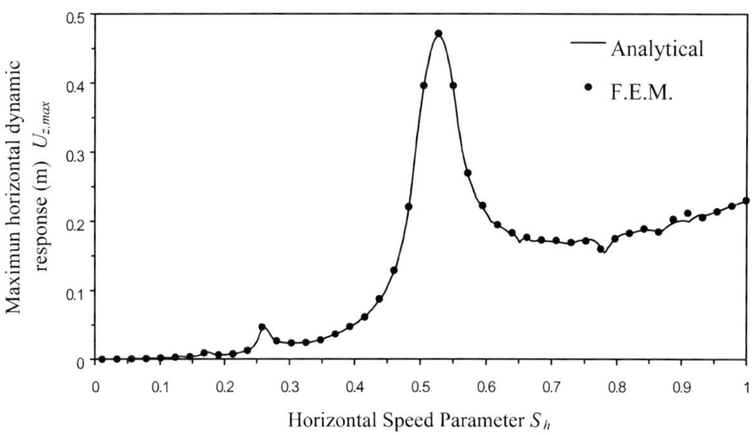


Fig. 5.15. The maximum response for midpoint horizontal deflection (eight moving masses).

In Fig. 5.14, it is confirmed that the present solutions agree very well with the finite element results. By substituting $d = 25$ m and $L = 24$ m into Eq. (5.54), the resonance speeds computed for the two directions are: $S_1 = 0.521,\ 0.260,\ 0.174,\ 0.130, \ldots$, which are close to the points where the peaks occur in Fig. 5.14. On the other hand, from Eq. (5.49) the speeds of cancellation can be computed as: $S_1 = 1.000, 0.333, 0.200, 0.143, \ldots$, which are also close to the points where the minimum values occur in Fig. 5.14.

As was noted, the impact response for the horizontal vibration of the curved beam is unreal, since there is no static centrifugal force. In Fig. 5.15, the maximum horizontal deflection for the midpoint of the curved beam has been plotted. As can be seen, the speeds for the maximum and minimum values to occur are generally consistent with those of Fig. 5.14, although they are not as clear.

## 5.8. Concluding Remarks

In this chapter, a general theory has been presented for treating the vibration of a horizontally-curved beam subjected to either a single or a series of moving masses, with each moving mass simulated as a set of gravitational and centrifugal forces. Unlike the gravitational force, the centrifugal force is not constant, but is proportional to the square of the speed of the moving vehicle along the curved beam. The problem has been solved in an analytical but approximate manner considering the contribution of the first mode of vibration. The accuracy of the present approach has been confirmed by an independent finite element analysis, which, by nature, considers the contribution of all modes of vibration of the beam. The present approach has the advantage that it provides clear physical insights into the various vibration phenomena induced by vehicles, in particular, the phenomena of resonance and cancellation, while allowing us to identify the key parameters involved. The solution established herein for the horizontal vibration of curved beams subjected to the centrifugal force is new in the literature.

# Part II

# Interaction Dynamics Problems

## Chapter 6

# Vehicle–Bridge Interaction Element Based on Dynamic Condensation

In this chapter, an element that is both accurate and efficient for modeling the vehicle–bridge interactions is presented, which is suitable for applications to high-speed railway bridges. The train is modeled as a sequence of lumped sprung masses, with the effects of suspension and energy dissipation of each vehicle taken into account, and a bridge sustaining the railway track with irregular elevations by a number of beam elements. Two sets of equations of motion are written, with the first set for the bridge element and the second set for the moving mass, which interact with each other through the contact forces. To resolve the problem of coupling between the two subsystems, the sprung mass equation is first discretized using Newmark's finite difference formulas, by which the sprung mass degrees of freedom (DOFs) are condensed to those of the bridge element in contact. The element derived is referred to as the *vehicle–bridge interaction* (VBI) element, which possesses the same number of DOFs as the parent element, while the properties of symmetry and bandedness in the element matrices are preserved.

## 6.1. Introduction

In the past two decades, partly due to the construction of high-speed railways worldwide, the vibration of bridges caused by the passage of trains is becoming a subject of increasing interest. In the survey paper by Diana and Cheli (1989), issues related to the train–bridge interactions have been discussed and a total of 90 papers were cited. By modeling a moving vehicle as a moving load, moving mass, or moving sprung mass considering the suspension and/or energy dissipation mechanisms, the dynamic response of bridges induced by moving vehicles has been studied by researchers from time to time (Timoshenko, 1922; Jeffcott, 1929; Lowan, 1935; Ayre *et al.*, 1950; Ayre and Jacobsen, 1950; Biggs, 1964; Frýba, 1972; Chu *et al.*, 1979; Stanišić, 1985; Sadiku and Leipholz, 1987; Chatterjee *et al.*, 1994). More sophisticated models that consider the various dynamic properties of vehicles or railroad cars have also been implemented in the study of vehicle–bridge interactions by researchers (Veletsos and Huang, 1970; Garg and Dukkipati, 1984; Yang and Lin, 1995).

In studying the dynamic response of a vehicle–bridge system, two sets of equations of motion can be written, one for the supporting bridge (i.e., the stationary subsystem) and the other for each of the moving vehicles (i.e., the moving subsystem) over the bridge. It is the interaction or contact forces existing at the contact points of the two subsystems that make the two sets of equations coupled and nonlinear. As the contact forces vary with respect both to time and space, the system matrices, which are functions of the contact forces, must be updated and factorized at each time step in a time-history analysis. To solve the two sets of differential equations, procedures of iterative nature are often adopted (Hwang and Nowak, 1991; Green and Cebon, 1994; Yang and Fonder, 1996). In this case, one may first start by assuming the displacements for the contact points, and then solve the vehicle equations to obtain the contact forces. By substituting the contact forces into the bridge equations, improved values of displacements for the contact points can be obtained. The advantage of such a procedure is that the responses of both the vehicles and bridge are simultaneously made available at each time step. However, the rate of convergence of iteration is likely to be low, when dealing with the situation of a bridge sustaining a large number of moving vehicles, for there exists twice the number of contact points if each vehicle is modeled as the composition of two lumped sprung masses.

Other approaches for solving the vehicle–bridge interaction problems include those based on the condensation method. Garg and

Dukkipati (1984) used the Guyan reduction scheme to condense the vehicle degrees of freedom (DOFs) to the associated DOFs of the bridge. Recently, Yang and Lin (1995) employed the dynamic condensation method to eliminate the vehicle DOFs on the element level. These condensation approaches were initially proposed for solving the response of the bridge (i.e., the larger or *master subsystem*), and they are good for this purpose. However, if the response of the moving vehicles (i.e., the smaller or *slave subsystem*) is desired, which serves as a reference for evaluating the riding comfort of passengers, the aforementioned two approaches cannot be relied to yield accurate solutions, since approximations have been made in relating the vehicle (slave) to the bridge (master) DOFs.

To resolve the dependency of system matrices on the wheel load positions, the condensation technique that eliminates the vehicle DOFs on the element level will be adopted in this chapter. First, two sets of equations of motion will be written, one for the bridge element and the other for the sprung mass lumped from the train car directly acting on the element. The sprung mass equation is then discretized using Newmark's finite difference formulas, and condensed to those of the bridge element in contact, which will result in the so-called *vehicle–bridge interaction* (VBI) element. Such an element has the advantage that it possesses exactly the same number of DOFs as the parent element, while the properties of *symmetry* and *bandedness* in element matrices are preserved. The materials presented in this chapter concerning the element formulation, i.e., in Secs. 6.2–6.7, have been revised mainly from the paper by Yang and Yau (1997), and those related to the parametric studies in Sec. 6.8 from the paper by Yau *et al.* (1999).

## 6.2. Equations of Motion for the Vehicle and Bridge

As shown in Fig. 6.1, a bridge is modeled as a beam-like structure, and the train traveling over the bridge with constant speed $v$ is idealized as a sequence of lumped sprung mass units of regular intervals. In this chapter, each sprung mass unit is used to represent either the front or rear half of a train car, which consists of two concentrated

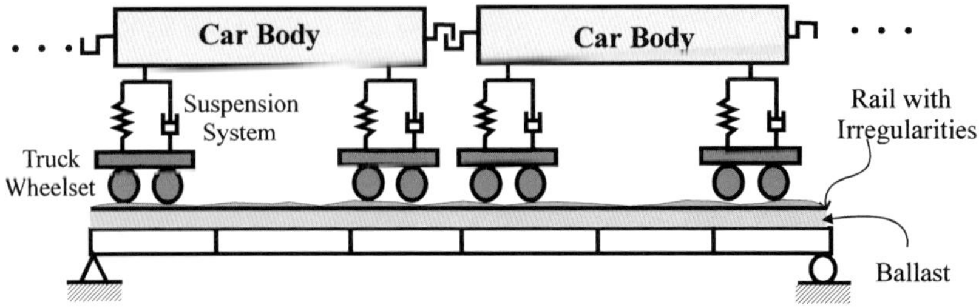


(a)

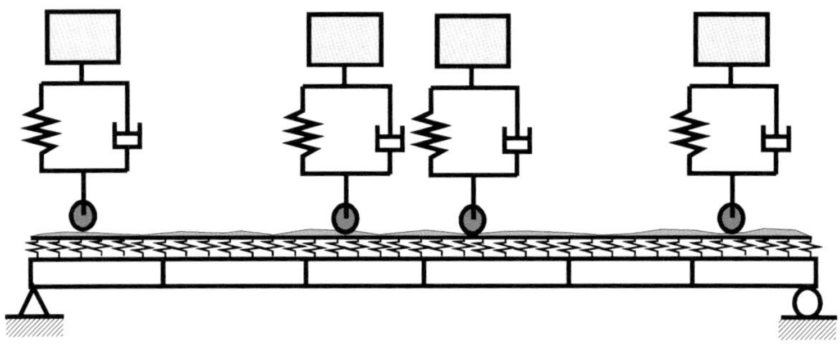

(b)

Fig. 6.1. The train–bridge system: (a) general model and (b) sprung mass model.

masses, with the top one representing the mass lumped from the car body, which is equal to half of the car body mass, and the bottom one the mass of the wheel assembly. The two masses are connected by a set of spring and dashpot that serves to represent the vehicle's suspension and energy dissipation mechanism.

For the present purposes, an interaction element is defined such that it consists of a bridge (beam) element and the sprung mass unit directly acting on it, as shown in Fig. 6.2, where the *ballast* with stiffness $k_B$ and the *rail irregularity* with a profile $r(x)$ are also indicated. The stiffness of the beam element may be computed as the summation of the stiffness of the bridge girder, rails and track structures. In other words, no consideration will be made specifically of the stiffness of the rails or track structures. For the parts of the bridge that are not directly acted upon by the moving vehicles or

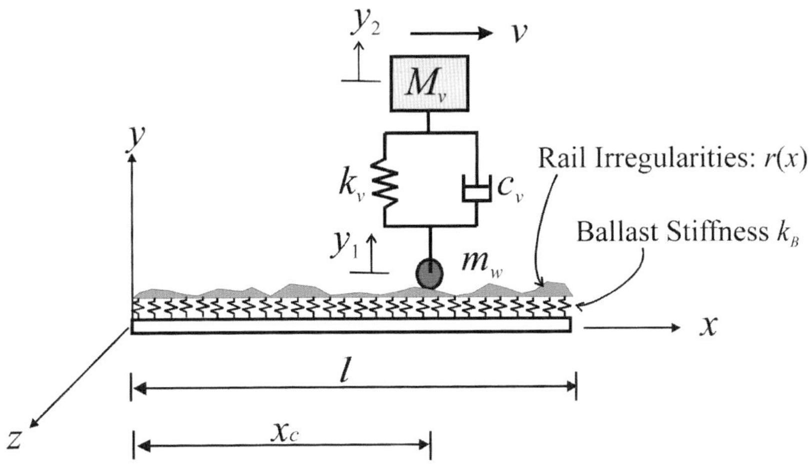


Fig. 6.2. The vehicle–bridge interaction element.

sprung mass units, they are modeled by traditional bridge elements. However, for the remaining parts that are in direct contact with the vehicles, interaction elements considering the effects of the sprung mass units have to be used instead. In this study, the notation [ ] is used for a square matrix, { } for a column vector, and $\langle\ \rangle$ for a row vector.

The vehicle model depicted in Fig. 6.2 will be grossly referred to the *sprung mass model* throughout this chapter. Let the stiffness and damping coefficients of the suspension unit be denoted by $k_v$ and $c_v$, respectively, the mass of the wheel assembly by $m_w$, and the mass lumped from the car body as $M_v$ (assumed to be equal to half of the mass of the car body). The sprung mass model is regarded as a two-node system, with one node associated with each of the two concentrated masses. Also, let the vertical displacements of the two nodes measured from the static equilibrium positions be denoted by the coordinates $\{z\}^T = \langle z_1, z_2\rangle$. Corresponding to the nodal displacements $\{z\}^T$ are the external forces $\{p_v\}^T = \langle p, 0\rangle$, where $p$ represents the total weight of the two masses, $p = -(M_v + m_w)g$, with $g$ denoting the acceleration of gravity.

The equations of motion for the sprung mass model shown in Fig. 6.2 can be written as follows (Frýba, 1972):

$$\begin{bmatrix} m_w & 0 \\ 0 & M_v \end{bmatrix} \begin{Bmatrix} \ddot{z}_1 \\ \ddot{z}_2 \end{Bmatrix} + \begin{bmatrix} c_v & -c_v \\ -c_v & c_v \end{bmatrix} \begin{Bmatrix} \dot{z}_1 \\ \dot{z}_2 \end{Bmatrix} + \begin{bmatrix} k_v & -k_v \\ -k_v & k_v \end{bmatrix} \begin{Bmatrix} z_1 \\ z_2 \end{Bmatrix} = \begin{Bmatrix} p + f_c \\ 0 \end{Bmatrix}, \tag{6.1}$$

where $f_c$ is the interaction or *contact force* existing between the wheel mass and the bridge element. Let $x_c$ denote the acting position of the sprung mass unit (see Fig. 6.2) and $\{N_c\}$ a vector containing the cubic Hermitian interpolation functions for the vertical displacement of the beam evaluated at the contact point $x_c$, i.e., $\{N_c\} = \{N(x_c)\}$ (see, for instance, p. 52 of Yang and Kuo (1994) for the interpolation functions of a two-dimensional beam element). The contact force $f_c$ can be expressed as

$$f_c = k_B(\langle N_c \rangle \{u_b\} + r_c - z_1) \geq 0\,, \tag{6.2}$$

where the condition of $f_c \geq 0$ is imposed for the VBI system to exclude the possibility of separation for the wheels from the bridge, $k_B$ is the ballast stiffness, $\{u_b\}$ the nodal displacements of the beam element, and $r_c$ the elevation of rail irregularity at the contact point $x_c$, which is assumed to be known throughout the element considered. Note that the interpolation function vector $\{N_c\}$ contains only four nonzero entries for those DOFs of the element related to the vertical displacements, the other entries not related to the vertical displacements are simply set to zero.

The equations of motion for the bridge element in contact with the sprung mass unit can be written as

$$[m_b]\{\ddot{u}_b\} + [c_b]\{\dot{u}_b\} + [k_b]\{u_b\} = \{p_b\} - \{N_c\} f_c\,, \tag{6.3}$$

where $[m_b]$, $[c_b]$, and $[k_b]$ denote the mass, damping, and stiffness matrices of the bridge element; and $\{p_b\}$ denotes the external nodal loads acting on the element. Regarding the bridge element as a three-dimensional solid beam element, one can assign six DOFs to each node of the element, with three for translations and the other three for rotations. The mass matrix $[m_b]$ and stiffness matrix $[k_b]$ for the three-dimensional beam element have been given in Appendix F. In this study, *Rayleigh damping* is assumed for the bridge, which means that the damping matrix of the entire bridge can be computed as a linear combination of the mass and stiffness matrices of the structure, for which the procedure has also been described in Appendix F. Thus, the element damping matrix $[c_b]$ appearing in Eq. (6.3) is required only for conceptual needs.

As can be seen from Eqs. (6.1) and (6.3), the sprung mass unit and the bridge element interact with each other through the contact force $f_c$, which varies as a function of both time and position. To ensure that the sprung mass unit is in full contact with the bridge element, the reactive force exerted by the bridge element on the sprung mass unit must be greater than zero. Whenever the contact force $f_c$ is less than zero, the wheel mass will jump upward, meaning that the contact condition between the sprung mass unit and the bridge is violated. In this chapter, no consideration will be made for separation of the moving vehicles from the bridge. From the first line of Eq. (6.1), together with the second line, the contact force $f_c$ can also be expressed as follows:

$$f_c = -p + m_w \ddot{z}_1 + M_v \ddot{z}_2 \,. \tag{6.4}$$

From Eqs. (6.3) and (6.4), it can be seen that the dynamic response of the beam element is affected not only by the static weight $p$, but also by the inertial effects of the moving vehicles in the vertical direction.

## 6.3. Element Equations in Incremental Form

The system equations as given in Eqs. (6.1)–(6.3) are coupled, nonlinear, and time-dependant in nature, due to existence of the contact force $f_c$ between the two subsystems, i.e., the moving vehicle and the sustaining bridge. In this chapter, an incremental method will be presented for solving the system equations of motion, with iterations performed at each incremental step for removing the unbalanced forces. Consider a typical incremental step for the system from time

$t$ to $t+\Delta t$. The component equations as presented in Eqs. (6.1)–(6.3) should be interpreted as those established for the deformed position of the system at time $t + \Delta t$. For the incremental time step, one may let

$$\{z\}_{t+\Delta t} = \{z\}_t + \{\Delta z\}\,, \tag{6.5}$$

where $\{\Delta z\}$ denotes the displacement increments of the sprung mass unit occurring during the time step considered. By using Eq. (6.2), the sprung mass equation (6.1) can be rewritten in an incremental form as

$$\begin{aligned} &\begin{bmatrix} m_w & 0 \\ 0 & M_v \end{bmatrix} \begin{Bmatrix} \ddot{z}_1 \\ \ddot{z}_2 \end{Bmatrix}_{t+\Delta t} \\ &\qquad + \begin{bmatrix} c_v & -c_v \\ -c_v & c_v \end{bmatrix} \begin{Bmatrix} \dot{z}_1 \\ \dot{z}_2 \end{Bmatrix}_{t+\Delta t} + \begin{bmatrix} k_v + k_B & -k_v \\ -k_v & k_v \end{bmatrix} \begin{Bmatrix} \Delta z_1 \\ \Delta z_2 \end{Bmatrix}_{t+\Delta t} \\ &\qquad = \begin{Bmatrix} p + k_B(\langle N_c \rangle \{u_b\} + r_c) \\ 0 \end{Bmatrix}_{t+\Delta t} - \begin{Bmatrix} q_{s1} \\ q_{s2} \end{Bmatrix}_t , \end{aligned} \tag{6.6}$$

where $\{q_s\}_t$ denotes the internal resistant forces of the sprung mass unit computed at the end of the last incremental step, i.e., at time $t$,

$$\begin{Bmatrix} q_{s1} \\ q_{s2} \end{Bmatrix}_t = \begin{bmatrix} k_v + k_B & -k_v \\ -k_v & k_v \end{bmatrix} \begin{Bmatrix} z_1 \\ z_2 \end{Bmatrix}_t . \tag{6.7}$$

It should be noted that the resistant forces $\{q_s\}_t$ and rail irregularities $r_{c,t+\Delta t}$ appearing on the right-hand side of Eq. (6.6) are known at the beginning of the current incremental step. However, the bridge displacements $\{u_b\}_{t+\Delta t}$ remain unknown. The vehicle equations are coupled with the bridge equations to be shown below.

Similarly, one may write the following for the bridge displacements at the current time step:

$$\{u_b\}_{t+\Delta t} = \{u_b\}_t + \{\Delta u_b\}\,, \tag{6.8}$$

where $\{\Delta u_b\}$ denotes the displacement increments of the bridge element during the current incremental step. By using Eqs. (6.2) and (6.8), the bridge equations of motion as given in Eq. (6.3) can be rewritten in an incremental form,

$$\begin{aligned}&[m_b]\{\ddot{u}_b\}_{t+\Delta t} + [c_b]\{\dot{u}_b\}_{t+\Delta t} + [[k_b] + k_B\{N_c\}\langle N_c\rangle]\{\Delta u_b\}\\ &\quad = \{p_b\}_{t+\Delta t} - \{N_c\}k_B(r_c - z_1)_{t+\Delta t} - [[k_b] + k_B\{N_c\}\langle N_c\rangle]\{u_b\}_t\,,\end{aligned} \tag{6.9}$$

where all the terms on the right-hand side of the equal sign are either given, such as the external loads $\{p_b\}_{t+\Delta t}$ and irregularity function $r_{c,t+\Delta t}$, or have been made available at the last time step, such as the bridge displacements $\{u_b\}_t$, except the wheel mass displacement $z_{1,t+\Delta t}$. This again represents another source of coupling between the bridge and moving vehicle.

## 6.4. Equivalent Stiffness Equation for Vehicles

As can be seen from Eqs. (6.6) and (6.9) for the sprung mass unit and the bridge element, the two system equations are coupled, since they depend on the solution of each other. In the following, the sprung mass equation (6.6) will first be transformed into an equivalent stiffness equation using Newmark's single-step finite difference scheme. From the equivalent stiffness equation, the displacements of the sprung mass unit can be solved and then substituted into the bridge equations. By doing so, the vehicle DOFs can be eliminated and condensed to those of the bridge element in contact.

Based on *Newmark's $\beta$ method* with constant average acceleration, i.e., with $\beta = 0.25$ and $\gamma = 0.5$ for its unconditional stability (see Appendix B),

$$\begin{aligned}\{\dot{z}\}_{t+\Delta t} &= \{\dot{z}\}_t + \{(1-\gamma)\{\ddot{z}\}_t + \gamma\{\ddot{z}\}_{t+\Delta t}\}\Delta t\,,\\ \{z\}_{t+\Delta t} &= \{z\}_t + \{\dot{z}\}_t\Delta t + \{(0.5-\beta)\{\ddot{z}\}_t\\ &\qquad + \beta\{\ddot{z}\}_{t+\Delta t}\}\Delta t^2\,,\end{aligned} \tag{6.10}$$

from which the accelerations and velocities of the vehicle at time $t + \Delta t$ can be solved,

$$\begin{aligned} \{\ddot{z}\}_{t+\Delta t} &= a_0\{\Delta z\} - a_2\{\dot{z}\}_t - a_3\{\ddot{z}\}_t\,, \\ \{\dot{z}\}_{t+\Delta t} &= \{\dot{z}\}_t + a_6\{\ddot{z}\}_t + a_7\{\ddot{z}\}_{t+\Delta t}\,, \end{aligned} \tag{6.11}$$

where the coefficients and those to appear later are defined as follows:

$$\begin{aligned} &a_0 = \frac{1}{\beta \Delta t^2}\,, \quad a_1 = \frac{\gamma}{\beta \Delta t}\,, \quad a_2 = \frac{1}{\beta \Delta t}\,, \\ &a_3 = \frac{1}{2\beta} - 1\,, \quad a_4 = \frac{\gamma}{\beta} - 1\,, \quad a_5 = \frac{\Delta t}{2}\left(\frac{\gamma}{\beta} - 2\right), \\ &a_6 = \Delta t(1-\gamma)\,, \quad a_7 = \gamma \Delta t\,. \end{aligned} \tag{6.12}$$

With the relations given in Eq. (6.11) for the vehicle accelerations and velocities, the sprung mass equation (6.6) can be manipulated to yield an *equivalent stiffness equation*:

$$\begin{aligned} &\begin{bmatrix} k_v + k_B + a_0 m_w + a_1 c_v & -k_v - a_1 c_v \\ -k_v - a_1 c_v & k_v + a_0 M_v + a_1 c_v \end{bmatrix} \begin{Bmatrix} \Delta z_1 \\ \Delta z_2 \end{Bmatrix} \\ &\quad = \begin{Bmatrix} p + k_B r_c + k_B \langle N_c \rangle \{u_b\} \\ 0 \end{Bmatrix}_{t+\Delta t} - \left( \begin{Bmatrix} q_{s1} \\ q_{s2} \end{Bmatrix}_t + \begin{Bmatrix} q_{e1} \\ q_{e2} \end{Bmatrix}_t \right), \end{aligned} \tag{6.13}$$

where

$$\begin{aligned} q_{e1,t} &= -m_w(a_2\dot{z}_1 + a_3\ddot{z}_1) - c_v[a_4(\dot{z}_1 - \dot{z}_2) + a_5(\ddot{z}_1 - \ddot{z}_2)]\,, \\ q_{e2,t} &= -M_v(a_2\dot{z}_2 + a_3\ddot{z}_2) - c_v[a_4(\dot{z}_2 - \dot{z}_1) + a_5(\ddot{z}_2 - \ddot{z}_1)]\,. \end{aligned} \tag{6.14}$$

Here, all the terms on the right-hand side of the equal sign of Eq. (6.13) are either given or already made known at the end of the last incremental step, except the bridge displacements $\{u_b\}_{t+\Delta t}$.

Accordingly, the displacement increments $\{\Delta z\}$ of the sprung mass unit can be solved from Eq. (6.13) and related to the bridge displacements $\{u_b\}_{t+\Delta t}$ as

$$\begin{Bmatrix} \Delta z_1 \\ \Delta z_2 \end{Bmatrix} = -\frac{1}{D}\begin{Bmatrix} k_v + a_0 M_v + a_1 c_v \\ k_v + a_1 c_v \end{Bmatrix}(p + k_B r_{c,t+\Delta t} + k_B \langle N_c \rangle \{u_b\}_{t+\Delta t}) - \frac{1}{D}\begin{Bmatrix} (q_{s1,t} + q_{e1,t}) a_0 M_v + (\tilde{q}_{s,t} + \tilde{q}_{e,t})(k_v + a_1 c_v) \\ (q_{s1,t} + q_{e1,t})(a_0 m_w + k_B) + (\tilde{q}_{s,t} + \tilde{q}_{e,t})(k_v + a_1 c_v) \end{Bmatrix}, \tag{6.15}$$

where $D$ represents the determinant of the vehicle equations in Eq. (6.13),

$$D = \begin{vmatrix} k_v + k_B + a_0 m_w + a_1 c_v & -k_v - a_1 c_v \\ -k_v - a_1 c_v & k_v + a_0 M_v + a_1 c_v \end{vmatrix} \tag{6.16}$$

and

$$\begin{aligned} \tilde{q}_{e,t} &= (q_{e1} + q_{e2})_t\,, \\ \tilde{q}_{s,t} &= (q_{s1} + q_{s2})_t = k_B z_{1,t}\,. \end{aligned} \tag{6.17}$$

Equation (6.15) represents the *master–slave relation* for condensing the sprung mass DOFs to the bridge DOFs. Compared with the approximate master–slave relation used by Yang and Lin (1995), the present relation in Eq. (6.15) has the advantage of being simple, accurate, and explicit. Since Eq. (6.15) has been derived through introduction of the finite difference formulas of the Newmark type, it possesses the same order of accuracy as that implied in solution of the equations for the entire vehicle–bridge interaction system, which will also be solved by Newmark's $\beta$ method. Because of this, the accuracy and convergence characteristics of the VBI element to be derived below can be assured with no difficulty.

## 6.5. Vehicle–Bridge Interaction Element

By the fact that $z_{1,t+\Delta t} = z_{1,t} + \Delta z_1$ and using the first line of Eq. (6.15) for $\Delta z_1$, one can derive from Eq. (6.9) the *condensed equations of motion* for the beam element at time $t + \Delta t$ with the interaction effect of the sprung mass unit taken into account, that is,

$$[m_b]\{\ddot{u}_b\}_{t+\Delta t} + [c_b]\{\dot{u}_b\}_{t+\Delta t} + [\hat{k}_b]\{\Delta u_b\}$$
$$= (\{p_b\}_{t+\Delta t} + \{p_w\}_{t+\Delta t}) - (\{f_s\}_t + [\hat{k}_b]\{u_b\}_t)\,, \quad (6.18)$$

where $[\hat{k}_b]$ is the stiffness matrix for the condensed system,

$$[\hat{k}_b] = [k_b] + k_B \frac{a_0}{D}[(M_v + m_w)(k_v + a_1 c_v) + a_0 M_v m_w]\{N_c\}\langle N_c\rangle\,. \quad (6.19)$$

$\{p_w\}_{t+\Delta t}$ denotes the load actions induced by the wheels, including the contributions from rail irregularities and ballast stiffness,

$$\{p_w\}_{t+\Delta t}$$
$$= -k_B \left[r_{c,t+\Delta t} - (p + k_B r_{c,t+\Delta t})\frac{1}{D}(k_v + a_0 M_v + a_1 c_v)\right]\{N_c\}$$
$$(6.20)$$

and $\{f_s\}_t$ denotes the resistant forces associated with the sprung mass unit,

$$\{f_s\}_t = k_B \left[\frac{1}{D}(q_{s1,t} + q_{e1,t})a_0 M_v \right.$$
$$\left. + (\tilde{q}_{s,t} + \tilde{q}_{e,t})(k_v + a_1 c_v) - z_{1,t}\right]\{N_c\}\,. \quad (6.21)$$

In Eq. (6.18), the term $[\hat{k}_b]\{u_b\}_t$ should be interpreted as the resistant forces exerted by the bridge element at the last time step.

The condensed stiffness matrix $[\hat{k}_b]$ as given in Eq. (6.19) contains essentially two major parts. The $[k_b]$ matrix represents the noninteraction part, which originates from the bridge element itself. The remainder in Eq. (6.19) represents the interaction part induced by the moving sprung mass, which is a function of the loading position $x_c$, and thus is time-dependent, as implied by the shape function $\{N_c\}$. In a time-history analysis, only the second (interaction) part must be updated at each incremental step. A convenient way is to assemble the $[k_b]$ matrices as if it was the case for the bridge with no moving vehicles, keeping this part of stiffness matrix unchanged throughout all the incremental steps, and to update only the second (interaction) part for each time step according to the acting positions of the sprung mass units.

The equations as given in Eq. (6.18) represent exactly the equations of motion for the *vehicle–bridge interaction* (VBI) element. Using such an element, only one sprung mass unit is allowed to be acting at one beam element. The maximum length that is allowed for each element should be no greater than the minimum spacing between any two wheel assemblies. Such a restriction can be removed, if the foregoing procedure has been generalized to include the case of two or more sprung mass units acting simultaneously on a single beam element.

Consider the special case of *moving loads*, i.e., with the effect of inertia of the moving vehicles neglected. By letting $k_v = 0$, $m_w = 0$, and $c_v = 0$, one can obtain from Eq. (6.16) the determinant as $D = a_0 M_v k_B$, from Eq. (6.7) the resistant force $q_{s1,t} = k_B z_{1,t}$, and from Eq. (6.14a) the sprung mass force $q_{e1,t} = 0$. Furthermore, from Eqs. (6.19)–(6.21), the following can be obtained: $[\hat{k}_b] = [k_b], \{p_s\}_{t+\Delta t} = p\{N_c\}$ and $\{f_s\}_t = \{0\}$. Consequently, the condensed equation as given in Eq. (6.18) reduces to

$$\begin{aligned}[m_b]\{\ddot{u}_b\}_{t+\Delta t} + [c_b]\{\dot{u}_b\}_{t+\Delta t} + [k_b]\{\Delta u_b\}\\ = \{p_b\}_{t+\Delta t} + p\{N_c\} - [k_b]\{u_b\}_t\,,\end{aligned} \tag{6.22}$$

in which the term $p\{N_c\}$ denotes the action of the moving load, as can be expected according to the concept of *consistent nodal loads*. On the other hand, by setting the damping coefficient $c_v$ and the mass $m_w$ of the wheel assembly equal to zero, and assigning a very large number to the stiffnesses of the vehicle and ballast, i.e., $k_v$ and $k_B$, one can arrive at the *moving mass* model as another special case.

Since the VBI element derived above possesses exactly the same number of DOFs as the parent element, while preserving the properties of *symmetry* and *bandedness* in element matrices, conventional element assembly procedure, which takes into account the conditions of equilibrium and compatibility of the structure at each nodal point, through transformation of all the element equations and associated variables to the same global coordinates, can be applied to constructing the equations of motion for the entire vehicle–bridge system. Namely, by looping over all the VBI elements, as represented by Eq. (6.18), and the elements not in contact with the moving vehicles, the *equations of motion for the structure* can be established as follows:

$$[M_b]\{\ddot{U}_b\}_{t+\Delta t} + [C_b]\{\dot{U}_b\}_{t+\Delta t} + [K_b]\{\Delta U_b\}_{t+\Delta t} \\ = \{P_b\}_{t+\Delta t} - \{F_b\}_t\,, \tag{6.23}$$

where $\{U_b\}_{t+\Delta t}$ denotes the total displacements of the bridge at time $t+\Delta t$, and $\{\Delta U_b\}$ the displacement increments of the bridge from time $t$ to $t+\Delta t$,

$$\{U_b\}_{t+\Delta t} = \{U_b\}_t + \{\Delta U_b\}\,. \tag{6.24}$$

In the system equations (6.23), each of the terms has been assembled by looping over all the corresponding terms of the beam elements, including the condensed ones, of the bridge. Thus, $[M_b]$ should be interpreted as the mass matrix and $[K_b]$ the stiffness matrix of the bridge, considering the contributions of both the VBI elements and the parent elements are not directly acted upon by the wheel loads. The external loads $\{P_b\}_{t+\Delta t}$ and resistant forces $\{F_b\}_t$ on the right-hand side of Eq. (6.23) are constructed as follows:

$$\{P_b\}_{t+\Delta t} = \sum_{elm=1}^{n.o.e} (\{p_b\}_{t+\Delta t} + \{p_w\}_{t+\Delta t})\,, \tag{6.25a}$$

$$\{F_b\}_t = \sum_{elm=1}^{n.o.e} (\{f_s\}_t + [\hat{k}_b]\{u_b\}_t)\,. \tag{6.25b}$$

One exception is the damping matrix $[C_b]$, which is constructed using the procedure described in Appendix F for bridges of which the damping property is assumed to be of the Rayleigh type.

## 6.6. Incremental Dynamic Analysis with Iterations

By Newmark's $\beta$ method, the system equations (6.23) can be rendered into a set of equivalent stiffness equations, from which the displacement increments $\{\Delta U_b\}$ of the bridge for a typical time step can be solved. By repeating such a procedure, the total bridge and vehicle responses can be computed as well. The accuracy of the incremental approach can be enhanced by inclusion of an iterative procedure to be presented below.

### 6.6.1. *Equivalent Stiffness Equations for VBI System*

First of all, the accelerations and velocities of the bridge at time $t+\Delta t$ can be related to the displacement increments $\{\Delta U_b\}$ and those of the bridge at time $t$ using finite difference formulas of the Newmark type similar to those given in Eq. (6.11) for the vehicles as

$$\begin{aligned} \{\ddot{U}_b\}_{t+\Delta t} &= a_0\{\Delta U_b\} - a_2\{\dot{U}_b\}_t - a_3\{\ddot{U}_b\}_t \,, \\ \{\dot{U}_b\}_{t+\Delta t} &= \{\dot{U}_b\}_t + a_6\{\ddot{U}_b\}_t + a_7\{\ddot{U}_b\}_{t+\Delta t} \,. \end{aligned} \tag{6.26}$$

By substitution of the expressions in Eq. (6.26), the system equations (6.23) can be manipulated to yield the following *equivalent stiffness equations*:

$$[\bar{K}_b]_{t+\Delta t}\{\Delta U_b\} = \{P_b\}_{t+\Delta t} - \{\bar{F}_b\}_t \,, \tag{6.27}$$

where the effective stiffness matrix $[\bar{K}_b]_{t+\Delta t}$ is

$$[\bar{K}_b]_{t+\Delta t} = a_0[M_b] + a_1[C_b] + [K_b] \tag{6.28}$$

and the effective resistant force vector $\{\bar{F}_b\}_t$ is

$$\begin{aligned} \{\bar{F}_b\}_t = \{F_b\}_t &+ [M_b](a_2\{\dot{U}_b\}_t + a_3\{\ddot{U}_b\}_t) \\ &+ [C_b](a_4\{\dot{U}_b\}_t + a_5\{\ddot{U}_b\}_t) \end{aligned} \tag{6.29}$$

Both the effective stiffness matrix $[\bar{K}_b]_{t+\Delta t}$ and the load vector $\{P_b\}_{t+\Delta t}$ can be regarded as constant within each time step, if sufficiently small time steps are used in the incremental analysis.

As was mentioned previously, the order of accuracy of the equivalent stiffness equations (6.27), including the condensation procedure involved in the derivation, is the same as that implied by the Newmark-type formulas given in Eq. (6.10). For most of the vehicle–bridge interaction problems encountered in practice, the accuracy of the present approach is satisfactory and no iterations are required, on the condition that reasonably small time increments $\Delta t$ have been selected. For this case, the displacement increments $\{\Delta U_b\}$ of the bridge can be solved directly from Eq. (6.27), which can then be substituted into Eq. (6.24) to yield the total displacements $\{U_b\}_{t+\Delta t}$ and into Eq. (6.26) to yield the accelerations $\{\ddot{U}_b\}_{t+\Delta t}$ and velocities $\{\dot{U}_b\}_{t+\Delta t}$ of the bridge at time $t + \Delta t$. Accordingly, the displacements, velocities and accelerations of each element of the bridge, i.e., $\{u_b\}_{t+\Delta t}$, $\{\dot{u}_b\}_{t+\Delta t}$ and $\{\ddot{u}_b\}_{t+\Delta t}$, can be computed. By backward substitution, the displacement increments $\{\Delta z\}$ of the sprung mass unit can be computed from Eq. (6.15), with which the total response of the sprung mass unit at time $t+\Delta t$ can be computed from Eqs. (6.5) and (6.11). This completes the cycle of analysis for the current time step from $t$ to $t + \Delta t$. As a side note, the vehicle acceleration $\{\ddot{z}\}$ serves as a measure of the riding comfort for passengers carried by the train.

As was stated previously, the condensation of the vehicle DOFs that leads to the VBI element is performed on the element level. It follows that the conventional procedure for assembling the structural matrices from the element ones can be directly applied. For this reason, the amount of effort required in programming and computation is minimal, compared with approaches that perform condensation on the structure level or with no condensation at all. Such an advantage becomes clearer in the study of bridges subjected to a sequence of moving vehicular loads. In such cases, what one needs is a proper book-keeping scheme to identify the acting position of each wheel load of the railroad cars composing the train at each time step. It should be noted that the VBI element derived herein is applicable not only for modeling the railroad cars with regular intervals and constant sizes, but also for vehicles of varying sizes that constitute a random traffic flow through the highway bridges.

### 6.6.2. *Procedure of Iterations*

During the passage of vehicles over a bridge, on the one hand, the vehicles excite the bridge through the *interaction* or *contact forces*, which change their action positions from time to time; on the other hand, the bridge affects the behavior of the vehicles by its deflection, as well as by the contact forces. Such a phenomenon is a typical nonlinear interaction problem, which can only be solved by procedures of incremental nature. For bridges whose displacements can be regarded as small under the moving vehicular loads, there is no need to perform iterations at each incremental step. However, for bridges whose displacements may not be regarded as small under the moving vehicular loads, iterations for removing the *unbalanced forces* at the structural nodes for each incremental step should be conducted. The following is a summary of the procedure for iterations.

For the purpose of performing iterations, the equivalent stiffness equations of the VBI system in Eq. (6.27), which has been presented in incremental form, should be modified to include the feature of iteration, that is,

$$[\bar{K}_b]_{t+\Delta t}\{\Delta U_b\}^i = \{P_b\}_{t+\Delta t} - \{\bar{F}_b\}^{i-1}_{t+\Delta t}, \tag{6.30}$$

in which the right superscript $i$ on each symbol indicates the current number of iterations. The right hand side of Eq. (6.30) should now be interpreted as the external load increments for the first iteration ($i = 1$) and as the system unbalanced forces for the following iterations ($i \geq 2$) (Yang and Kuo, 1994). The philosophy for modifying an equation originally presented in incremental form into one incorporating the iterative measure is that all the terms originally associated with time $t$ be interpreted as those for the last, i.e., the $(i-1)$th, iterative step of the current incremental step, and all the terms originally associated with time $t + \Delta t$ as those for the current, i.e., the $i$th, iterative step. For instance, the resistant force vector in

Eq. (6.29) should be modified for the iterative steps with $i \geq 2$ as follows:

$$\{\bar{F}_b\}_{t+\Delta t}^{i-1} = \{F_b\}_{t+\Delta t}^{i-1} + [M_b](a_2\{\dot{U}_b\}_{t+\Delta t}^{i-1} + a_3\{\ddot{U}_b\}_{t+\Delta t}^{i-1}) \\ + [C_b](a_4\{\dot{U}_b\}_{t+\Delta t}^{i-1} + a_5\{\ddot{U}_b\}_{t+\Delta t}^{i-1}) . \quad (6.31)$$

Equation (6.30) represents a typical nonlinear equation encountered in the study of a great number of nonlinear problems. To solve problems of this sort, the *modified Newton–Raphson method* that performs iterations at constant loads can be employed (Yang and Kuo, 1994).

The initial conditions (i.e., for $i = 1$) to the system equations (6.30) are

$$\{\bar{F}_b\}_{t+\Delta t}^{0} = \{\bar{F}_b\}_{t}^{l} , \\ \{U_b\}_{t+\Delta t}^{0} = \{U_b\}_{t}^{l} , \quad (6.32)$$

where $\{\bar{F}_b\}_t^l$ has been given in Eq. (6.29) and the right superscript $l$ denotes the last iteration of the previous incremental or time step. Accordingly, the system equations (6.30) reduce to the following for the first iteration (i.e., with $i = 1$) of the current incremental step:

$$[\bar{K}]_{t+\Delta t}\{\Delta U_b\}^1 = \{P_b\}_{t+\Delta t} - \{\bar{F}_b\}_t^l . \quad (6.33)$$

Here, it should be noted that the effective stiffness matrix $[\bar{K}]_{t+\Delta t}$ remains constant within each incremental step, which need not be updated for the iterations involved using the modified Newton–Raphson method.

For each iterative step, the displacement increments $\{\Delta U_b\}^i$ of the bridge can be solved from Eq. (6.30). Accordingly, the total displacements of the bridge for the current iterative step can be accumulated as

$$\{U_b\}_{t+\Delta t}^{i} = \{U_b\}_{t+\Delta t}^{i-1} + \{\Delta U_b\}^i . \quad (6.34)$$

The acceleration and velocity of the bridge can be computed from Eq. (6.26) with due account taken of the feature of iteration, that is,

$$\begin{aligned} \{\ddot{U}_b\}^i_{t+\Delta t} &= a_0\{\Delta U_b\}^i - a_2\{\dot{U}_b\}^{i-1}_{t+\Delta t} - a_3\{\ddot{U}_b\}^{i-1}_{t+\Delta t}, \\ \{\dot{U}_b\}^i_{t+\Delta t} &= \{\dot{U}_b\}^{i-1}_{t+\Delta t} + a_6\{\ddot{U}_b\}^{i-1}_{t+\Delta t} + a_7\{\ddot{U}_b\}^i_{t+\Delta t}. \end{aligned} \tag{6.35}$$

Once the displacements $\{U_b\}^i_{t+\Delta t}$ of the bridge in global coordinates are made available, the displacements $\{u_b\}^i_{t+\Delta t}$ for each element of the bridge can be computed accordingly. It follows that the vehicle displacement increments $\{\Delta z\}^i$ can be computed from Eq. (6.15), by treating the terms with subscript $t+\Delta t$ as those associated with the $i$th iterative step, and the terms with subscript $t$ as those associated with the $(i-1)$th iterative step. The total responses of the sprung mass unit for the $i$th iterative step can be determined from Eqs. (6.5) and (6.11) as follows:

$$\{z\}^i_{t+\Delta t} = \{z\}^{i-1}_{t+\Delta t} + \{\Delta z\}^i, \tag{6.36a}$$

$$\{\ddot{z}\}^i_{t+\Delta t} = a_0\{\Delta z\}^i - a_2\{\dot{z}\}^{i-1}_{t+\Delta t} - a_3\{\ddot{z}\}^{i-1}_{t+\Delta t}, \tag{6.36b}$$

$$\{\dot{z}\}^i_{t+\Delta t} = \{\dot{z}\}^{i-1}_{t+\Delta t} + a_6\{\ddot{z}\}^{i-1}_{t+\Delta t} + a_7\{\ddot{z}\}^i_{t+\Delta t}, \tag{6.36c}$$

with the following initial condition: $\{z\}^0_{t+\Delta t} = \{z\}^l_t$.

The following is a summary of the procedure for performing the incremental-iterative analysis based on the modified Newton–Raphson algorithm:

(1) Read in all the structural and vehicle data.
(2) Start with time $t = 0$ and adopt the following initial conditions: $\{F_b\}^l_0 = \{0\}$, $\{U_b\}^l_0 = \{\dot{U}_b\}^l_0 = \{\ddot{U}_b\}^l_0 = \{0\}$, and $\{z_b\}^l_0 = \{\dot{z}_b\}^l_0 = \{\ddot{z}_b\}^l_0 = \{0\}$. Calculate the mass matrices $[m_b]$ for all elements of the bridge and assemble the mass matrix $[M_b]$ for the bridge that is free of any vehicles. Select a proper time increment $\Delta t$ for the Newmark integration scheme.
(3) For each incremental or time step, check if the total duration $t$ for which the solution has been made available is greater than the duration desired. If yes, stop the incremental procedure.

Otherwise, let $t = t + \Delta t$ and use $i = 1$ for the first iteration. Determine the acting position $x_c$ of each wheel load and the rail irregularity $r_c$ for each contact point.

(4) For the elements carrying no wheel loads, calculate the element matrix $[k_b]$; and for the elements in direct contact with the wheel loads, calculate the modified element matrix $[\hat{k}_b]$ using Eq. (6.19), in which the shape function $\{N_c\}$ is known to be function of the contact position $x_c$.

(5) Assemble the bridge matrix $[K_b]$ and load vector $\{P_b\}_{t+\Delta t}$, using Eq. (6.25a) for the latter. The damping matrix $[C_b]$ of the bridge is assumed to be of the Rayleigh type, which can be computed as a linear combination of the mass matrix $[M_b]$ and stiffness matrix $[K_b]$ of the bridge free of the vehicles (see Appendix F). Compute the equivalent stiffness matrix $[\bar{K}_b]_{t+\Delta t}$ using Eq. (6.28). For the present problem, the system matrices $[\bar{K}_b]_{t+\Delta t}$ and $\{P_b\}_{t+\Delta t}$ are treated as constant for each incremental or time step.

(6) Determine the resistant force vector $\{\bar{F}_b\}^{i-1}_{t+\Delta t}$ using Eq. (6.31). For $i \geq 2$, check if the unbalanced forces $(\{P_b\}_{t+\Delta t} - \{\bar{F}_b\}^{i-1}_{t+\Delta t})$ are less than a given tolerance. If yes and if the contact condition $f_c \geq 0$ as given in Eq. (6.2) is satisfied, go to step 3 for the next increment or time step. Otherwise, perform the following.

(7) Solve the displacement increments $\{\Delta U_b\}^i$ from the system equations (6.30). Determine the vehicle displacement increments $\{\Delta z\}^i$ from Eq. (6.15), where the terms with subscript $t + \Delta t$ should be interpreted as those for the $i$th iterative step, and the terms with subscript $t$ as those for the $(i-1)$th step.

(8) Update the total displacements $\{U_b\}^i_{t+\Delta t}$ of the bridge using Eq. (6.34) and $\{z\}^i_{t+\Delta t}$ of the vehicle using Eq. (6.36a). Compute the displacement derivatives $\{\dot{U}_b\}^i_{t+\Delta t}$ and $\{\ddot{U}_b\}^i_{t+\Delta t}$ for the bridge using Eq. (6.35) and $\{\dot{z}\}^i_{t+\Delta t}$ and $\{\ddot{z}\}^i_{t+\Delta t}$ for each sprung mass unit using Eqs. (6.36b) and (6.36c).

(9) Let $i = i + 1$ and go to step 6 for the next iterative step.

As was stated in Sec. 6.5, a more efficient way for updating the stiffness matrix $[K_b]$ for the VBI system at each time step is as follows. First, one assembles the stiffness $[K_b]$ for the bridge that is free of any vehicles, using the element matrices $[k_b]$ for all the bridge elements. Then, at each incremental step, one checks the acting position $x_c$ of each sprung mass unit and adds only the interaction (i.e., the second) part of the $[\hat{k}_b]$ matrix of the VBI element in Eq. (6.19) to the entries in $[K_b]$ corresponding to the bridge element on which the sprung mass unit is acting. By watching the movement of each sprung mass unit, one can either update the interaction part of the VBI element to account for the change in the acting position of the sprung mass unit, or remove the interaction part from the $[K_b]$ matrix, if the sprung mass unit is no longer acting on the element.

With the above procedure of iteration, the coupling effect between the bridge and moving vehicles is considered through the condensed VBI elements. Because the same order of accuracy is maintained both in establishing the *master–slave relations* leading to the VBI element and in discretizing the structural equations of motion, the number of iterations required for each time step is generally small, compared with approaches that do not rely on the condensation technique. In addition to the bridge response, the present approach allows us to compute the vehicle response as a by-product, which serves as an indicator for the *riding comfort* of passengers carried by the train.

## 6.7. Numerical Verification

Three examples are prepared to verify the VBI element and the procedure of solution presented in this chapter. First, the dynamic responses solved by the present method for a simple beam subjected to a moving sprung mass will be compared with those obtained by considering only the contribution of the first mode of vibration of the beam. Second, by modeling a train as a sequence of moving lumped loads or lumped masses, the impact responses of bridges excited by the train will be investigated and compared with the existing solutions. Finally, the dynamic responses of a cantilever subjected to moving vehicles that are simulated by different models will be studied. In each case, the beam is represented by ten elements.

### 6.7.1. *Simple Beam Subjected to Moving Sprung Mass*

As shown in Fig. 6.3, a simple beam of span length $L = 25$ m is subjected to a moving sprung mass. The following data are adopted: Young's modules $E = 2.87$ GPa, Poisson's ratio $v = 0.2$, moment of inertia $I = 2.90\ \mathrm{m}^4$, mass per unit length $m = 2303$ kg/m, suspended mass $M_v = 5750$ kg, suspension stiffness $k_v = 1595$ kN/m, and speed $v = 100$ km/h. For this problem, the mass ratio of the vehicle to the bridge is $M_v/mL = 0.1$. The frequency of vibration computed of the bridge is $\omega_1 = 30.02$ rad/s, and the one for the sprung mass is $\omega_v = 16.66$ rad/s. By representing the vertical displacement of the beam as $u_b = q_b(t)\sin(\pi x/L)$, the displacement of the sprung mass as $q_v(t)$, and neglecting the effect of damping of the beam, the equations of motion for the beam and the sprung mass moving at speed $v$ can be given as (Biggs, 1964):

$$\begin{Bmatrix} \ddot{q}_b \\ \ddot{q}_v \end{Bmatrix} + \begin{bmatrix} 2\omega_v^2 \dfrac{M_v}{mL} \sin^2 \dfrac{\pi v t}{L} + \omega_1^2 & -2\omega_v^2 \dfrac{M_v}{mL} \sin \dfrac{\pi v t}{L} \\ -\omega_v^2 \dfrac{M_v}{mL} \sin \dfrac{\pi v t}{L} & \omega_v^2 \end{bmatrix} \begin{Bmatrix} q_b \\ q_v \end{Bmatrix}$$

$$= \begin{Bmatrix} -2\dfrac{M_v g}{mL} \sin \dfrac{\pi v t}{L} \\ 0 \end{Bmatrix}. \tag{6.37}$$

The dynamic responses of the midpoint vertical displacement of the beam subjected to a moving load or sprung mass have been plotted in Fig. 6.4. As can be seen, the response obtained by the present procedure using the VBI element for the special case of sprung mass agrees very well with the single mode solution to Eq. (6.37).

From the response of vertical acceleration for the midpoint of the beam shown in Fig. 6.5, one observes that inclusion of the higher modes can result in drastic oscillation of the acceleration response, which was neglected in the solution to Eq. (6.37). The deflection and vertical acceleration computed of the sprung mass have been plotted in Figs. 6.6 and 6.7, respectively. In these two figures, the difference between the present solution and that of Eq. (6.37) can be attributed mainly to the omission of *higher vibration modes* in the latter. A comparison of Figs. 6.4 and 6.6 indicates that the sprung mass is more sensitive than the beam to the omission of higher modes. As was noted previously, the vertical acceleration of the sprung mass serves as a measure of the *riding comfort* for passengers carried by the vehicle.

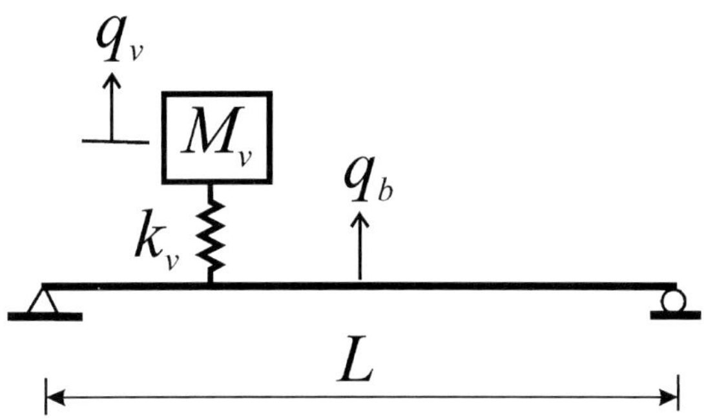


Fig. 6.3. Beam with moving sprung mass.

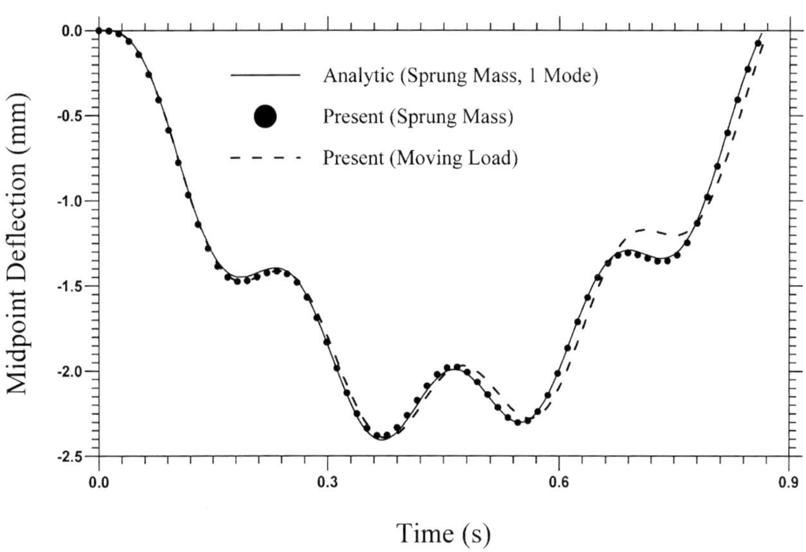


Fig. 6.4. Midpoint vertical deflection of beam.

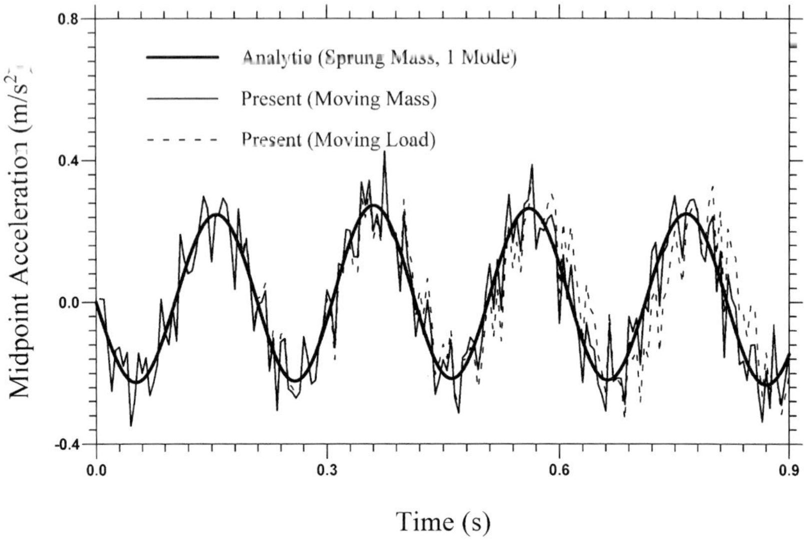


Fig. 6.5. Midpoint vertical acceleration of beam.

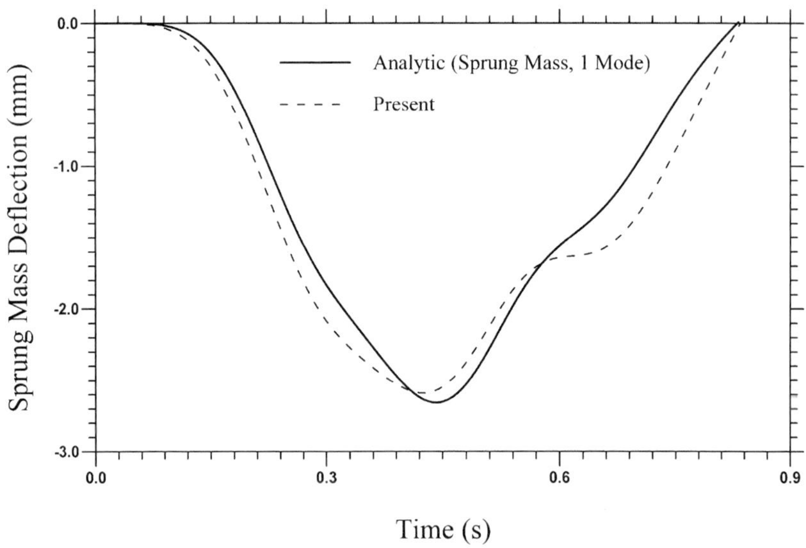


Fig. 6.6. Deflection of sprung mass.

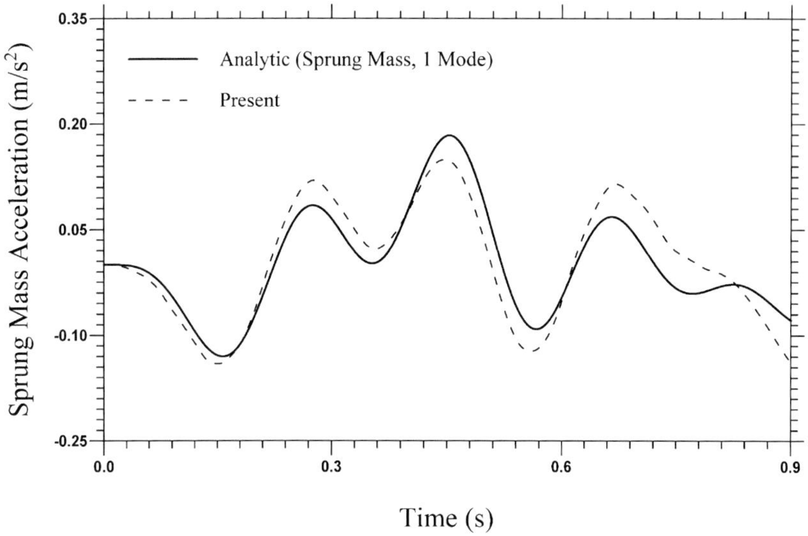


Fig. 6.7. Vertical acceleration of sprung mass.

### 6.7.2. *Simple Beam Subjected to Moving Train*

Consider a simply-supported beam with the following properties: $L = 20$ m, $I = 3.81$ m$^4$, $E = 29.43$ GPa, $m = 34\,088$ kg/m, and damping ratio = 2.5%. The train traveling over the bridge contains 10 wheel assemblies, which can be modeled as a sequence of moving lumped loads with regular nonuniform intervals arranged as follows: $\downarrow\leftarrow L_c \rightarrow\downarrow\leftarrow L_d \rightarrow\downarrow \ldots \downarrow\leftarrow L_d \rightarrow\downarrow\leftarrow L_c \rightarrow\downarrow$, where $L_c = 18$ m, $L_d = 6$ m, and "$\downarrow$" indicates a lumped load. For this example, the lumped load is assumed as $p = 215.6$ kN, the car body mass as $M_v = 22\,000$ kg and the wheel mass as $m_w = 0$ kg. The impact factor $I$ as defined in Eq. (1.1) is adopted.

Two cases are considered herein. In the first case, the moving load model is conceived by setting the suspension stiffness $k_v$, damping $c_v$, and ballast stiffness $k_B$ all equal to zero. In the second case, the moving mass model is approximated by setting the vehicle stiffness $k_v$ and ballast stiffness $k_B$ equal to a very large number, say, with $k_v = k_B = 9.0 \times 10^6$ kN/m. The impact factors $I$ calculated for the midpoint displacement of the beam subjected to the moving loads

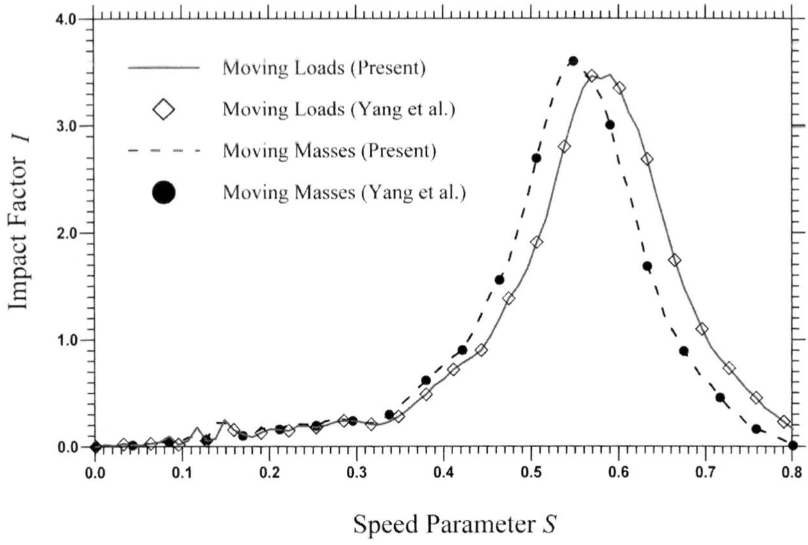


Fig. 6.8. Impact response for bridge sustaining a moving train.

using the two models have been plotted in Fig. 6.8, with respect to the (first) speed parameter $S$, which, according to Eq. (1.4), is defined as the ratio of the exciting frequency of the moving vehicle, $\pi v/L$, to the fundamental frequency $\omega$ of the bridge. Also shown in the figure are the results based on the analytical work of Yang *et al.* (1997b) or those presented in Sec. 2.6. Clearly, the present solutions correlate very well with those of the analytical study. The moving mass model tends to reduce the frequency of vibration of the vehicle–bridge system, in the sense that the critical speed for the peak response to occur shifts to a smaller value.

### 6.7.3. *Free-Fixed Beam with Various Models for Moving Vehicles*

Figure 6.9 shows a cantilever subjected to a moving lumped mass. The following data are assumed: length $L = 300$ in (7.62 m); velocity $v = 2000$ in/s (50.8 m/s); flexural rigidity $EI = 3.3 \times 10^9$ lbf-sq in ($9.474 \times 10^6$ N-m$^2$); mass of the beam per unit length

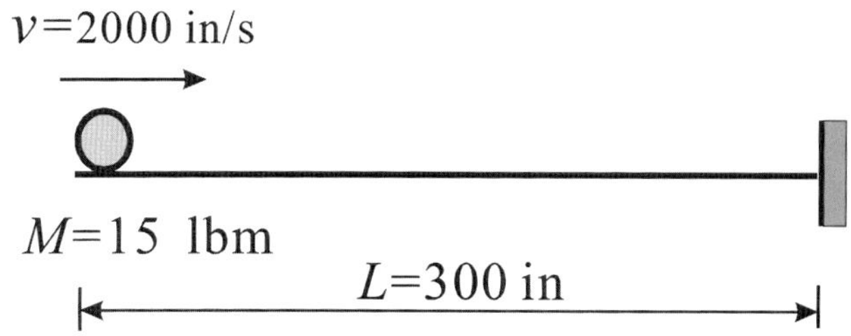


Fig. 6.9. Cantilever with mass moving at constant speed.

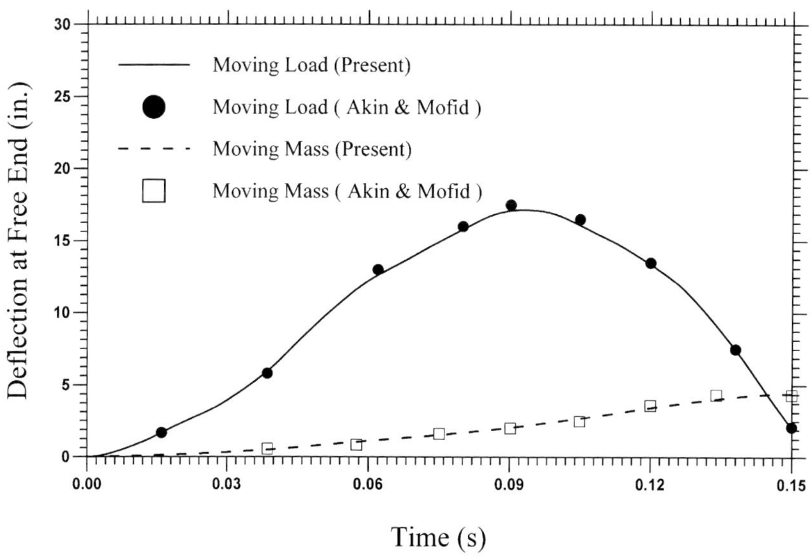


Fig. 6.10. Comparison of results for the cantilever.

$m = 0.006667$ lbm/in (0.1192 kg/m); lumped mass $M_v = 15$ lbm (6.81 kg); and moving load = 5793 lbf (66.74 N). The results obtained for the free-end deflection of the cantilever using both the moving mass and moving load models have been plotted in Fig. 6.10, along with those of Akin and Mofid (1989). As can be seen, good agreement has been achieved between the present solutions and those of Akin and Mofid (1989). It can be seen that throughout most of the acting period of the vehicle on the beam, i.e., for $t < 0.14$ s, the moving load model tends to produce larger response, compared with the moving mass model.

## 6.8. Parametric Studies

In the design of bridges for highways and railways, it is required that the stresses caused by the static live (vehicular) loads on the bridge be increased by a dynamic allowance factor or impact factor to ensure the capability of the bridge in resisting the impact loads. Some additional requirements are placed on the design of high-speed railway bridges concerning the running safety of trains and the *riding comfort* of passengers. For instance, a limit of 0.05 g ($= 0.49$ m/s$^2$) was previously used for the vertical acceleration of train cars traveling over bridges by the Bureau of Taiwan High Speed Railways. Due to the relatively stringent requirements, the design of high-speed railway bridges is governed generally by the conditions of serviceability, rather than by the strength or yielding of materials, as encountered in most highway bridges.

In this section, parametric studies will be carried out for both a simply-supported beam and a three-span continuous beam traveled by trains moving at high speeds, using the VBI element and procedure developed in the early part of this chapter, based primarily on the paper by Yau *et al.* (1999). By high speeds, we mean that the train is allowed to travel in a range of speeds from 250 to 400 km/h, to reflect the current advances in power and control technologies on operation of high-speed trains.

The numerical results to be discussed in the following subsections indicate that the moving load model is generally accurate if only the bridge response is desired. However, the use of the sprung mass model is necessary whenever the riding comfort of passengers carried by the train is of concern. Noteworthy is the fact that if the *characteristic length*, rather than the span length, is used in computing the first speed parameter $S$ for the *continuous beam* (see Eq. (1.4) for definition), then both the simple and continuous beams will attain their peaks at the same critical speed $S$, when subjected to the same sequence of axle loads. Factors such as the rail irregularity, ballast stiffness, suspension stiffness and suspension damping can affect drastically the riding comfort of the train cars traveling over simple beams. Their influence appears to be comparatively small for continuous beams.

### 6.8.1. *Models for Bridge, Train and Rail Irregularities*

A simply-supported beam (SB) and a three-span continuous beam (CB) are considered in this section, which are made of prestressed concrete with elastic constant $E = 29.43$ GPa and Poisson's ratio $v = 0.2$. The damping of the bridge is assumed to be of the Rayleigh type, with a damping ratio of 2.5% assumed for the first two modes. The properties of the beams have been given in Table 6.1, where $l =$ the span length, $L =$ characteristic length, $A =$ cross sectional area, $I$ = moment of inertia, $m =$ mass per unit length, and $\omega =$ frequency of vibration. Using the characteristic length $L$ for the continuous beam and the fundamental frequency $\omega$ of vibration for the simple and continuous beams given in Table 6.1, the vehicle velocity $v$ can be related to the first speed parameter $S$ using Eq. (1.4) as $v = 1007.3\ S$ km/h for the simple beam, and as $v = 1017.6\ S$ km/h for the continuous beam.

In this section, a train is modeled as a sequence of sprung masses of regular intervals, as shown in Fig. 6.1(b). Two models of train similar to those commercially available are considered herein, of which the dynamic properties are given in Table 6.2, in which $M_v$ denotes the mass lumped from one half of the car body, $m_w$ the wheel mass, $k_v$ and $c_v$ respectively denote the stiffness and damping of the suspension devices, as defined in Fig. 6.2. The S25 model consists of 10 axles with regular nonuniform intervals and the T18 model consists of eight equidistantly-spaced axles.

Track irregularities may be caused by factors such as small imperfections in materials, imperfections in manufacturing of rails and rail joints, terrain irregularities, and errors in surveying during design and construction. The irregularity function proposed by Nielsen

Table 6.1. Properties of beams.

| Beams | $l$ (m) | $L$ (m) | $A$ (m$^2$) | $I$ (m$^4$) | $m$ (kg/m) | $\omega$ (rad/s) | | |
|---|---|---|---|---|---|---|---|---|
| Simple | 30 | 30 | 7.94 | 8.72 | 36 056 | 29.3 | 117.0 | 263.3 |
| Continuous | 25-40-25 | 30 | 8.66 | 14.05 | 37 784 | 29.6 | 60.4 | 70.7 |

Table 6.2. Dynamic properties of trains.

| Model | Axle arrangement (m) | $M_v$ (kg) | $m_w$ (kg) | $k_v$ (kN/m) | $c_v$ (kN-s/m) |
|---|---|---|---|---|---|
| S25 | ↓18 ↓7 ↓18 ↓7 ↓18 ↓7 ↓18 ↓7 ↓18 ↓ | 24 000 | 5000 | 1500 | 85 |
| T18 | ↓18 ↓18 ↓18 ↓18 ↓18 ↓18 ↓18 ↓ | 30 000 | 5000 | 1700 | 90 |

and Abrahamsson (1992) for the vertical profile of the rails will be adopted in this section:

$$r(x) = -r_0 \left[ 1 - \exp\left( -\frac{x}{x_0} \right)^3 \right] \sin \frac{2\pi x}{\gamma_0}, \tag{6.38}$$

where $x$ is the along-track distance (in m), $x_0 = 1.0$ m, $r_0$ ($= 0.5$ mm) is the amplitude of irregularities, and $\gamma_0$ ($= 1.0$ m) the wavelength of the corrugation. As for the ballast stiffness $k_B$, the value of 20 MN/m per rail has been used by Nielsen and Abrahamsson (1992). In this section, the ballast stiffness $k_B$ is taken as 40 MN/m for two rails.

### 6.8.2. *Moving Load versus Sprung Mass Model*

To investigate the effect of different vehicle modelings on the bridge response, the train is modeled either as a series of moving loads or sprung masses, with no consideration made of rail irregularities. The impact factors $I$ solved for the midpoint displacement of the simple beam and three-span continuous beams have been plotted against the first speed parameter $S$ in Figs. 6.11 and 6.12, for the two vehicle models T18 and S25, respectively. Evidently, the moving load model can be reliably used to predict the bridge responses, as little difference exists between the solutions obtained for the two types of

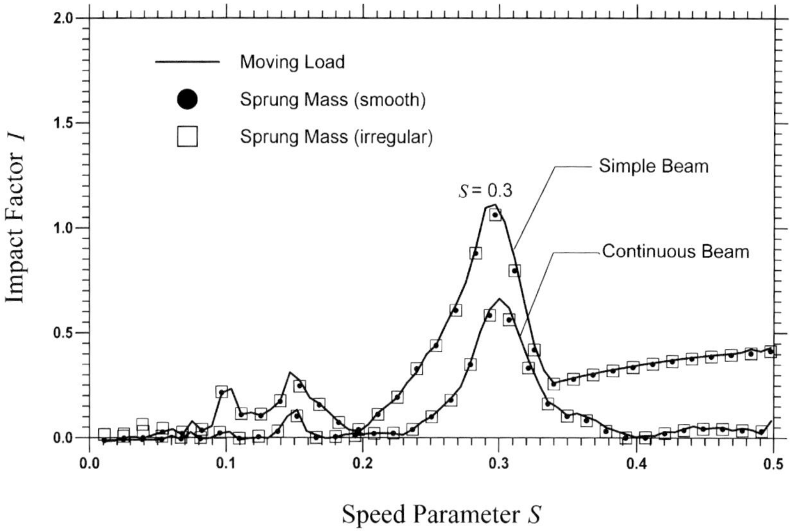


Fig. 6.11. Impact factor for midpoint deflection of beam (T18)-different bridge models.

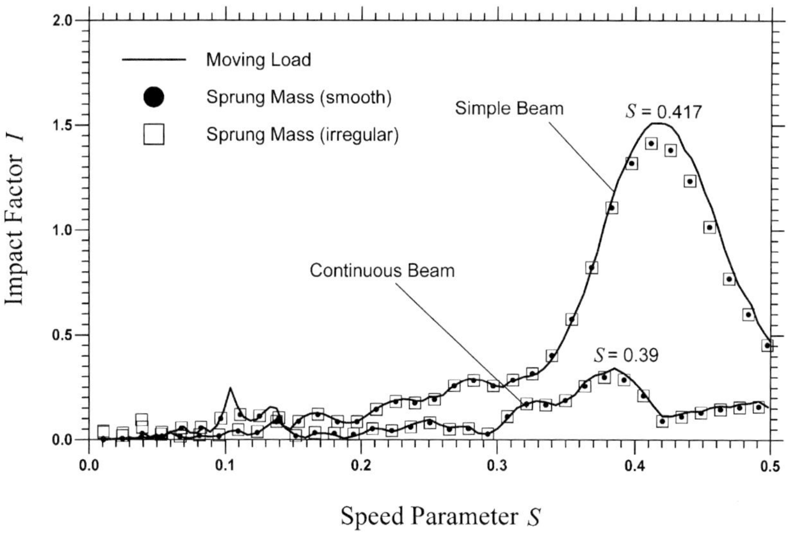


Fig. 6.12. Impact factor for midpoint deflection of beam (S25)-different bridge models.

vehicle modeling. Moreover, the inclusion of the inertial effect of the moving vehicles, as represented by the sprung mass model, has resulted in slight reduction of the peak response of the bridge, due to the fact that the sprung mass behaves in some sense like a *tuned mass* to the bridge.

Figures 6.11 and 6.12 indicate that the simple beam reaches its peak responses at the speeds $S = 0.300$ and 0.417 (or equivalently at $v = 302$ and 420 km/h for the case considered) under the moving action of the two trains T18 and S25, respectively. These are exactly the *critical speeds* (nondimensionalized) for the bridge to resonate under the passage of the two trains, which correspond very well to the values predicted from the formula given by Yang *et al.* (1997b) for the resonance of simple beams, i.e., the one given in Eq. (2.59) or $S = d/(2L)$, where $d$ is the train car length, which equals 25 m for the S25 model and 18 m for the T18 model.

From Fig. 6.11, one observes that for the T18 train, whose axle loads are equidistantly-spaced, the continuous beam attains its peak response at exactly the same critical speed $S$ as that for the simple beam, although the peak amplitude has been drastically reduced due to the restraint effect of the multi supports. No similar behavior is observed from Fig. 6.12 for the S25 train, partly due to the fact that the axles are not uniformly-spaced for this case. Moreover, the impact response appears to be much smaller for the continuous beam than for the simple beam. Such an observation is consistent with the observation made by Yang *et al.* (1995) for continuous beams under the passage of a single HS20-44 truck.

### 6.8.3. *Effect of Rail Irregularities*

In general, the rail irregularities have little influence on the impact response of the bridges (see Figs. 6.11 and 6.12). However, the same is not true with the moving vehicles or sprung masses. For the present purposes, let us adopt the rail irregularities given in Eq. (6.38). For both the simple beam (SB) and continuous beam (CB), the vertical acceleration of the sprung masses have been plotted in Figs. 6.13 and 6.14 for the two train models T18 and S25, respectively. From

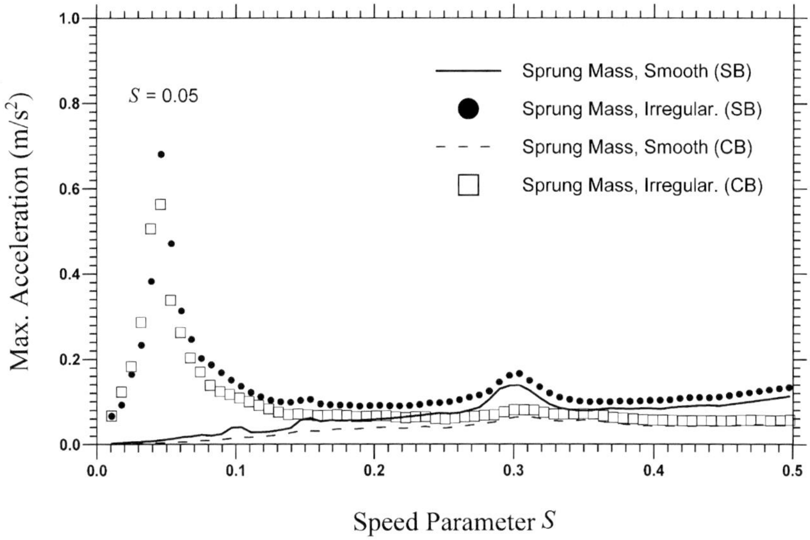


Fig. 6.13. Maximum acceleration of sprung mass (T18)-effect of irregularities.

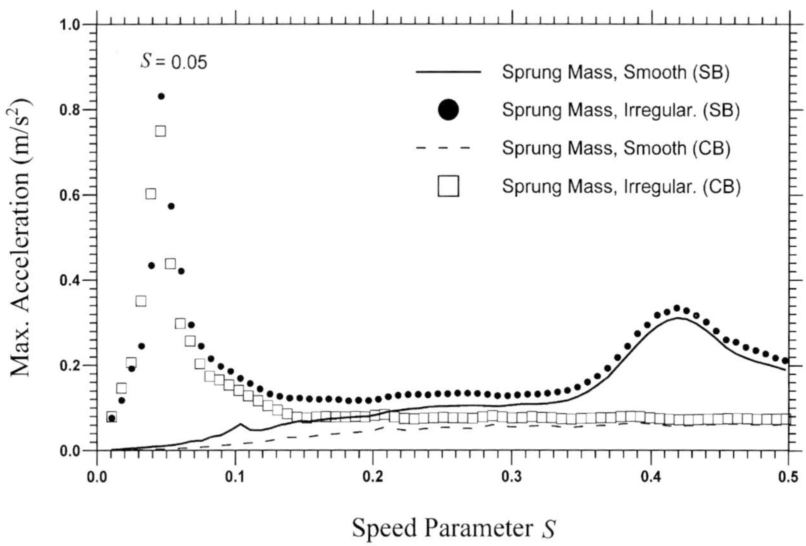


Fig. 6.14. Maximum acceleration of sprung mass (S25)-effect of irregularities.

both figures, a critical speed of $S = 0.05$ can be identified, which is equivalent to $v = 50.4$ km/h for the simple beam and to 50.9 km/h for the continuous beam. Let $\omega_B$ denote the frequency of vibration of the ballast, i.e.,

$$\omega_B = \sqrt{\frac{k_B}{m_w}} = 89.44 \text{ rad/s}. \tag{6.39}$$

The speed at which the peak vehicle response occurs can be explained as a phenomenon of resonance caused by the coincidence of the ballast frequency $\omega_B$ with the frequency of rail irregularities as represented by $2\pi v/\gamma_0$, where $\gamma_0$ is the wavelength of corrugation (= 1.0 m). For instance, by letting $2\pi v/\gamma_0 = \omega_B$, one obtains $v = \omega_B \gamma_0 / 2\pi$. Hence,

$$S = \frac{\pi v}{\omega L} = \frac{\omega_B \gamma_0}{2\omega L}. \tag{6.40}$$

By substitution of the values of $\omega$, $\omega_B$, $\gamma_0$, and $L$ for the simple and continuous beams, the speeds at which the peak response occurs can be computed, which turn out to be $S = 0.05$, coincident with that observed from Figs. 6.13 and 6.14. From Eq. (6.40), it is obvious that the larger the ballast stiffness, the greater is the speed for resonance to occur. It should be added that the peak responses plotted in Figs. 6.13 and 6.14 for the vehicles exceed significantly the tolerance limit of 0.49 m/s$^2$ (= 0.05 g) previously used by the Bureau of Taiwan High Speed Railways.

### 6.8.4. *Effect of Ballast Stiffness*

To investigate the influence of ballast stiffness on the bridge response, three different values of ballast stiffness, $0.25k_B$, $k_B$ and $2.5k_B$, are used. For both the simple and continuous beams, the corresponding impact factors solved for the midpoint displacement have been plotted in Figs. 6.15 and 6.16, for the two train models T18 and S25, respectively. As can be seen, softer ballast tends to reduce the impact response of the bridges, although the degree of reduction is marginal. One possible explanation for this is that the ballast layer serves to some extent as an energy dissipating mechanism for the bridge under the action of moving trains.

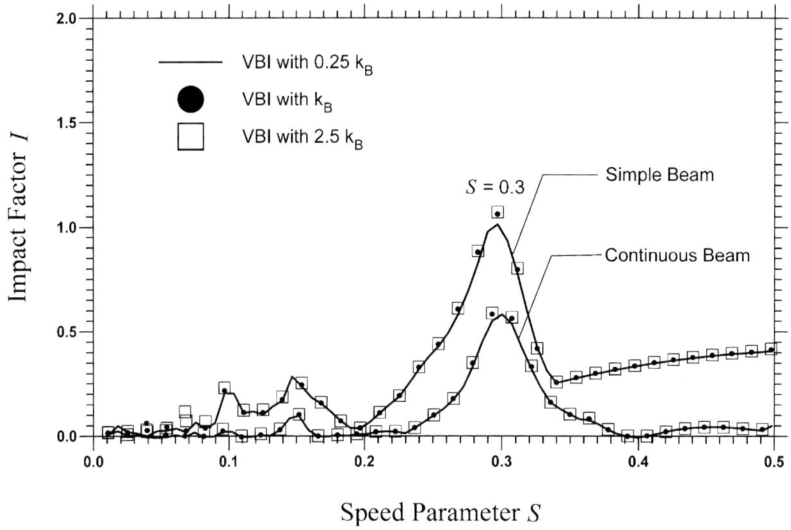


Fig. 6.15. Impact factor for midpoint deflection of beam (T18)-effect of ballast stiffness.

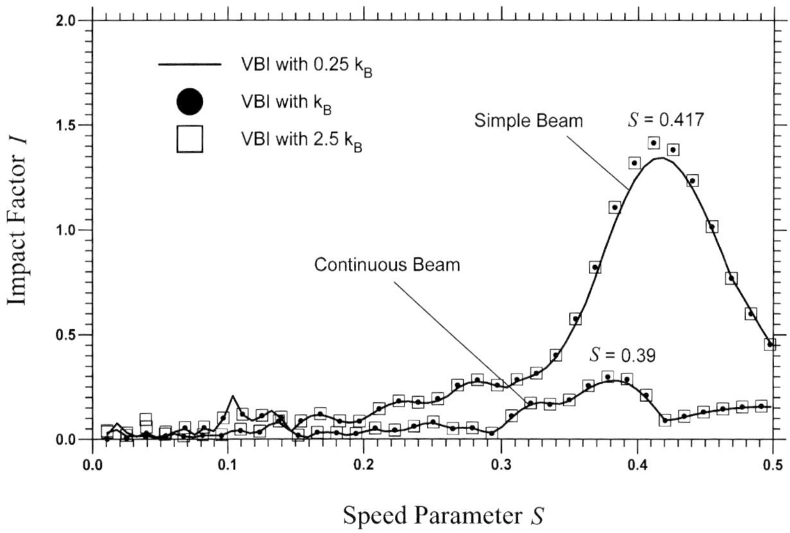


Fig. 6.16. Impact factor for midpoint deflection of beam (S25)-effect of ballast stiffness.

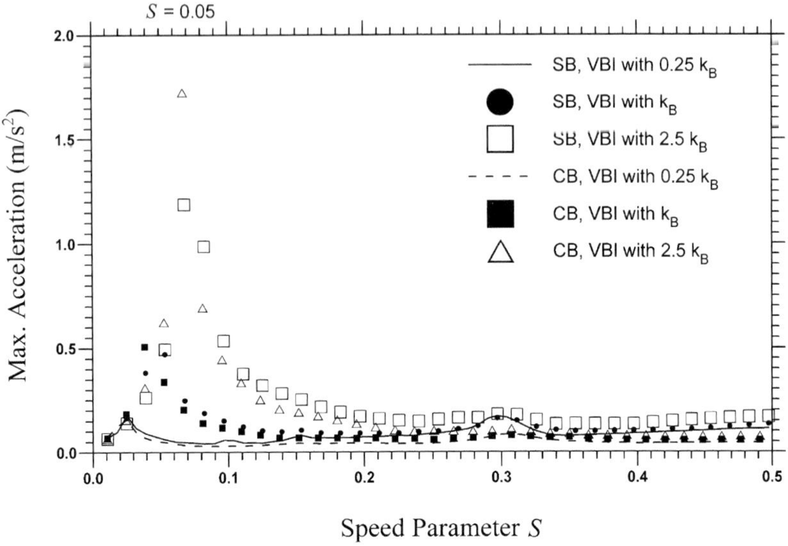


Fig. 6.17. Maximum acceleration of sprung mass (T18)-effect of ballast stiffness.

Unlike the case with the bridge response, the effect of ballast on the vehicle response can be quite significant. For instance, the maximum vertical accelerations of the sprung masses have been plotted in Figs. 6.17 and 6.18 for trains moving over the simple and continuous beams with different ballast stiffnesses, but no rail irregularities, for the two train models T18 and S25. The following observations can be drawn from the two figures. First, for both simple and continuous beams, resonant peaks exceeding the tolerance limit of 0.49 $m/s^2$ (= 0.05 g) will be induced for train cars moving over harder ballast (i.e., with stiffness $2.50k_B$) in the low speed range with $S < 0.15$ (or $v < 151.1$ km/h). Second, softer ballast increases significantly the acceleration level of the sprung masses moving over simple beams in the high-speed range, say, with $S = 0.34 \sim 0.50$ (or $v = 342.5 \sim 503.7$ km/h), especially for the S25 model. Finally, the dynamic response of the train cars moving over continuous beams is generally small for $S > 0.20$ (or $v > 203.5$ km/h) regardless of the stiffness of ballast. For the continuous beam, the use of softer ballast appears to be very effective for reducing the sprung mass response.

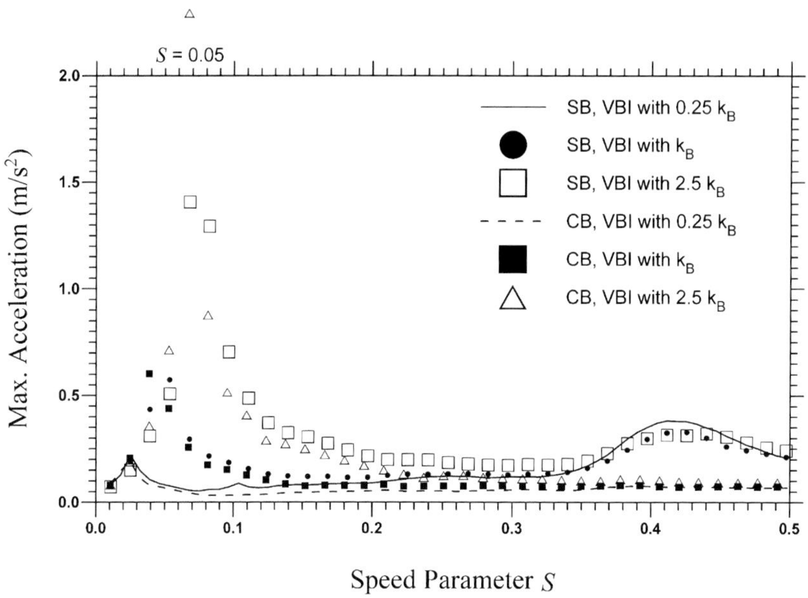


Fig. 6.18. Maximum acceleration of sprung mass (S25)-effect of ballast stiffness.

### 6.8.5. *Effect of Vehicle Suspension Stiffness*

Three values of suspension stiffness $0.25k_v$, $k_v$ and $2.5k_v$ are assumed for the two train models. As can be seen from Figs. 6.19 and 6.20, the influence of suspension stiffness of the train on the bridge response is generally small for the simple beam and even invisible for the continuous beam, although a stiffer suspension system can slightly reduce the bridge response. The conclusion here is that the effect of suspension stiffness of the moving train on the bridge response can be generally ignored in a practical design.

However, the same is not true with the vertical acceleration of the train cars running over the bridge. As can be seen from Fig. 6.21 for the T18 model and Fig. 6.22 for the S25 model, in addition to the peak response caused by resonance with the ballast at $S = 0.05$ (or $v = 50.4$ km/h), the use of stiffer suspension systems increases significantly the vertical acceleration of the train cars in running. This is certainly harmful concerning the *riding comfort* of passengers.

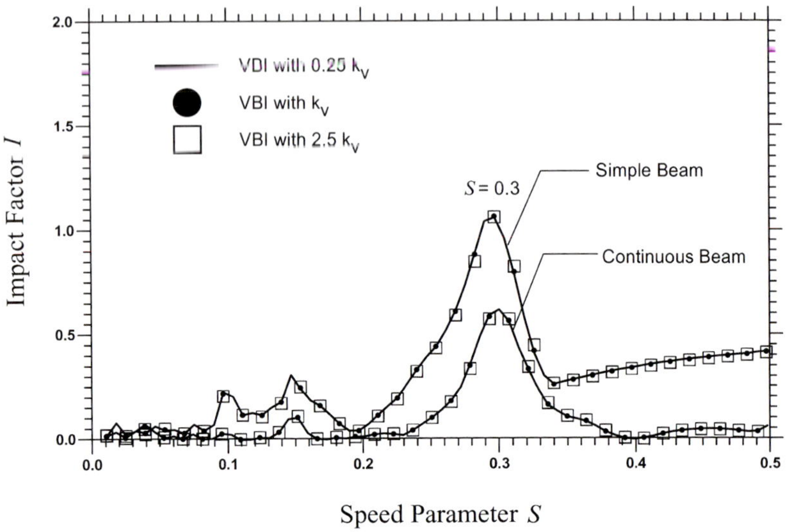


Fig. 6.19. Impact factor for midpoint deflection of beam (T18)-effect of suspension stiffness.

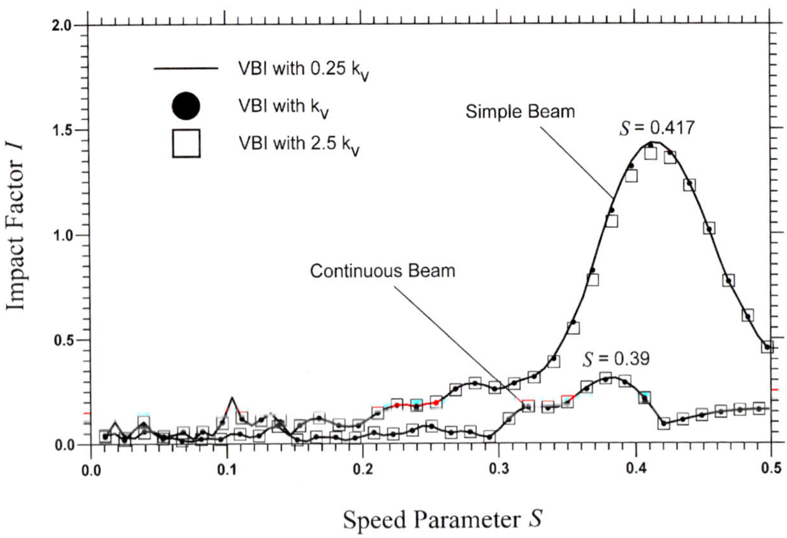


Fig. 6.20. Impact factor for midpoint deflection of beam (S25)-effect of suspension stiffness.

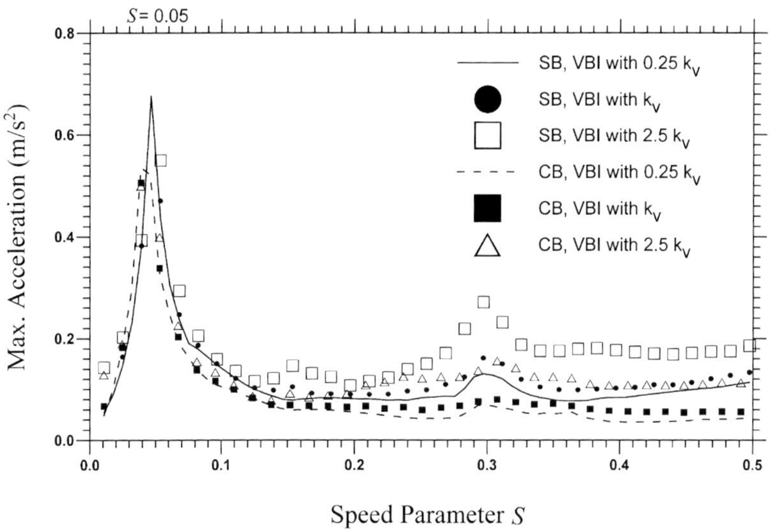


Fig. 6.21. Maximum acceleration of sprung mass (T18)-effect of suspension stiffness.

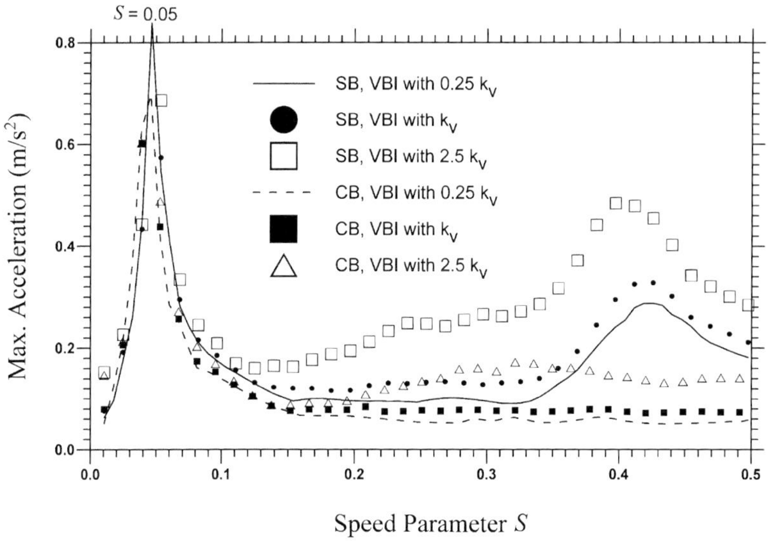


Fig. 6.22. Maximum acceleration of sprung mass (S25)-effect of suspension stiffness.

The reason behind this is that by making the suspension systems stiffer, the frequency of the sprung mass, as represented by $\sqrt{k_v/M_v}$, tends to increase and approach that of the bridge. As the two frequencies become closer, the sprung mass will attain its maximum response, in a manner similar to that of the *tuned mass* for the bridge. For the continuous beam, although the same tendency can be observed for the suspension stiffness from both figures, the level of influence is comparatively small. Note that there also exists a peak around $S = 0.05$ (or $v = 50.9$ km/h) for the continuous beam.

### 6.8.6. *Effect of Vehicle Suspension Damping*

Three values of suspension damping are considered, that is, $0.25c_v$, $c_v$, and $2.5c_v$. As can be seen from Figs. 6.23 and 6.24, for the two train models, by increasing the damping of the suspension systems, the response of the bridge will be reduced, although the degree of influence is only marginal. However, as far as the dynamic response of the train cars (or sprung masses) is concerned, it can be seen from

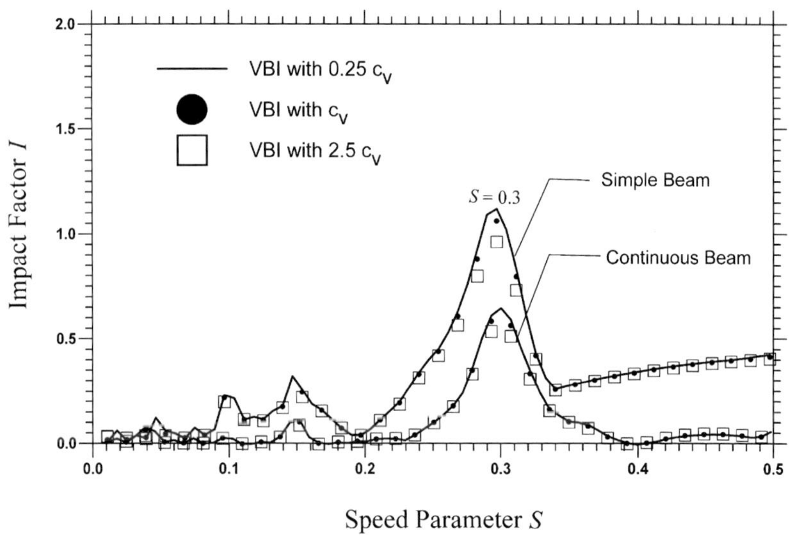


Fig. 6.23. Impact factor for midpoint deflection of beam (T18)-effect of suspension damping.

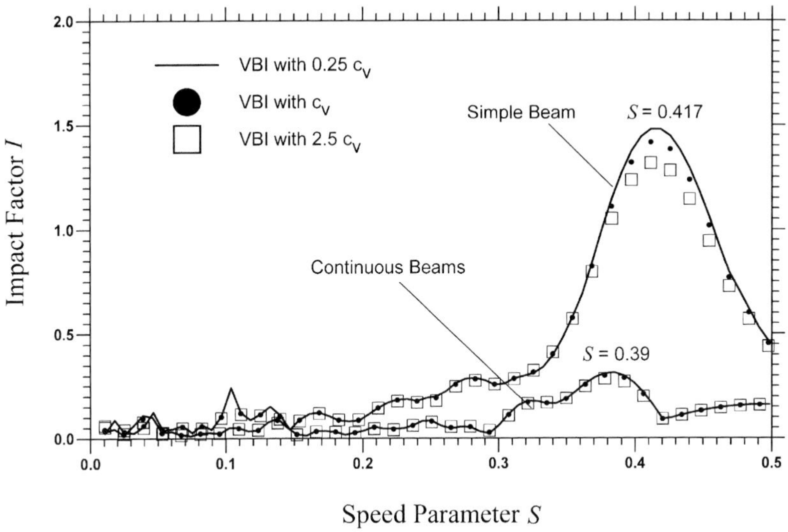


Fig. 6.24. Impact factor for midpoint deflection of beam (S25)-effect of suspension damping.

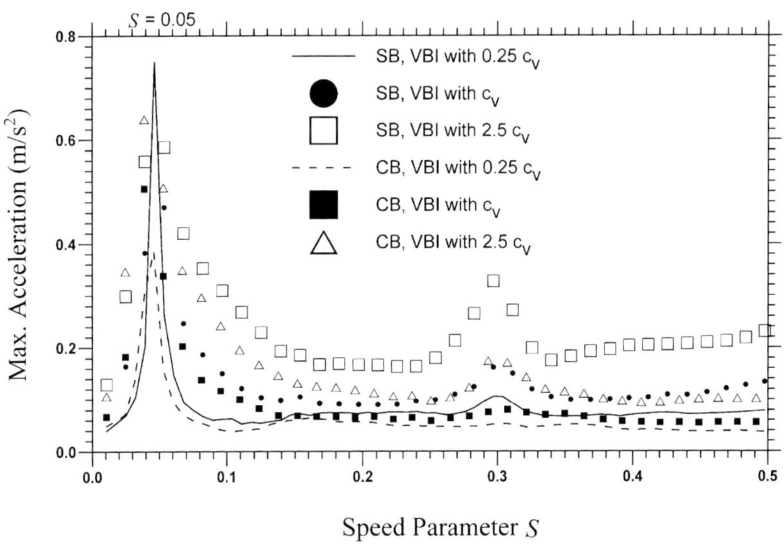


Fig. 6.25. Maximum acceleration of sprung mass (T18)-effect of suspension damping.

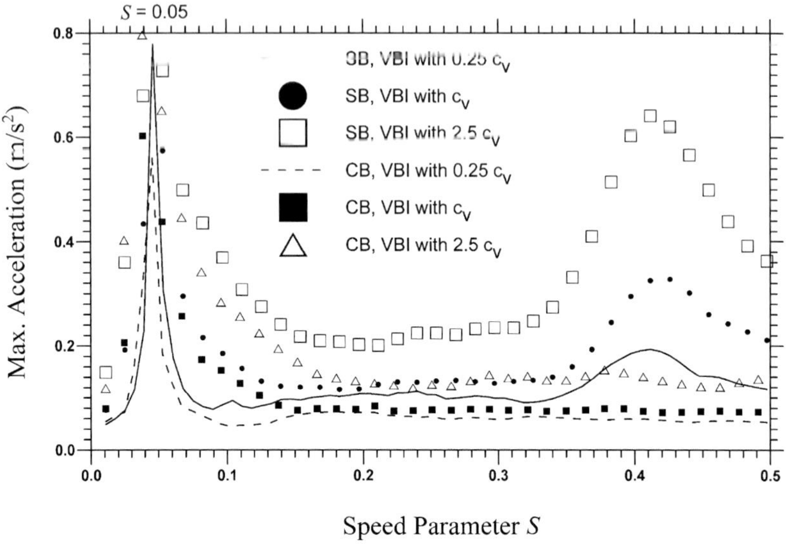


Fig. 6.26. Maximum acceleration of sprung mass (S25)-effect of suspension damping.

Figs. 6.25 and 6.26 for the T18 and S25 models, respectively, that larger values of suspension damping tend to drastically increase the maximum vertical acceleration of the train cars traveling over the simple beam for a wide range of speeds that is of interest. Such a fact must be carefully taken into account in the design of train cars. In contrast, the effect of suspension damping on the vehicle response is comparatively small for the case of continuous beams, as can be seen from both figures.

## 6.9. Concluding Remarks

In this chapter, the equations of motion for the vehicle is first discretized using Newmark's finite difference formulas and then condensed to the bridge equations, considering the condition of contact between the bridge and moving vehicles. The vehicle–bridge interaction (VBI) element derived has the advantage that it possesses the same number of DOFs as the parent element, while preserving the properties of symmetry and bandedness in element matrices. Because of this, conventional assembly process can be applied to forming the equations of motion for the entire VBI system. The present approach allows us to compute not only the bridge response, but also the vehicle response. The applicability of the derived element has been demonstrated in the numerical examples.

The following conclusions can be extracted from the parametric studies for the simple and three-span continuous beams using the VBI element derived: (1) The moving load model can be reliably used to predict the bridge response. However, the use of the sprung mass model enables us to compute the vehicle response, in addition to the bridge response. (2) Resonant response may be induced on the bridge by the train within the range of speeds of operation. The critical speed for resonance solved by the present approach coincides very well with that predicted by the analytical formula, as given in Eq. (2.59). (3) If the characteristic length, rather than the span length, is used for a continuous beam, then both the simple and continuous beams will attain their peaks at the same critical speed $S$, when subjected to moving loads of constant intervals. (4) The effect of rail irregularities is generally small on the bridge response. However, it can affect drastically the vertical acceleration of the sprung masses, which serves as a measure of the riding comfort for passengers carried by the train. (5) Softer ballast tends to reduce the impact response of the bridges, although the degree of reduction is marginal. For simple beams, the use of ballast that is too hard or too soft is not good for running vehicles concerning the riding comfort. (6) The influence of suspension stiffness on the bridge response is generally small, which can be neglected in practice. However, the use of stiffer suspension devices can dramatically increase the vertical acceleration of the train cars. (7) Increasing the suspension damping can only result in marginal reduction of the bridge response, but can adversely magnify the dynamic response of the train cars. (8) In general, much smaller response can be expected of a continuous beam, compared with that of the simple beam.

Chapter 7

# Vehicle–Bridge Interaction Element Considering Pitching Effect

Vehicle–bridge interaction (VBI) elements that were derived by treating a vehicle as discrete lumped masses, such as the one shown in Chapter 6, has the apparent drawback that the *pitching effect* of the car body caused by the differential deflections of the suspension systems was not duly taken into account. To overcome this drawback, a vehicle with two suspension systems, each for the front and rear wheel assemblies, is modeled as a rigid beam supported by two spring-dashpot units in this chapter. The equations of motion written for the vehicle and the bridge (beam) elements are coupled due to the existence of two interacting forces acting through the contact points. Following the procedure presented in the preceding chapter, the vehicle equations are first reduced to a set of equivalent stiffness equations using Newmark's finite difference scheme. The vehicle degrees of freedom (DOFs) are then condensed to those of the beam elements in contact. The rigid vehicle–bridge interaction elements derived is good for computing not only the bridge response, but also the vehicle response, with the vehicle's pitching effect duly taken into account.

## 7.1. Introduction

Previous research on the interaction between a bridge and the vehicles traveling over it has been abundant, which continues to grow in recent years due to the booming demands of high-speed railways in several European and Asian countries. Some of the related research works have been reviewed in Chapter 1. When studying the dynamic response of a vehicle–bridge interaction system, two sets of second-order differential equations can be written, one for the moving vehicles and the other for the bridge. It is the interaction forces existing at the contact points of the two subsystems that make the two sets of equations coupled. As the contact points are time-dependent, whose acting positions move from time to time, so are the system matrices, which therefore must be updated and factorized at each time step. To solve these two sets of second-order differential equations, procedures of an iterative nature are often adopted. For instance, by first assuming a trial solution for the displacements of the contact points, the contact forces can be solved from the vehicle equations. Then, by substituting these forces into the bridge equations, an improved solution for the displacements of the contact points can be solved. One drawback with iterative approaches of this type is that the convergence rate of iteration is likely to be low, when dealing with the more realistic case of a bridge traveled by a series of vehicles that make up a train, for there exists a great number of contact points.

The condensation method has been demonstrated to be an efficient method for solving the vehicle–bridge interaction problems. Garg and Dukkipati (1984) used the Guyan reduction scheme to condense the degrees of freedom (DOFs) of the vehicles to the DOFs of the bridge. In the paper by Yang and Lin (1995), the dynamic condensation method was used instead. However, because of the approximation made in relating the vehicle (slave) to the bridge (master) DOFs, this approach is good only for computing the bridge response, but not for the vehicle response; the latter serves as the indicator of riding comfort useful to the design of high-speed railroad bridges. To overcome this drawback, the vehicle equations were first reduced to a set of equivalent stiffness equations using Newmark's finite difference scheme, by which the vehicle DOFs are condensed to those of the bridge element in contact. The vehicle–bridge interaction (VBI) element derived is featured by the fact that the symmetry and bandedness of the element matrices are retained, while the vehicle response can be computed simply by back substitution (Yang and Yau, 1997). Also, it possesses the same order of accuracy as that implied by the Newmark scheme used in the solution of the entire vehicle–bridge system.

The aforementioned VBI element is not perfect, however, since no account has been made of the pitching effect of the car body with respect to the front and rear wheels, which appears to be of paramount importance in computation of the vehicle response and may be aggravated in the presence of track irregularities. Previously, a series expansion approach was conducted by Wen (1960) to analyze the dynamic response of beams traversed by two-axle loads. This chapter can be regarded as an extension of the theory presented in Chapter 6 toward the development of accurate and efficient VBI elements. It surpasses most existing VBI elements, in that the vehicle is modeled as a rigid beam supported by two suspension units, rather than as one or two discrete sprung masses. Further extension of the condensation technique presented herein is not impossible, say, to include the rolling motion or other three-dimensional effects of the moving vehicle.

Theoretically, more sophisticated models that contain dozens of DOFs can be developed for the vehicle, similar to those presented in Chu *et al.* (1986), and Huang and Novak (1991). However, due to their complexities and the large amount of computation required, these models are most suitable for the simulation of a *single vehicle* traveling over bridges or for the case where the vehicle response is of major concern. Since we are interested in the simulation of the more realistic case of a bridge traveled by a series of vehicles that constitute a train, the use of too complicated a vehicle model will make the derivation of the VBI element a formidable task. It is with these considerations that the vehicle model presented in this chapter, along with the special condensation technique, is considered appropriate. It should be noted that the materials presented in this chapter are based generally on the paper by Yang *et al.* (1999), except those in Secs. 7.5.4–7.5.6, which are based on the thesis by Chang (1997).

## 7.2. Equations of Motion for the Vehicle and Bridge

Figure 7.1 shows a train moving with speed $v$ over a bridge, in which a bridge is modeled as a number of beam elements and each car of the train as a uniform rigid beam supported by two suspension (spring-dashpot) units, as an improvement over the sprung mass model used in the preceding chapter. A schematic of the VBI system considered is given in Fig. 7.2, in which $r(x)$ denotes the track irregularity and $m_w$ the (unsprung) mass of the bogie. The motion of the rigid beam is described by the generalized coordinates $\{y\}^T = \{y_v \theta_v\}$, with $y_v$ denoting the vertical displacement and $\theta_v$ the rotation about the center point. In this chapter, the subscripts $i$ and $j$ respectively are

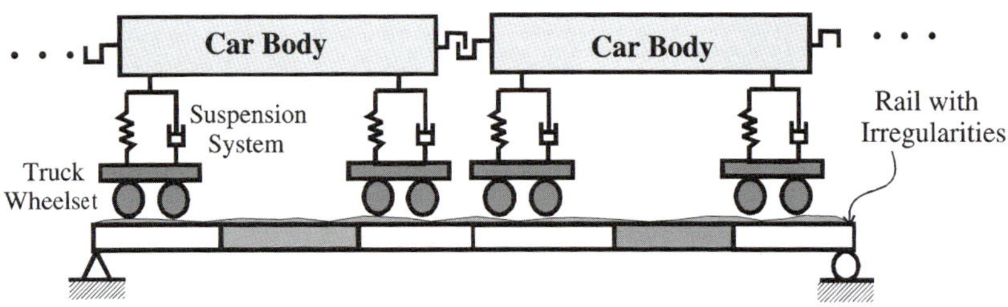


Fig. 7.1. Train–bridge system.

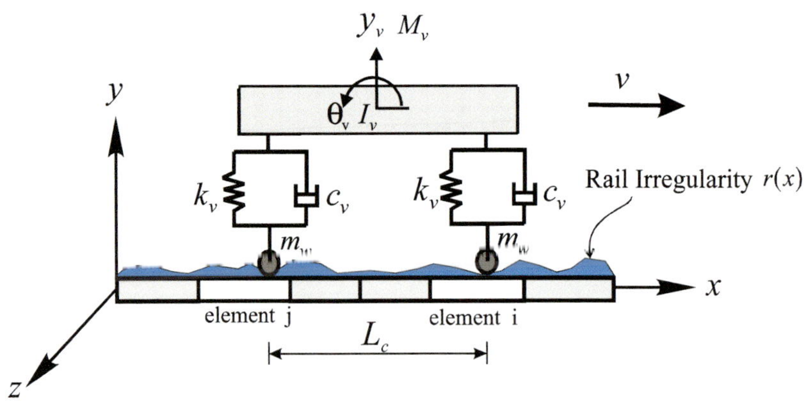


Fig. 7.2. Typical VBI system.

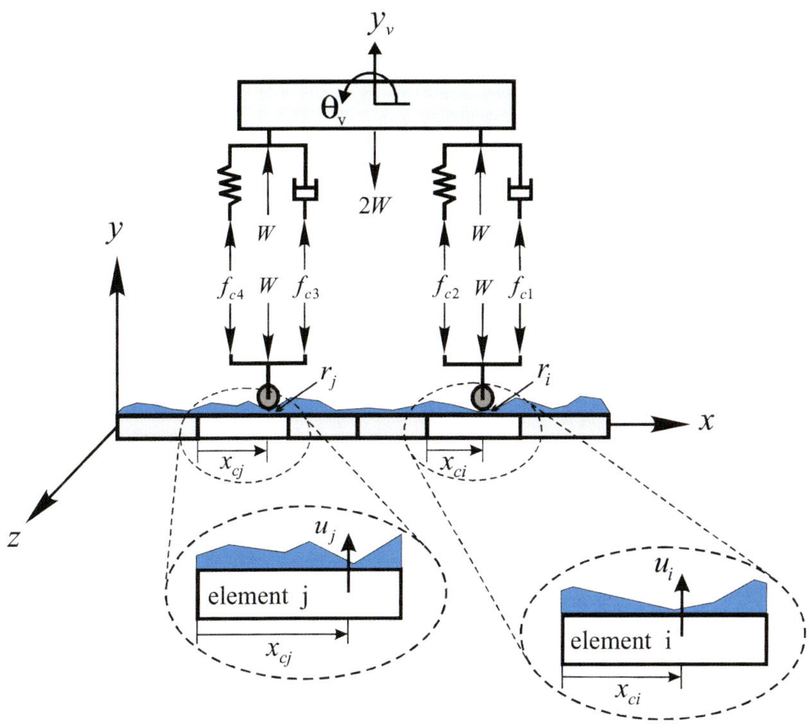


Fig. 7.3. Free body diagrams for components of the VBI system.

used to denote quantities associated with the front and rear wheels (or wheel assemblies) of the vehicle under consideration. As can be seen from Fig. 7.2, both elements $i$ and $j$ of the bridge, on which the vehicle is acting will be affected by the pitching motion of the car body via the two suspension units. The reverse is also true.

Figure 7.3 shows the free body diagrams for the components of the VBI system that are of interest, in which $2W$ denotes the weight of the car body (i.e., $W = 0.5\ M_v g$, with $M_v$ indicating the total mass of the car body, and $g$ the acceleration of gravity), $x_c$ the contact position of each set of wheels (or wheel assemblies) on the beam element, and $u$ the vertical displacement of the beam element. As can be seen, the (rigid) car body is acted upon by the contact forces $f_{c1}$, $f_{c2}$, $f_{c3}$, $f_{c4}$, and the front and rear wheels by the contact forces

$(f_{c1} + f_{c2} + W)$ and $(f_{c3} + f_{c4} + W)$, respectively. For the present case, all the component forces can be given as follows:

$$\begin{aligned} f_{c1} &= c_v(\dot{u}_i - \dot{y}_v - 0.5L_c\dot{\theta}_v)\,, \\ f_{c2} &= k_v(u_i + r_i - y_v - 0.5L_c\theta_v)\,, \\ f_{c3} &= c_v(\dot{u}_j - \dot{y}_v + 0.5L_c\dot{\theta}_v)\,, \\ f_{c4} &= k_v(u_j + r_j - y_v + 0.5L_c\theta_v)\,, \end{aligned} \tag{7.1}$$

where an overdot denotes differentiation with respect to time $t$; $k_v$, $c_v$ = the spring constant and damping coefficient of the suspension unit; $L_c$ = the axle distance of the vehicle; $r_i$, $r_j$ = the track irregularities of elements $i$ and $j$ at the contact points; and $u_i$, $u_j$ = the displacements of elements $i$ and $j$ evaluated at the contact points. The equations of equilibrium for the rigid car body are

$$\begin{aligned} f_{c1} + f_{c2} + f_{c3} + f_{c4} &= M_v\ddot{y}_v\,, \\ 0.5d(f_{c1} + f_{c2} - f_{c3} - f_{c4}) &= I_v\ddot{\theta}_v\,, \end{aligned} \tag{7.2}$$

where $I_v$ = the rotatory inertia of the car body. Substituting the component forces in Eq. (7.1) into the preceding two equations in Eq. (7.2) yields the equations of motion for the vertical and rotational displacements of the car body as:

$$\begin{bmatrix} M_v & 0 \\ 0 & I_v \end{bmatrix} \begin{Bmatrix} \ddot{y}_v \\ \ddot{\theta}_v \end{Bmatrix} + \begin{bmatrix} 2c_v & 0 \\ 0 & 0.5c_vL_c^2 \end{bmatrix} \begin{Bmatrix} \dot{y}_v \\ \dot{\theta}_v \end{Bmatrix} + \begin{bmatrix} 2k_v & 0 \\ 0 & 0.5k_vL_c^2 \end{bmatrix} \begin{Bmatrix} y_v \\ \theta_v \end{Bmatrix} = \begin{Bmatrix} f_{ver} \\ f_{rot} \end{Bmatrix}, \tag{7.3}$$

where the term on the right-hand side denotes the acting forces resulting from the two suspension (spring-dashpot) units,

$$\begin{Bmatrix} f_{ver} \\ f_{rot} \end{Bmatrix} = \begin{Bmatrix} c_v(\dot{u}_i + \dot{u}_j) + k_v(u_i + u_j) + k_v(r_i + r_j) \\ 0.5L_c[c_v(\dot{u}_i - \dot{u}_j) + k_v(u_i - u_j) + k_v(r_i - r_j)] \end{Bmatrix}. \tag{7.4}$$

As can be seen, the vertical force $f_{ver}$ and rotational moment $f_{rot}$ experienced by the car body are caused by the velocity $\dot{u}$, displacement $u$, and track irregularity $r$ of the bridge and transmitted through the suspension units at the contact points.

In this chapter, the most commonly used 12-DOF beam element will be used to simulate the bridge structure, of which the axial displacement is interpolated by linear functions, and the transverse displacements by cubic *interpolation* (*Hermitian*) *functions*. For instance, the $y$-direction displacement $u$ of the element at section $x$ (in local coordinate) can be related to the nodal DOFs of the beam element as $u = \langle N \rangle \{\bar{u}_y\}$, where $\{\bar{u}_y\}$ denotes the $y$-direction related nodal DOFs, $\{\bar{u}_y\}^T = \langle u_{yA}\ l\theta_{zA}\ u_{yB}\ l\theta_{zB} \rangle$, and the associated interpolation function $\{N\}$ is

$$\{N\} = \left\langle 1 - 3\left(\frac{x}{l}\right)^2 + 2\left(\frac{x}{l}\right)^3,\quad x\left(1 - \frac{x}{l}\right)^2,\quad 3\left(\frac{x}{l}\right)^2 - 2\left(\frac{x}{l}\right)^3,\quad \left(\frac{x}{l} - 1\right)\frac{x^2}{l} \right\rangle^T, \tag{7.5}$$

where $l$ denotes the length of the element. See Appendix F for notation of the nodal DOFs of the beam element, and Yang and Kuo (1994), Chapter 2, for more details on the interpolation functions used by the beam element. Assuming that Rayleigh damping is valid for the bridge, the equations of motion can be written for elements $i$ and $j$ of the bridge as

$$\begin{aligned} [m_i]\{\ddot{u}_i\} + [c_i]\{\dot{u}_i\} + [k_i]\{u_i\} &= -p_i\{N_{ci}\}, \\ [m_j]\{\ddot{u}_j\} + [c_j]\{\dot{u}_j\} + [k_j]\{u_j\} &= -p_j\{N_{cj}\}, \end{aligned} \tag{7.6}$$

where $\{u\}$ = the displacement vector; $[m]$, $[c]$, and $[k]$ = the mass, damping, and stiffness matrices of the element [see for instance Appendix F for details of these matrices]; $\{N_c\}$ = the Hermitian functions evaluated for the DOFs associated with the $y$-direction displacement at the contact point (i.e., at $x = x_c$), which can be obtained by substituting the contact point position $x_c$ into Eq. (7.5), but with those associated with the $x$- and $z$-direction displacements set to zero;

and $p$ = the contact force for each of the two elements considered,

$$\begin{aligned} p_i &= f_{c1} + f_{c2} + 0.5M_v g + m_w(g + \ddot{u}_i)\,, \\ p_j &= f_{c3} + f_{c4} + 0.5M_v g + m_w(g + \ddot{u}_j)\,, \end{aligned} \tag{7.7}$$

with $m_w$ denoting the mass of the wheel assemblies.

In Eq. (7.6), the terms on the right-hand side should be recognized as the *consistent nodal loads* acting on the two elements due to the contact forces. Besides, a comparison of the vehicle equations, i.e., Eq. (7.2), with the bridge equations, i.e., Eq. (7.6), together with the contact force expressions in Eq. (7.7), indicates that the two subsystems are coupled via the spring and damping forces, i.e., $f_{c1}$, $f_{c2}$, $f_{c3}$, $f_{c4}$, of the suspension units.

By substitution of the spring and damping forces in Eq. (7.1), the contact forces in Eq. (7.7) can be rewritten,

$$\begin{aligned} p_i &= c_v(\dot{u}_i - \dot{y}_v - 0.5L_c\theta_v) + k_v(u_i + r_i - y_v - 0.5L_c\theta_v) \\ &\quad + 0.5M_v g + m_w(g + \ddot{u}_i)\,, \\ p_j &= c_v(\dot{u}_j - \dot{y}_v + 0.5L_c\dot{\theta}_v) + k_v(u_j + r_j - y_v + 0.5L_c\theta_v) \\ &\quad + 0.5M_v g + m_w(g + \ddot{u}_j)\,. \end{aligned} \tag{7.8}$$

As can be seen, the contact forces acting on the two elements $i$ and $j$ are composed of four components: (1) the static weights associated with the car body and wheel assemblies, as represented by $M_v g$ and $m_w g$; (2) the damping forces resulting from the relative velocity of the car body to the bridge elements, as indicated by the terms containing $c_v$; (3) the elastic forces resulting from the relative displacement of the car body to the bridge elements, as indicated by the terms involving $k_v$; and (4) the inertial forces due to the vertical acceleration of the bridge elements, as indicated by the term $m_w\ddot{u}$. It is these coupling forces existing between the two subsystems that make the vehicle–bridge interaction a problem difficult to solve. In the preceding chapter, it was demonstrated that such a problem can be effectively resolved by first reducing the vehicle equations into a set of equivalent stiffness equations using Newmark's finite difference scheme, and then by condensing the vehicle DOFs to those of the bridge element in contact, which will result in the so-called VBI element. Such an approach will be generally followed in this chapter.

## 7.3. Rigid Vehicle–Bridge Interaction Element

The system equations as given in Eqs. (7.3) and (7.6) for the car body of the moving vehicle and the associated bridge elements, along with the acting forces in Eq. (7.4) and the contact forces in Eq. (7.8), are nonlinear in nature, as the contact points move from time to time, which can only be solved by incremental methods. Consider a typical incremental step from time $t$ to $t + \Delta t$. The system equations, i.e., Eqs. (7.3) and (7.6), together with the contact forces in Eqs. (7.4) and (7.8), should now be interpreted as those established for the structure in the current configuration at time $t + \Delta t$.

Accordingly, the equations of motion for the car body, i.e., Eq. (7.3), can be rewritten for the current time step, with $t + \Delta t$ clearly inserted as the subscript as:

$$\begin{bmatrix} M_v & 0 \\ 0 & I_v \end{bmatrix} \begin{Bmatrix} \ddot{y}_v \\ \ddot{\theta}_v \end{Bmatrix}_{t+\Delta t} + \begin{bmatrix} 2c_v & 0 \\ 0 & 0.5c_v L_c^2 \end{bmatrix} \begin{Bmatrix} \dot{y}_v \\ \dot{\theta}_v \end{Bmatrix}_{t+\Delta t} + \begin{bmatrix} 2k_v & 0 \\ 0 & 0.5k_v L_c^2 \end{bmatrix} \begin{Bmatrix} y_v \\ \theta_v \end{Bmatrix}_{t+\Delta t} = \begin{Bmatrix} f_{ver} \\ f_{rot} \end{Bmatrix}_{t+\Delta t}, \tag{7.9}$$

where the acting forces are given as

$$\begin{Bmatrix} f_{ver} \\ f_{rot} \end{Bmatrix}_{t+\Delta t} = \begin{Bmatrix} c_v(\dot{u}_i + \dot{u}_j) + k_v(u_i + u_j) + k_v(r_i + r_j) \\ 0.5L_c[c_v(\dot{u}_i - \dot{u}_j) + k_v(u_i - u_j) + k_v(r_i - r_j)] \end{Bmatrix}_{t+\Delta t}. \tag{7.10}$$

Similarly, the equations of motion for the bridge elements $i$ and $j$, as given in Eq. (7.6), can be rewritten for the current time step $t+\Delta t$ as

$$\begin{aligned}
&[m_i]\{\ddot{u}_i\}_{t+\Delta t} + [c_i]\{\dot{u}_i\}_{t+\Delta t} + [k_i]\{u_i\}_{t+\Delta t} = -p_{i,t+\Delta t}\{N_{ci}\}\,, \\
&[m_j]\{\ddot{u}_j\}_{t+\Delta t} + [c_j]\{\dot{u}_j\}_{t+\Delta t} + [k_j]\{u_j\}_{t+\Delta t} = -p_{j,t+\Delta t}\{N_{cj}\}\,,
\end{aligned} \tag{7.11}$$

where the associated contact forces are

$$\begin{aligned}
p_{i,t+\Delta t} &= [c_v(\dot{u}_i - \dot{y}_v - 0.5L_c\dot{\theta}_v) + k_v(u_i + r_i - y_v - 0.5L_c\theta_v) \\
&\quad + 0.5M_v g + m_w(g + \ddot{u}_i)]_{t+\Delta t}\,, \\
p_{j,t+\Delta t} &= [c_v(\dot{u}_j - \dot{y}_v + 0.5L_c\dot{\theta}_v) + k_v(u_j + r_j - y_v + 0.5L_c\theta_v) \\
&\quad + 0.5M_v g + m_w(g + \ddot{u}_j)]_{t+\Delta t}\,.
\end{aligned} \tag{7.12}$$

Following the procedures of the preceding chapter, the vehicle equations, Eq. (7.9), which are of second order, will first be reduced to a set of equivalent stiffness equations using Newmark's finite difference scheme. Then, the vehicle DOFs will be condensed to those of the bridge elements in contact.

Let $\{y\}_t$ denote the car body displacements at time $t$. The following are the finite difference equations proposed by Newmark:

$$\begin{aligned}
\{\dot{y}\}_{t+\Delta t} &= \{\dot{y}\}_t + [(1-\gamma)\{\ddot{y}\}_t + \gamma\{\ddot{y}\}_{t+\Delta t}]\Delta t\,, \\
\{y\}_{t+\Delta t} &= \{y\}_t + \{\dot{y}\}_t\Delta t + [(0.5-\beta)\{\ddot{y}\}_t + \beta\{\ddot{y}\}_{t+\Delta t}]\Delta t^2\,,
\end{aligned} \tag{7.13}$$

where $\beta = 0.25$ and $\gamma = 0.5$ will be adopted, for they imply an integration scheme of constant average acceleration within the time step, which has the advantage of being unconditionally stable. From Eq. (7.13), the following can be solved,

$$\begin{aligned}
\{\ddot{y}\}_{t+\Delta t} &= a_0(\{y\}_{t+\Delta t} - \{y\}_t) - a_2\{\dot{y}\}_t - a_3\{\ddot{y}\}_t\,, \\
\{\dot{y}\}_{t+\Delta t} &= a_1(\{y\}_{t+\Delta t} - \{y\}_t) - a_4\{\dot{y}\}_t - a_5\{\ddot{y}\}_t\,,
\end{aligned} \tag{7.14}$$

where the coefficients and those to appear are defined as

$$a_0 = \frac{1}{\beta \Delta t^2}, \quad a_1 = \frac{\gamma}{\beta \Delta t}, \quad a_2 = \frac{1}{\beta \Delta t},$$
$$a_3 = \frac{1}{2\beta} - 1, \quad a_4 = \frac{\gamma}{\beta} - 1, \quad a_5 = \frac{\Delta t}{2}\left(\frac{\gamma}{\beta} - 2\right), \tag{7.15}$$
$$a_6 = \Delta t(1-\gamma), \quad a_7 = \gamma \Delta t .$$

By the relations in Eq. (7.14), the car body equations, i.e., Eq. (7.9), can be manipulated to yield a set of *equivalent stiffness equations* as follows:

$$\begin{bmatrix} D_{11} & 0 \\ 0 & D_{22} \end{bmatrix} \begin{Bmatrix} y_v \\ \theta_v \end{Bmatrix}_{t+\Delta t} = \begin{Bmatrix} f_{ver} \\ f_{rot} \end{Bmatrix}_{t+\Delta t} + \begin{Bmatrix} f_{ver} \\ f_{rot} \end{Bmatrix}_t , \tag{7.16}$$

where

$$\begin{aligned} D_{11} &= a_0 M_v + 2a_1 c_v + 2k_v , \\ D_{22} &= a_0 I_v + 0.5a_1 c_v L_c^2 + 0.5k_v L_c^2 , \end{aligned} \tag{7.17}$$

and the last term on the right-hand side denotes the equivalent nodal loads resulting from the vehicle responses at time $t$,

$$\begin{Bmatrix} f_{ver} \\ f_{rot} \end{Bmatrix}_t = \begin{Bmatrix} M_v(a_0 y_v + a_2 \dot{y}_v + a_3 \ddot{y}_v) + 2c_v(a_1 y_v + a_4 \dot{y}_v + a_5 \ddot{y}_v) \\ I_v(a_0 \theta_v + a_2 \dot{\theta}_v + a_3 \ddot{\theta}_v) + 0.5c_v L_c^2(a_1 \theta_v + a_4 \dot{\theta}_v + a_5 \ddot{\theta}_v) \end{Bmatrix}_t . \tag{7.18}$$

Solving Eq. (7.16) yields the car body displacements $\{y_v \ \theta_v\}^T$ at time $t + \Delta t$ as

$$\begin{Bmatrix} y_v \\ \theta_v \end{Bmatrix}_{t+\Delta t} = \frac{1}{D} \begin{Bmatrix} D_{22} f_{ver} \\ D_{11} f_{rot} \end{Bmatrix}_{t+\Delta t} + \frac{1}{D} \begin{Bmatrix} D_{22} f_{ver} \\ D_{11} f_{rot} \end{Bmatrix}_t , \tag{7.19}$$

where $D$ is the determinant of the system matrix in Eq. (7.16), $D = D_{11}D_{22}$. Once the vehicle displacements $\{y_v\ \theta_v\}^T$ are made available at time $t+\Delta t$, the car body velocities at the same moment can be computed using Eq. (7.14) as

$$\left\{\begin{array}{c}\dot{y}_v\\ \dot{\theta}_v\end{array}\right\}_{t+\Delta t} = \frac{a_1}{D}\left\{\begin{array}{c}D_{22}f_{ver}\\ D_{11}f_{rot}\end{array}\right\}_{t+\Delta t} + \left\{\begin{array}{c}f_1\\ f_2\end{array}\right\}_t, \tag{7.20}$$

where

$$\left\{\begin{array}{c}f_1\\ f_2\end{array}\right\}_t = \frac{1}{D}\left\{\begin{array}{c}D_{22}(-2a_1k_vy_v + a_0M_v\dot{y}_v - 2a_4k_v\dot{y}_v\\ +a_0a_6M_v\ddot{y}_v - 2a_5k_v\ddot{y}_v)\\ D_{11}(-0.5a_1k_vL_c^2\theta_v + a_0I_v\dot{\theta}_v - 0.5a_4k_vL_c^2\dot{\theta}_v\\ +a_0a_6I_v\ddot{\theta}_v - 0.5a_5k_vL_c^2\ddot{\theta}_v)\end{array}\right\}_t. \tag{7.21}$$

In the meantime, the car body accelerations can be computed as

$$\left\{\begin{array}{c}\ddot{y}_v\\ \ddot{\theta}_v\end{array}\right\}_{t+\Delta t} = \frac{a_0}{D}\left\{\begin{array}{c}D_{22}f_{ver}\\ D_{11}f_{rot}\end{array}\right\}_{t+\Delta t} - \left\{\begin{array}{c}f_3\\ f_4\end{array}\right\}_t, \tag{7.22}$$

where

$$\left\{\begin{array}{c}f_3\\ f_4\end{array}\right\}_t = \frac{1}{D}\left\{\begin{array}{c}2D_{22}(a_0k_vy_v + a_0c_v\dot{y}_v + a_2k_v\dot{y}_v\\ +a_0a_6c_v\ddot{y}_v + a_3k_v\ddot{y}_v)\\ 0.5L_c^2D_{11}(a_0k_v\theta_v + a_0c_v\dot{\theta}_v + a_2k_v\dot{\theta}_v\\ +a_0a_6c_v\ddot{\theta}_v + a_3k_v\ddot{\theta}_v)\end{array}\right\}_t. \tag{7.23}$$

As can be seen from Eqs. (7.19), (7.20) and (7.22), one feature with Newmark's finite difference equations is that all the quantities to be solved at time $t+\Delta t$ can be related exclusively to those existing at the beginning and ending points, i.e., at time $t$ and $t+\Delta t$, of the current time step.

By substitution of Eqs. (7.19) and (7.20), the car body displacements $y_v$ and $\theta_v$ can be eliminated from the associated bridge element equations, Eq. (7.11), resulting in

$$\begin{aligned}
&\begin{bmatrix} [m_i]+[m_{ii}] & [0] \\ [0] & [m_j]+[m_{jj}] \end{bmatrix} \begin{Bmatrix} \{\ddot{u}_i\} \\ \{\ddot{u}_j\} \end{Bmatrix}_{t+\Delta t} \\
&\quad + \begin{bmatrix} [c_i]+[c_{ii}] & [c_{ij}] \\ [c_{ji}] & [c_j]+[c_{jj}] \end{bmatrix} \begin{Bmatrix} \{\dot{u}_i\} \\ \{\dot{u}_j\} \end{Bmatrix}_{t+\Delta t} \\
&\quad + \begin{bmatrix} [k_i]+[k_{ii}] & [k_{ij}] \\ [k_{ji}] & [k_j]+[k_{jj}] \end{bmatrix} \begin{Bmatrix} \{u_i\} \\ \{u_j\} \end{Bmatrix}_{t+\Delta t} \\
&= -\begin{Bmatrix} \bar{p}_i\{N_{ci}\} \\ \bar{p}_j\{N_{cj}\} \end{Bmatrix}.
\end{aligned} \tag{7.24}$$

These are exactly the equations of motion for the two elements $i$ and $j$ on which the front and rear wheels (or wheel assemblies) of the vehicle are directly acting. The elements as presented in Eq. (7.24) are referred to as the *rigid vehicle–bridge interaction elements*, which are characterized by the fact that the car body dynamic properties have been included through the condensation process, and that the element has the same order of accuracy as that implied by the Newmark finite difference scheme. In Eq. (7.24), [0] is a $12 \times 12$ matrix containing only zero entries, the vehicle-induced matrices appearing on the first row of Eq. (7.24) associated with element $i$ can be given as

$$\begin{aligned}
[m_{ii}] &= m_w\{N_{ci}\}\langle N_{ci}\rangle\,, \\
[c_{ii}] &= c_v\xi\{N_{ci}\}\langle N_{ci}\rangle\,, \\
[c_{ij}] &= c_v\eta\{N_{ci}\}\langle N_{cj}\rangle\,, \\
[k_{ii}] &= k_v\xi\{N_{ci}\}\langle N_{ci}\rangle\,, \\
[k_{ij}] &= k_v\eta\{N_{ci}\}\langle N_{cj}\rangle\,,
\end{aligned} \tag{7.25}$$

and those on the second row associated with element $j$ as

$$\begin{aligned}
[m_{jj}] &= m_w\{N_{cj}\}\langle N_{cj}\rangle\,,\\
[c_{jj}] &= c_v\xi\{N_{cj}\}\langle N_{cj}\rangle\,,\\
[c_{ji}] &= c_v\eta\{N_{cj}\}\langle N_{ci}\rangle\,,\\
[k_{jj}] &= k_v\xi\{N_{cj}\}\langle N_{cj}\rangle\,,\\
[k_{ji}] &= k_v\eta\{N_{cj}\}\langle N_{ci}\rangle\,.
\end{aligned} \tag{7.26}$$

It is easy to see that the terms on the right-hand side of the equal sign of Eq. (7.24) represent the *consistent nodal loads* acting on elements $i$ and $j$, resulting from the following contact force $\bar{p}_i$ for element $i$:

$$\begin{aligned}
\bar{p}_i = &-[k_v(\varsigma r_i + \eta r_j) + 0.5M_v g + m_w g]_{t+\Delta t}\\
&+ \left[c_v(f_1 + 0.5L_c f_2) + \frac{k_v}{D}(D_{22} f_{ver} + 0.5D_{11} L_c f_{rot})\right]_t
\end{aligned} \tag{7.27}$$

and the following for element $j$:

$$\begin{aligned}
\bar{p}_j = &-[k_v(\varsigma r_j + \eta r_i) + 0.5M_v g + m_w g]_{t+\Delta t}\\
&+ \left[c_v(f_1 - 0.5L_c f_2) + \frac{k_v}{D}(D_{22} f_{ver} - 0.5D_{11} L_c f_{rot})\right]_t\,,
\end{aligned} \tag{7.28}$$

where

$$\begin{aligned}
\varsigma &= \frac{a_0}{D}[(a_1 c_v + k_v)(0.25M_v L_c^2 + I_v) + a_0 M_v I_v]\,,\\
\eta &= \frac{a_0}{D}(a_1 c_v + k_v)(0.25M_v L_c^2 - I_v)\,.
\end{aligned} \tag{7.29}$$

As can be seen from Eq. (7.24), the mass, damping and stiffness matrices derived for the rigid vehicle–bridge interaction elements differ from those of the parent (beam) elements in the addition of the matrices given in Eq. (7.25) for element $i$, those in Eq. (7.26) for element $j$, to account for the effect of interaction from the moving vehicle. In particular, the matrices $[c_{ij}]$ and $[k_{ij}]$ represent the pitching action of the car body via the front wheels on element $i$, and the matrices $[c_{ji}]$ and $[k_{ji}]$ via the rear wheels on element $j$. It should be noted that because of the existence of the transposition relations:

$[c_{ij}] = [c_{ji}]^T$ and $[k_{ij}] = [k_{ji}]^T$, the system matrices $[M]$, $[C]$, $[K]$ assembled from all the element matrices, including those that are free of moving vehicles, retain the desired property of *symmetry*.

## 7.4. Equations of Motion for the VBI System

When a train is traveling over a bridge, the elements of the bridge that are directly acted upon by the wheels (or wheel assemblies) of the vehicles constituting the train at the instant considered should be modeled by the VBI elements derived, and the remaining by the conventional (parent) beam elements. Here, two things need be noted. First, the vehicle loads can only transmit through the nodal points of the VBI elements, but not of the conventional beam elements. Second, the vehicle-induced matrices that are special for the VBI elements, when summed up all over the entire VBI system, turn out to be symmetric on the global level. Based on these considerations, the first step in analyzing a VBI system is to construct the mass matrix, damping matrix, and stiffness matrix for the primary bridge structure with no vehicles acting on it. The next step is to determine the acting positions ($x_c$) of the front and rear wheels (or wheel assemblies) of each vehicle, given its speed and acceleration. The vehicle-induced matrices and the contact forces $\bar{p}$ of the VBI elements, which are considered "extra" to those of the primary structure, can then be determined using Eqs. (7.25)–(7.26) and Eqs. (7.27)–(7.28), respectively, depending on whether they are acted upon by the front or rear wheels. In particular, the contact forces $\bar{p}$ can be transformed into the (consistent) nodal loads of the VBI elements as given on the right-hand side of Eq. (7.24). All these "extra" terms, which are to account for the VBI effect, including the pitching actions, should be added to the entries in the system matrices corresponding to the nodal DOFs of each of the VBI elements considered.

For illustration, let us consider a simple beam traveled by a two-axle vehicle. By dividing the beam into three elements, the equations of motion, in terms of $[M]$, $[C]$ and $[K]$, are first established for the primary structure that is free of any vehicles, with the load vector

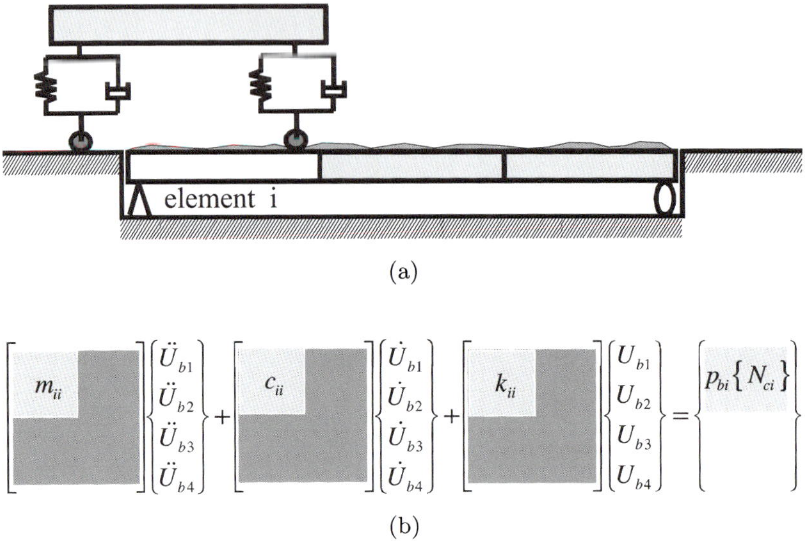


Fig. 7.4. Vehicle with front wheels only acting on the bridge.

$\{P\}$ initialized as $\{0\}$. For the case when only the front wheels of the vehicle are acting on the bridge, say, with coordinate $x_{ci}$ known for the acting position on element $i$ (which becomes now a VBI element), the vehicle nodal loads $\bar{p}_i\{N_{ci}\}$, where $\{N_{ci}\}$ is evaluated using Eq. (7.5) by setting $x = x_{ci}$, and the vehicle-induced matrices $[m_{ii}]$, $[c_{ii}]$, $[k_{ii}]$ should be added to the rows and columns corresponding to the nodal DOFs of element $i$ in the system equations, as was depicted in Fig. 7.4. Several things should be noted here. First, for this case, the pitching actions vanish, i.e., $[c_{ij}] = [k_{ij}] = [0]$. Second, the vehicle load $\bar{p}_i\{N_{ci}\}$ and the vehicle-induced matrices $[m_{ii}]$, $[c_{ii}]$, $[k_{ii}]$ should be adjusted according to the acting position $x_{ci}$ of the front wheels at each time step. Finally, all these "extra" terms should be removed from the system matrices associated with element $i$ once the front wheels move to the neighboring element.

For the case when both the front and rear wheels of the vehicle are acting on the bridge, say, with the front wheels acting at coordinate $x_{ci}$ on element $i$ and the rear wheels at coordinate $x_{cj}$ on element $j$ (here both elements $i$ and $j$ are the VBI elements), as shown in

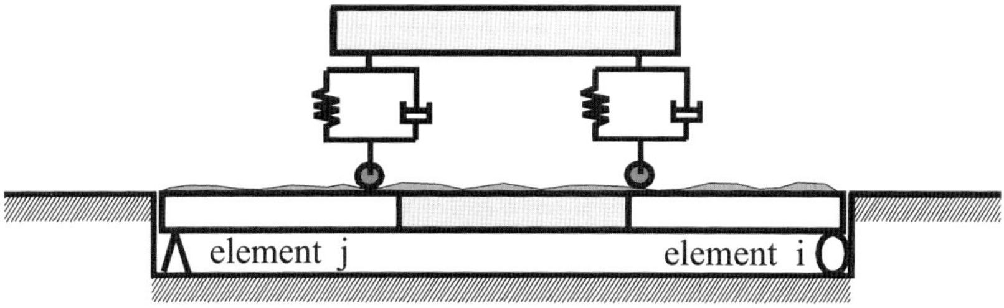


(a)

$$\begin{bmatrix} m_{jj} & \\ & m_{ii} \end{bmatrix}\begin{Bmatrix} \ddot{U}_{b1} \\ \ddot{U}_{b2} \\ \ddot{U}_{b3} \\ \ddot{U}_{b4} \end{Bmatrix} + \begin{bmatrix} c_{jj} & c_{ji} \\ c_{ij} & c_{ii} \end{bmatrix}\begin{Bmatrix} \dot{U}_{b1} \\ \dot{U}_{b2} \\ \dot{U}_{b3} \\ \dot{U}_{b4} \end{Bmatrix} + \begin{bmatrix} k_{jj} & k_{ji} \\ k_{ij} & k_{ii} \end{bmatrix}\begin{Bmatrix} U_{b1} \\ U_{b2} \\ U_{b3} \\ U_{b4} \end{Bmatrix} = \begin{Bmatrix} p_{bj}\{N_{cj}\} \\ p_{bi}\{N_{ci}\} \end{Bmatrix}$$

(b)

Fig. 7.5. Vehicle with both front and rear wheels acting on the bridge.

Fig. 7.5, the vehicle nodal loads $\bar{p}_i\{N_{ci}\}$ and $\bar{p}_j\{N_{cj}\}$, which are evaluated by setting $x = x_{ci}$ and $x_{cj}$ for elements $i$ and $j$, respectively, and the vehicle-induced matrices $[m_{ii}]$, $[m_{jj}]$, $[c_{ii}]$, $[c_{ij}]$, $[c_{ji}]$, $[c_{jj}]$, $[k_{ii}]$, $[k_{ij}]$, $[k_{ji}]$ and $[k_{jj}]$ should be added to the rows and columns corresponding to the nodal DOFs of elements $i$ and $j$ in the system equations. As was stated, all the load vectors and vehicle-induced matrices should be adjusted at each time step according to the acting positions of the vehicle axles, and removed from the columns and rows in the system matrices corresponding to each element if that element is no longer acted upon by the front or rear wheels.

The third case occurs when only the rear wheels of the vehicle are acting on the bridge, as shown in Fig. 7.6. For this case, there is no pitching actions, i.e., $[c_{ij}] = [k_{ij}] = [0]$. By setting the acting position to be $x = x_{cj}$, the vehicle nodal loads $\bar{p}_j\{N_{cj}\}$ and the vehicle-induced matrices $[m_{jj}]$, $[c_{jj}]$, $[k_{jj}]$ can be evaluated and substituted into the rows and columns corresponding to the nodal DOFs of element $j$ in the system equations. The other treatments are similar to those stated above.

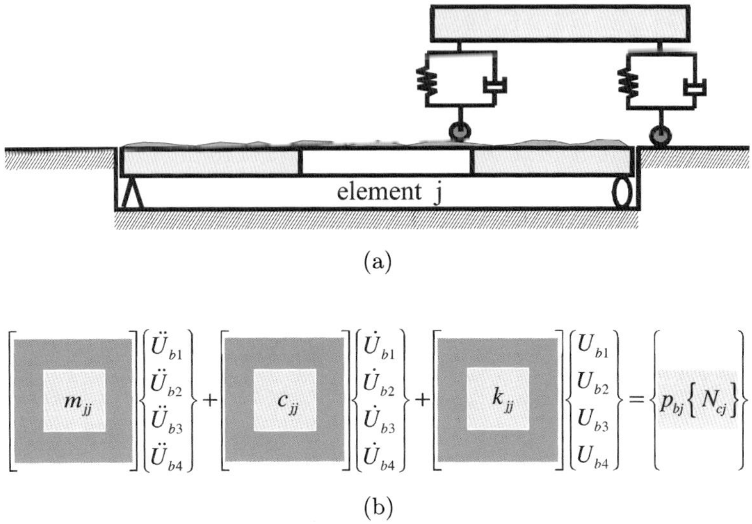


Fig. 7.6. Vehicle with rear wheels only acting on the bridge.

Based on the above considerations, the system equations can be established for the VBI system at time $t + \Delta t$ as follows:

$$[M]\{\ddot{U}\}_{t+\Delta t} + [C]\{\dot{U}\}_{t+\Delta t} + [K]\{U\}_{t+\Delta t} = \{P\}_{t+\Delta t}\,, \tag{7.30}$$

in which $\{U\}$ = the displacement vector, $\{P\}$ = the external load vector, and $[M]$, $[C]$ and $[K]$ = the mass, damping and stiffness matrices of the VBI system. By substitution of the Newmark-type equations similar to those given in Eq. (7.14) for the velocity and acceleration vectors, $\{\dot{U}\}_{t+\Delta t}$ and $\{\ddot{U}\}_{t+\Delta t}$, the preceding second-order equations can be reduced to a set of equivalent stiffness equations as

$$[\bar{K}]\{U\}_{t+\Delta t} = \{\bar{P}\}\,, \tag{7.31}$$

where the effective stiffness matrix $[\bar{K}]$ is

$$[\bar{K}] = a_0[M] + a_1[C] + [K] \tag{7.32}$$

and the effective load vector $[\bar{P}]$ is

$$\begin{aligned}\{\bar{P}\} = \{P\}_{t+\Delta t} &+ [M](a_0\{U\}_t + a_2\{\dot{U}\}_t + a_3\{\ddot{U}\}_t)\\ &+ [C](a_1\{U\}_t + a_4\{\dot{U}\}_t + a_5\{\ddot{U}\}_t)\,.\end{aligned} \tag{7.33}$$

From Eq. (7.31), the displacements $\{U\}_{t+\Delta t}$ for the bridge can be solved. Consequently, the velocities and accelerations of the bridge, i.e., $\{\dot{U}\}_{t+\Delta t}$ and $\{\ddot{U}\}_{t+\Delta t}$, can be computed using equations similar to Eq. (7.14). After the bridge responses are made available, the car body responses can be computed simply by back substitution. In this regard, one may first compute the car body forces $f_{ver}$ and $f_{rot}$ using Eq. (7.10), with which the car body displacements, velocities and accelerations can be computed using Eqs. (7.19), (7.20), and (7.22).

## 7.5. Numerical Studies

In this section, the VBI element derived in this chapter and the associated computer program developed will be verified through the study of some examples. Thus, all the examples presented in this chapter are characterized by the fact that the pitching motion of the moving vehicles is duly taken into account. The time step size used throughout all time histories of the examples is $\Delta t = 0.001$ s.

### 7.5.1. *Simple Beam Traveled by a Two-Axle System*

Consider a simple beam traveled by a two-axle system in Fig. 7.7. The frequencies of the vertical and rotational vibration of the car body are as follows (Yau and Yang, 1998):

$$\omega_v = \sqrt{\frac{2k_v}{M_v}}, \qquad \omega_\theta = \sqrt{\frac{k_v d^2}{2I_v}}, \tag{7.34}$$

where $M_v$ = the mass, $I_v$ = rotatory inertia, $k_v$ = spring constant of the car body; and $L_c$ = the axle distance. The following data are assumed: $M_v = 180$ t, $I_v = 4600$ t-m$^2$, $k_v = 13\,783$ kN/m, $L_c = 17.5$ m, and $v = 100$ km/h ($= 27.78$ m/s). Correspondingly, the frequencies for the two-axle system are: $\omega_v = 12.38$ rad/s; $\omega_\theta = 21.42$ rad/s. The following data are assumed for the beam: $E$ (Young's modulus)

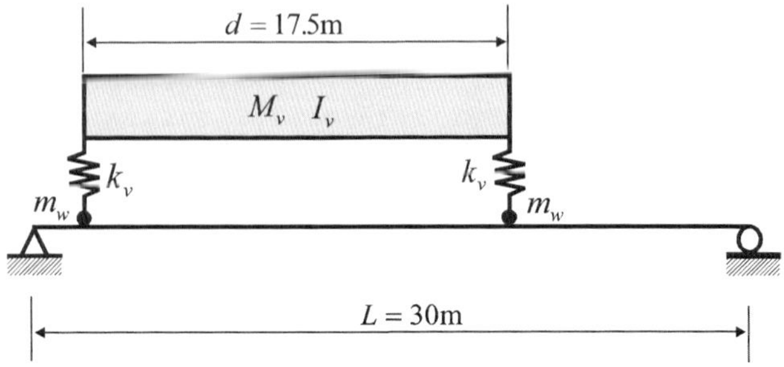


Fig. 7.7. Simple beam traveled by a two-axle system.

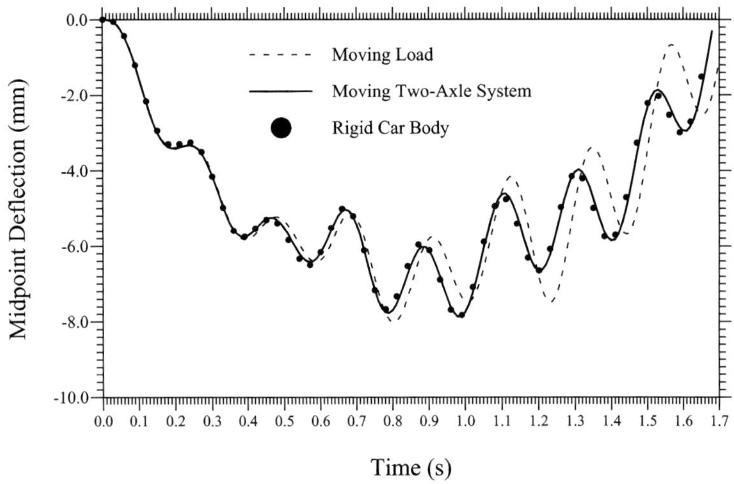


Fig. 7.8. Midpoint deflection for the two-axle system.

= 29.43 GPa; $v$ (Poisson's ratio) = 0.2; $L = 30$ m; $A$ (cross-sectional area) = 2.14 m$^2$; $I_y$ (moment of inertia) = 2.88 m$^4$; $m$ (mass per unit length) = 12 t/m; and zero damping. The first frequency computed of the bridge is $\omega_1 = 29.15$ rad/s. For the present case, the vehicle-to-bridge mass ratio is $M_v/mL = 0.5$. In the finite element analysis, 10 elements were used for the beam.

The midpoint displacement of the beam calculated using the present VBI element has been plotted in Fig. 7.8, along with the analytical solution given by Yau and Yang (1998) for the two-axle system and the solution of the moving load model, obtained by setting the suspension stiffness $k_v$ of the sprung mass model equal to a very large value. It is observed that the present solution agrees very well with the analytical one, from which the accuracy of the present VBI element in simulating the vehicle response is confirmed. However, slight deviation exists between the present solution and the one based on the moving load model. This can be attributed to the fact that the moving load model loses its accuracy for large vehicle-to-bridge mass ratios; for the present case, the mass ratio is 0.5.

### 7.5.2. *Simple Beam Traveled by a Train Consisting of Five Identical Cars*

Figure 7.9 shows a train consisting of five identical cars, which travels over a simple beam represented by 10 elements. The following data are assumed for each car: $M_v = 48$ t, $I_v = 2500$ t-m$^2$, $k_v = 1500$ kN/m, $c_v = 85$ kN-s/m, $m_w = 5$ t, $L_c$ (axle distance) $=$ 18 m, and $d$ (car length) $= 25$ m. The frequencies of vibration calculated for the car body are: $\omega_v = 7.9$ rad/s; $\omega_\theta = 9.86$ rad/s. The data assumed below for the simple beam are close to those used in the high-speed railway bridges under construction in Taiwan: $E = 29.43$ GPa; $v = 0.2$; $L = 30$ m; $A = 7.94$ m$^2$; $I_z = 8.72$ m$^4$; $m = 36$ t/m; and damping coefficient $= 2.5\%$. The first frequency computed of the bridge is $\omega_1 = 29.30$ rad/s.

For illustration, the time-history response of the midpoint displacement of the bridge traveled by the train at speeds $v = 140$ and

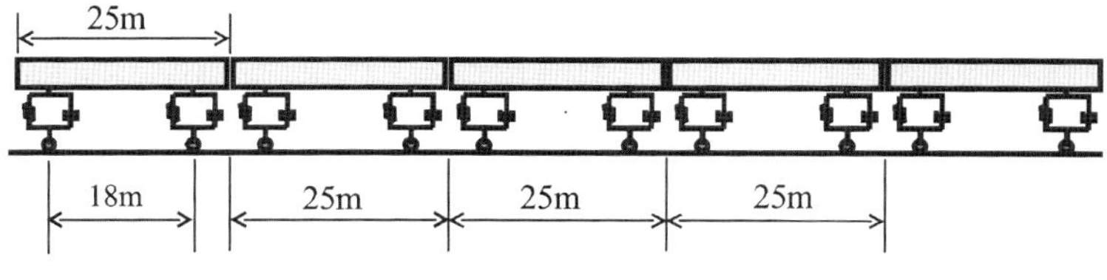


Fig. 7.9. A train consisting of five identical cars.

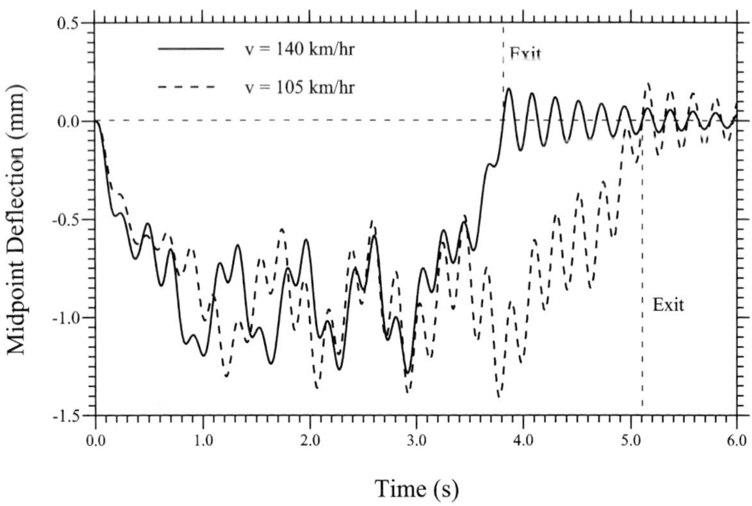


Fig. 7.10. Time-history responses of the bridge excited by moving trains.

105 km/h have been plotted in Fig. 7.10. Assuming that the trains enter the bridge at $t = 0$ s, they will leave the bridge at $t = 3.81$ s and 5.07 s for the speeds 140 and 105 km/h, respectively, which were also indicated in the figure. It is interesting to note that the maximum response of the bridge excited by the train moving at the lower speed, i.e., $v = 105$ km/h, is higher than that at the higher speed, i.e., $v = 140$ km/h. The reason can be given as follows. From Eq. (2.59) of Chapter 2, it is known that the dimensionless resonant speed is $S_r = d/2nL$, where $d$ is the car length, $L$ the bridge span length, and $n$ a positive integer. With $d = 25$ m, $L = 30$ m, $\omega_1 = 29.30$ rad/s, one can obtain the resonant speed from Eq. (1.4) as $v_r = 116.6/n$ (m/s) $= 419.7/n$ (km/h). For $n = 3$, 4, the resonant speed $v_r$ turns out to be 140, 105 km/h, implying that the two speeds analyzed equal the resonant speeds. On the other hand, from Eq. (2.61) of Chapter 2, it is also known that the dimensionless speed for the waves generated by the continuously moving axles loads to cancel each other is $S_c = 1/(2i - 1)$, where $i$ is a positive integer. Equivalently, the cancellation speed is $v_c = 279.8/(2i - 1)$ (m/s) $= 1007/(2i - 1)$ (km/h). Here, it can be seen that the traveling speed 140 km/h of the train

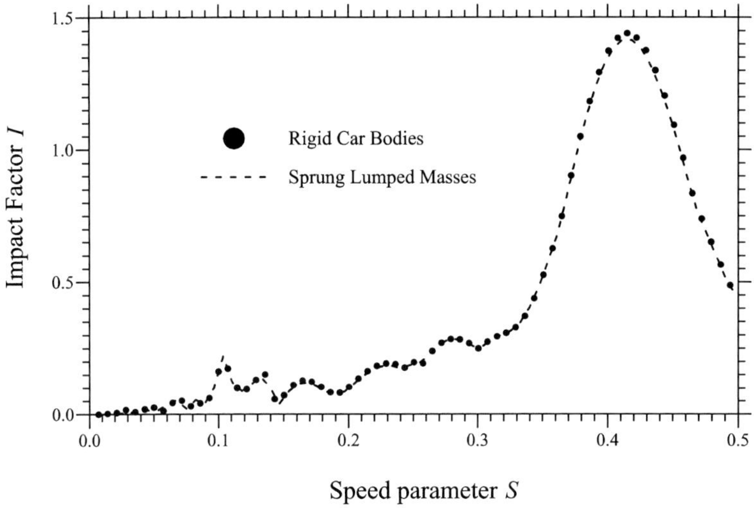


Fig. 7.11. Impact factor for midpoint displacement of simple beam.

is close to the cancellation speed of 144 km/h obtained by letting $i = 4$, implying that the bridge response excited by the train moving at 140 km/h will be significantly suppressed, as the cancellation condition is approximately met. However, the response of the bridge traversed by the train at 105 km/h will not be suppressed, since it is not close or equal to any cancellation speed. This explains why the bridge response is greater for the train traveling at speed 105 km/h, rather than at 140 km/h.

The deflection response computed for the midpoint of the beam has been plotted with respect to the speed parameter $S$ in Fig. 7.11, in which the solution obtained alternatively by modeling each vehicle as two discrete sprung masses was also shown. One observation herein is that for the case with no track irregularities, the influence of the vehicle's pitching effect on the bridge response can basically be neglected. From Fig. 7.11, the critical speed parameter for the resonant response to occur is found as $S_r = 0.417$, which is consistent with that predicted analytically using Eq. (2.59) of Chapter 2: $S = d/2nL$ with $n = 1$. Correspondingly, the resonant velocity of the train can be computed from Eq. (1.4) as $v_r = 120$ km/h.

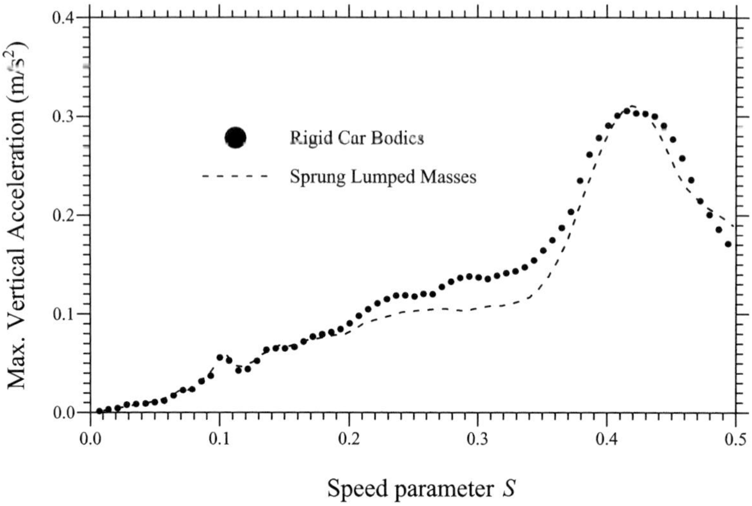


Fig. 7.12. Maximum vertical acceleration of train with no track irregularities.

The vertical acceleration of the moving vehicles has been regarded as an indicator of the *riding comfort* or runnability of high-speed trains. Due to the stringent requirements placed for the bridge response and riding comfort, the design of high-speed railroad bridges is generally governed by the conditions of serviceability, rather than by strength. For the present example, the maximum vertical acceleration computed for the train versus the speed parameter $S$ has been plotted in Fig. 7.12, along with the solution based on the sprung mass model. As can be seen, the omission of the effect of vehicle pitching or the interaction between the front and rear wheels, as implied by the sprung mass model, may result in significant underestimate of the vehicle response even in the absence of track irregularities, which is not conservative from the design point of view. The effect of pitching of the car body appears to be especially important in the range from $S = 0.2$ to $S = 0.4$, which may be encountered by most high-speed trains.

In this subsection and those to follow, the maximum vertical acceleration of the train denotes the maximum of the vertical accelerations computed for all parts of each vehicle constituting the train.

Therefore, it should be interpreted as the absolute vertical acceleration of the train.

### 7.5.3. *Riding Comfort in the Presence of Track Irregularities*

For the present purposes, the irregularity function proposed by Nielsen and Abrahamsson (1992) for the vertical profile of the railroad is adopted:

$$r(x) = -r_0 \left[1 - \exp\left(-\frac{x}{x_0}\right)^3\right] \sin\frac{2\pi x}{\gamma_0}, \tag{7.35}$$

where $x$ is the along-track distance (in m), $x_0 = 1.0$ m, $r_0$ (= 0.5 mm) = amplitude of irregularities, and $\gamma_0$ (= 1.0 m) = wavelength of the corrugation. The train model and bridge model are identical to those of the preceding example. The maximum vertical accelerations of the train computed for the cases with and with no track irregularities have been compared in Fig. 7.13. As can be seen, even for the case with very small track irregularities, i.e., with $r_0 = 0.5$ mm, significant amplification on the vehicle acceleration can be observed, which is harmful to the riding comfort and, in some cases, to the running safety of the high-speed trains. Obviously, the importance of maintaining a smooth track surface in high-speed railroad engineering cannot be overstressed.

### 7.5.4. *Effect of Elasticity of the Suspension System*

Let us consider again the problem of a train consisting of five identical cars traveling over a simply-supported bridge. The data for the train and bridge are identical to those used in Sec. 7.5.2. To investigate the effect of elasticity of the suspension system, three values of suspension stiffness, i.e., $k_v$, $2k_v$, $3k_v$, are assumed. The results computed for the impact factor of the midpoint displacement of the bridge, the maximum vertical acceleration and maximum rotational acceleration of the train have been plotted against the speed parameter $S$ in Figs. 7.14–7.16. From Fig. 7.14, one observes that the impact

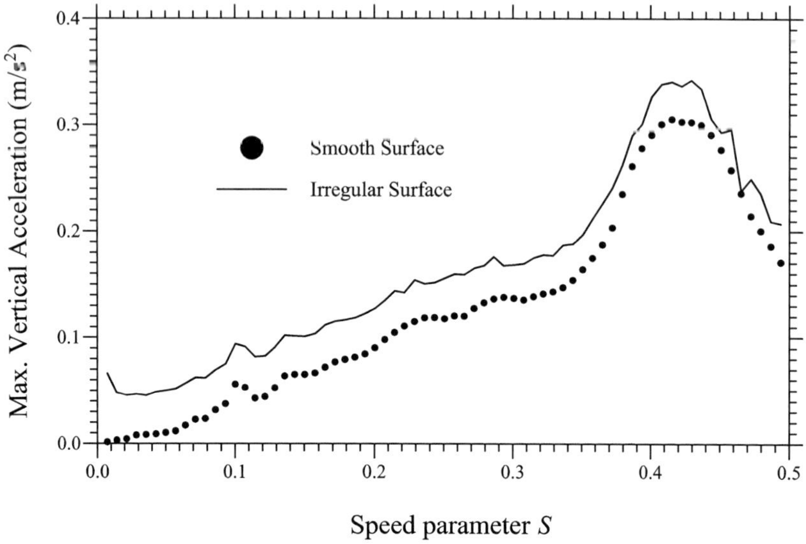


Fig. 7.13. Maximum vertical acceleration of train with track irregularities.

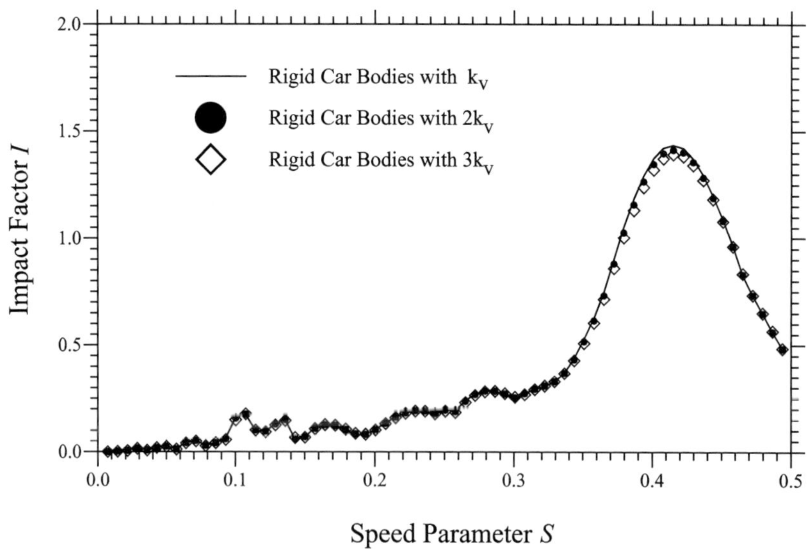


Fig. 7.14. Impact factor for the midpoint displacement of the bridge.

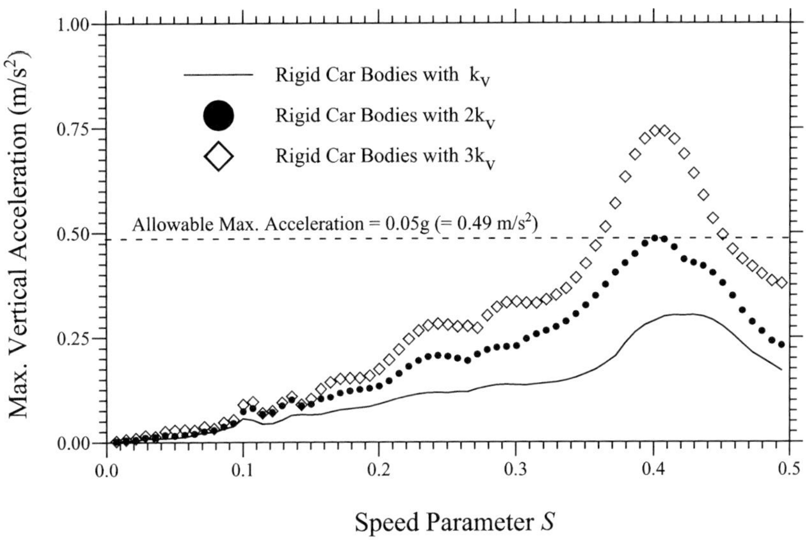


Fig. 7.15. Maximum vertical acceleration of the train.

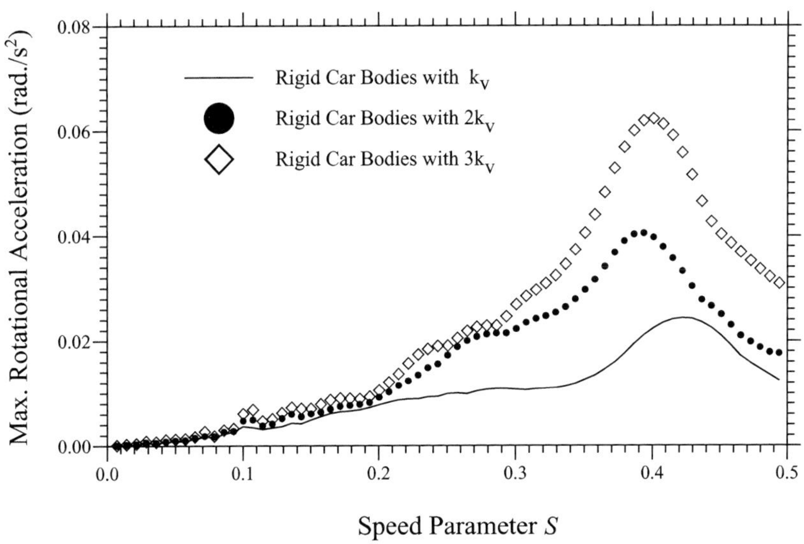


Fig. 7.16. Maximum rotational acceleration of the train.

factor for the midpoint displacement of the bridge increases generally following the increase in the speed parameter $S$ of the train, and reaches the peak when $S = 0.417$ for all the three values of suspension stiffness considered. In general, the influence of the suspension stiffness on the bridge response is quite little, although larger values of suspension stiffness can induce marginally smaller bridge response.

Concerning the train vibrations, it can be seen from Figs. 7.15 and 7.16 that the suspension stiffness can drastically affect the train response. The larger the suspension stiffness, the greater is the train response, which is consistent with our common understanding that it is uncomfortable to ride on a vehicle with flat tires. The speed parameter for the peak response to occur is around $S = 0.417$ and becomes smaller for larger suspension stiffness. Such a trend is valid for both the vertical acceleration and rotation acceleration of the train. In fact, the allowable maximum acceleration of 0.05 g (= 0.49 m/s$^2$) previously used by the Taiwan High-Speed Railway was exceeded in the case with $3k_v$.

### 7.5.5. *Effect of Damping of the Suspension System*

Same data as those used in Sec. 7.5.2 for the train and bridge are adopted herein. Let us consider three values of suspension damping, i.e., $c_v$, $2c_v$, $3c_v$. The results computed for the impact response of the midpoint displacement of the bridge, the maximum vertical acceleration and maximum rotational acceleration of the train were shown in Figs. 7.17–7.19. As can be seen from Fig. 7.17, the general trend for the bridge midpoint displacement impact is similar to the one shown in Fig. 7.14, except that larger suspension damping will result in slightly smaller impact response for the bridge. From Figs. 7.18 and 7.19, one observes that the suspension damping can affect significantly the train response. The larger the damping coefficient of the suspension system, the greater is the train response, whether it is vertical or rotational acceleration. The magnification effect of suspension damping on the train response seems to be coupled by the resonance effect, as indicated by the drastic increase in train response for the speed parameter range with $S$ greater than 0.32.

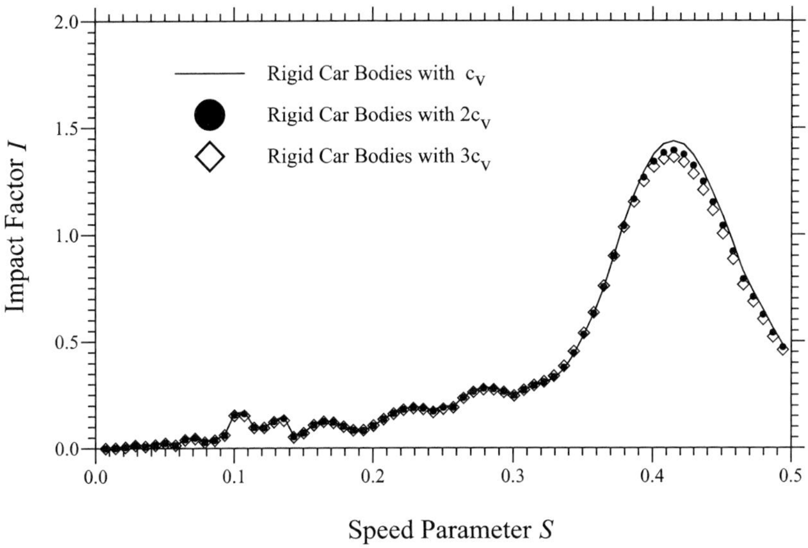


Fig. 7.17. Impact factor of the midpoint displacement of the bridge.

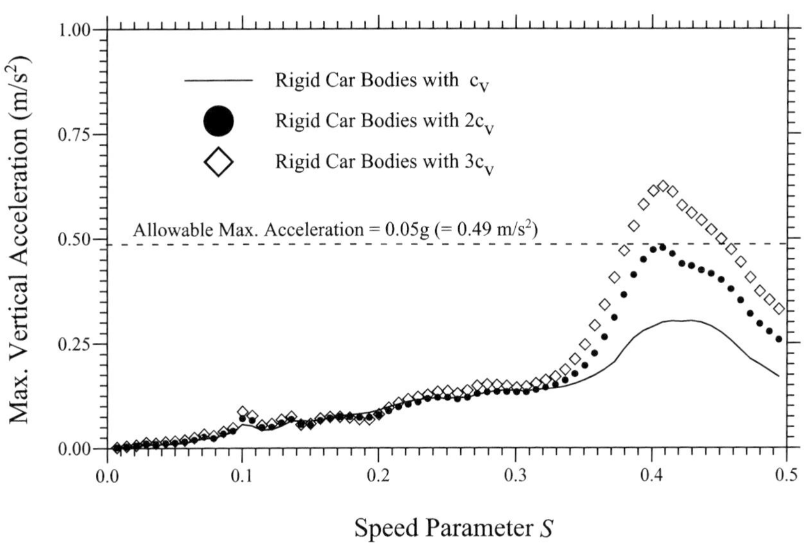


Fig. 7.18. Maximum vertical acceleration of the train.

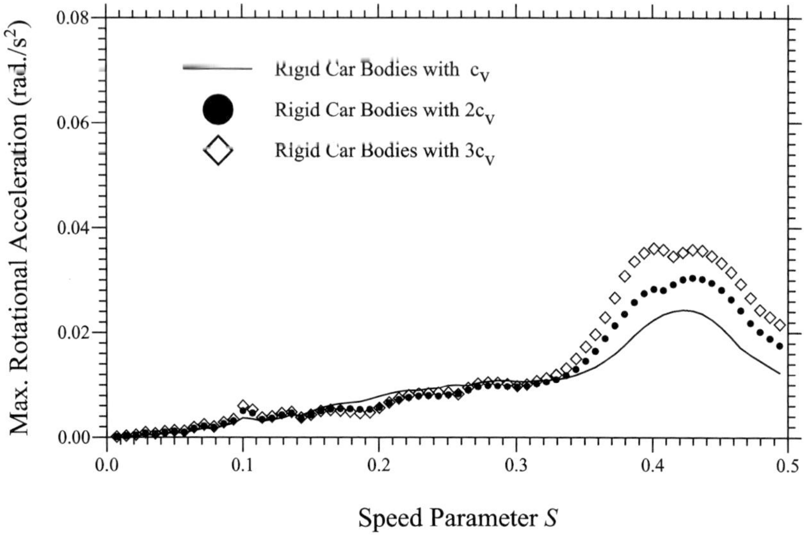


Fig. 7.19. Maximum rotational acceleration of the train.

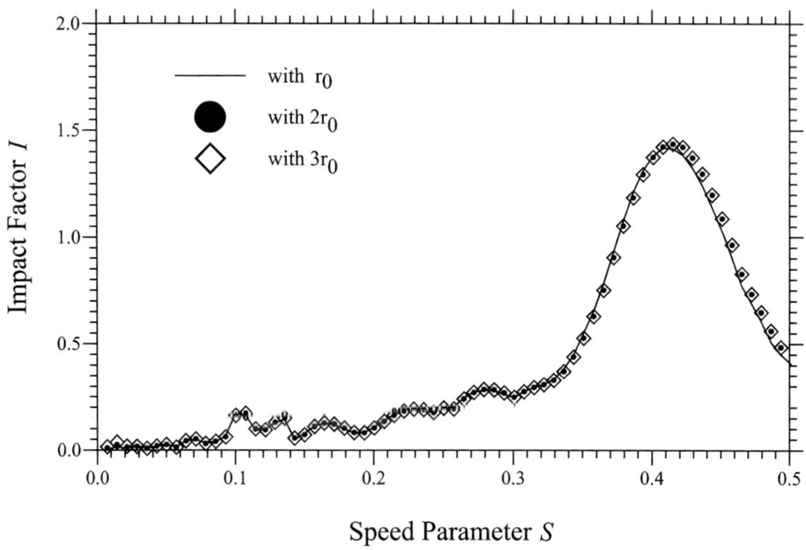


Fig. 7.20. Impact factor for the midpoint displacement of the bridge.

### 7.5.6. *Effect of Track Irregularity*

Again, the same data as those used in Sec. 7.5.2 for the train and bridge are adopted herein. To investigate the effect of track irregularity, we shall consider three values of irregularity amplitude $r_0$, $2r_0$, $3r_0$. In Figs. 7.20–7.22, the impact factor of the midpoint displacement of the bridge, the maximum vertical acceleration and the maximum rotational acceleration of the train have been plotted with respect to the speed parameter $S$. As can be seen from Fig. 7.20, the influence of track irregularity on the bridge response is so small that it can virtually be neglected. On the other hand, from Figs. 7.21 and 7.22, one observes that the increase in the magnitude of track irregularity will result in significant magnification of the train response, no matter it is vertical or rotational acceleration.

## 7.6. Concluding Remarks

In this chapter, a vehicle is modeled as a rigid beam supported by two suspension units, and a bridge by a number of beam elements. The equations of motion written for the rigid beam (actually the car body) are first reduced to a set of equivalent stiffness equations using Newmark's finite difference equations, by which the car body DOFs are condensed to those of the bridge elements in contact, following basically the procedures in Chapter 6. The VBI elements derived are characterized by the fact that the pitching or linking action of the car body is duly taken into account, while the property of symmetry is retained in the system matrices. The applicability of the derived element has been demonstrated in the numerical studies concerning both the bridge and the vehicle responses.

Some conclusions can be made from the numerical studies conducted in this chapter. (1) For the case with no track irregularities, the pitching effect of vehicles on the bridge response can be generally neglected. (2) Wherever the vehicle response is of major concern, it is necessary to consider the effect of pitching of the car body. The use of sprung mass model for the vehicles is inadequate and not conservative from the point of view of design. (3) The resonant speed

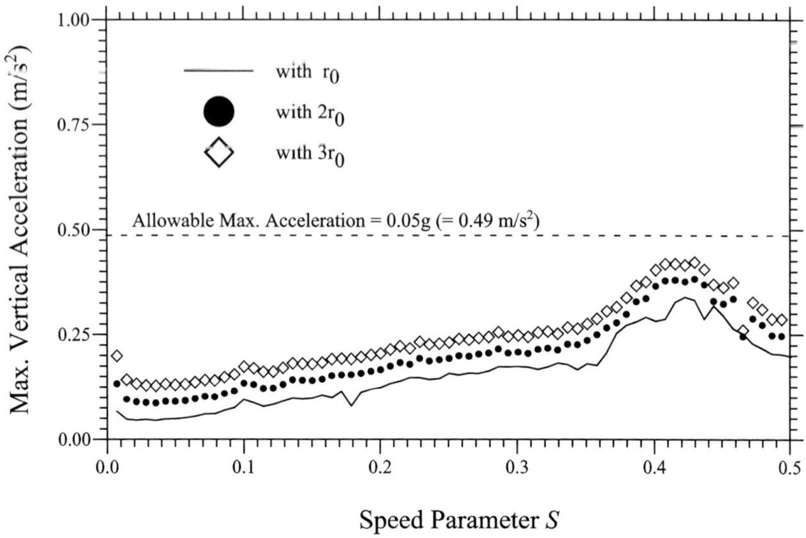


Fig. 7.21. Maximum vertical acceleration of the train.

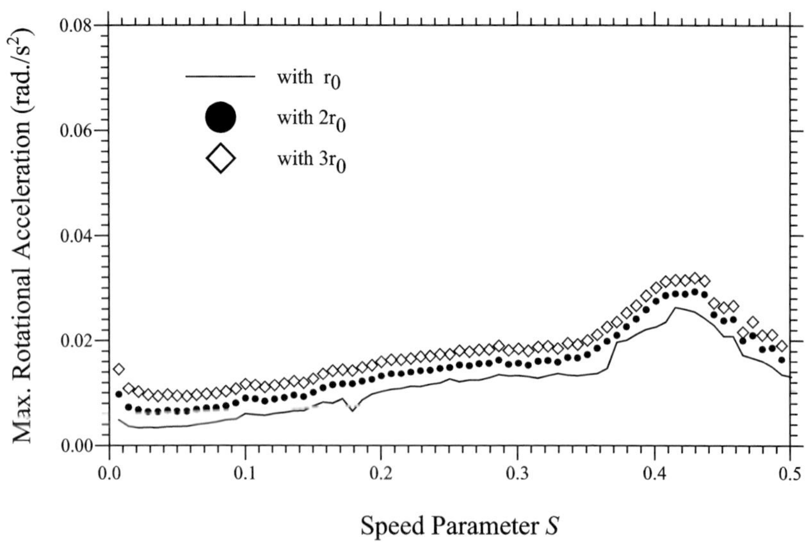


Fig. 7.22. Maximum rotational acceleration of the train.

parameter predicted using the present finite element model agrees very well with that predicted analytically. (4) The vehicle response will be drastically amplified in the presence of track irregularities and through consideration of the pitching effect (i.e., using the present VBI elements), which is harmful to both the riding comfort and running safety of high-speed trains. (5) The effects of both elasticity and damping of the suspension system on the bridge response are so small that they can be virtually neglected. However, increasing the stiffness or damping coefficient of the suspension system tends to increase largely the maximum train response. (6) The trend for the track irregularity is similar to that for the suspension stiffness or damping concerning the bridge and train responses. Naturally, greater magnitude of track irregularity will induce larger train response.

## Chapter 8

# Modeling of Vehicle–Bridge Interactions by the Concept of Contact Forces

Unlike Chapter 6 that deals with the vehicle–bridge interactions (VBIs) by condensing the vehicle degrees of freedom (DOFs) to the bridge DOFs, in this chapter we present a procedure that utilizes the contact forces as the interface for treating the same problem. Central to the present formulation is the adoption of Newmark's finite difference scheme for discretizing the vehicle equations of motion, by which the contact forces are solved and expressed in terms of the wheel displacements. Through the use of no-jump condition for vehicles, the contact forces can then be related to the displacements of the contact points of the bridge. As such, a VBI element that considers all the interaction effects can be derived from the bridge equations, with which the vehicle response, contact forces and bridge response can be computed with no iterations required. The present procedure is versatile in that there is virtually no limit on the level of complexity of the vehicles simulated, which may range from the moving load, moving mass, sprung mass, to suspended rigid bar, and so on. The capability of the present procedure is demonstrated in the study of a number of VBI problems, including those caused by vehicles in braking.

## 8.1. Introduction

The vibration of bridges caused by the moving vehicles or trains has been a subject of continuous research since the nineteenth century, as was revealed by the review presented in Chapter 1. In the past three decades, due to the construction of high-speed railways and the upgrading of existing railways worldwide, the problem of train–bridge interactions has received more attention from engineers and researchers than ever (Richardson and Wormley, 1974; Matsuura, 1976; Cherchas, 1979; Chu *et al.*, 1979; Aida *et al.*, 1990; Wakui *et al.*, 1995; Yang *et al.*, 1997b). Previously, vehicles have often been approximated as moving loads, which in many cases offer a feasible means for obtaining solutions in closed form. To consider the inertial effect of the moving vehicles, the moving mass model has been adopted instead (Stanišić and Hardin, 1969; Ting *et al.*, 1974; Sadiku and Leipholz, 1987; Akin and Mofid, 1989). However, for the case where the *riding comfort* or vehicle response is of concern, it is necessary to consider the effect of suspension systems of the vehicles as well. The simplest model that can be conceived in this regard is a lumped mass supported by a spring-dashpot unit, referred to as the sprung mass model (Tan and Shore, 1968b; Genin *et al.*, 1975; Blejwas *et al.*, 1979). Although more sophisticated models can still be devised for simulating the vehicle structures, the efficiency of solution of the vehicle–bridge interaction (VBI) system becomes an issue of great concern, especially when there exists a large number of vehicles, e.g., for the case of a train consisting of a number of vehicles in connection. By taking into account the VBI, it is realized that all the contact forces vary not only in magnitudes, but also in acting positions.

In analyzing the VBI systems, two sets of second-order differential equations can be written each for the vehicles and for the bridge. It is the *interaction forces* or *contact forces* existing at the contact points that make the two subsystems coupled. As the contact points move from time to time, the system matrices are time-dependent and must be updated and factorized at each time step in an incremental time-history analysis. To solve these two sets of equations, procedures of iterative nature are often adopted (Hwang and Nowak, 1991; Yang and Fonder, 1996; Yang and Yau, 1998). For instance, by first assuming the displacements for the contact points, one can solve the vehicle equations to obtain the interaction (contact) forces and then proceed to solve the bridge equations for improved values of displacements for the contact points. This completes the first cycle of iteration. One drawback with methods of iterative nature is that the convergence rate is likely to be low when dealing with the more realistic case of a bridge sustaining a large number of vehicles in motion. In the literature, Lagrange's equation with multipliers and constraint equations has also been used (Blejwas *et al.*, 1979). However, the use of Lagrange multipliers increases the number of unknowns and thus the effort of computation, especially for problems involving a large number of moving vehicles.

Still, another category of methods exists for solving the VBI problems, e.g., those based on the condensation method. Garg and Dukkipati (1984) used the Guyan reduction technique to condense the vehicle degrees of freedom (DOFs) to the associated bridge DOFs. In the paper by Yang and Lin (1995), the dynamic condensation method was used to eliminate all the vehicle DOFs on the element level. These methods have been demonstrated to be efficient for computing the bridge response. However, because of the approximations made in relating the vehicle (slave) DOFs to the bridge (master) DOFs, they are not adequate for computing the vehicle response, which serves as an indicator of the riding comfort generally required in the design of high-speed railway bridges. By using the Newmark finite difference scheme to discretize the vehicle equations, rather accurate *master–slave relations* have been established and used in eliminating the vehicle DOFs from the bridge equations. Such relations have been employed in deriving the VBI elements in Chapters 6 and 7, which are good for computing both the vehicle and bridge responses.

The objective of this chapter is to develop a more general approach for solving both the vehicle and bridge responses that are of particular interest in high-speed railways, but not to study solely the bridge impact response, as was the case for highway bridges (Nassif and Nowak, 1995; Kim and Nowak, 1997). First of all, the second-order differential equations for the moving vehicles will be rendered into a set of equivalent stiffness equations using Newmark's finite difference scheme, as presented in Chapter 6. However, instead of establishing the master–slave relations, we shall concentrate on computation of the contact forces from the equivalent vehicle equations, which are first expressed in terms of the wheel displacements, and then, through enforcement of the constraint equations, related to the bridge displacements at contact points. With the contact forces made available, the VBI element can be derived from the bridge equations by treating the contact forces as consistent nodal loads. The advantage of the present procedure is its *versatility* in dealing with vehicle structures with practically no limit on the level of complexity, which may range from the moving loads, moving masses, sprung masses, to rigid car bodies supported by spring-dashpot units, and so on. The capability and reliability of the present procedure will be demonstrated in the study of a number of examples close to those encountered in practice. The materials presented in this chapter follow basically those of Yang and Wu (2001).

## 8.2. Vehicle Equations and Contact Forces

In this study, matrices, column and row vectors will be represented by quantities enclosed by [ ], { } and ⟨ ⟩, respectively. Consider a simply-supported beam that is under the action of a moving vehicle in Fig. 8.1. The vehicle is assumed to be composed of two parts.

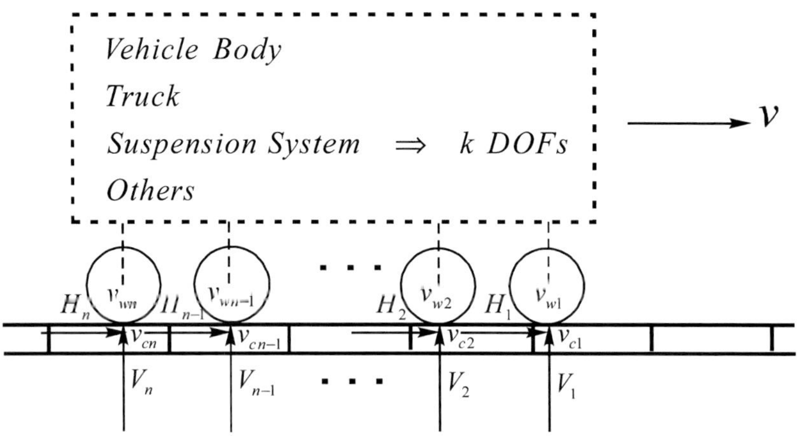


Fig. 8.1. Schematic of vehicle–bridge interaction.

The *upper* or noncontact part consists of the car body, suspension systems, and bogies, which has a total of $k$ DOFs, as indicated by the vector $\{d_u\}$. The *wheel* or contact part consists of $n$ wheelsets. Assuming that each wheelset is represented by one vertical DOF, the wheel part can be denoted as $\{d_w\} = \langle v_{w1} v_{w2} \ldots v_{wi} \ldots v_{wn}\rangle^T$, where $v_{wi}$ denotes the displacement of the $i$th wheel. Correspondingly, there exist $n$ contact points on the bridge, of which the displacement may be denoted as $\{d_c\} = \langle v_{c1} v_{c2} \ldots v_{ci} \ldots v_{cn}\rangle^T$, where $v_{ci}$ denotes the displacement at the $i$th contact point.

Let $[m_v]$, $[c_v]$, and $[k_v]$ respectively denote the mass, damping and stiffness matrices of the vehicle, and $\{d_v\}$ the displacement vector of the vehicle, i.e., $\{d_v\} = \langle\{d_u\}\{d_w\}\rangle^T$. The following is the equation of motion for the vehicle:

$$[m_v]\{\ddot{d}_v\} + [c_v]\{\dot{d}_v\} + [k_v]\{d_v\} = \{f_v\}, \tag{8.1}$$

where $\{f_v\}$ is the force vector, which can be decomposed into two parts,

$$\{f_v\} = \{f_e\} + [l]\{f_c\}. \tag{8.2}$$

Here, $\{f_e\}$ denotes the external force components excluding the contact forces, $\{f_c\}$ the contact forces acting through the wheels, $\{f_c\} = \langle V_1 V_2 \ldots V_i \ldots V_n\rangle^T$, where $V_i$ is the force acting at the $i$th contact point of the bridge, and $[l]$ is a transformation matrix. The wheel displacements $\{d_w\}$ can be related to the contact displacements $\{d_c\}$ of the bridge by the *constraint conditions*,

$$\{d_w\} = [\Gamma]\{d_c\} + \{r\}, \tag{8.3}$$

where $[\Gamma]$ should be interpreted as a unit matrix for the case where *no jumps* occur between the vehicle's wheels and the bridge, as is considered herein, and $\{r\}$ is a vector representing the rail irregularity (or pavement roughness for highway bridges) at the contact points. The irregularity or roughness vector $\{r\}$ depends on the condition of contact between the wheels and bridge and the level of complexity of the problem considered, that is, whether the problem is 2D or 3D in nature, as will be studied in depth in Chapters 9 and 10 to follow.

In this section and the section to follow, only contact forces that act along the vertical (gravity) direction will be considered. The procedure presented in these two sections will be generalized in Sec. 8.5 to include the horizontal contact forces, which may be generated as the frictional forces between the wheels and rails.

The VBI system will be analyzed in an incremental manner in time domain. Assuming that all the information of the system at time $t$ is known and $\Delta t$ is a small time increment, we are interested in the behavior of the system at time $t + \Delta t$. The vehicle equations given in Eqs. (8.1) can be rewritten for time $t + \Delta t$ with partitions in matrices for the *upper* and *wheel* parts as

$$\begin{bmatrix} [m_{uu}] & [m_{uw}] \\ [m_{wu}] & [m_{ww}] \end{bmatrix} \begin{Bmatrix} \{\ddot{d}_u\} \\ \{\ddot{d}_w\} \end{Bmatrix}_{t+\Delta t} + \begin{bmatrix} [c_{uu}] & [c_{uw}] \\ [c_{wu}] & [c_{ww}] \end{bmatrix} \begin{Bmatrix} \{\dot{d}_u\} \\ \{\dot{d}_w\} \end{Bmatrix}_{t+\Delta t} + \begin{bmatrix} [k_{uu}] & [k_{uw}] \\ [k_{wu}] & [k_{ww}] \end{bmatrix} \begin{Bmatrix} \{d_u\} \\ \{d_w\} \end{Bmatrix}_{t+\Delta t} = \begin{Bmatrix} \{f_{ue}\} \\ \{f_{we}\} \end{Bmatrix}_{t+\Delta t} + \begin{bmatrix} [l_u] \\ [l_w] \end{bmatrix} \{f_c\}_{t+\Delta t} , \tag{8.4}$$

where $\{f_{ue}\}$ and $\{f_{we}\}$ respectively denote the external forces acting on the upper and wheel parts of the vehicle. The first row in Eq. (8.4) relates to the behavior of the upper or non-contact part of the vehicle, and the second row of the wheels or contact part. Since only the wheels will be acted upon by the contact forces, the submatrix $[l_u]$ should be set to $[0]$. Expanding the first row of Eq. (8.4) yields

$$[m_{uu}]\{\ddot{d}_u\}_{t+\Delta t} + [c_{uu}]\{\dot{d}_u\}_{t+\Delta t} + [k_{uu}]\{d_u\}_{t+\Delta t} = \{f_{ue}\}_{t+\Delta t} - \{q_{uc}\}_{t+\Delta t} , \tag{8.5}$$

where

$$\{q_{uc}\}_{t+\Delta t} = [m_{uw}]\{\ddot{d}_w\}_{t+\Delta t} + [c_{uw}]\{\dot{d}_w\}_{t+\Delta t} + [k_{uw}]\{d_w\}_{t+\Delta t} . \tag{8.6}$$

Let $\{\Delta d_u\}$ denote the increment in the upper-part vehicle displacement $\{d_u\}$ occurring during the time step from $t$ to $t+\Delta t$. By Newmark's finite difference scheme (see Appendix B), the vector $\{d_u\}$ and its derivatives at the instant $t+\Delta t$ can be related to those at time $t$ as

$$\begin{aligned}
\{\ddot{d}_u\}_{t+\Delta t} &= b_0\{\Delta d_u\} - b_1\{\dot{d}_u\}_t - b_2\{\ddot{d}_u\}_t\,,\\
\{\dot{d}_u\}_{t+\Delta t} &= \{\dot{d}_u\}_t + b_3\{\ddot{d}_u\}_t + b_4\{\ddot{d}_u\}_{t+\Delta t}\,,\\
\{d_u\}_{t+\Delta t} &= \{d_u\}_t + \{\Delta d_u\}\,,
\end{aligned} \tag{8.7}$$

where the variables with subscript $t$ denote the quantities occurring at time $t$, which are assumed to be known. Using Newmark's parameters $\beta$ and $\gamma$, the coefficients and those to be used later can be given as

$$\begin{gathered}
b_0 = \frac{1}{\beta\Delta t^2}\,, \quad b_1 = \frac{1}{\beta\Delta t}\,, \quad b_2 = \frac{1}{2\beta} - 1\,,\\
b_3 = (1-\gamma)\Delta t\,, \quad b_4 = \gamma\Delta t\,, \quad b_5 = \frac{\gamma}{\beta\Delta t}\,,\\
b_6 = \frac{\gamma}{\beta} - 1\,, \quad b_7 = \frac{\Delta t}{2}\left(\frac{\gamma}{\beta} - 2\right),
\end{gathered} \tag{8.8}$$

which are equivalent to the coefficients $a_0 \sim a_7$ presented in Appendix B. Substituting Eqs. (8.7) into the differential equations for the upper part of the vehicle in Eq. (8.5), one obtains after some manipulations the following,

$$[\Psi_{uu}]\{\Delta d_u\} = \{f_{ue}\}_{t+\Delta t} - \{q_{uc}\}_{t+\Delta t} + \{q_u\}_t\,, \tag{8.9}$$

where

$$[\Psi_{uu}] = b_0[m_{uu}] + b_5[c_{uu}] + [k_{uu}]\,, \tag{8.10a}$$

$$\begin{aligned}
\{q_u\}_t = {} & [m_{uu}](b_1\{\dot{d}_u\}_t + b_2\{\ddot{d}_u\}_t)\\
& + [c_{uu}](b_6\{\dot{d}_u\}_t + b_7\{\ddot{d}_u\}_t) - [k_{uu}]\{d_u\}_t\,.
\end{aligned} \tag{8.10b}$$

From Eq. (8.9), the displacement increments $\{\Delta d_u\}$ for the upper part of the vehicle can be solved as

$$\{\Delta d_u\} = [\Psi_{uu}]^{-1}(\{f_{ue}\}_{t+\Delta t} - \{q_{uc}\}_{t+\Delta t} + \{q_u\}_t)\,. \tag{8.11}$$

With the use of Eq. (8.11), the displacement vector $\{d_u\}_{t+\Delta t}$ and its derivatives for the upper part of the vehicle can be obtained as

$$\begin{aligned}
\{\ddot{d}_u\}_{t+\Delta t} &= b_0[\Psi_{uu}]^{-1}(\{f_{ue}\}_{t+\Delta t} - \{q_{uc}\}_{t+\Delta t} + \{q_u\}_t) \\
&\quad - b_1\{\dot{d}_u\}_t - b_2\{\ddot{d}_u\}_t\,, \\
\{\dot{d}_u\}_{t+\Delta t} &= b_5[\Psi_{uu}]^{-1}(\{f_{ue}\}_{t+\Delta t} - \{q_{uc}\}_{t+\Delta t} + \{q_u\}_t) \\
&\quad - b_6\{\dot{d}_u\}_t - b_7\{\ddot{d}_u\}_t\,, \\
\{d_u\}_{t+\Delta t} &= [\Psi_{uu}]^{-1}(\{f_{ue}\}_{t+\Delta t} - \{q_{uc}\}_{t+\Delta t} + \{q_u\}_t) + \{d_u\}_t\,,
\end{aligned} \tag{8.12}$$

of which the order of accuracy is the same as that implied by the Newmark finite difference equations presented in Eq. (8.7).

## 8.3. Solution of Contact Forces from Vehicle Equations

One key step in solution of the VBI systems is to solve for the contact forces existing between the two subsystems, i.e., the moving vehicles and the bridge. By substituting Eq. (8.12) into the second row of the vehicle equations in Eq. (8.4), one obtains the *contact forces* $\{f_c\}_{t+\Delta t}$ as

$$\begin{aligned}
\{f_c\}_{t+\Delta t} &= [m_c]\{\ddot{d}_w\}_{t+\Delta t} + [c_c]\{\dot{d}_w\}_{t+\Delta t} + [k_c]\{d_w\}_{t+\Delta t} \\
&\quad + \{p_c\}_{t+\Delta t} + \{q_c\}_t\,,
\end{aligned} \tag{8.13}$$

where the contact matrices $[m_c]$, $[c_c]$ and $[k_c]$ are

$$\begin{aligned}
[m_c] &= [l_w]^{-1}([m_{ww}] - [\Psi_{wu}][\Psi_{uu}]^{-1}[m_{uw}])\,, \\
[c_c] &= [l_w]^{-1}([c_{ww}] - [\Psi_{wu}][\Psi_{uu}]^{-1}[c_{uw}])\,, \\
[k_c] &= [l_w]^{-1}([k_{ww}] - [\Psi_{wu}][\Psi_{uu}]^{-1}[k_{uw}])\,,
\end{aligned} \tag{8.14}$$

the load vectors $\{p_c\}_{t+\Delta t}$ and $\{q_c\}_t$ are

$$\begin{aligned}\{p_c\}_{t+\Delta t} &= [l_w]^{-1}([\Psi_{wu}][\Psi_{uu}]^{-1}\{f_{ue}\}_{t+\Delta t} - \{f_{we}\}_{t+\Delta t}),\\ \{q_c\}_t &= [l_w]^{-1}([\Psi_{wu}][\Psi_{uu}]^{-1}\{q_u\}_t - \{q_w\}_t),\end{aligned} \tag{8.15}$$

and

$$[\Psi_{wu}] = b_0[m_{wu}] + b_5[c_{wu}] + [k_{wu}], \tag{8.16a}$$

$$\begin{aligned}\{q_w\}_t &= [m_{wu}](b_1\{\dot{d}_u\}_t + b_2\{\ddot{d}_u\}_t)\\ &\quad + [c_{wu}](b_6\{\dot{d}_u\}_t + b_7\{\ddot{d}_u\}_t) - [k_{wu}]\{d_u\}_t.\end{aligned} \tag{8.16b}$$

From Eq. (8.13), it can be seen that the contact forces $\{f_c\}_{t+\Delta t}$ depend not only on the wheel response and the forces acting on the vehicle at time $t+\Delta t$, but also on those at time $t$. By the constraint condition of no jumps for the vehicles in Eq. (8.3), i.e., $\{d_w\} \equiv \{d_c\}$, the contact forces $\{f_c\}_{t+\Delta t}$ can be expressed in terms of the contact displacements $\{d_c\}$ of the bridge as follows:

$$\begin{aligned}\{f_c\}_{t+\Delta t} &= [m_c]\{\ddot{d}_c\}_{t+\Delta t} + [c_c]\{\dot{d}_c\}_{t+\Delta t} + [k_c]\{d_c\}_{t+\Delta t}\\ &\quad + \{p_c\}_{t+\Delta t} + \{q_c\}_t,\end{aligned} \tag{8.17}$$

from which each of the contact forces $V_{i,t+\Delta t}$, with $i = 1, \ldots, n$, can be given as

$$\begin{aligned}V_{i,t+\Delta t} &= p_{ci,t+\Delta t} + q_{ci,t}\\ &\quad + \sum_{j=1}^{n}(m_{cij}\ddot{d}_{cj,t+\Delta t} + c_{cij}\dot{d}_{cj,t+\Delta t} + k_{cij}d_{cj,t+\Delta t}),\end{aligned} \tag{8.18}$$

where $m_{cij}$, $c_{cij}$ and $k_{cij}$ represent the entry located at the $i$th row and $j$th column of the matrices $[m_c]$, $[c_c]$, and $[k_c]$, respectively, and $p_{ci,t+\Delta t}$, $q_{ci,t}$ are the $i$th entry of the vectors $\{p_c\}_{t+\Delta t}$ and $\{q_c\}_t$, respectively.

## 8.4. VBI Element Considering Vertical Contact Forces Only

Consider that at time $t+\Delta t$, there are $n$ wheels acting simultaneously on the $e_1, e_2, \ldots, e_n$th elements of the bridge, which will be referred to as the *VBI elements* as they are directly under the action of the wheel loads. For the time being, we shall consider only the vertical components of the contact forces and assume that the $e_i$th element is acted upon by the $i$th contact force $V_{i,t+\Delta t}$. Consequently, the equation of motion for the $e_i$th element of the bridge at time $t+\Delta t$ can be written as follows:

$$\begin{aligned}[m_{bi}]\{\ddot{d}_{bi}\}_{t+\Delta t} + [c_{bi}]\{\dot{d}_{bi}\}_{t+\Delta t} + [k_{bi}]\{d_{bi}\}_{t+\Delta t} \\ = \{f_{bi}\}_{t+\Delta t} - \{f_{bci}\}_{t+\Delta t}\,, \end{aligned} \tag{8.19}$$

where $[m_{bi}]$, $[c_{bi}]$ and $[k_{bi}]$ denote the mass, damping and stiffness matrices of the $e_i$th element of the bridge, $\{d_{bi}\}$ the nodal displacement vector, $\{f_{bi}\}$ the vector of external forces directly acting on the nodal points, and $\{f_{bci}\}$ the vector of *consistent nodal forces* resulting from action of the $i$th vertical contact force $V_{i,t+\Delta t}$,

$$\{f_{bci}\}_{t+\Delta t} = \{N^v_{ci}\}V_{i,t+\Delta t}\,, \tag{8.20}$$

where $\{N^v_{ci}\}$ denotes the interpolation vector of the $e_i$th bridge element in which all entries are set to zero except for those associated with the vertical displacements, which are represented by cubic *interpolation* (*Hermitian*) *functions.* The subscript $c$ indicates that the interpolation vector $\{N^v_{ci}\}$ is evaluated at the contact point, i.e.,

$$\{N^v_{ci}\} = \{N^v(x_i)\}\,, \tag{8.21}$$

where $x_i$ is the local coordinate of the $i$th contact point on the $e_i$th element. Note that in Eq. (5.19) the *centrifugal* and *Coriolis force* effects induced on the beam by the moving vehicle loadings have been neglected, as they are usually small in realistic bridge structures. By using Eqs. (8.18) and (8.20), one can rewrite the bridge equations, Eq. (8.19), as follows:

$$
\begin{aligned}
&[m_{bi}]\{\ddot{d}_{bi}\}_{t+\Delta t} + [c_{bi}]\{\dot{d}_{bi}\}_{t+\Delta t} + [k_{bi}]\{d_{bi}\}_{t+\Delta t} \\
&\quad = \{f_{bi}\}_{t+\Delta t} - \sum_{j=1}^{n}([m^*_{cij}]\{\ddot{d}_{bj}\} + [c^*_{cij}]\{\dot{d}_{bj}\} + [k^*_{cij}]\{d_{bj}\}) \\
&\qquad - \{p^*_{ci}\}_{t+\Delta t} - \{q^*_{ci}\}_t \,,
\end{aligned}
\tag{8.22}
$$

where the matrices with an asterisk represent the interaction effect of the VBI elements due to the *interlocking action* of the moving vehicle via the front and rear wheelsets,

$$
\begin{aligned}
[m^*_{cij}] &= \{N^v_{ci}\} m_{cij} \langle N^v_{cj}\rangle \,, \\
[c^*_{cij}] &= \{N^v_{ci}\} c_{cij} \langle N^v_{cj}\rangle \,, \\
[k^*_{cij}] &= \{N^v_{ci}\} k_{cij} \langle N^v_{cj}\rangle \,,
\end{aligned}
\tag{8.23}
$$

and the equivalent nodal loads are

$$
\begin{aligned}
\{p^*_{ci}\}_{t+\Delta t} &= \{N^v_{ci}\} p_{ci,t+\Delta t} \,, \\
\{q^*_{ci}\}_t &= \{N^v_{ci}\} q_{ci,t} \,.
\end{aligned}
\tag{8.24}
$$

Evidently, the effect of interaction with the moving vehicles has been considered in Eq. (8.22) for the $e_i$th element of the bridge through the asterisked matrices and vectors. The equations of motion as given in Eq. (8.22) will be referred to as the *condensed equation of motion for the VBI element*, as all the vehicle DOFs in contact with the bridge element have been eliminated. It should be noted that all the asterisked matrices and vectors in Eqs. (8.23) and (8.24) are time-dependent, since they are all functions of the contact positions, as implied by the shape function $\{N^v_{ci}\}$. Because of this, all these matrices and vectors should be updated at each time step in an incremental analysis. Besides, the accuracy of the VBI element derived, as given in Eq. (8.22), is of the same order as that of the Newmark equations given in Eq. (8.7). Because of this, the VBI element derived herein is quite accurate and can be used to solve the time-history response of most VBI systems with no iterations required.

## 8.5. VBI Element Considering General Contact Forces

The equations of motion as derived in Eq. (8.22) for the VBI element consider only the action of the vertical contact forces $V_{i,t+\Delta t}$. For the case where *acceleration* or *deceleration* are involved in the movement of the vehicles over the bridge, it is necessary to consider the action of the *horizontal contact forces* $H_{i,t+\Delta t}$ along the bridge axis, as shown in Fig. 8.1. To this end, the contact force vector $\{f_{bci}\}_{t+\Delta t}$ in Eq. (8.20) should be generalized as

$$\{f_{bci}\}_{t+\Delta t} = \{N_{ci}^h\}H_{i,t+\Delta t} + \{N_{ci}^v\}V_{i,t+\Delta t}\,, \qquad (8.25)$$

where $\{N_{ci}^h\}$ denotes the interpolation vector of the $e_i$th (VBI) element in which all entries are set to zero except those associated with the axial displacements, which are represented by linear functions and evaluated at the contact point $x_i$. The horizontal components $H_{i,t+\Delta t}$ of the contact forces may be generated by the rolling, accelerating or braking action of the wheels. All these actions can be referred to as the variational forms of the frictional force, which vary according to the cohesion existing between the wheels and rails. For the present purposes, one may relate the $i$th horizontal contact force $H_i$ to the corresponding vertical contact force $V_i$ simply as

$$H_i = \mu_i V_i\,, \qquad (8.26)$$

where $\mu_i$ is the *frictional coefficient* for the $i$th wheel, which equals either the coefficient of braking or the coefficient of acceleration; the frictional coefficient in rolling is so small that it can be neglected in practice.

Considering the more general expression for the contact forces in Eq. (8.25), along with Eq. (8.26), one can proceed to derive a set of condensed equations of motion for the $e_i$th element of the bridge identical in the form of Eq. (8.22), but with the asterisked matrices modified as

$$
\begin{aligned}
[m^*_{cij}] &= \{N^h_{ci}\}\mu_i m_{cij}\langle N^v_{cj}\rangle + \{N^v_{ci}\}m_{cij}\langle N^v_{cj}\rangle\,, \\
[c^*_{cij}] &= \{N^h_{ci}\}\mu_i c_{cij}\langle N^v_{cj}\rangle + \{N^v_{ci}\}c_{cij}\langle N^v_{cj}\rangle\,, \\
[k^*_{cij}] &= \{N^h_{ci}\}\mu_i k_{cij}\langle N^v_{cj}\rangle + \{N^v_{ci}\}k_{cij}\langle N^v_{cj}\rangle\,,
\end{aligned}
\tag{8.27}
$$

and the asterisked vectors as

$$
\begin{aligned}
\{p^*_{ci}\}_{t+\Delta t} &= \{N^h_{ci}\}\mu_i p_{ci,t+\Delta t} + \{N^v_{ci}\}p_{ci,t+\Delta t}\,, \\
\{q^*_{ci}\}_t &= \{N^h_{ci}\}\mu_i q_{ci,t} + \{N^v_{ci}\}q_{ci,t}\,.
\end{aligned}
\tag{8.28}
$$

Again, it should be noted that all the matrices and vectors with an asterisk in Eqs. (8.27) and (8.28) are time-dependent. In this chapter, the general equations as given in Eqs. (8.27) and (8.28), which consider the effect of both the vertical and horizontal contact forces, will be used in studying the vehicle–bridge interactions whenever the acceleration or braking of the moving vehicles are involved. The two equations given in Eqs. (8.23) and (8.24), which consider only the vertical contact forces, will be used for the case when the vehicles are allowed to travel at a constant speed.

## 8.6. System Equations and Structural Damping

As was stated previously, a bridge element that is directly under the action of wheel loads is referred to as a VBI element. Now, let us consider a bridge that is traveled by a connected line of vehicles, as the case with railway bridges. At some instant, say, at time $t + \Delta t$, only parts of the bridge will be directly acted upon by the wheel loads, which should be modeled by the VBI elements. However, for the remaining parts of the bridge that are not in direct touch with the wheel loads, they should be modeled by the original (parent) bridge element. Following the conventional finite element procedure, all the VBI elements and bridge elements can be assembled to yield the following equations:

$$[M]\{\ddot{D}\}_{t+\Delta t} + [C]\{\dot{D}\}_{t+\Delta t} + [K]\{D\}_{t+\Delta t} = \{F_b\}_{t+\Delta t} - \{P_c^*\}_{t+\Delta t} - \{Q_c^*\}_t\,, \tag{8.29}$$

where $\{D\}_{t+\Delta t}$ denotes the displacements of the entire VBI system, $[M]$, $[C]$ and $[K]$ the mass, damping and stiffness matrices, $\{F_b\}_{t+\Delta t}$ the external loads, and $\{P_c^*\}_{t+\Delta t}$ and $\{Q_c^*\}_t$ the equivalent contact forces in global coordinates. A convenient way to construct the system matrices $[M]$, $[C]$ and $[K]$ is first to assemble the matrices $[M_b]$, $[C_b]$ and $[K_b]$ for the bridge that is free of any wheel actions, and then to add to them the interaction effects of vehicles contributed by the VBI elements, as represented by the terms with an asterisk in Eqs. (8.23) or (8.27), i.e.,

$$\begin{aligned} [M] &= [M_b] + [M_c^*] = \sum[m_{bi}] + \sum\sum[m_{cij}^*]\,, \\ [C] &= [C_b] + [C_c^*] = \sum[c_{bi}] + \sum\sum[c_{cij}^*]\,, \\ [K] &= [K_b] + [K_c^*] = \sum[k_{bi}] + \sum\sum[k_{cij}^*]\,. \end{aligned} \tag{8.30}$$

Similarly, the equivalent contact forces $\{P_c^*\}_{t+\Delta t}$ and $\{Q_c^*\}_t$ are

$$\begin{aligned} \{P_c^*\}_{t+\Delta t} &= \sum\{p_{ci}^*\}_{t+\Delta t}\,, \\ \{Q_c^*\}_t &= \sum\{q_{ci}^*\}_t\,. \end{aligned} \tag{8.31}$$

In Eqs. (8.30) and (8.31), all the terms or components with an asterisk should be interpreted as those generated by the interaction effect of the moving vehicles. Therefore, they should be looped over the VBI elements only. As the wheel loads move from time to time, it is necessary to check at each time step whether a bridge element becomes a VBI element and vice versa, and to update (i.e., to add or delete) the entries of the system matrices and vectors, concerning the contribution of the terms or components identified by an asterisk in Eqs. (8.30) and (8.31), for the DOFs that are directly affected by the vehicle actions, according to the variation of the contact positions.

One feature with the present formulation is that the total number of DOFs of the VBI system remains identical to that of the original bridge, regardless of the consideration of vehicles interaction effects.

The *symmetry* property of the system matrices is also preserved. Furthermore, the present procedure can be adopted to deal with the vehicle models of various complexities, which may vary from the simplest case of moving load to models with dozens of DOFs. As will be demonstrated in the numerical examples, with the present approach there is virtually no limit on the number of DOFs used to describe the vehicle structure.

In this chapter, *Rayleigh damping* is assumed for the bridge, namely, the damping matrix $[C_b]$ of the bridge that is free of any vehicle actions can be computed as a linear combination of the mass matrix $[M_b]$ and stiffness matrix $[K_b]$,

$$[C_b] = \alpha_0[M_b] + \alpha_1[K_b]\,, \tag{8.32}$$

where, given the damping ratio $\xi$, the two coefficients $\alpha_0$ and $\alpha_1$ are

$$\begin{aligned} \alpha_0 &= \frac{2\xi\omega_1\omega_2}{\omega_1+\omega_2}\,, \\ \alpha_1 &= \frac{2\xi}{\omega_1+\omega_2}\,. \end{aligned} \tag{8.33}$$

Here, $\omega_1$ and $\omega_2$ are the first and second frequencies of vibration of the bridge.

Also, the system equations as given in Eq. (8.29) will be solved in an incremental sense using *Newmark's $\beta$ method.* The parametric values of $\beta = 0.25$ and $\gamma = 0.5$ are used throughout, implying that the marching scheme is unconditionally stable. By this method, the system accelerations $\{\ddot{D}\}$, velocities $\{\dot{D}\}$ and displacements $\{D\}$ at time $t + \Delta t$ can first be discretized and related to those at time $t$ in exactly the same manner as in Eq. (8.7), with which the system equation (8.29) can be reduced to an equivalent stiffness equation and be solved (see Appendix B for more details).

## 8.7. Procedure of Time-History Analysis for VBI Systems

The following is a summary of the procedure for incremental analysis of the VBI system based on Newmark's $\beta$ method:

(1) Input the fundamental data of the bridge and vehicles.

(2) Construct the mass matrix $[M_b]$ and stiffness matrix $[K_b]$ for the bridge that is free of any vehicles. Perform an eigenvalue analysis to obtain the first two frequencies $\omega_1$ and $\omega_2$. Given the damping ratio $\xi$, determine the damping matrix $[C_b]$ of the bridge using Eqs. (8.32) and (8.33).

(3) Specify the initial conditions for the bridge and the vehicles, and the position $x_{i0}$ of each wheel at time $t = 0$, i.e., $\{\ddot{D}\}_0 = \{\dot{D}\}_0 = \{D\}_0 = \{0\}$ for the bridge and $\{\ddot{D}_v\}_0 = \{\dot{D}_v\}_0 = \{D_v\}_0 = \{0\}$ for all the vehicles. Specify the following data: (a) the initial velocity $v_0$ and acceleration $a$ of the vehicles; (b) the position $x_{bs0}$ for the first wheel to start acceleration or braking; (c) the position $x_{bsf}$ for the first wheel to stop acceleration or braking, $x_{bsf} > x_{bs0}$; (d) the frictional coefficient for each wheel $\mu_i = \mu$ ($\mu = \mu_b$ for braking and $\mu = \mu_s$ for acceleration); (e) the ending time $t_{\text{end}}$ for analysis or ending position $x_{\text{end}}$ of the first wheel; and (g) the time increment $\Delta t$.

(4) Construct the matrices and vectors for the vehicles equations, i.e., Eq. (8.4). Compute $[\Psi_{uu}]$ and $[\Psi_{wu}]$, using Eqs. (8.10a) and (8.16a), and the contact matrices $[m_c]$, $[c_c]$ and $[k_c]$ using Eq. (8.14).

(5) For a new time with $t' = t + \Delta t$, check if $t' > t_{\text{end}}$ or $x_1 > x_{\text{end}}$ for the first wheel. If yes, stop analysis; otherwise, proceed to Step 6.

(6) Calculate the global position $x_i$ of each wheel, identify the element $e_i$ in contact, and compute the values of the interpolation functions $\{N_{ci}^v\}$ and $\{N_{ci}^h\}$ at the contact point.

(7) Compute $\{q_u\}_t$ and $\{q_w\}_t$ using Eqs. (8.10b) and (8.16b), and $\{p_c\}_{t+\Delta t}$ and $\{q_c\}_t$ using Eq. (8.15).

(8) If $x_1 < x_{bs0}$ or $x_1 > x_{bsf}$, which means that the vehicles are moving with constant speed, use Eqs. (8.23) and (8.24) to compute the vehicle-related matrices $[m_{cij}^*]$, $[c_{cij}^*]$, $[k_{cij}^*]$ and vectors $\{p_{ci}^*\}_{t+\Delta t}$, $\{q_{ci}^*\}_t$ for each element $e_i$ in contact. If $x_{bs0} \le x_1 \le x_{bsf}$, which means that the vehicles are in acceleration or in braking, then use Eqs. (8.27) and (8.28) to compute the matrices $[m_{cij}^*]$, $[c_{cij}^*]$, $[k_{cij}^*]$ and vectors $\{p_{ci}^*\}_{t+\Delta t}$, $\{q_{ci}^*\}_t$ for

each element $e_i$ in contact. In each case, update the global matrices $[M_c^*]$, $[C_c^*]$, $[K_c^*]$ and vectors $\{P_c^*\}_{t+\Delta t}$, $\{Q_c^*\}_t$ accordingly. Note that for vehicles with acceleration $a < 0$, use $\mu = \mu_b < 0$, and for acceleration $a > 0$, use $\mu = \mu_s > 0$.

(9) Summing up $[M_b]$, $[C_b]$ and $[K_b]$ respectively with $[M_c^*]$, $[C_c^*]$, $[K_c^*]$, one can obtain the matrices $[M]$, $[C]$ and $[K]$ for the entire VBI system following Eq. (8.30).

(10) Solve Eq. (8.29) by Newmark's $\beta$ method for the displacement increments $\{\Delta D\}$. Using equations identical in form to Eq. (8.7), compute the displacements $\{D\}$, velocities $\{\dot{D}\}$ and accelerations $\{\ddot{D}\}$ of the bridge.

(11) Compute the response for all the contact points of the bridge, i.e., $\{d_c\}$, $\{\dot{d}_c\}$ and $\{\ddot{d}_c\}$. Compute the contact force $V_{i,t+\Delta t}$ for each wheel using Eq. (8.18). Based on the no-jump condition for vehicles, $\{d_w\} \equiv \{d_c\}$, one can compute $\{q_{uc}\}_{t+\Delta t}$ from Eq. (8.6) and $\{\Delta d_u\}$ from Eq. (8.9). Consequently, the response of the upper or noncontact part of the vehicle, i.e., $\{d_u\}$, $\{\dot{d}_u\}$ and $\{\ddot{d}_u\}$, can be computed from Eq. (8.7).

(12) Go to Step 5 to perform the next time increment of analysis.

## 8.8. Numerical Examples and Verification

In order to demonstrate the capability and reliability of the present procedure in dealing with various vehicle models, six typical examples will be studied in this section. In each case, the beam is modeled as 10 elements.

### 8.8.1. *Cantilever Beam Subjected to a Moving Load*

As shown in Fig. 8.2(a), a cantilever beam of length $L = 300$ in. is subjected to a moving load $p$. The following data are adopted: flexural rigidity $EI = 3.3 \times 10^9$ lb-in.$^2$, mass per unit length $m = 0.00667$ lbm/in., damping ratio $\xi = 0$, moving load $p = 5793$ lb and velocity $v = 2000$ in./s. For this case, the moving load $p$ should be interpreted as a vehicle with only one wheel ($n = 1$) of zero mass which has a single DOF denoted as $v_w$ and is subjected to an external

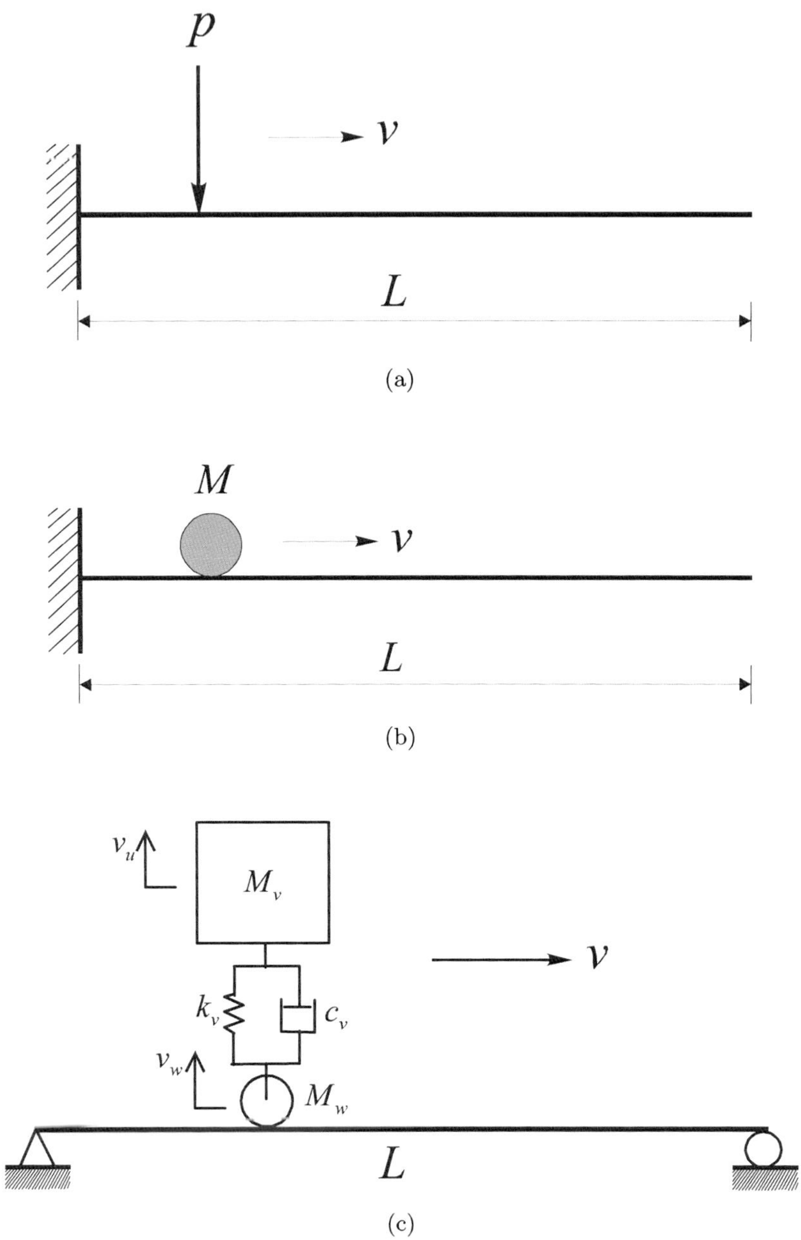


Fig. 8.2. VBI models: (a) moving load, (b) moving mass, (c) sprung mass, and (d) suspended rigid beam.

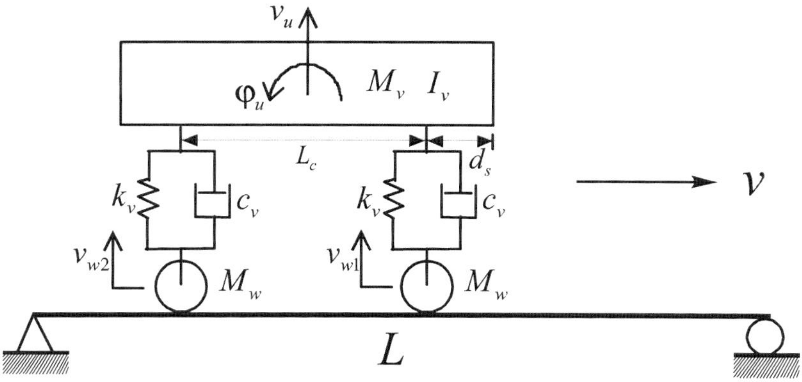


(d)

Fig. 8.2. (*Continued*).

load of magnitude $p$. The partitioned matrices corresponding to the upper part of the vehicle have to be eliminated from the relevant equations, and the remaining matrices corresponding to the wheel can be given as follows:

$$\begin{aligned} [m_{ww}] &\equiv 0\,, \\ [c_{ww}] &\equiv 0\,, \\ [k_{ww}] &\equiv 0\,, \\ \{f_{we}\}_{t+\Delta t} &\equiv -p\,, \\ [l_w] &\equiv 1\,, \\ [\Gamma] &\equiv 1\,. \end{aligned} \tag{8.34}$$

By using Eq. (8.34), together with Eqs. (8.14) and (8.15), we can obtain from Eq. (8.18) the contact force $V_{1,t+\Delta t} = p$, as it should be for the moving load case. The free end displacements of the cantilever caused by load $p$ moving from the left to the right and in the reversed direction were plotted in Fig. 8.3. As can be seen, the results obtained by the present procedure agree well with those of Akin and Mofid (1989).

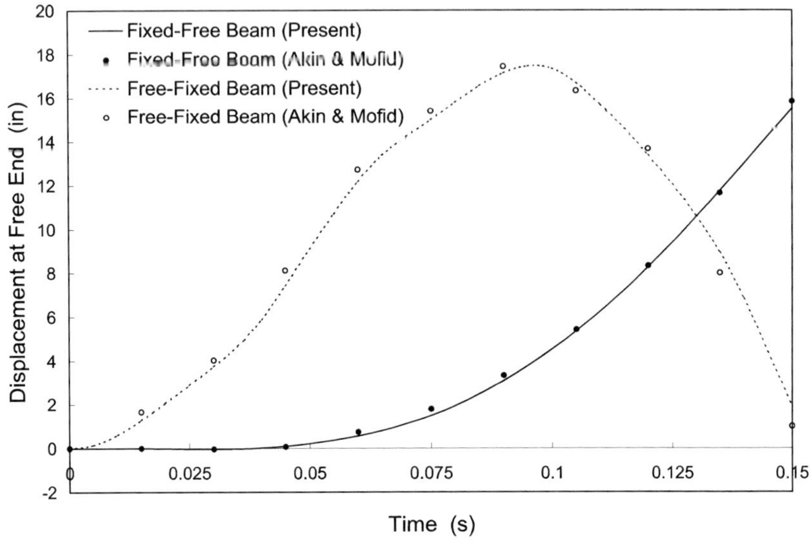


Fig. 8.3. Free-end responses of cantilever (moving load model).

### 8.8.2. *Cantilever Beam Subjected to a Moving Mass*

Consider a cantilever beam subjected to a lumped mass of $M = 15$ lbm moving at $v = 2000$ in./s, as shown in Fig. 8.2(b). The properties of the beam are identical to those used in the preceding example. The moving mass should now be treated as a vehicle with nothing but a single wheel of mass $M$, which has a single DOF denoted as $v_w$, and one contact point. The corresponding partitioned matrices for the wheel are

$$\begin{aligned} [m_{ww}] &\equiv M\,, \\ [c_{ww}] &\equiv 0\,, \\ [k_{ww}] &\equiv 0\,, \end{aligned} \tag{8.35}$$

$$\begin{aligned} \{f_{we}\}_{t+\Delta t} &\equiv -Mg\,, \\ [l_w] &\equiv 1\,, \\ [\Gamma] &\equiv 1\,. \end{aligned} \tag{8.36}$$

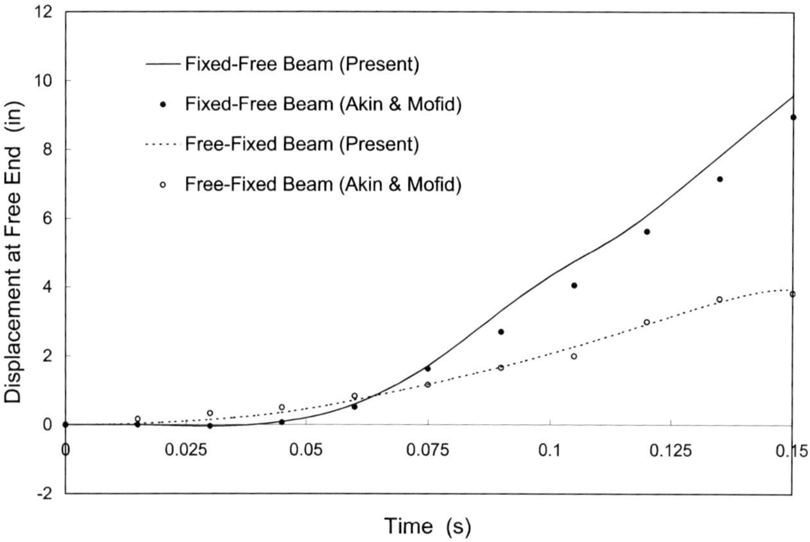


Fig. 8.4. Free-end responses of cantilever (moving mass model).

Similarly, by using Eqs. (8.35), (8.36) and relevant equations, one can obtain the contact force as $V_{1,t+\Delta t} = M\ddot{v}_{w,t+\Delta t} + Mg$, of which the first term represents the inertial effect and the second term the moving load effect. The dynamic responses of the free end of the cantilever caused by the mass $M$ moving from the fixed to the free end and in the reverse direction have been plotted in Fig. 8.4. Again, the responses obtained by the present procedure agree well with those of Akin and Mofid (1989). From the contact force response plotted in Fig. 8.5, one observes that the contact force fluctuates and does not remain equal to the static weight of the mass, implying that the effect of inertia is significant and cannot be neglected for the present case. Moreover, for the case with the mass moving from the free to the fixed end, the contact force encounters a serious drop at the beginning, then escalates and eventually drops again to negative values near the fixed end. The occurrence of negative contact forces should be recognized as part of the effect of no-jump assumption, which enforces the mass to keep in contact with the beam at all times.

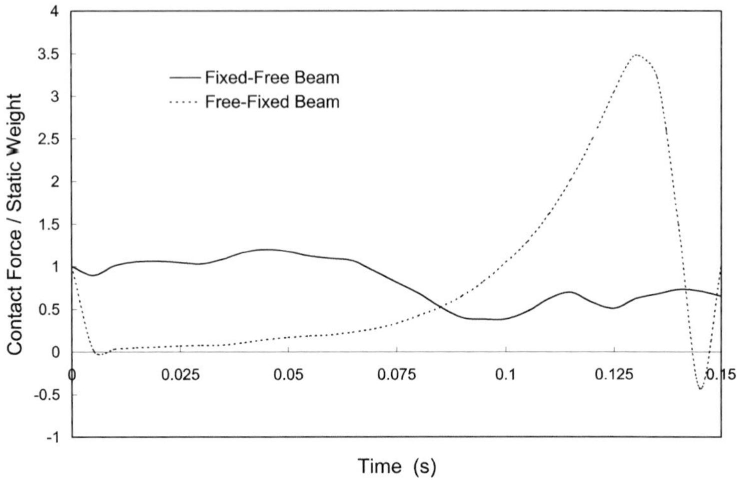


Fig. 8.5. Contact force response (moving mass model).

### 8.8.3. *Simple Beam Subjected to a Moving Sprung Mass*

Figure 8.2(c) shows a simply-supported beam subjected to a moving sprung mass system, of which the vehicle (or sprung) mass $M_v$ is supported by a dashpot-spring unit of spring constant $k_v$ and damping $c_v$, which is further supported by a wheel mass of $M_w$. The data identical to those of Sec. 6.7.1 are adopted herein: Young's modules $E = 2.87$ GPa kN/m$^2$, Poisson's ratio $\upsilon = 0.2$, moment of inertia $I = 2.90$ m$^4$, per-unit-length mass $m = 2.303$ t/m, length of beam $L = 25$ m; sprung mass $M_v = 5.75$ t, and suspension spring constant $k_v = 1595$ kN/m. For illustration, the effect of the dashpot damping and the wheel mass will be neglected, i.e., by letting $c_v = 0$ kN-s/m, $M_w = 0$ t. By using $v_u$ and $v_w$ to denote the DOF associated with the car body and wheel masses, respectively, the partitioned matrices corresponding to the two DOFs can be given as follows:

$$\begin{aligned} [m_{uu}] &\equiv M_v\,, \\ [m_{uw}] &= [m_{wu}]^T \equiv 0\,, \\ [m_{ww}] &\equiv M_w\,, \end{aligned} \tag{8.37}$$

$$\begin{aligned} [c_{uu}] &\equiv c_v \,, \\ [c_{uw}] = [c_{wu}]^T &\equiv -c_v \,, \\ [c_{ww}] &\equiv c_v \,, \end{aligned} \tag{8.38}$$

$$\begin{aligned} [k_{uu}] &\equiv k_v \,, \\ [k_{uw}] = [k_{wu}]^T &\equiv -k_v \,, \\ [k_{ww}] &= k_v \,, \end{aligned} \tag{8.39}$$

$$\begin{aligned} \{f_{ue}\}_{t+\Delta t} &\equiv 0 \,, \\ \{f_{we}\}_{t+\Delta t} &\equiv -M_v g - M_w g \,, \\ [l_w] &\equiv 1 \,, \\ [\Gamma] &\equiv 1 \,. \end{aligned} \tag{8.40}$$

For this particular problem, an approximate analytical solution can be established by considering only the first mode of vibration, i.e., by representing the deflection of the beam as $v_b(x,t) = q_b(t)\sin(\pi x/L)$. The general procedure of solution was outlined in Biggs (1964).

For the sprung mass with speed $v = 27.78$ m/s (= 100 km/h), the midpoint displacement of the beam obtained based either on the sprung mass or moving load assumption has been plotted in Fig. 8.6. As can be seen, the response obtained by the present VBI element based on the sprung mass assumption agrees well with the analytical solution. Moreover, from the vertical acceleration of the midpoint of the beam shown in Fig. 8.7, one observes that inclusion of the *higher vibration modes* can result in drastic, but local oscillation around the acceleration obtained by the analytical approach that considers only the first mode. On the other hand, the response of the vertical acceleration of the sprung mass has been plotted in Fig. 8.8, from which the effect of higher modes on the general trend can be appreciated. This figure also indicates that the acceleration of the vehicle or sprung mass is quite sensitive to the omission of higher-order terms, which appears to be reasonable as the vehicle can in some sense be regarded as a tuned mass to the bridge. It is interesting

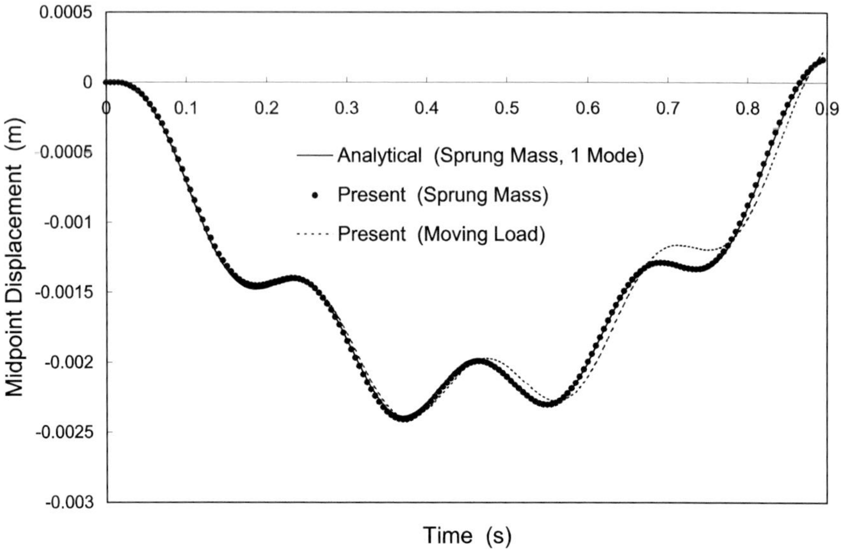


Fig. 8.6. Midpoint displacement of simple beam (sprung mass model).

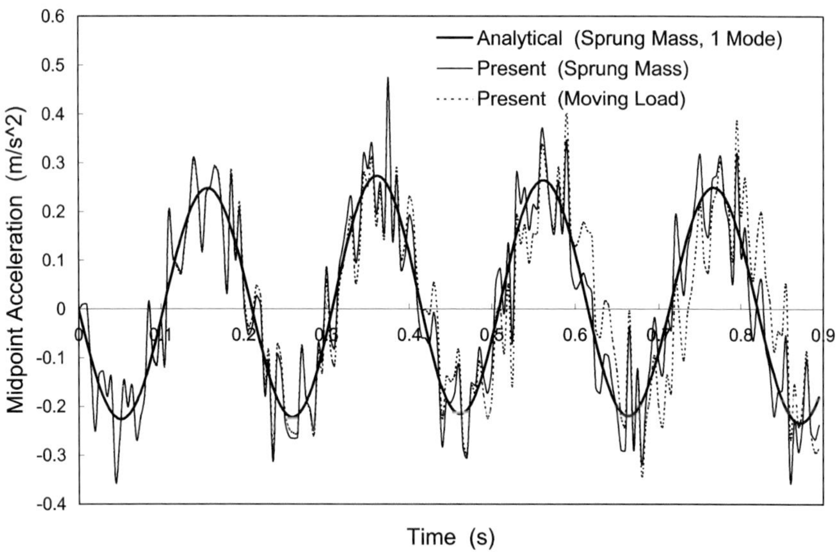


Fig. 8.7. Midpoint acceleration of simple beam (sprung mass model).

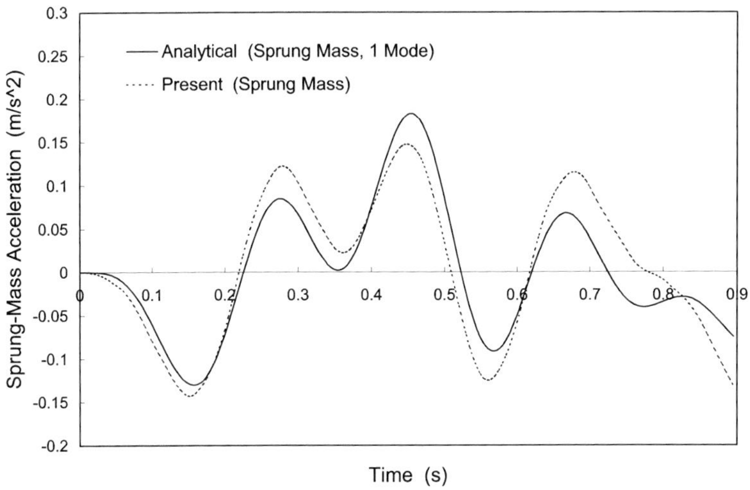


Fig. 8.8. Vertical acceleration of sprung mass (sprung mass model).

to note that the acceleration of the sprung mass has been taken as a measure of the passengers' *riding comfort* in the design of high-speed railways. Clearly, as far as the riding comfort is concerned, it is necessary to take into account the higher mode vibrations of the beam.

### 8.8.4. *Simple Beam Subjected to a Moving Rigid Bar Supported by Spring-Dashpot Units*

Let us consider a simple beam with the following properties: $L = 30$ m, $E = 2.943 \times 10^7$ kN/m$^2$, Poisson's ratio $v = 0.2$, $I = 8.65$ m$^4$, $m = 36$ t/m, $\xi = 0$, and cross-sectional area $A = 5.16$ m$^2$. The vehicle moving over the beam is modeled as a rigid bar supported by two identical spring-dashpot units, as shown in Fig. 8.2(d). With this model, the interlocking effect of the car body on the front and rear wheels can be duly taken into account. The following data are adopted for the vehicle: rigid bar mass $M_v = 540$ t, mass moment of inertia $I_v = 13\,800$ t-m$^2$, spring stiffness $k_v = 41\,350$ kN/m, dashpot coefficient $c_v = 0$ kN-s/m, wheel mass $M_w = 0$ t, wheel-to-wheel distance $L_c = 17.5$ m, and vehicle speed $v = 27.78$ m/s. For this example, the DOFs of the rigid bar are denoted as $v_u$ and $\varphi_u$ and the DOFs of the front and rear wheels as $v_{w1}$ and $v_{w2}$, respectively. Using the present notation, the dynamic matrices and relevant terms for the vehicle model considered can be written as follows:

$$
\begin{aligned}
[m_{uu}] &= \begin{bmatrix} M_v & 0 \\ 0 & M_v \end{bmatrix}, \\
[m_{uw}] &= [m_{wu}]^T = [0], \\
[m_{ww}] &= \begin{bmatrix} M_w & 0 \\ 0 & M_w \end{bmatrix},
\end{aligned}
\tag{8.41}
$$

$$
\begin{aligned}
[c_{uu}] &= \begin{bmatrix} 2c_v & 0 \\ 0 & 0.5d^2 c_v \end{bmatrix}, \\
[c_{uw}] = [c_{wu}]^T &= \begin{bmatrix} -c_v & -c_v \\ -0.5dc_v & 0.5dc_v \end{bmatrix}, \\
[c_{ww}] &= \begin{bmatrix} c_v & 0 \\ 0 & c_v \end{bmatrix},
\end{aligned}
\tag{8.42}
$$

$$
\begin{aligned}
[k_{uu}] &= \begin{bmatrix} 2k_v & 0 \\ 0 & 0.5d^2 k_v \end{bmatrix}, \\
[k_{uw}] = [k_{wu}]^T &= \begin{bmatrix} -k_v & -k_v \\ -0.5dk_v & 0.5dk_v \end{bmatrix}, \\
[k_{ww}] &= \begin{bmatrix} k_v & 0 \\ 0 & k_v \end{bmatrix},
\end{aligned}
\tag{8.43}
$$

$$
\begin{aligned}
\{f_{ue}\}_{t+\Delta t} &= \{0\}, \\
\{f_{we}\}_{t+\Delta t} &= \begin{Bmatrix} -0.5M_v g - M_w g \\ -0.5M_v g - M_w g \end{Bmatrix}, \\
[l_w] &= \begin{bmatrix} 1 & 0 \\ 0 & 1 \end{bmatrix}, \\
[\Gamma] &= \begin{bmatrix} 1 & 0 \\ 0 & 1 \end{bmatrix}.
\end{aligned}
\tag{8.44}
$$

The deflection and acceleration of the midpoint of the beam and the vertical acceleration of the center of gravity of the rigid bar have been plotted in Figs. 8.9–8.11, along with analytical solutions given by Yau and Yang (1998) based on the first mode approximation. As can be seen, good agreement has been achieved between the present solutions and the analytical ones. Like the sprung mass example, the differences in these figures between the present solutions and the analytical ones can be attributed mainly to the omission of higher modes in the latter.

For comparison, the rigid bar model has also been approximated by two identical moving loads each of 2648.7 kN ($= 0.5 \times 540 \times 9.81$) or as two suspended masses each of 270 t ($= 0.5 \times 540$) with identical interval $L_c$. The results obtained for the midpoint displacement of the simple beam by the moving load, sprung mass and suspended rigid bar models have been shown in Fig. 8.12. Here, one observes that the maximum response obtained by the moving load model is similar to that of the suspended rigid bar model and both are greater than that of the sprung mass model. In Fig. 8.13, the results

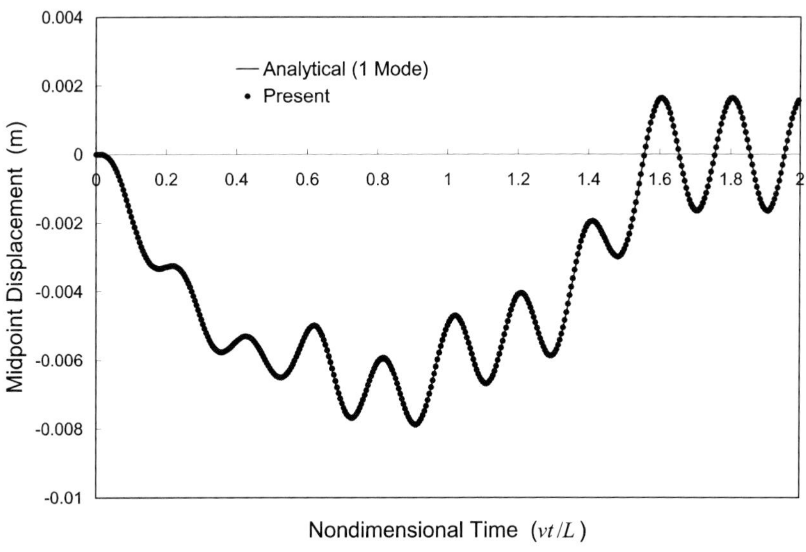


Fig. 8.9. Midpoint displacement of simple beam (rigid beam model).

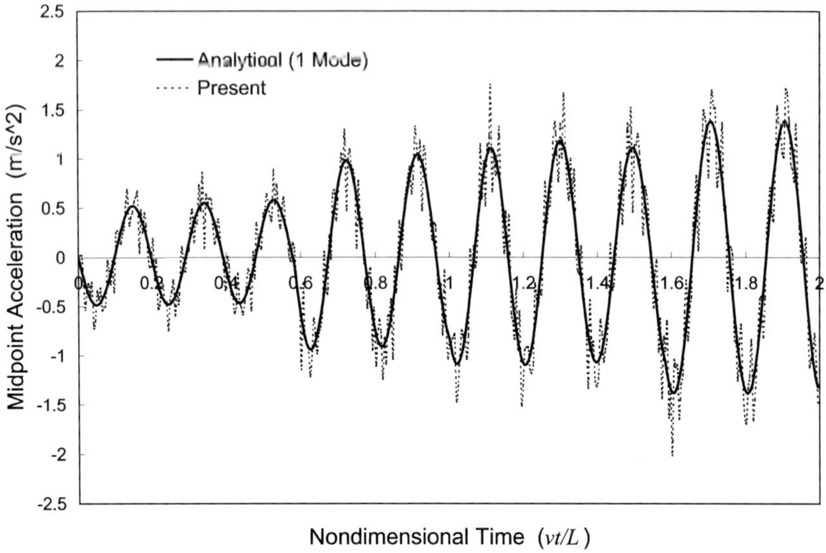


Fig. 8.10. Midpoint acceleration of simple beam (rigid beam model).

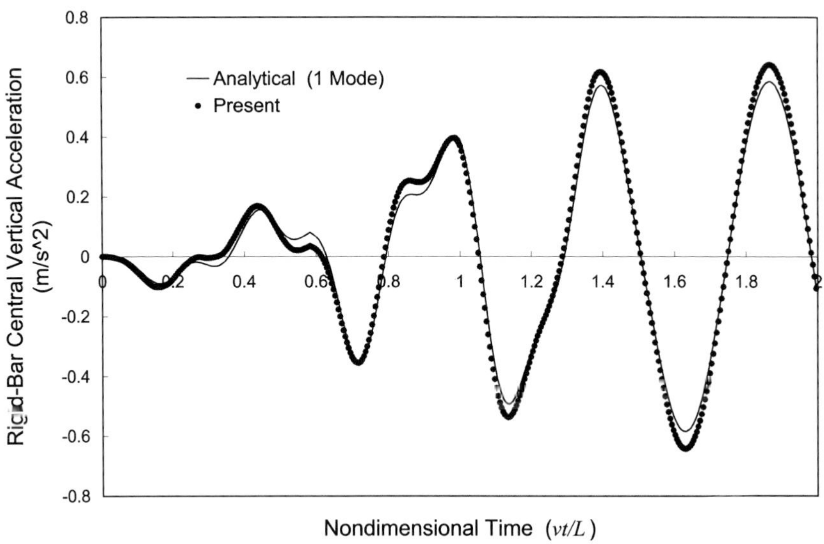


Fig. 8.11. Vertical acceleration of rigid beam.

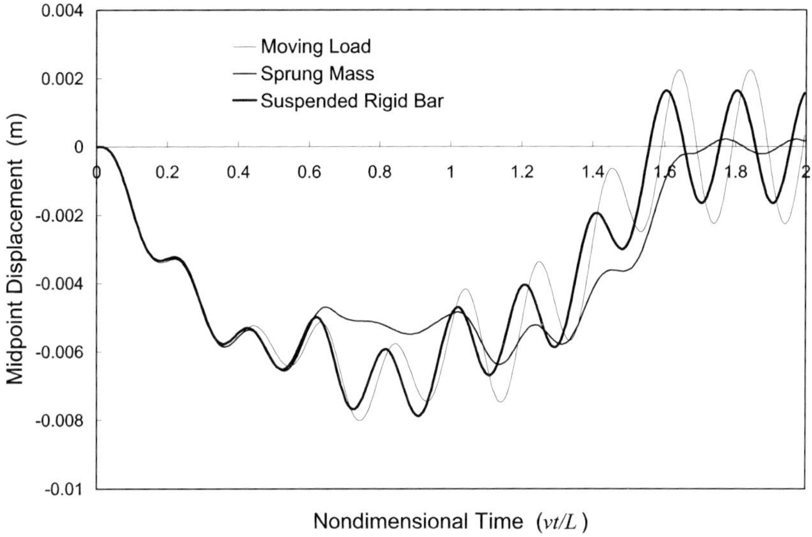


Fig. 8.12. Comparison of midpoint responses of simple beam.

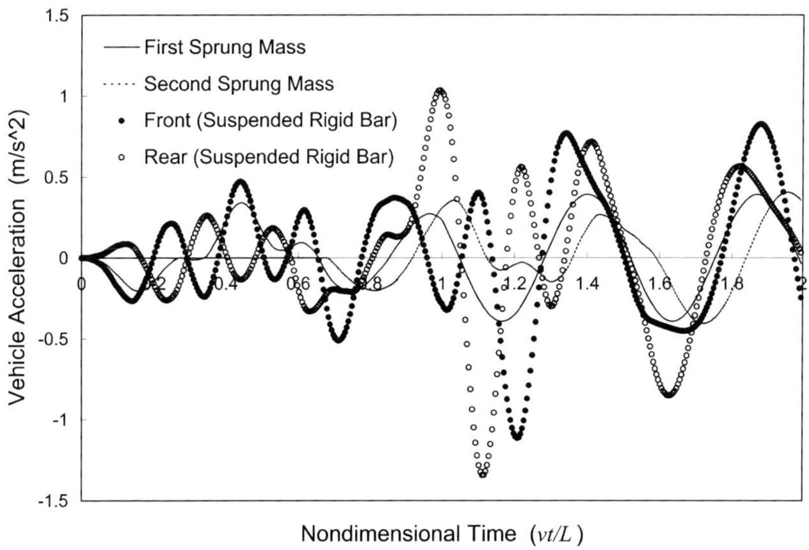


Fig. 8.13. Responses of vehicle based on rigid beam model.

obtained for the rigid bar at points corresponding to the front and rear wheels using the suspended rigid bar model and the discrete sprung mass model have been plotted. As can be seen, the maximum acceleration of the vehicle computed using the suspended rigid bar model appears to be much larger than those using the sprung mass models, indicating that the *interlocking effect* of the rigid bar tends to increase the vehicle response, and that using the sprung mass model to compute the vehicle response is generally nonconservative.

### 8.8.5. *Bridge Subjected to a Vehicle in Deceleration*

Let us now proceed to investigate the behavior of a bridge caused by a vehicle in deceleration. The bridge is modeled as a simple beam and the vehicle as a suspended rigid bar. The data adopted for the beam and the suspended rigid bar are the same as those used in the preceding example, except for inclusion of the damping ratio $\xi = 0.025$. Here, we shall use $V$ to denote the initial velocity of the rigid bar, $a$ the acceleration, $x_{bs0}$ the initial position of deceleration, $x_{bsf}$ the final position of deceleration, and $\mu$ the frictional coefficient. The following three cases are considered: (1) $V = 50$ m/s, $a = -10$ m/s$^2$, (2) $V = 100$ m/s, $a = -10$ m/s$^2$, (3) $V = 100$ m/s, $a = -20$ m/s$^2$, where in each case the following are adopted: $x_{bs0} = 0$ m, $x_{bsf} = 60$ m and $\mu = -0.26$.

The midpoint deflection and acceleration respectively of the beam for the three cases have been plotted in Figs. 8.14 and 8.15, and the vertical acceleration of the vehicle was plotted in Fig. 8.16. As can be seen, the responses for Cases (2) and (3) are close to each other, which are larger than those for Case (1). This has the indication that the response of the bridge is generally not influenced by the action of braking, but by the initial velocity before braking is applied. The reason is that the change in speed is quite limited during the passage of the vehicle over the beam, as the acting time is so short. The *effect of braking* will be evident, however, in the case of a beam traveled by a train of which the speed can change significantly during a longer acting time.

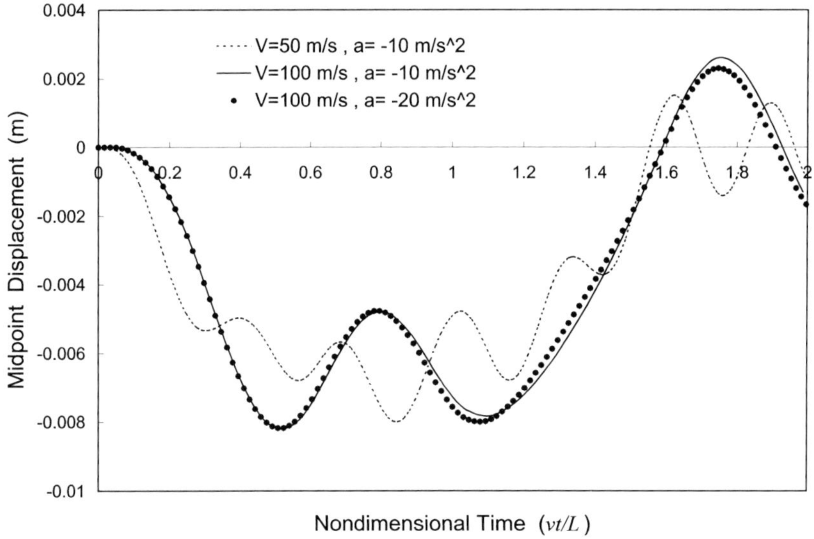


Fig. 8.14. Midpoint displacement of simple beam for vehicles in braking.

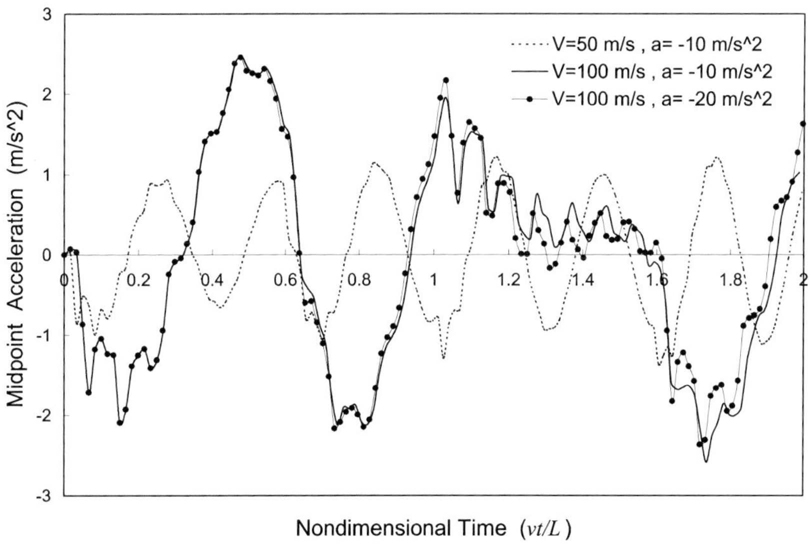


Fig. 8.15. Midpoint acceleration of simple beam for vehicles in braking.

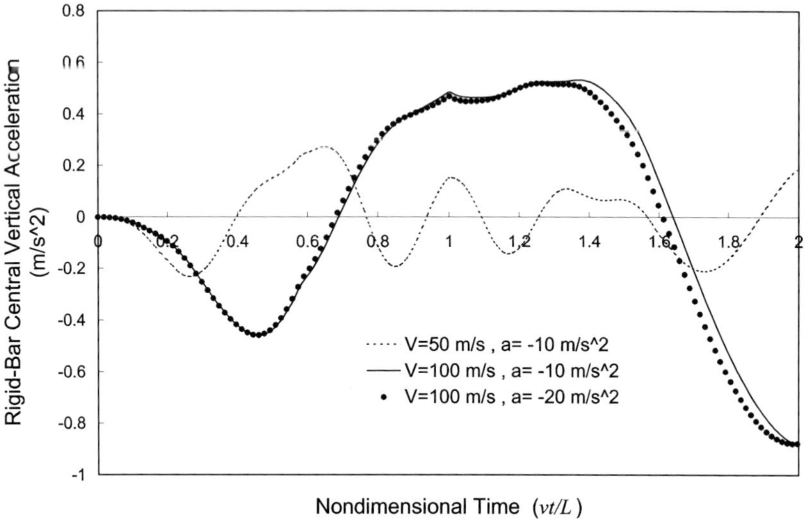


Fig. 8.16. Vertical acceleration of vehicles in braking.

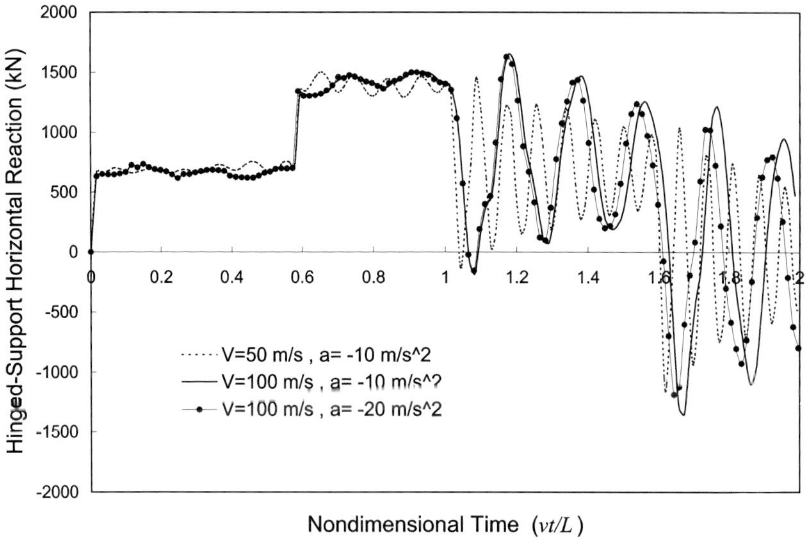


Fig. 8.17. Horizontal reaction at hinged support of simple beam.

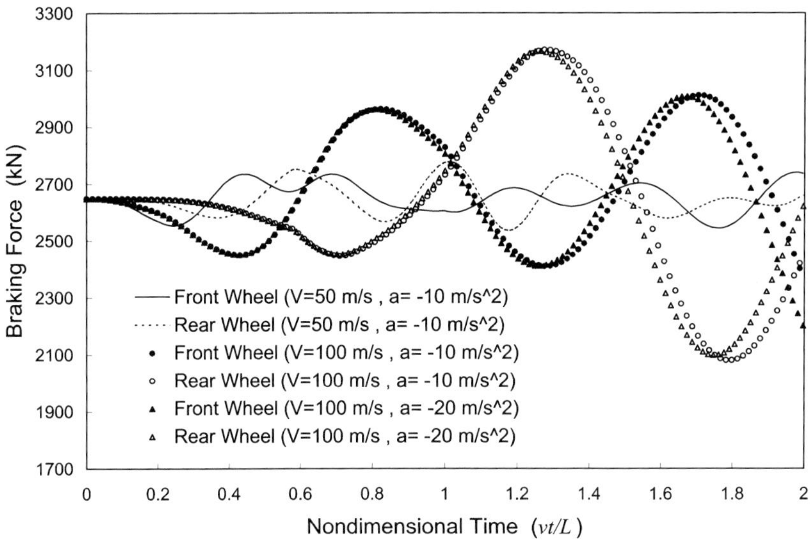


Fig. 8.18. Braking forces of vehicles.

Figure 8.17 depicts the *horizontal reaction force* occurring at the hinged support for the above three cases of deceleration. As can be seen, the reaction force for each case increases significantly at the instant when the front wheel or rear wheel enters the beam. It decreases rapidly at the instant when the front wheel leaves the beam and rebounds immediately to the maximum value. Thereafter it fluctuates around the value of force caused only by the rear wheel. When the rear wheel leaves the beam, it oscillates around the value of zero. Besides, it can be observed that the reaction forces for the three cases are nearly the same before the front wheel leaves the beam.

The braking forces existing between each of the two wheels and the beam have been plotted in Fig. 8.18. It indicates that the braking forces for the cases with higher initial speed, i.e., Cases (2) and (3), are much larger than those for the case with lower initial speed, i.e., Case (1). The maximum braking force of the rear wheel is also larger than the front wheel for each case and will occur at some time after the front wheel leaves the beam.

### 8.8.6. *Bridges Subjected to a Train Consisting of 10 Identical Cars*

The train considered herein consists of 10 identical cars, each of which is modeled as a suspended rigid bar as shown in Fig. 8.2(d). The bridge is modeled as a simple beam with the following properties: $E = 2.87 \times 10^7$ kN/m$^2$, $v = 0.2$, $I = 6.635$ m$^4$, $m = 32.4$ t/m, $L = 30$ m, and $\xi = 0.025$. Ten elements are used for the beam. The data adopted for the cars are extracted from Iwnick (1998) for the Manchester vehicle model, i.e., $M_v = 32$ t, $I_v = 1970$ t-m$^2$, $k_v = 430$ kN/m, $c_v = 20$ kN-s/m, $M_w = 6.241$ t, $L_c = 19$ m, and $d_s = 3$ m (the distance from the front or rear wheels of a car to its nearest end, see Fig. 8.2(d)). Thus, the total length $d$ of each car is $19 + 2 \times 3 = 25$ m. Two different speeds are considered for the train, i.e., $v = 60$ and 100 km/h. In each case, the train is considered entering the bridge when the train head reaches $x = 0$ m, i.e., left end of the bridge, and departing from the bridge when it reaches $x = 300$ m.

The response of the midspan displacement and acceleration of the bridge for the two speeds were plotted in Figs. 8.19 and 8.20. As can

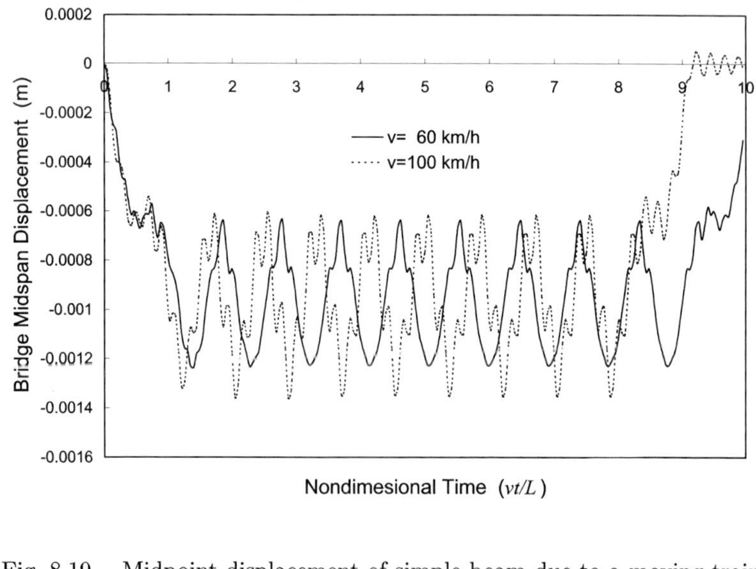


Fig. 8.19. Midpoint displacement of simple beam due to a moving train.

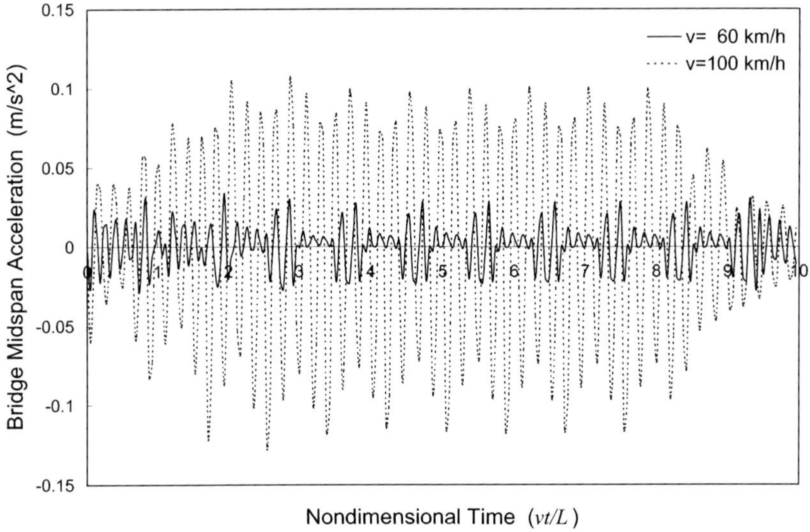


Fig. 8.20. Midpoint acceleration of simple beam due to a moving train.

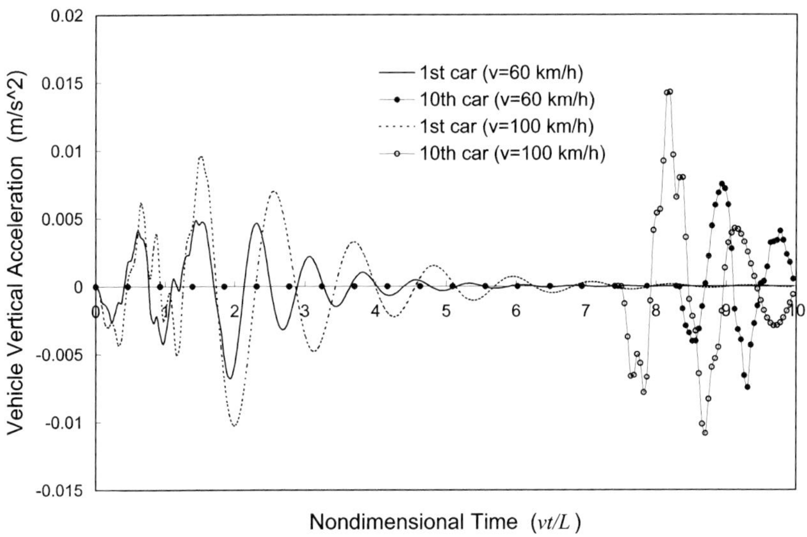


Fig. 8.21. Vertical acceleration of train cars

be seen, the bridge vibrates in a periodic manner as a result of the passage of each car of the train. Moreover, the response of the bridge for the train with higher speeds is larger than that with lower speeds. This is especially true for the acceleration response. The results for the vertical acceleration of the first and tenth (last) cars of the train were shown in Fig. 8.21. It can be observed that the train vibrates more severely at higher speeds, and that the response of the last car is larger than that of the first car at the same speed.

## 8.9. Concluding Remarks

In this chapter, an effective procedure hinging on computation of the contact forces is proposed for solving the interaction equations of general vehicle–bridge systems. One key step herein is the discretization of the second-order equations of motion for the vehicles using Newmark's finite difference scheme, which enables us to relate the contact forces to the wheel displacements, and then to the contact displacements of the bridge through enforcement of the no-jump condition for the vehicles. As the contact forces have been made available, the VBI element can be directly derived from the bridge equations by the concept of consistent nodal forces. The VBI element so derived possesses the same number of DOFs as its parent element, while possessing the property of symmetry. Since the VBI effect has been duly taken into account, the derived element can be used to compute not only the bridge response, the contact force, but also the vehicle response, the latter serves as a measure of passengers' riding comfort.

The versatility and applicability of the proposed procedure have been demonstrated in the numerical examples. It is concluded that the high modes of vibration of the bridge can affect more significantly the vehicles than the bridge response. The vehicle response is underestimated when using the sprung mass model, in comparison with those using the suspended rigid bar model. Using a single moving vehicle, it was demonstrated that the action of braking has no substantial influence on the response of the bridge due to the relatively small change in vehicle speed and relatively short acting time. The vertical response of the bridge is determined mainly by the vehicle speed before braking is applied. However, the horizontal reaction at the hinged support of the bridge is affected mainly by the entrance and departure of each of the two wheels from the beam. For a bridge traveled by a train with multi cars, periodic response can be observed following the passage of each car. Higher accelerations can be expected for the cars of a train moving with a higher speed, and the last car of a train tends to vibrate more seriously than the first car.

## Chapter 9

# Vehicle–Rails–Bridge Interaction — Two-Dimensional Modeling

This chapter deals with the steady-state response and riding comfort of a train moving over a series of simply-supported railway bridges, together with the impact response of the rails and bridges, using the two-dimensional model. The dynamic response of the vehicle–rails–bridge interaction system is solved by the condensation technique presented in Chapter 8. For the moving train to achieve the steady-state response, a bridge segment consisting of a minimal number of bridge units should be considered. Track irregularity with random nature is considered through the use of a power spectral density (PSD) function. The steady-state response of the train, rails and bridges, together with the fast Fourier transform (FFT) of the response, are computed and discussed. The impact responses of the rails and bridges under different train speeds are investigated using the impact factor. The maximum response of the train caused by the train–rails–bridge resonance is identified. Finally, the riding comfort of the trains moving over tracks of different classes of irregularities is assessed using Sperling's ride index.

## 9.1. Introduction

In regions where the ground traffic is congested, elevated bridges consisting of a series of simply-supported beams are often adopted as the supporting structure for railways, particularly for the high-speed railways, to offer unobstructed right of way, while minimizing the total area of lands used in construction. For this kind of problems, the bridge units are considered separate, but the track structure supported by the bridge units that carries the trains are treated as continuous. Thus, a train traveling over such a bridge system can excite certain interactions between the various components of the bridge. The dynamic behavior of a train traveling over a multi-unit bridge system as mentioned here is different from that over a short bridge. For the latter case, the responses of the train and bridge are generally of the transient nature, as the duration of the train traveling over the bridge is too short for the response to accumulate. Most existing researches in the literature are in this category (Frýba, 1972; Matsuura, 1976; Chu *et al.*, 1979; Bhatti *et al.*, 1985; Wakui *et al.*, 1995; Yau *et al.*, 1999). For the former case, however, due to the repetitive nature of the span units and the continuity nature of the track, the response of the train moving over the bridge can reach a *steady state* if the travel time of the train is long enough. To the authors' knowledge, such a problem was not thoroughly studied in the literature.

Smith *et al.* (1975) analyzed the response of the moving vehicles interacting with single, multiple and continuous span elevated structures, where the operating conditions for the occurrence of multi-span resonance were identified, along with the resonant amplitudes computed. Chen and Li (2000) studied the dynamic response of single and three-span uniform railway bridges subjected to three types of high-speed trains. In their work, only the response of the bridge was obtained, as the train was simplified as a series of moving loads. Based on the Lagrangian approach, Cheung *et al.* (1999) analyzed the vibration of multi-span nonuniform bridges due to the moving vehicles and trains, and presented the dynamic magnification factor for the bridge under different velocity ratios. Dukkipati and Dong (1999) studied the idealized steady state interaction between the railway vehicle and track, with the track deflection and contact force computed for the steady state.

In the aforementioned works, none has dealt with the steady-state vibration of the train moving over multi-unit railway bridges using realistic vehicle and bridge models that take into account the dynamic effect of the track system, which is assumed to be infinitely continuous. For the purpose of assessing the *riding comfort* of passengers, while enhancing the maneuverability of the train, it is necessary to investigate the dynamic response of the train, track and bridge as a whole, considering the interaction among the various components of the train–railway bridge system, including in particular the steady-state response.

The interaction between the moving train and the sustaining bridge makes the equations of motion for the two subsystems coupled and nonstationary. To solve this problem, a dynamic condensation technique has been developed by Yang and Wu (2001) for decoupling the two sets of equations of motion, which is basically the technique presented in Chapter 8. This chapter has been rewritten primarily from the work by Wu and Yang (2003), which can be regarded as an extension of the theoretical framework established by Yang and Wu (2001) in dealing with the two-dimensional steady-state response of a train moving over a bridge system that consists of a number of separate bridge units, which are connected to the track structure. This is a problem commonly encountered in railway engineering, but not very well studied previously. The riding comfort of the train will be evaluated using a comfort index based on the vertical acceleration response of the train.

## 9.2. Train and Bridge Models and Minimal Bridge Segment

Figure 9.1 shows the two-dimensional view of a train traveling over a multi-unit railway bridge with constant speed $v$. The train is simplified as a sequence of identical vehicles, each of which comprises one car body, two bogies and four wheels, as shown in Fig. 9.2. Each of the suspension systems of the vehicle is represented by a spring-dashpot unit. The bridge consists of many identical units, each of which is made of a concrete box-girder of constant section and is simply-supported at both ends. The track system lying on the bridges, particularly the rails, is simplified as an *infinite beam* resting on uniformly-distributed spring-dashpot units. Because the track structure is continuous, it serves as a medium for

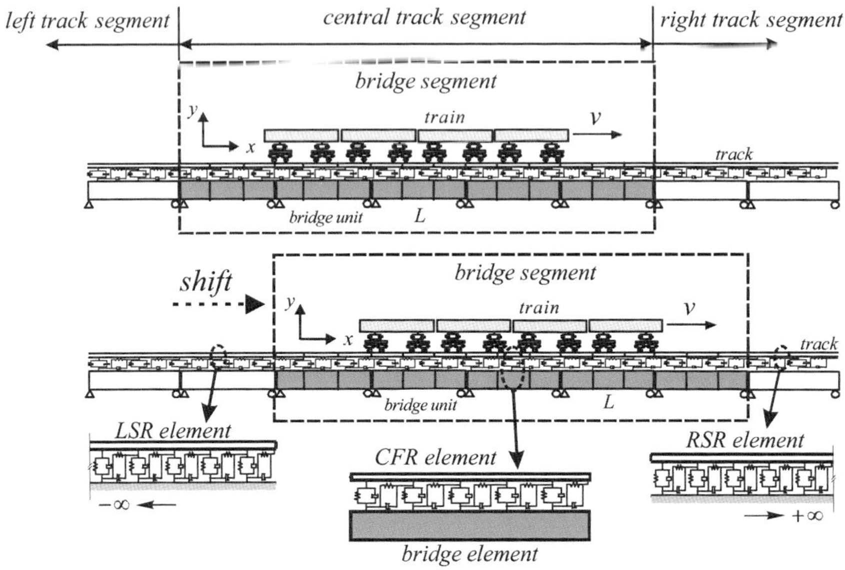


Fig. 9.1. A train moving on a multi-unit railway bridge.

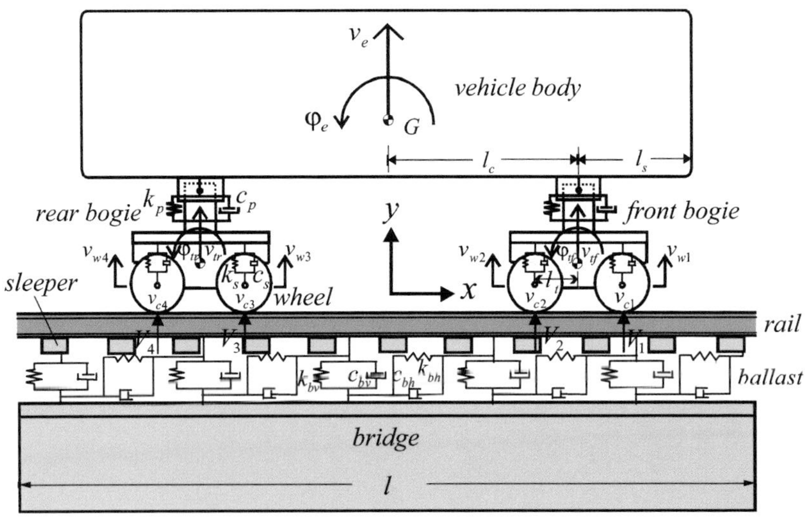


Fig. 9.2. Vehicle, track and bridge models.

transmitting the kinetic energy brought by the moving trains to the various components of the multi-unit bridge. All the wheels are assumed to be in close contact with the rails as they roll over. The physical parameters of the vehicle, track and bridge interaction system were indicated in Fig. 9.2 and summarized in Tables 9.1 and 9.2.

The car body of the vehicle is represented by a rigid bar, while the rails and bridges by beam elements of the Bernoulli–Euler type. It is realized that the equations of motion of the vehicle–rails–bridge interaction system are coupled and time-dependent, which will be solved using the dynamic condensation technique presented in the preceding chapter. By this technique all the degrees of freedom (DOFs) of the vehicle (i.e., the moving subsystem) are condensed into those of the rail elements in contact to form a condensed set of equations of motion, which are expressed in terms of only the DOFs of the rails and bridge units (i.e., the supporting subsystem).

For a moving train to reach the *steady-state response*, the minimal number $N_{\min}$ of the bridge units that should be considered for the

Table 9.1. The properties of vehicle model.

| Item | Notation | Value* |
|---|---|---|
| Mass of vehicle body | $M_c$ | 41.75 t |
| Mass moment of inertia of vehicle body | $I_c$ | 2080 t-m$^2$ |
| Mass of bogie | $M_t$ | 3.04 t |
| Mass moment of inertia of bogie | $I_t$ | 3.93 t-m$^2$ |
| Mass of wheel | $M_w$ | 1.78 t |
| Stiffness of primary suspension system | $k_p$ | 1180 kN/m |
| Damping of primary suspension system | $c_p$ | 39.2 kN-s/m |
| Stiffness of secondary suspension system | $k_s$ | 530 kN/m |
| Damping of secondary suspension system | $c_s$ | 90.2 kN-s/m |
| Half of longitudinal distance between centers of gravity of front and rear bogies | $l_c$ | 8.75 m |
| Half of wheelbase | $l_t$ | 1.25 m |
| Longitudinal distance between center of gravity of bogie and nearest side of vehicle body | $l_s$ | 3.75 m |

*For SKS series 300 vehicle model, extracted from Wakui *et al.* (1995) with some modifications.

Table 9.2. The properties of track and bridge unit.

| Item | Notation | Value* |
|---|---|---|
| **Track** | | |
| Young's modulus of rail | $E_r$ | 210 GPa |
| Poisson's ratio of rail | $v_r$ | 0.3 |
| Per-unit-length mass of rail (including mass of sleeper) | $m_r$ | 0.2995 t |
| Sectional area of rail | $A_r$ | $7.686 \times 10^{-3}$ m$^2$ |
| Flexural moment of inertia of rail | $I_r$ | $3.055 \times 10^{-5}$ m$^4$ |
| Per-unit-length vertical stiffness of ballast layer | $k_{bv}$ | $1.04 \times 10^5$ kN/m |
| Per-unit-length vertical damping of ballast layer | $c_{bv}$ | 50 kN-s/m |
| Per-unit-length horizontal stiffness of ballast layer | $k_{bh}$ | $1.04 \times 10^4$ kN/m |
| Per-unit-length horizontal damping of ballast layer | $c_{bh}$ | 50 kN-s/m |
| **Bridge unit** | | |
| Young's modulus of concrete | $E_b$ | 28.2 GPa |
| Poisson's ratio of concrete | $v_b$ | 0.2 |
| Per-unit-length mass (including mass of ballast layer) | $m_b$ | 31.4 t |
| Sectional area | $A_b$ | 7.94 m$^2$ |
| Flexural moment of inertia | $I_b$ | 8.72 m$^4$ |
| Length of bridge unit | $L$ | 30 m |

*The data for the bridge unit were modified from those used by the Taiwan-HSR.

bridge segment included in analysis is:

$$N_{\min} = \left\| \frac{n_v \times d}{L} \right\| + 2\,, \tag{9.1}$$

where $n_v$ denotes the number of vehicles comprising the train, $d$ is the length of the vehicle, i.e., $d = 2(l_c + l_s)$ as indicated in Fig. 9.2, $L$ is the length of each bridge unit, and $\|a\|$ represents the integer that is less than but nearest to $a (a \geqq 0)$. As can be seen from Fig. 9.1, the number $N_{\min}$ given in Eq. (9.1) is the maximum number of bridge units that can be occupied by the train at any instant. On the other hand, since all the vehicles constituting the train can affect each other through the bridge units, which are connected by the track structure, it is necessary to consider a minimum of $N_{\min}$ bridge units so that the response of all the vehicles can be properly captured. The part of the bridge formed integrally by these $N_{\min}$ bridge units is referred to as the *minimal bridge segment* that should be included in analysis in order for the moving train to reach the steady-state response. In reality, the minimal bridge segment moves following the motion of the train but its size remains unchanged.

## 9.3. Vehicle's Equations of Motion and Contact Forces

Let $[m_v]$, $[c_v]$, and $[k_v]$ respectively denote the mass, damping and stiffness matrices of the vehicle, and $\{d_v\}$ the displacement vector of the vehicle. The equation of motion for the vehicle can be written as

$$[m_v]\{\ddot{d}_v\} + [c_v]\{\dot{d}_v\} + [k_v]\{d_v\} = \{f_v\}, \tag{9.2}$$

where $\{f_v\}$ is the external force vector, which can be decomposed into two parts,

$$\{f_v\} = \{f_e\} + [l]\{f_c\}. \tag{9.3}$$

Here, $\{f_e\}$ denotes the external force components excluding the contact forces, $[l]$ is a transformation matrix, and $\{f_c\}$ represents the contact forces acting through the wheels. As shown in Fig. 9.2, for a vehicle with four wheels, the contact force vector is $\{f_c\} = \langle V_1\ V_2\ V_3\ V_4\rangle^T$, where $V_i$ is the $i$th contact force.

For the present purposes, we shall decompose the vehicle structure in Fig. 9.2 into two parts, i.e., the *upper* (or *noncontact*) and *wheel* (or *contact*) parts. The upper part consists of the car body, the primary and secondary suspension systems and the front and rear bogies, which has a total of six DOFs as denoted by $\{d_u\}$. The wheel part consists of four wheels, which can be represented as $\{d_w\} = \langle v_{w1}\ v_{w2}\ v_{w3}\ v_{w4}\rangle^T$ with $v_{wi}$ denoting the vertical DOF of the $i$th wheel. Correspondingly, the vertical displacements of the four contact points on the rails are denoted as $\{d_c\} = \langle v_{c1}\ v_{c2}\ v_{c3}\ v_{c4}\rangle^T$. The total displacement DOFs of the vehicle can therefore be presented as $\{d_v\} = \langle\{d_u\}\{d_w\}\rangle^T$.

The vehicle–bridge interaction (VBI) system will be analyzed in an incremental manner in time domain. Assuming that all the information of the system at time $t$ is known and $\Delta t$ is a small time increment, we are interested in the behavior of the system at time

$t+\Delta t$. The equations of motion as given in Eq. (9.2) for the vehicle can be rewritten for time $t+\Delta t$ with proper partitions in the system matrices to account for the *upper* and *wheel* parts as

$$\begin{bmatrix} [m_{uu}] & [m_{uw}] \\ [m_{wu}] & [m_{ww}] \end{bmatrix} \begin{Bmatrix} \{\ddot{d}_u\} \\ \{\ddot{d}_w\} \end{Bmatrix}_{t+\Delta t} + \begin{bmatrix} [c_{uu}] & [c_{uw}] \\ [c_{wu}] & [c_{ww}] \end{bmatrix} \begin{Bmatrix} \{\dot{d}_u\} \\ \{\dot{d}_w\} \end{Bmatrix}_{t+\Delta t} + \begin{bmatrix} [k_{uu}] & [k_{uw}] \\ [k_{wu}] & [k_{ww}] \end{bmatrix} \begin{Bmatrix} \{d_u\} \\ \{d_w\} \end{Bmatrix}_{t+\Delta t} = \begin{Bmatrix} \{f_{ue}\} \\ \{f_{we}\} \end{Bmatrix}_{t+\Delta t} + \begin{bmatrix} [l_u] \\ [l_w] \end{bmatrix} \{f_c\}_{t+\Delta t}\,, \tag{9.4}$$

where $[m_{uu}]$, $[m_{uw}]$, $[m_{wu}]$ and $[m_{ww}]$ denote the partitioned mass matrices of the vehicle, $[c_{uu}]$, $[c_{uw}]$, $[c_{wu}]$ and $[c_{ww}]$ the damping matrices, and $[k_{uu}]$, $[k_{uw}]$, $[k_{wu}]$ and $[k_{ww}]$ the stiffness matrices; $\{f_{ue}\}$ and $\{f_{we}\}$ the external forces acting on the upper and wheel parts of the vehicle, respectively; and $[l_u]$ and $[l_w]$ the transformation matrices for the two parts. Since only the wheels are acted upon by the contact forces, the matrix $[l_u]$ is set to $[0]$. All the partitioned matrices and vectors were listed in Appendix G, in which $W = (M_w+0.5M_t+0.25M_c)g$ represents the static axle load and $g$ is the acceleration of gravity. Here, we use $M_w$, $M_t$, and $M_c$ to denote the mass of the wheel, bogie, and car body, respectively.

The wheel displacements $\{d_w\}$ can be related to the contact displacements $\{d_c\}$ of the rails by the *constraint equation*,

$$\{d_w\} = [\Gamma]\{d_c\} + \{r\}\,, \tag{9.5}$$

where $[\Gamma]$ is the constraint matrix and $\{r\}$ is the vector of *track irregularity*, $\{r\} = \langle r_1\ r_2\ r_3\ r_4\rangle^T$, where $r_i = r(x_i)$ is the track profile evaluated at the $i$th contact point. Assuming the wheels to be in full contact with the rails, $[\Gamma]$ reduces to a unit matrix [I]. By the procedure presented in Chapter 8, the $i$th contact force $V_i$ of the vehicle can be solved from Eqs. (9.4) and (9.5) as:

$$V_{i,t+\Delta t} = p_{ci,t+\Delta t} + q_{ci,t} + \sum_{j=1}^{4} (m_{cij}\ddot{v}_{cj,t+\Delta t} + c_{cij}\dot{v}_{cj,t+\Delta t} + k_{cij}v_{cj,t+\Delta t}), \quad (9.6)$$

where $m_{cij}$, $c_{cij}$ and $k_{cij}$ represent the entries located at the $i$th row and $j$th column of the contact matrices $[m_c]$, $[c_c]$, and $[k_c]$, respectively, as given in Eq. (8.14), and $p_{ci,t+\Delta t}$ and $q_{ci,t}$ the $i$th entries of the load vectors $\{p_c\}_{t+\Delta t}$ and $\{q_c\}_t$ resulting from the contact forces. Both the load vectors $\{p_c\}_{t+\Delta t}$ and $\{q_c\}_t$ are available in Eq. (8.15).

## 9.4. Rails and Bridge Element Equations

With reference to Fig. 9.1, the track is divided into three typical parts, i.e., the *central*, *left* and *right track segments*. The *central track segment* refers to the part of the track within the bridge range, in which the rails interact with the bridge through the ballast layer. The *left* and *right track segments* correspond to the semi-infinite parts of the track outside the bridge range. The track structure in the central track segment is modeled as a set of rail elements supported by spring-dashpot units and in turn by bridge elements. Each of the rail elements is of identical length and referred to as the *central finite rail* (CFR) element. The track structures in the two side segments are modeled as semi-infinite rail elements supported by distributed spring-dashpot units, which have been referred to as the *left semi-infinite rail* (LSR) element and *right semi-infinite rail* (RSR) element.

### 9.4.1. *Central Finite Rail (CFR) Element and Bridge Element*

As shown in Fig. 9.3(a), a CFR element and underlying bridge element are connected by a uniformly-distributed spring-dashpot-unit layer, both of which are of length $l$ and are modeled as conventional beam elements. The nodal DOF's of the rail element can be denoted as $\{d_r\} = \langle u_{r1}\ v_{r1}\ \theta_{r1}\ u_{r2}\ v_{r2}\ \theta_{r2}\rangle^T$, and those of the bridge element as $\{d_b\} = \langle u_{b1}\ v_{b1}\ \theta_{b1}\ u_{b2}\ v_{b2}\ \theta_{b2}\rangle^T$. By the principle of virtual work, the equation of equilibrium for the rail element can be written as

$$\begin{aligned}
&\int_0^l E_r(2A_r)u_r'\delta u_r' dx + \int_0^l E_r(2I_r)v_r''\delta v_r'' dx \\
&\quad = -\int_0^l (2m_r)\ddot{u}_r\delta u_r dx - \int_0^l (2c_r)\dot{u}_r\delta u_r dx \\
&\qquad + \int_0^l k_{bh}(u_b - u_r)\delta u_r dx + \int_0^l c_{bh}(\dot{u}_b - \dot{u}_r)\delta u_r dx \\
&\qquad - \int_0^l (2m_r)\ddot{v}_r\delta v_r dx - \int_0^l (2c_r)\dot{v}_r\delta v_r dx \\
&\qquad + \int_0^l k_{bv}(v_b - v_r)\delta v_r dx + \int_0^l c_{bv}(\dot{v}_b - \dot{v}_r)\delta v_r dx \\
&\qquad + \langle \delta d_r\rangle\{f_r\}\,,
\end{aligned} \tag{9.7}$$

where $x$ is the local coordinate, $0 \leqq x \leqq l$, $\{\delta d_r\}$ denotes the variation of the rail element displacement vector $\{d_r\}$, $\{f_r\}$ the corresponding nodal loads acting on the rail element, i.e., $\{f_r\} = \langle H_{r1}\ V_{r1}\ M_{r1}\ H_{r2}\ V_{r2}\ M_{r2}\rangle^T$, $(u_r, v_r)$ the axial and vertical displacements of the rail element, and $(u_b, v_b)$ the corresponding displacements of the bridge element. The displacement fields of the rail and bridge elements can be expressed in terms of the nodal DOFs as

$$\begin{aligned}
u_r(x) &= \langle N_u\rangle\{d_r\}\,, \\
v_r(x) &= \langle N_v\rangle\{d_r\}\,, \\
u_b(x) &= \langle N_u\rangle\{d_b\}\,, \\
v_b(x) &= \langle N_v\rangle\{d_b\}\,,
\end{aligned} \tag{9.8}$$

where $\langle N_u\rangle = \langle N_1\ 0\ 0\ N_2\ 0\ 0\rangle$ denotes the interpolation vector for the axial displacement, $\langle N_v\rangle = \langle 0\ N_3\ N_4\ 0\ N_5\ N_6\rangle$ the interpolation vector for the vertical displacement, $N_1$ and $N_2$ are linear, and $N_3$,

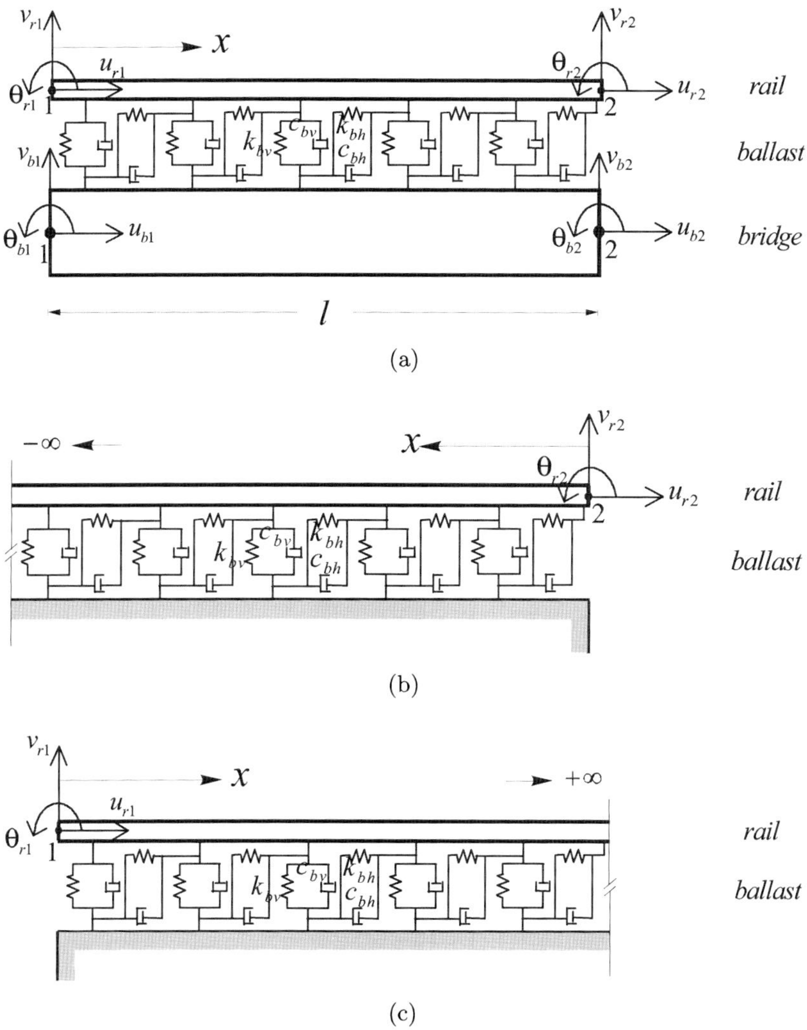


Fig. 9.3. Rail element: (a) central, (b) left, and (c) right.

$N_4$, $N_5$ and $N_6$ are cubic Hermitian functions. Substituting the preceding displacement fields into Eq. (9.7), one can obtain the equation of motion for the *CFR element* as follows:

$$[m_r]\{d_r\} + [c_r]\{\dot{d}_r\} + [k_r]\{d_r\} = \{f_r\} + [c_d]\{\dot{d}_b\} + [k_g]\{d_b\}, \quad (9.9)$$

where $\{d_r\}$ and $\{d_b\}$ denote the nodal DOFs of the CFR element and bridge element, respectively, $\{f_r\}$ the external nodal forces. The mass, damping and stiffness matrices, $[m_r]$, $[c_r]$ and $[k_r]$, are

$$\begin{aligned}
[m_r] &= 2m_r[\psi_u^0] + 2m_r[\psi_v^0]\,, \\
[c_r] &= \underline{2c_r[\psi_u^0] + 2c_r[\psi_v^0]} + \underline{c_{bh}[\psi_u^0] + c_{bv}[\psi_v^0]} = [c_{r0}] + [c_{rb}]\,, \\
[k_r] &= 2E_rA_r[\psi_u^1] + 2E_rI_r[\psi_v^2] + k_{bh}[\psi_u^0] + k_{bv}[\psi_v^0]\,.
\end{aligned} \tag{9.10}$$

The damping matrix $[c_r]$ is composed of the rail material damping matrix $[c_{r0}]$ and the ballast damping matrix $[c_{rb}]$. The other parameters in Eq. (9.10) have been defined in Table 9.2. The damping and stiffness matrices, $[c_d]$ and $[k_s]$, due to interaction with the bridge element, are

$$\begin{aligned}
[c_d] &= c_{bh}[\psi_u^0] + c_{bv}[\psi_v^0]\,, \\
[k_s] &= k_{bh}[\psi_u^0] + k_{bv}[\psi_v^0]\,,
\end{aligned} \tag{9.11}$$

where the matrices $[\psi_u^0]$, $[\psi_u^1]$, $[\psi_v^0]$ and $[\psi_v^2]$ are defined as

$$\begin{aligned}
[\psi_u^0] &= \int_0^l \{N_u\}\langle N_u\rangle dx\,, \\
[\psi_u^1] &= \int_0^l \{N_u'\}\langle N_u'\rangle dx\,, \\
[\psi_v^0] &= \int_0^l \{N_v\}\langle N_v\rangle dx\,, \\
[\psi_v^2] &= \int_0^l \{N_v''\}\langle N_v''\rangle dx\,.
\end{aligned} \tag{9.12}$$

The results for all these matrices have been listed in Appendix H.

Similarly, the equation of motion for the *bridge element* can be derived as

$$[m_b]\{\ddot{d}_b\} + [c_b]\{\dot{d}_b\} + [k_b]\{d_b\} = \{f_b\} + [c_d]\{\dot{d}_r\} + [k_s]\{d_r\}\,. \tag{9.13}$$

Here, $\{f_b\}$ denotes the external nodal forces. The mass, damping and stiffness matrices, $[m_b]$, $[c_b]$ and $[k_b]$, are defined as follows:

$$
\begin{aligned}
[m_b] &= m_b[\psi_u^0] + m_b[\psi_v^0]\,, \\
[c_b] &= \underline{c_b[\psi_u^0] + c_b[\psi_v^0]} + \underline{c_{bh}[\psi_u^0] + c_{bv}[\psi_v^0]} = [c_{b0}] + [c_{br}]\,, \\
[k_b] &= E_b A_b[\psi_u^1] + E_b I_b[\psi_v^2] + k_{bh}[\psi_u^0] + k_{bv}[\psi_v^0]\,,
\end{aligned}
\tag{9.14}
$$

where the damping matrix $[c_b]$ consists of the material damping matrix $[c_{b0}]$ and the ballast damping matrix $[c_{br}]$, with $[c_{br}] = [c_{rb}]$. The two matrices $[c_d]$ and $[k_s]$ in Eqs. (9.9) and (9.13) are to account for the interaction between the rails and the bridge, which were given in Eq. (9.11).

### 9.4.2. *Left Semi-Infinite Rail (LSR) Element*

The LSR element is idealized as a semi-infinite beam with a single node, as shown in Fig. 9.3(b). The nodal displacements may be denoted as $\{d_{rl}\} = \langle u_{r2}\ v_{r2}\ \theta_{r2}\rangle^T$ and the nodal external forces as $\{f_{rl}\} = \langle H_{r2}\ V_{r2}\ M_{r2}\rangle^T$. By the virtual work principle, one can write

$$
\begin{aligned}
&\int_{-\infty}^{0} E_r(2A_r)u_r'\delta u_r'\,dx + \int_{-\infty}^{0} E_r(2I_r)v_r''\delta v_r''\,dx \\
&\quad = -\int_{-\infty}^{0} (2m_r)\ddot{u}_r\delta u_r\,dx - \int_{-\infty}^{0} (2c_r)\dot{u}_r\delta u_r\,dx \\
&\qquad - \int_{-\infty}^{0} k_{bh}u_r\delta u_r\,dx - \int_{-\infty}^{0} c_{bh}\dot{u}_r\delta u_r\,dx \\
&\qquad - \int_{-\infty}^{0} (2m_r)\ddot{v}_r\delta v_r\,dx - \int_{-\infty}^{0} (2c_r)\dot{v}_r\delta v_r\,dx \\
&\qquad - \int_{-\infty}^{0} k_{bv}v_r\delta v_r\,dx - \int_{-\infty}^{0} c_{bv}\dot{v}_r\delta v_r\,dx + \langle\delta d_{rl}\rangle\{f_{rl}\}\,,
\end{aligned}
\tag{9.15}
$$

where $\langle\delta d_{rl}\rangle = \langle\delta u_{r2}\ \delta v_{r2}\ \delta\theta_{r2}\rangle$ denotes the nodal virtual displacements, and $x$ is the local coordinate, $-\infty < x \leqq 0$. The horizontal and vertical displacement fields of the element can be described using the relations such as: $u_r = \langle N_u\rangle\{d_{rl}\}$, $v_r = \langle N_v\rangle\{d_{rl}\}$, where

$\langle N_u \rangle = \langle N_2\ 0\ 0 \rangle$ and $\langle N_v \rangle = \langle 0\ N_5\ N_6 \rangle$ denote the interpolation vectors. Because of the semi-infinite nature of the LSR element, the interpolation functions should be obtained from the static solution to the problem of a beam resting on the *Winkler foundation* subjected to a unit load (Hetényi, 1979), that is,

$$\begin{aligned} N_2 &= e^{\lambda_u x}\,, \\ N_5 &= e^{\lambda_v x}(\cos \lambda_v x - \sin \lambda_v x)\,, \\ N_6 &= \frac{1}{\lambda_v} e^{\lambda_v x} \sin \lambda_v x\,, \end{aligned} \tag{9.16}$$

where $\lambda_u$ and $\lambda_v$ denote the horizontal and vertical *characteristic* numbers of the beam-Winkler foundation system,

$$\begin{aligned} \lambda_u &= \sqrt{\frac{k_{bh}}{E_r(2A_r)}}\,, \\ \lambda_v &= \sqrt[4]{\frac{k_{bv}}{4E_r(2I_r)}}\,. \end{aligned} \tag{9.17}$$

By using Eq. (9.16) and the definition of the interpolation functions, Eq. (9.15) can be manipulated to yield the equation of motion for the LSR element,

$$[m_{rl}]\{\ddot{d}_{rl}\} + [c_{rl}]\{\dot{d}_{rl}\} + [k_{rl}]\{d_{rl}\} = \{f_{rl}\}\,. \tag{9.18}$$

Here, the mass, damping and stiffness matrices $[m_{rl}]$, $[c_{rl}]$ and $[k_{rl}]$ are

$$\begin{aligned} [m_{rl}] &= 2m_r[\psi_u^0]_l + 2m_r[\psi_v^0]_l\,, \\ [c_{rl}] &= \underline{2c_r[\psi_u^0]_l + 2c_r[\psi_v^0]_l} + \underline{c_{bh}[\psi_u^0]_l + c_{bv}[\psi_v^0]_l} \\ &= [c_{rl0}] + [c_{rlb}]\,, \\ [k_{rl}] &= 2E_rA_r[\psi_u^1]_l + 2E_rI_r[\psi_v^2]_l + k_{bh}[\psi_u^0]_l + k_{bv}[\psi_v^0]_l\,, \end{aligned} \tag{9.19}$$

where $[\psi_u^0]_l$, $[\psi_u^1]_l$, $[\psi_v^0]_l$, and $[\psi_v^2]_l$ can be found in Appendix H.

### 9.4.3. *Right Semi-Infinite Rail (RSR) Element*

As shown in Fig. 9.3(c), the RSR element is also idealized as a semi-infinite beam with a single node. The nodal displacements are $\{d_{rr}\} = \langle u_{r1}\ v_{r1}\ \theta_{r1}\rangle^T$ and the nodal forces are $\{f_{rr}\} = \langle H_{r1}\ V_{r1}\ M_{r1}\rangle^T$. The horizontal and vertical displacements in the element can be related to the nodal DOFs as $u_r = \langle N_u\rangle\{d_{rr}\}$, $v_r = \langle N_v\rangle\{d_{rr}\}$, where $\langle N_u\rangle = \langle N_1\ 0\ 0\rangle$ and $\langle N_v\rangle = \langle 0\ N_3\ N_4\rangle$. Here the interpolation functions can also be determined from the static solution as

$$
\begin{aligned}
N_1 &= e^{-\lambda_u x}\,,\\
N_3 &= e^{-\lambda_v x}(\cos\lambda_v x + \sin\lambda_v x)\,,\\
N_4 &= \frac{1}{\lambda_v}e^{-\lambda_v x}\sin\lambda_v x\,,
\end{aligned}
\tag{9.20}
$$

where $\lambda_u$ and $\lambda_v$ are the same parameters as those given in Eq. (9.17). Following the same procedure as that for the LSR element, the equation of motion for the RSR element can be derived as

$$
[m_{rr}]\{\ddot{d}_{rr}\} + [c_{rr}]\{\dot{d}_{rr}\} + [k_{rr}]\{d_{rr}\} = \{f_{rr}\}\,. \tag{9.21}
$$

The mass, damping and stiffness matrices $[m_{rr}]$, $[c_{rr}]$ and $[k_{rr}]$ are

$$
\begin{aligned}
[m_{rr}] &= 2m_r[\psi_u^0]_r + 2m_r[\psi_v^0]_r\,,\\
[c_{rr}] &= \underline{2c_r[\psi_u^0]_r + 2c_r[\psi_v^0]_r} + \underline{c_{bh}[\psi_u^0]_r + c_{bv}[\psi_v^0]_r}\\
&= [c_{rr0}] + [c_{rrb}]\,,\\
[k_{rr}] &= 2E_rA_r[\psi_u^1]_r + 2E_rI_r[\psi_v^2]_r + k_{bh}[\psi_u^0]_r + k_{bv}[\psi_v^0]_r\,,
\end{aligned}
\tag{9.22}
$$

where the matrices $[\psi_u^0]_r$, $[\psi_u^1]_r$, $[\psi_v^0]_r$, and $[\psi_v^2]_r$ have been listed in Appendix H. In a step-by-step time-history analysis, the equations of motion in Eqs. (9.9), (9.13), (9.18), and (9.21) should be interpreted as those established for the deformed position of the system at time $t + \Delta t$.

## 9.5. VRI Element Considering Vertical Contact Forces Only

Assume that at time $t+\Delta t$, the four wheelsets of the vehicle are acting simultaneously at the $e_1$, $e_2$, $e_3$ and $e_4$th rail elements, which will be grossly referred to as the *vehicle–rails interaction* (VRI) elements, as they are directly under the action of the wheel loads. Consider the $e_i$th element that is acted upon only by the vertical component of the $i$th contact force $V_{i,t+\Delta t}$. The equation of motion for the $e_i$th rail element at time $t + \Delta t$ can be written as follows:

$$[m_{ri}]\{\ddot{d}_{ri}\}_{t+\Delta t} + [c_{ri}]\{\dot{d}_{ri}\}_{t+\Delta t} + [k_{ri}]\{d_{ri}\}_{t+\Delta t} = \{f_{ri}\}_{t+\Delta t} + \varepsilon_i([c_d]\{\dot{d}_{bi}\} + [k_s]\{d_{bi}\}) - \{f_{rci}\}_{t+\Delta t}\,, \tag{9.23}$$

where $[m_{ri}] = [m_r]$, $[c_{ri}] = [c_r]$, $[k_{ri}] = [k_r]$, $\{d_{ri}\} = \{d_r\}$ and $\varepsilon_i = 1$ for the case with the contact force acting on the CFR element; $[m_{ri}] = [m_{rl}]$, $[c_{ri}] = [c_{rl}]$, $[k_{ri}] = [k_{rl}]$, $\{d_{ri}\} = \{d_{rl}\}$ and $\varepsilon_i = 0$ for the case with the LSR element; $[m_{ri}] = [m_{rr}]$, $[c_{ri}] = [c_{rr}]$, $[k_{ri}] = [k_{rr}]$, $\{d_{ri}\} = \{d_{rr}\}$ and $\varepsilon_i = 0$ for the case with the RSR element; and $\{f_{rci}\}$ denotes the vector of equivalent nodal forces resulting from the action of the $i$th vertical contact force $V_{i,t+\Delta t}$,

$$\{f_{rci}\}_{t+\Delta t} = \{N_{vi}\}V_{i,t+\Delta t}\,, \tag{9.24}$$

where $\{N_v\}$ denotes the interpolation vector for the vertical displacement of the $e_i$th element, which varies according to the type of elements, i.e., CFR, RSR or LSR, to which the contact force is acting. The subscript $i$ indicates that the vector $\{N_{vi}\}$ is evaluated at the $i$th contact point, i.e.,

$$\{N_{vi}\} = \{N_v(x_i)\}\,, \tag{9.25}$$

where $x_i$ is the local coordinate of the $i$th contact point on the $e_i$th element. By using Eqs. (9.24) and (9.6), Eq. (9.23) can be rewritten as follows:

$$
\begin{aligned}
&[m_{ri}]\{\ddot{d}_{ri}\}_{t+\Delta t} + [c_{ri}]\{\dot{d}_{ri}\}_{t+\Delta t} + [k_{ri}]\{d_{ri}\}_{t+\Delta t} \\
&\quad = \{f_{ri}\}_{t+\Delta t} + \varepsilon_i([c_d]\{\dot{d}_{bi}\}_{t+\Delta t} + [k_s]\{d_{bi}\}_{t+\Delta t}) \\
&\qquad - \sum_{j=1}^{4}([m^*_{cij}]\{\ddot{d}_{rj}\}_{t+\Delta t} + [c^*_{cij}]\{\dot{d}_{rj}\}_{t+\Delta t} \\
&\qquad + [k^*_{cij}]\{d_{rj}\}_{t+\Delta t}) - \{p^*_{ci}\}_{t+\Delta t} - \{q^*_{ci}\}_t \,,
\end{aligned}
\tag{9.26}
$$

where the asterisked matrices represent the *linking action* transmitted through the car body by the $e_j$th element (under the $j$th wheel load) on the $e_i$th element (under the $i$th wheel load),

$$
\begin{aligned}
[m^*_{cij}] &= \{N_{vi}\}m_{cij}\langle N_{vj}\rangle \,, \\
[c^*_{cij}] &= \{N_{vi}\}c_{cij}\langle N_{vj}\rangle \,, \\
[k^*_{cij}] &= \{N_{vi}\}k_{cij}\langle N_{vj}\rangle \,,
\end{aligned}
\tag{9.27}
$$

and the equivalent nodal loads resulting from the contact forces are

$$
\begin{aligned}
\{p^*_{ci}\}_{t+\Delta t} &= \{N_{vi}\}p_{ci,t+\Delta t} \,, \\
\{q^*_{ci}\}_t &= \{N_{vi}\}q_{ci,t} \,.
\end{aligned}
\tag{9.28}
$$

The equation of motion as given in Eq. (9.26) will be referred to as the *condensed equation of motion* for the VRI element, as all the relevant vehicle DOFs have been eliminated. Note that all the asterisked matrices and load vectors involved in Eqs. (9.27) and (9.28), which are representative of the interaction effects, are time-dependent, since $\{N_{vi}\}$ varies as the contact point moves. They should be updated at each time step in an incremental analysis.

## 9.6. VRI Element Considering General Contact Forces

The equation of motion as derived in Eq. (9.26) for the VRI element considers only the action of the vertical contact forces $V_{i,t+\Delta t}$. For the case where acceleration or deceleration are involved in the motion of vehicles over the bridge, it is necessary to consider the horizontal contact forces $H_{i,t+\Delta t}$ as well. In accordance, the contact force $\{f_{rci}\}_{t+\Delta t}$ in Eq. (9.23) should be modified as

$$\{f_{rci}\}_{t+\Delta t} = \{N_{ui}\}H_{i,t+\Delta t} + \{N_{vi}\}V_{i,t+\Delta t}\,, \tag{9.29}$$

where $\{N_u\}$ denotes the interpolation vector for the axial displacement of the $e_i$th element, which varies according to the type of rail elements, i.e., CFR, RSR or LSR elements, and $\{N_{ui}\} = \{N_u(x_i)\}$ is evaluated at the $i$th contact point. The horizontal contact forces $H_{i,t+\Delta t}$ may be generated by the rolling, accelerating or braking action of the wheelsets. All these actions can be regarded as parts of the frictional force, depending on the cohesion between the wheels and the rails. For convenience, the $i$th horizontal contact force $H_i$ may be related to the vertical force $V_i$ as

$$H_i = \mu_i V_i\,, \tag{9.30}$$

where $\mu_i$ is the *frictional coefficient* for the $i$th wheelset, which equals the coefficient of braking or acceleration. The frictional coefficient in rolling is neglected in this study, since it is quite small.

Considering the more general expression in Eq. (9.29) for the contact forces, along with Eq. (9.30), one can derive a condensed equation of motion for the $e_i$th rail element that is identical in form to Eq. (9.26), but with the asterisked matrices given as

$$\begin{aligned}
[m^*_{cij}] &= \{N_{ui}\}\mu_i m_{cij}\langle N_{uj}\rangle + \{N_{vi}\}m_{cij}\langle N_{vj}\rangle\,,\\
[c^*_{cij}] &= \{N_{ui}\}\mu_i c_{cij}\langle N_{uj}\rangle + \{N_{vi}\}c_{cij}\langle N_{vj}\rangle\,,\\
[k^*_{cij}] &= \{N_{ui}\}\mu_i k_{cij}\langle N_{uj}\rangle + \{N_{vi}\}k_{cij}\langle N_{vj}\rangle\,,
\end{aligned} \tag{9.31}$$

and the following for the nodal loads resulting from the contact forces,

$$\begin{aligned}
\{p^*_{ci}\}_{t+\Delta t} &= \{N_{ui}\}\mu_i p_{ci,t+\Delta t} + \{N_{vi}\}p_{ci,t+\Delta t}\,,\\
\{q^*_{ci}\}_t &= \{N_{ui}\}\mu_i q_{ci,t} + \{N_{vi}\}q_{ci,t}\,.
\end{aligned} \tag{9.32}$$

Again, all the asterisked matrices and vectors in Eqs. (9.31) and (9.32) are time-dependent and should be updated at each time increment.

## 9.7. System Equations and Structural Damping

Let us consider a train moving on a multi-unit railway bridge. At instant $t+\Delta t$, the parts of the rails directly acted upon by the wheel loads should be modeled by the VRI elements, while the remaining parts of the rails by the rail (beam) elements. All the VRI elements, rail elements and bridge elements can be assembled to yield the system equations:

$$\begin{aligned}[M]\{\ddot{D}\}_{t+\Delta t} + [C]\{\dot{D}\}_{t+\Delta t} + [K]\{D\}_{t+\Delta t} \\ = \{F\}_{t+\Delta t} - \{P_c^*\}_{t+\Delta t} - \{Q_c^*\}_t\,, \end{aligned} \tag{9.33}$$

where $\{D\} = \langle D_r \ D_b \rangle^T$ denotes the nodal DOFs of the entire railway bridge, with $\{D_r\}$, $\{D_b\}$ denoting those of the rails and bridge, respectively; $[M]$, $[C]$ and $[K]$ the system mass, damping and stiffness matrices; $\{F\} = \langle \{F\}_r \{F_b\} \rangle^T$ the external nodal loads, with $\{F_r\}$ and $\{F_b\}$ denoting those acting on the rails and on the bridge; $\{P_c^*\}_{t+\Delta t}$ and $\{Q_c^*\}_{t+\Delta t}$ the equivalent nodal (contact) forces in global coordinates.

A convenient way to construct the system matrices $[M]$, $[C]$ and $[K]$ is first to assemble the matrices $[M_0]$, $[C_0]$ and $[K_0]$ for the railway bridge that is free of any vehicle actions, i.e.,

$$\begin{aligned}[M_0] &= \begin{bmatrix} [M_r] & 0 \\ 0 & [M_b] \end{bmatrix} \\ &= \begin{bmatrix} [m_{rl}] + [m_{rr}] + \sum_{m=1}^{n_b} [m_r]_m & 0 \\ 0 & \sum_{m=1}^{n_b} [m_b]_m \end{bmatrix}, \end{aligned} \tag{9.34}$$

$$
\begin{aligned}
[C_0] &= [C_{00}] + [C_{0b}] \\
&= \begin{bmatrix} [C_{r0}] & 0 \\ 0 & [C_{b0}] \end{bmatrix} + \begin{bmatrix} [C_{rb}] & -[C_d] \\ -[C_d]^T & [C_{bb}] \end{bmatrix} \\
&= \begin{bmatrix} [c_{rl0}] + [c_{rr0}] + \sum_{m=1}^{n_b} [c_{r0}]_m & 0 \\ 0 & \sum_{m=1}^{n_b} [c_{b0}]_m \end{bmatrix} \\
&\quad + \begin{bmatrix} [c_{rlb}] + [c_{rrb}] + \sum_{m=1}^{n_b} [c_{rb}]_m & -\sum_{m=1}^{n_b} [c_d]_m \\ -\sum_{m=1}^{n_b} [c_d]_m^T & \sum_{m=1}^{n_b} [c_{bb}]_m \end{bmatrix},
\end{aligned} \tag{9.35}
$$

$$
\begin{aligned}
[K_0] &= \begin{bmatrix} [K_r] & -[K_s] \\ -[K_s]^T & [K_b] \end{bmatrix} \\
&= \begin{bmatrix} [k_{rl}] + [k_{rr}] + \sum_{m=1}^{n_b} [k_r]_m & -\sum_{m=1}^{n_b} [k_s]_m \\ -\sum_{m=1}^{n_b} [k_s]_m^T & \sum_{m=1}^{n_b} [k_b]_m \end{bmatrix},
\end{aligned} \tag{9.36}
$$

where $n_b$ is the number of the CFR (or bridge) elements considered within the minimal bridge segment. Then we can add to the preceding matrices the interaction effects of vehicles contributed through the VRI elements, as represented by the asterisked terms in Eq. (9.27) or Eq. (9.31), depending on the type of acting forces, namely,

$$
\begin{aligned}
[M] &= \begin{bmatrix} [M_r] + [M_c^*] & 0 \\ 0 & [M_b] \end{bmatrix} \\
&= \begin{bmatrix} [M_r] + \sum_{k=1}^{n_v} \left( \sum_{i=1}^{4} \sum_{j=1}^{4} [m_{cij}^*] \right)_k & 0 \\ 0 & [M_b] \end{bmatrix},
\end{aligned} \tag{9.37}
$$

$$[C] = [C_{00}] + \begin{bmatrix} [C_{rb}] + [C_c^*] & -[C_d] \\ -[C_d]^T & [C_{bb}] \end{bmatrix}$$

$$= [C_{00}] + \begin{bmatrix} [C_{rb}] + \sum_{k=1}^{n_v} \left( \sum_{i=1}^{4} \sum_{j=1}^{4} [c_{cij}^*] \right)_k & -[C_d] \\ -[C_d]^T & [C_{bb}] \end{bmatrix}, \quad (9.38)$$

$$[K] = \begin{bmatrix} [K_r] + [K_c^*] & -[K_s] \\ -[K_s]^T & [K_b] \end{bmatrix}$$

$$= \begin{bmatrix} [K_r] + \sum_{k=1}^{n_v} \left( \sum_{i=1}^{4} \sum_{j=1}^{4} [k_{cij}^*] \right)_k & -[K_s] \\ [K_s]^T & [K_b] \end{bmatrix}, \quad (9.39)$$

where $n_v$ is the number of vehicles. The equivalent nodal forces $\{P_c^*\}_{t+\Delta t}$ and $\{Q_c^*\}_{t+\Delta t}$ resulting from the contact forces are

$$\{P_c^*\}_{t+\Delta t} = \begin{Bmatrix} \sum_{k=1}^{n_v} \left( \sum_{i=1}^{4} \{p_{ci}^*\}_{t+\Delta t} \right)_k \\ 0 \end{Bmatrix}, \quad (9.40)$$

$$\{Q_c^*\}_t = \begin{Bmatrix} \sum_{k=1}^{n_v} \left( \sum_{i=1}^{4} \{q_{ci}^*\}_t \right)_k \\ 0 \end{Bmatrix}. \quad (9.41)$$

In Eqs. (9.37)–(9.39), all the asterisked terms or components represent exactly the interaction effects caused by the *linking action* of the car bodies. In particular, the subscript $k$ that represents the $k$th vehicle should be looped over from 1 to $n_v$. As the wheel loads move from time to time, it is necessary to check at each time step whether a rail element changes into a VRI element and vice versa, and to update the entries of the system matrices and vectors, concerning the contribution of the asterisked terms in Eqs. (9.37)–(9.39) and (9.40)–(9.41), according to the acting positions of the contact forces. One feature with the present procedure is that the total number of DOFs of the system remains *unchanged*, regardless of the interaction effects of moving vehicles. The *symmetry* property of the original system is also preserved.

Conventionally, the structural damping has been computed on the structure level. Based on the definition of *Rayleigh damping*, the damping matrix $[C_{00}]$ of the railway bridge in Eq. (9.38) is computed as follows:

$$[C_{00}] = \alpha_0[M_0] + \alpha_1[K_0]\,. \tag{9.42}$$

Given the damping ratio $\xi$, the two coefficients $\alpha_0$ and $\alpha_1$ can be determined as

$$\alpha_0 = \frac{2\xi\omega_1\omega_2}{\omega_1 + \omega_2}\,, \qquad \alpha_1 = \frac{2\xi}{\omega_1 + \omega_2}\,, \tag{9.43}$$

where $\omega_1$ and $\omega_2$ are the first two frequencies of vibration of the railway bridge.

The system equations as given in Eq. (9.33) will be solved using *Newmark's $\beta$ method* with $\beta = 0.25$ and $\gamma = 0.5$. By this method, the system responses $\{\ddot{D}\}$, $\{\dot{D}\}$ and $\{D\}$ at time $t + \Delta t$ can first be discretized and related to those at time $t$, with which the system equations in Eq. (9.33) can be reduced to an equivalent stiffness equation and solved. Details of such a procedure are available in Appendix B.

## 9.8. Shift of Bridge Segment and Renumbering of Nodal Degrees of Freedom

Once the train moves out of the right boundary of the bridge segment considered, a *shift* in the element mesh is performed, i.e., by adding a new bridge unit to the segment on the right-hand side, and by deleting the leftmost unit from the segment (see Fig. 9.1). By doing so, the train can interact continuously with the rails and bridge, while the size of the bridge segment remains unchanged. On the other hand, since the bridge segment has shifted one unit rightward, the nodal DOFs of the rails and bridge elements have to be renumbered such that consistence is maintained for the nodal response before and after shifting. Accordingly, the global coordinates of the element nodes and the positions of the contact points between the wheels and rails should also be updated. Besides, the irregular track profile, if there is any, has to be introduced for the newly-added bridge unit. Note that the system matrices of the rails–bridge model remain the same regardless of the element shifting, as the same properties are used for each bridge unit, which therefore need not be updated.

## 9.9. Verification of Proposed Procedure

For the purpose of verification, let us consider a simply-supported beam of length $L = 25$ m subjected to a moving sprung mass (Fig. 9.4), with the following properties: Young's modulus $E = 2.87$ GPa, Poisson's ratio $v = 0.2$, moment of inertia $I = 2.90$ m$^4$, mass per unit length $m = 2.303$ t/m, suspended mass $M_v = 5.75$ t, suspension stiffness $k_v = 1595$ kN/m, and speed $v = 27.78$ m/s. The first frequency computed of the beam is $\omega_1 = 30.02$ rad/s, and the frequency of the sprung mass is $\omega_v = 16.66$ rad/s. The damping of the beam is neglected. To make use of the present procedure, the following data are assumed for the vehicle and bridge: $M_c = M_v = 5.75$ t, $I_c = 1$ t-m$^4$, $M_t = 10^{-5}$ t, $I_t = 1$ t-m$^4$, $M_w = 10^{-5}$ t, $k_p = 10^{10}$ kN/m, $k_s = 0.5k_v = 797.5$ kN/m, $c_p = c_s = 0$ kN-s/m, and $l_c = l_t = l_s = 0$ m; $E_r = 210$ GPa, $v_r = 0.3$, $I_r = 10^{-10}$ m$^4$, $m_r = 10^{-10}$ t/m, $k_{bv} = 10^{10}$ kN/m/m, $c_{bv} = 0$ kN-s/m/m; $E_b = E = 2.87$ GPa, $v_b = v = 0.3$, $I_b = I = 2.90$ m$^4$, $m_b = m = 3.303$ t/m, $\xi = 0$. The bridge is modeled as 10 elements. The frequency $\omega_1^*$ of the vehicle obtained from the eigenvalue analysis is the same as the frequency $\omega_v$ of the sprung mass,

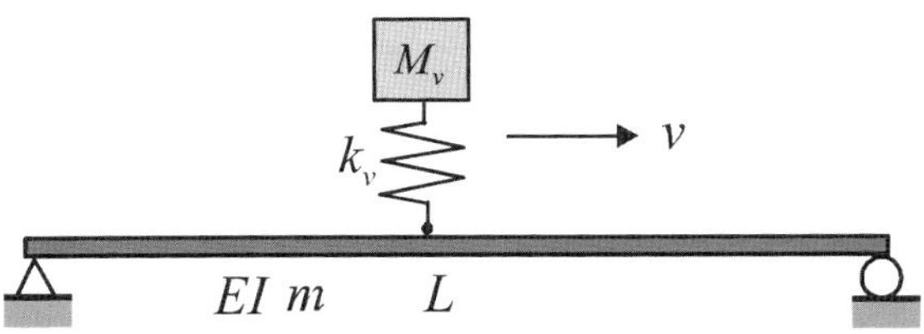


Fig. 9.4. A simple beam subjected to a moving sprung mass.

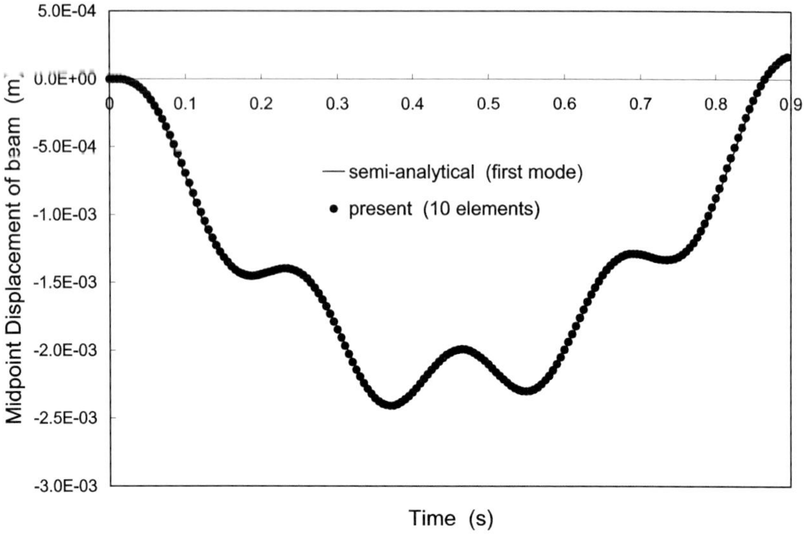


Fig. 9.5. The midpoint displacement of beam.

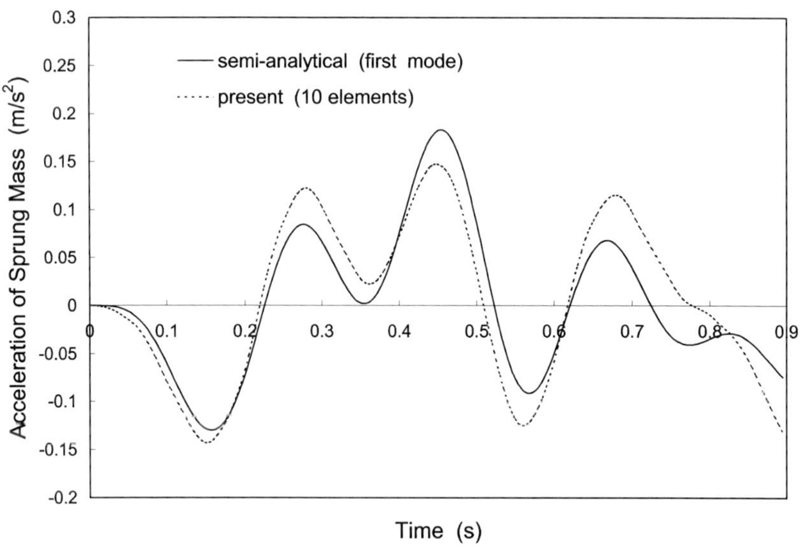


Fig. 9.6. The acceleration response of sprung mass.

i.e., $\omega_1^* = \omega_v = 16.66$ rad/s, indicating that the data assumed for the vehicle are equivalent to those adopted for the sprung mass model.

The dynamic response of the midpoint displacement of the beam has been plotted in Fig. 9.5, along with the analytical solution considering only the first mode (Biggs 1964). Evidently, good agreement has been achieved between the present and analytical solutions. From Fig. 9.6, one observes that the acceleration of the mass computed by the present approach agrees generally well with the analytical one, with slight deviation due to consideration of only the first mode of vibration of the beam in the latter. This example serves to illustrate the reliability of the present procedure.

## 9.10. Numerical Studies

The procedure presented above is applied to studying: (1) the steady-state behavior and riding comfort of a train moving over a multi-unit railway bridge, and (2) the impact response of the bridge during the passage of the train. The data adopted for the train cars are extracted from those for the SKS Series 300 rail cars (Wakui *et al.*, 1995) (see Table 9.1). The rails are assumed to be of the UIC-60 type. The bridge units comprising the track supporting system are assumed to be simply-supported and of constant cross sections. The properties of the rails, ballast layer and bridge units have been listed in Table 9.2. Ten vehicles are considered for the train, i.e., $n_v = 10$. Each bridge unit, assumed to be of length $L = 30$ m, is modeled by 10 elements, i.e., with element length $l = 3$ m. According to Eq. (9.1), ten bridge units are used for the bridge segment in analysis, i.e., $N_{\min} = 10$. In addition, a time increment of $\Delta t = 0.005$ s and damping ratio of $\xi = 0.025$ are used. No braking or accelerating actions are considered for the train ($\mu = 0$). The first and second natural frequencies of the bridge segment obtained from the eigenvalue analysis are: $\omega_1 = 26.87$ rad/s (4.28 Hz) and $\omega_2 = 107.42$ rad/s (17.10 Hz). The fundamental frequency is close to the mean value of those measured from several concrete railway bridges in service (Frýba, 1996), indicating the adequacy of the data used in the present analysis.

### 9.10.1. *Steady-State Responses of the Train, Rails and Bridge*

Figure 9.7 shows the steady-state vertical accelerations of the car body of the 5th vehicle of the train at speeds $v = 300$ and 375 km/h. Obviously, the train undergoes a periodic vibration at the steady state. The dominant frequencies of the acceleration response for $v = 300$ km/h obtained by the fast Fourier transform (FFT) are 2.83 and 5.56 Hz and those for $v = 375$ km/h are 3.52 and 7.03 Hz. It is found that the vibration frequency $f_v$ of the vehicle, the length $L$ (= 30 m) of the bridge unit and the train speed $v$ satisfy the relation: $f_v = n \times v/L$, $n = 1, 2, 3, \ldots, \infty$, where relatively small contributions are made by harmonic components with $n \geq 3$. In addition, the response for $v = 375$ km/h appears to be much larger than that for $v = 300$ km/h, the reason for which will be explained later.

The midspan responses of the bridge unit to the passage of the train have been plotted in Fig. 9.8. As can be seen from Fig. 9.8(a) for the displacement, the bridge unit vibrates periodically during

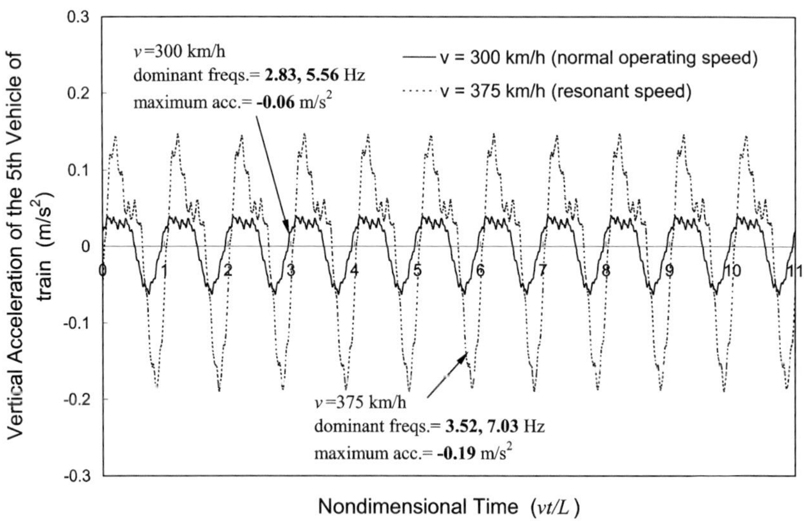


Fig. 9.7. The vertical acceleration of the 5th vehicle of train.

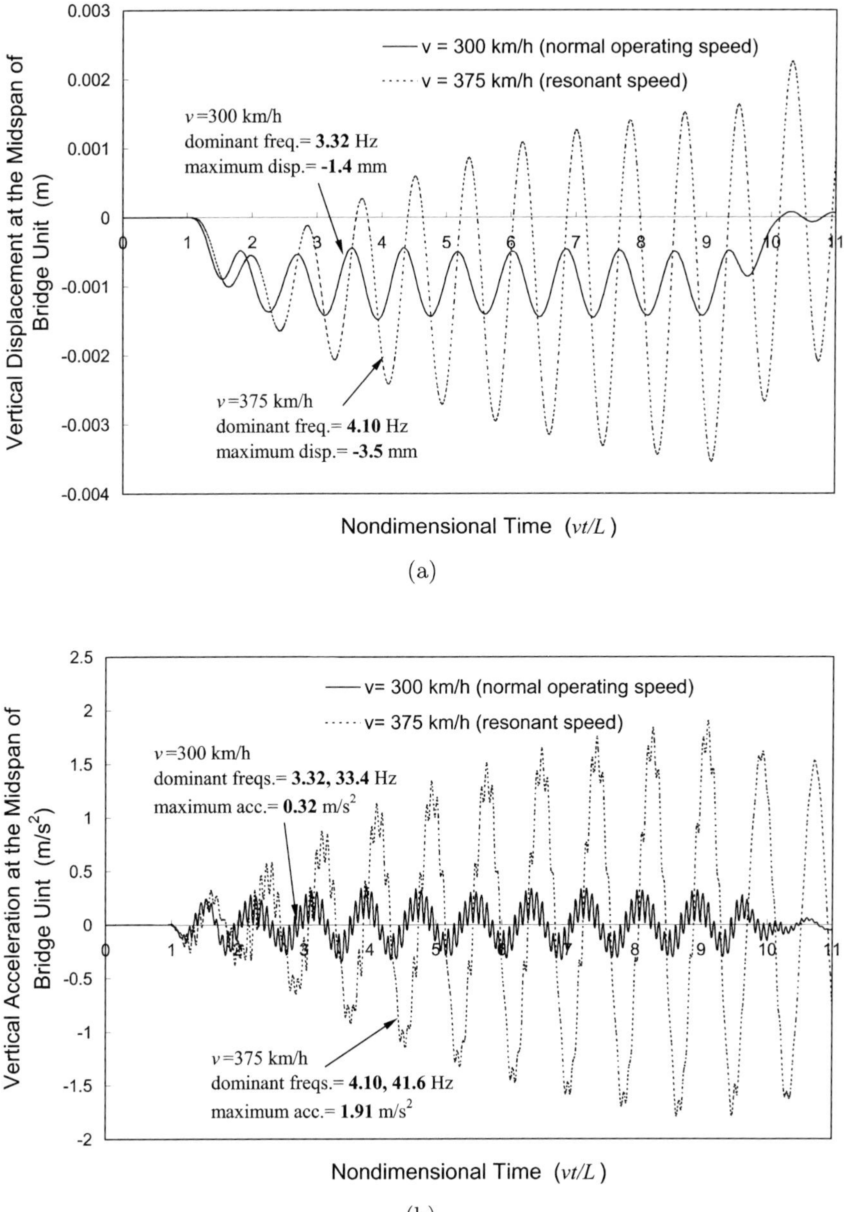


Fig. 9.8. The vertical response at the midspan of bridge unit: (a) displacement and (b) acceleration.

the passage of the train, implying a dominant frequency of 3.32 Hz for $v = 300$ km/h and 4.10 Hz for $v = 375$ km/h, both of which satisfy the relation for the frequency $f_b$ of the bridge unit: $f_b = n \times v/d$, $n = 1, 2, 3, \ldots, \infty$, where $d$ denotes the vehicle length ($= 25$ m). The contribution from the higher harmonic components, i.e., with $n > 2$, to the displacement can virtually be ignored. Similar trend exists for the acceleration response in Fig. 9.8(b), except that the contribution of the second dominant frequency can be readily observed, which is equal to 33.4 Hz for $v = 300$ km/h and 41.6 Hz for $v = 375$ km/h. The second frequency is about 10 times the value of the first frequency, which satisfies the following relation for the frequency $f_b'$ of the bridge unit: $f_b' = n \times v/d_{wb}$, $n = 1, 2, 3, \ldots, \infty$, where $d_{wb}$ denotes the wheelbase of each wheelset ($= 2l_t = 2.5$ m), as indicated in Fig. 9.2.

Evidently, the response of the bridge unit for $v = 375$ km/h is much larger than that for $v = 300$ km/h and increases as there are more vehicles passing the bridge unit. This is an indication of the occurrence of resonance on the bridge unit for the train moving at speed $v = 375$ km/h. The same is not true for $v = 300$ km/h. It should be added that due to its interaction with the bridge units the train also exhibits resonance at $v = 375$ km/h, as indicated by the rather large response in Fig. 9.7. This type of resonance, referred to as the *train–rails–bridge resonance*, occurs if the train speed $v$ (m/s), the vehicle length $d$ (m) and the loaded fundamental frequency of the bridge unit $\omega_1^*$ (rad/s) satisfy the following conditions (Li and Su, 1999):

$$\frac{\omega_1^* d}{2v} = n\pi\,, \qquad n = 1, 2, 3, \ldots\,, \tag{9.44}$$

where $\omega_1^* = \omega_1 \times [m/(m + M/L)]^{1/2}$, $M = M_c + 2M_t + 4M_w$ (see Tables 9.1 and 9.2 for definition of symbols). According to Eq. (9.44), the primary resonant speed for the present case can be predicted as 374 km/h ($= 103.9$ m/s) with $\omega_1^* = 26.1$ rad/s, $d = 25$ m and $n = 1$, which is very close to the value of 375 km/h indicated above. Besides, it is noted that the bridge unit exhibits certain positive displacements at the resonant speed of 375 km/h, implying the occurrence of negative moments in the bridge unit. This is an issue that should be taken into account in the design of bridges. No similar phenomenon can be observed for the train moving at speed $v = 300$ km/h.

Figure 9.9 shows the vertical response of the rails at the point right above the midspan of the bridge unit. The displacement of the rails is generally similar to that of the bridge unit due to the strong constraint effect of the ballast layer, where pronounced positive displacements (implying negative moments) are also observed for the resonant speed. Unlike that for the response of the bridge unit, the contribution of the second mode to the rails response is not negligible, which equals three times the fundamental one, as revealed by a FFT analysis. Because of the filtering effect of the ballast layer, the second frequency does not contribute to the response of the bridge unit. In contrast, the acceleration of the rails in Fig. 9.9(b) shows even much larger contribution from the higher modes, in comparison with the bridge response. Note that the dominant frequency of the acceleration of the rails at $v = 375$ km/h is lower than that at $v = 300$ km/h, contrary to the case for the displacement.

As shown in Fig. 9.10, the contact force between the rails and the first wheel of the 5th vehicle is quite different for the two speeds. Due to occurrence of the train–rails–bridge resonance, the maximum and mean values of the contact force for $v = 375$ km/h are greater than those for $v = 300$ km/h. The dominant frequency and maximum value of the responses of the 5th vehicle, rails, bridge unit and contact force have been indicated in the figure.

### 9.10.2. *Impact Response of Rails and Bridge Under Various Train Speeds*

The impact response of the rails or bridge is represented by the *impact factor* $I$ as defined in Eq. (1.1), i.e., $I = [R_d(x) - R_s(x)]/R_s(x)$, where $R_d(x)$ and $R_s(x)$ denote the maximum dynamic and static responses, respectively, of the rails or the bridge at section $x$. The impact factors computed for the bridge, including the displacements and internal forces, traveled by the train moving at speeds $0 \sim 400$ km/h have been plotted in Fig. 9.11. As can be

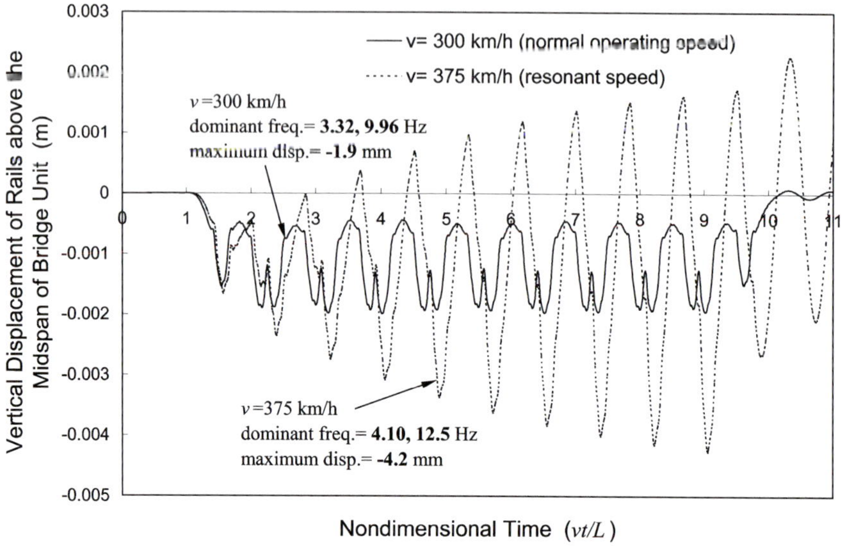


(a)

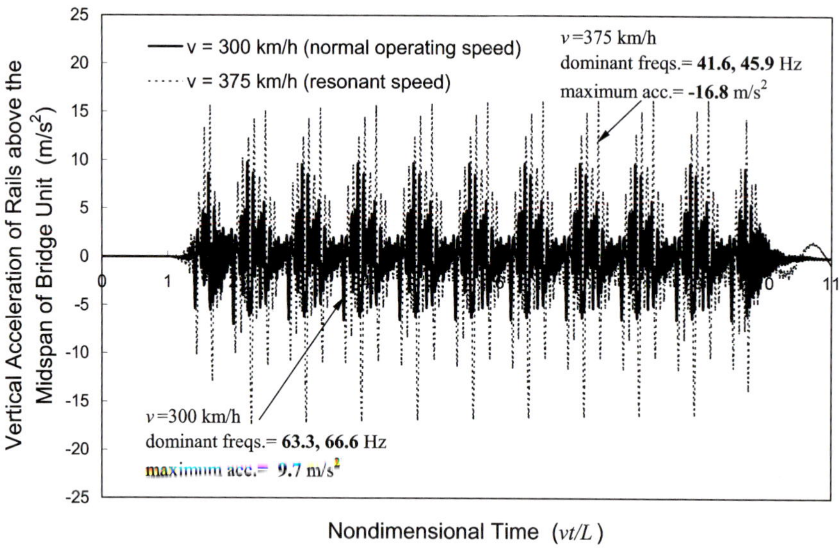


(b)

Fig. 9.9. The vertical response of rails above the midspan of bridge unit: (a) displacement and (b) acceleration.

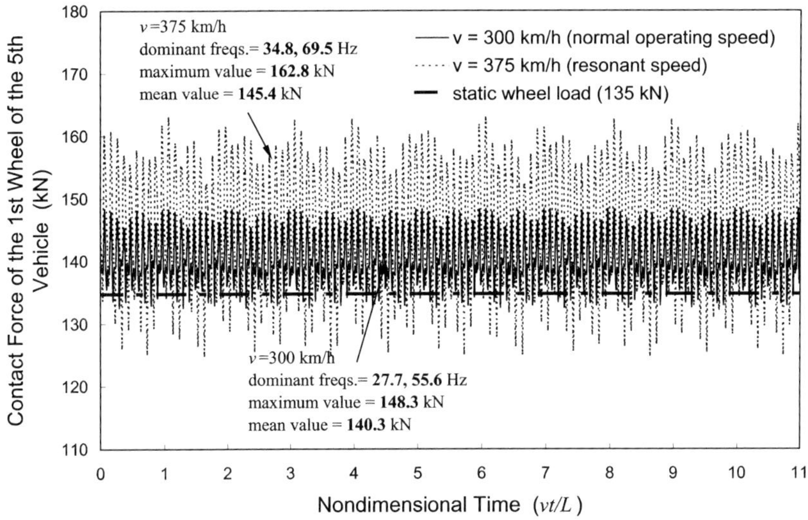


Fig. 9.10. The contact force of the 1st wheel of the 5th vehicle.

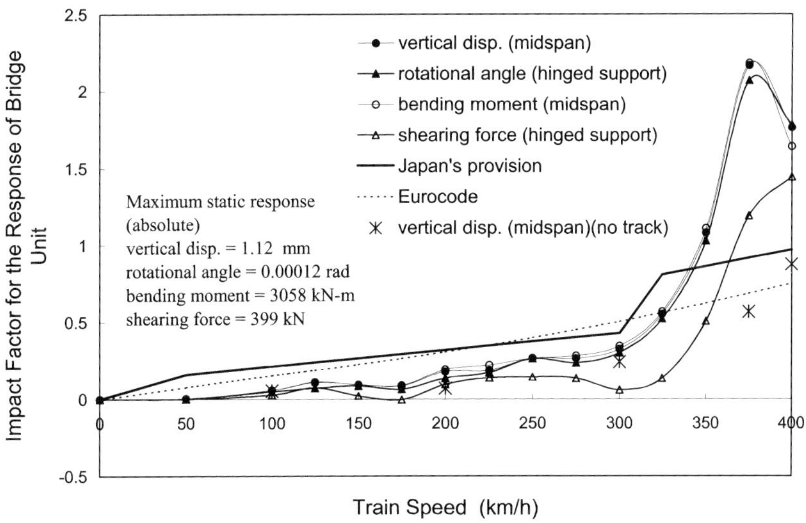


Fig. 9.11. The impact factor for the response of bridge unit under various train speeds.

seen, the impact factor increases as the train speed increases and exhibits a peak value at a speed around 375 km/h, indicating the occurrence of the train–bridge resonance at that speed. Little difference can be observed between the impact factors for the vertical displacement, rotational angle and bending moment. The impact factor for the shearing force is much smaller than those for the bending moment and displacement. A comparison of the present results with existing specifications, such as Japan's provision and Eurocode, indicates that the impact factor exceeds the allowable limit set by the specifications as the train speed exceeds 350 km/h or reaches the resonant speed, where the impact factor can be as high as 2.2.

Figure 9.12 shows the impact factor for the response of the rails at the points right above the midspan and hinged support of the bridge unit under different train speeds. In general, the impact factor increases as the train speed increases. The impact factor for the displacements is higher than that for the internal forces. The displacement impact factor reaches a peak at the resonant speed of 375 km/h, while such a peak does not exist for the internal forces.

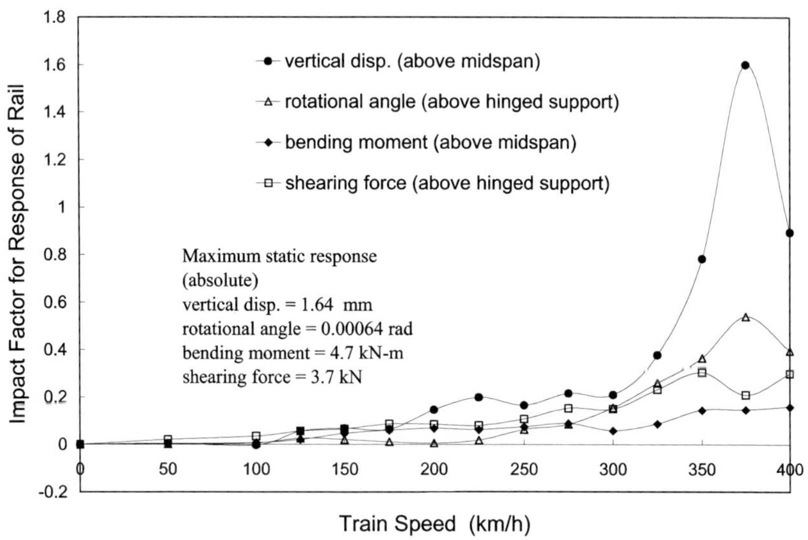


Fig. 9.12. The impact factor for the response of rails under various train speeds.

The implication is that the extent of amplification for the internal forces (including the moment and shear) is smaller than that for the displacement at resonant speeds.

### 9.10.3. *Response of Train to Track Irregularity and Riding Comfort of Train*

Track irregularity is one of the factors that can amplify the vibration response of the train. In this study, the track irregularity is assumed to be of random nature and characterized by the following *power spectral density* (PSD) function $S(\Omega)$ (White *et al.*, 1978):

$$S(\Omega) = \frac{A_v \Omega_c^2}{(\Omega^2 + \Omega_r^2)(\Omega^2 + \Omega_c^2)}, \tag{9.45}$$

where $\Omega(= 2\pi/\lambda_r)$ denotes the spatial frequency (rad/m), $\lambda_r$ the wavelength of the irregularity (m), and $A_v$ (m), $\Omega_r$ (rad/m) and $\Omega_c$ (rad/m) are relevant parameters. Table 9.3 contains the values of the parameters in Eq. (9.45) for track classes 4, 5 and 6 designated by the *Federal Railroad Administration* (FRA), with class 6 indicating the best quality (Fries and Coffey, 1990). By using PSD function in Eq. (9.45) with 0.209 rad/m $\leq \Omega \leq$ 209.441 rad/m (0.03 m $\leq \lambda_r \leq$ 30 m) and the method proposed in Chen and Zhai (1999), the irregular profiles $r(x)$ of the three classes of tracks are computed and plotted in Fig. 9.13 with the maximum deviations indicated.

The maximum accelerations of the train for the three class of track irregularity under different train speeds have been plotted in Fig. 9.14. As was expected, the maximum response of the train moving over irregular tracks appears to be greater for rougher surface, and is the worst for FRA class 4. The maximum response of the

Table 9.3. Track PSD model parameters.

| FRA class | 4 | 5 | 6 |
|---|---|---|---|
| $Av$ (m) | $2.39 \times 10^{-5}$ | $9.35 \times 10^{-6}$ | $1.50 \times 10^{-6}$ |
| $\Omega_r$ (rad/m) | $2.06 \times 10^{-2}$ | $2.06 \times 10^{-2}$ | $2.06 \times 10^{-2}$ |
| $\Omega_c$ (rad/m) | 0.825 | 0.825 | 0.825 |

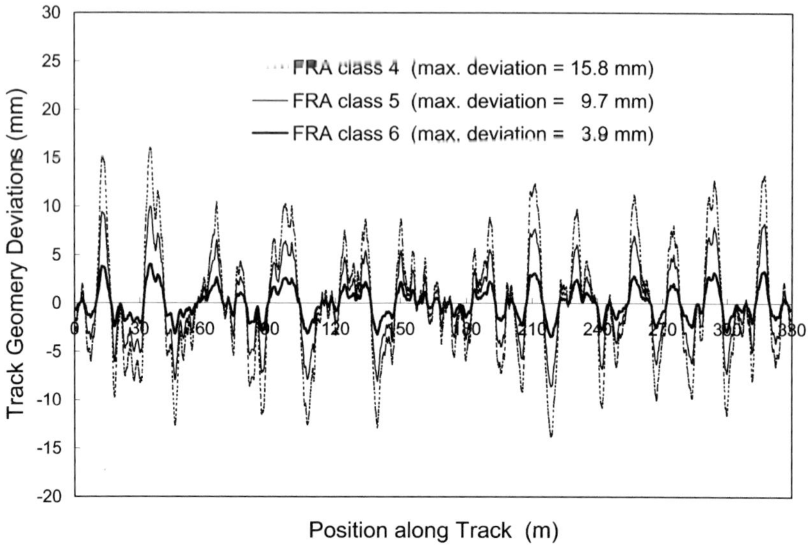


Fig. 9.13. The irregular track profiles for FRA classes 4, 5 and 6.

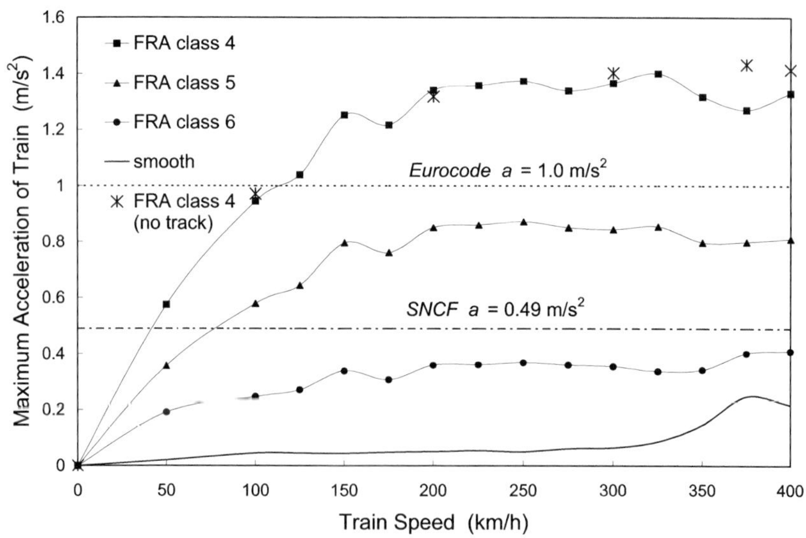


Fig. 9.14. Maximum acceleration responses of train under various train speeds.

train moving over irregular tracks approaches a limit value for train speeds $v$ over 200 km/h. The asymptotic values for FRA classes 4, 5 and 6 are estimated as 1.4, 0.9 and 0.4 m/s$^2$, respectively. The reason for existence of such a limit is that the exciting frequencies induced by track irregularities at higher speeds are filtered out by the suspension system of each vehicle in transmission to the car body. Moreover, the maximum acceleration of the train for FRA class 6 is below the allowable limit of 0.49 m/s$^2$ (= 0.05 g) imposed by *France-SNCF* (Grandil and Ramondenc, 1990) or *Taiwan-HSR* concerning passengers' *riding comfort*, while those for the FRA classes 4 and 5 are not. If a less strict limit of 1.0 m/s$^2$ is imposed, as suggested by *Eurocode* (1995), only FRA class 4 remains unqualified.

The amplification factor and decrement ratio for the maximum contact forces have been presented in Fig. 9.15 for the three different classes of track quality, which shows a trend of increase for higher train speeds. Of interest is the fact that the decrement ratios for all the three track classes are below the limit of 0.25 set in Japan's

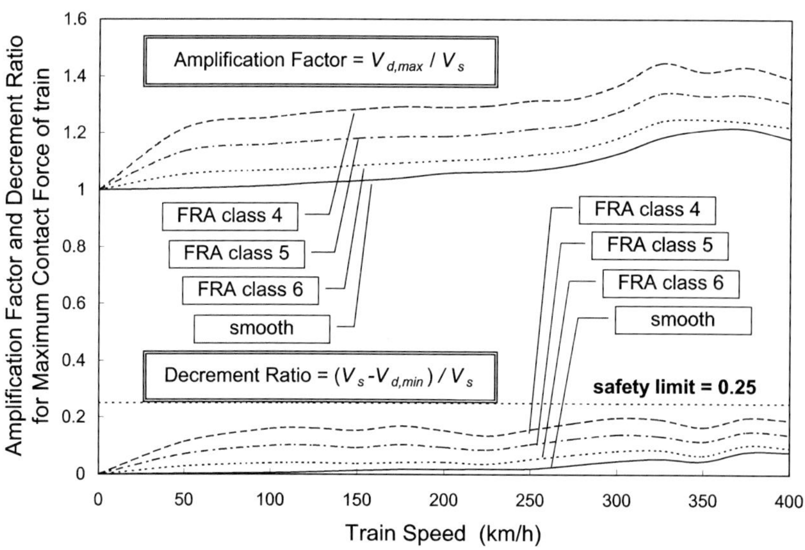


Fig. 9.15. The amplification factor and decrement ratio for maximum contact force of train.

provision for the running safety of trains (Ma and Zhu, 1998), indicating that the train can run safely over the bridge under the track conditions specified for speeds up to 400 km/h.

The riding comfort of trains is an issue of great concern for ground vehicles, especially for high-speed trains. In this study, the riding comfort of the train is assessed by *Sperling's ride index* $W_z$ (Garg and Dukkipati, 1984) defined as:

$$W_z = \left( \sum_{i=1}^{n_f} W_{z_i}^{10} \right)^{1/10} . \tag{9.46}$$

Here, $n_f$ is the total number of the discrete frequencies of the acceleration response of the train identified by the FFT and $W_{z_i}$ is the *comfort index* corresponding to the $i$th discrete frequency, computed as

$$W_{z_i} = [a_i^3 B(f_i)^3]^{1/10} , \tag{9.47}$$

where $a_i$ denotes the amplitude of the acceleration response (m/s$^2$) of the $i$th frequency identified by the FFT and $B(f_i)$ a weighting factor,

$$B(f_i) = 0.588 \left[ \frac{1.911 f_i^2 + (0.25 f_i^2)^2}{(1 - 0.277 f_i^2)^2 + (1.563 f_i - 0.0368 f_i^3)^2} \right]^{1/2} . \tag{9.48}$$

Figure 9.16 shows Sperling's ride index computed for the train under different speeds and track qualities. As can be seen, Sperling's index remains generally constant for trains moving at moderate to high speeds (100 km/h $\leq v \leq$ 300 km/h), and increases slightly at speeds over 300 km/h. For poorer track quality (i.e., for lower FRA class), there exists larger Sperling's index. Compared with the limit values set for different comfort levels in Fig. 9.16 (Garg and Dukkipati, 1984), the following observations can be made for train speeds in the range of 100 km/h $\leq v \leq$ 300 km: (1) for smooth track and tracks of FRA class 6, the level of riding comfort of the train is considered between "*Just noticeable* ($W_z = 1$)" and "*Clearly noticeable* ($W_z = 2$)"; (2) for tracks of FRA class 5, the comfort level becomes worse and falls within the limits of "*Clearly noticeable* ($W_z = 2$)"

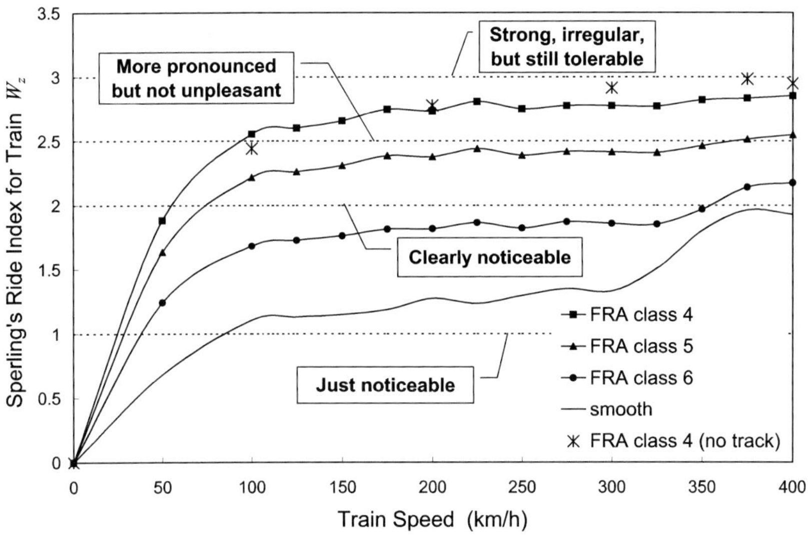


Fig. 9.16. Sperling's ride index for train under various train speeds.

and "*More pronounced but not unpleasant* ($W_z = 2.5$)"; and (3) for tracks with FRA class 4, i.e., the worst quality, the riding comfort deteriorates to a level between "*More pronounced but not unpleasant* ($W_z = 2.5$)" and "*Strong, irregular, but still tolerable* ($W_z = 3$)". Note that the comfort level for the smooth track is the best level that can be achieved by the train under the present conditions set for the train, track and bridge. Figure 9.17 shows the relation between Sperling's ride index $W_z$ and the maximum deviation of the track irregularity, which can be described by a cubic polynomial. The maximum track deviations corresponding to the riding comfort levels stated above are also shown in the figure.

## 9.10.4. *Effect of the Track System*

To evaluate the effect of the track system on the response of the bridge and moving train, a parallel analysis was performed with the track system omitted. The impact factor computed for the bridge unit by such an analysis was also shown in Fig. 9.11 (indicated as the solution with "no track"), together with the maximum acceleration

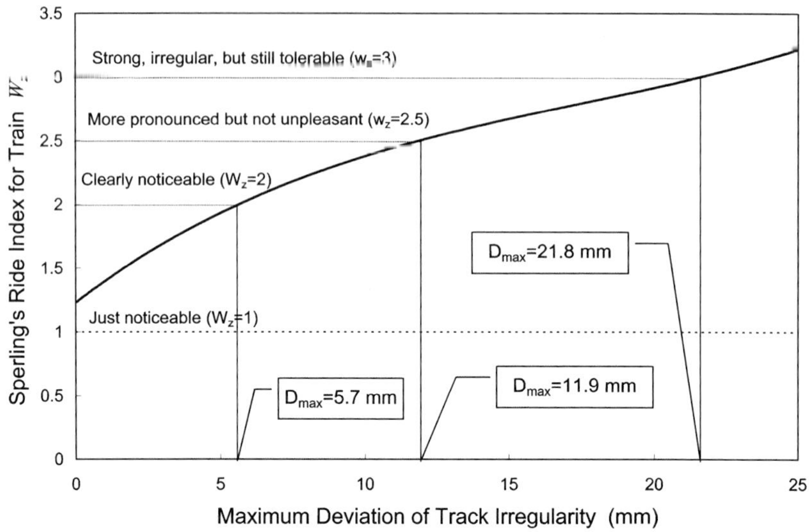


Fig. 9.17. Sperling's ride index with relation to maximum track deviation.

and Sperling's index of the train in Figs. 9.14 and 9.16. In general, the impact factor computed for such a case for the bridge is less than that for the case with the track system taken into account. Besides, the impact factor reaches its peak at a higher speed due to the relatively higher loaded fundamental frequency of the train–bridge system. For FRA class 4, the maximum acceleration and Sperling's index of the train computed for the case with no track are slightly larger than those for the case with track for train speeds over 200 km/h. They are nearly the same for both cases for train speeds below 200 km/h. It can be concluded that the presence of the track system can increase the train-induced impact effect on the bridge, while reducing the vibration amplitude (i.e., raising the riding comfort) of the traveling train.

## 9.11. Concluding Remarks

A procedure for analyzing the vehicle–rails–bridge system was presented, by which the dynamic response of each subsystem can be computed. The procedure was then applied to investigating the steady-state response and riding comfort of the train moving over a railway bridge comprising a series of simply-supported beams. The following are the conclusions made from the numerical studies: (1) The steady-state response of the train moving over a smooth track displays periodic behaviors, of which the dominant frequencies relate to the train speed and the bridge length. (2) The vibrations of the rails and bridge caused by the moving train also possess some characteristic frequencies related to the train speed and the vehicle length. (3) Resonance can occur in the interaction system of the train, rails, and bridge, as the vehicle length, frequencies of the bridge and train speed meet some specific relations, which can result in dramatic amplification of the response of each component. (4) The resonant effect on the rails and bridge responses was underestimated by existing specifications for the high-speed range, which should be considered in the design of railway bridges involving flexible bridge structures or for trains moving at high speeds. (5) The riding comfort of the train can be significantly affected by the presence of track irregularity. It remains nearly independent of the train speed in the moderate to high-speed range. (6) The effect of the track system should be taken into account in the analyzing the train–bridge system. Otherwise, one may underestimate the impact effect of the bridge and overestimate the vibration response of the train.

# Chapter 10

# Vehicle–Rails–Bridge Interaction — Three-Dimensional Modeling

The procedure presented in Chapters 8 and 9 is extended in this chapter to deal with the three-dimensional aspects of the vehicle–rails–bridge interaction (VRBI) system. In this chapter, the vehicle is idealized as an assembly of a car body, two bogies and four wheelsets, which has a total of 27 degrees of freedom. The bridge is assumed to carry two parallel tracks, each consisting of two rails of infinite length supported by the spring-dashpot units that are uniformly-distributed. The bridge is modeled by the central finite rail (CFR) element, left semi-infinite rail (LSR) element, and right semi-infinite rail (RSR) element, depending on the parts of rails concerned. Through elimination of the contact forces, three types of vehicle–rails interaction (VRI) elements were derived, with due account taken of the vehicle properties. The equation of motion for the entire VRBI model was then constructed by assembling the VRI elements, ordinary rail elements and bridge elements. The present procedure allows us to analyze a wide range of three-dimensional VRBI problems, including the crossing of two trains on the bridge, the variation of wheel/rail contact forces, the risk of derailment of moving trains, and so on.

## 10.1. Introduction

The dynamic response of bridge structures to the moving loads exerted by trains has been widely investigated by researchers, as was reviewed in Chapter 1. For the works that dealt with the train–bridge interactions, the scope of study was restricted mainly to the two-dimensional aspects of the train–bridge system, aimed at analyzing the vertical vibration of the bridge. For the relatively small amount of research works that dealt with the three-dimensional aspects of the train–bridge system, e.g., those by Bhatti *et al.* (1985), Wakui *et al.* (1995), and Schupp and Jaschinski (1996), among others, only partial considerations were made for the dynamic effects of the railway bridge and moving trains, which is not sufficient if focus is to be extended to some advanced aspects of the *vehicle–rails–bridge interaction* (VRBI) systems, such as the lateral and rotational or torsional vibrations, the responses induced by two trains in crossing, the risk of derailment of moving trains, and so on. All these problems are crucial and have to be considered in the design of high-speed railway bridges, concerning the maneuverability of the train at high speeds. Besides, the track system directly involved in the interaction with the moving train was either ignored or only partially considered in most previous studies. Since the track system is a flexible medium vibrating with the train (moving subsystem) and the bridge (stationary subsystem), it can affect seriously the extent of interaction between the two subsystems, especially for trains moving in the high-speed range. For this reason, the track system will be included in the analysis of the train–bridge system, with which the contact forces between the wheels and rails can be computed.

In this chapter, the three-dimensional characteristics of the train–track–bridge system, together with assumptions made for modeling such a system, will first be summarized. A three-dimensional VRBI model will then be constructed based on these assumptions. Next, the equations of motion for the major components of the model, i.e., the vehicle, rails and bridge, will be formulated, based on which equations of motion for the entire VRBI system are assembled. The procedure for performing the nonlinear dynamic analysis is also presented. The theory proposed herein will be verified through comparison of the results for a typical example obtained by the present 3D procedure with those by the 2D procedure presented in Chapter 8. The dynamic interactions between the moving trains and railway bridge under various conditions of track irregularities will be investigated.

The *risk of derailment* of a train running over the bridge will also be investigated. The research results presented in this chapter are based largely on the paper by Wu *et al.* (2001), but with substantial additions and revisions.

## 10.2. Three-Dimensional Models for Train, Track and Bridge

Figure 10.1 shows a simply-supported railway bridge traveled by a train of speed $v$. In this study, the train is idealized as a series of identical moving vehicles, each composed of one car body, two identical bogies and four identical wheelsets, as depicted in Fig. 10.2. All these components are assumed to be rigid. The suspension system, whether vertical or lateral, of each wheelset of the vehicle is modeled as a linear spring-dashpot unit. As shown in Fig. 10.3(a), the bridge is made of a box girder carrying two parallel tracks, which allows two trains to move in the same or opposite directions. Each of the two tracks is simplified as a set of two infinite rails lying on a single-layer ballast foundation. The two rails of each track are identical and connected by uniformly-distributed rigid sleepers. Each set of the two rails will be represented collectively as a Bernoulli–Euler beam of constant sections. By neglecting the nonlinear effects and interlock shear, the ballast foundation will be represented by uniformly-distributed linear spring-dashpot units.

Due to the flexibility of the ballast foundation, each track is allowed to move longitudinally ($x$ direction), vertically ($y$ direction) and laterally ($z$ direction), as well as to rotate about the longitudinal axis. The two rails of each track have the same longitudinal and lateral displacements due to the constraint of rigid sleepers. The vertical contact forces between the wheels and the two rails of each track are assumed to act along the centerline of the cross sections of the two rails. Therefore, each individual rail has no torsional deformations, although the track to which the rail belongs may undergo some torsional or rotational motion. The torques produced by the lateral forces acting through the torsional center of the two rails will be neglected. The torsional resistance of the two rails against the

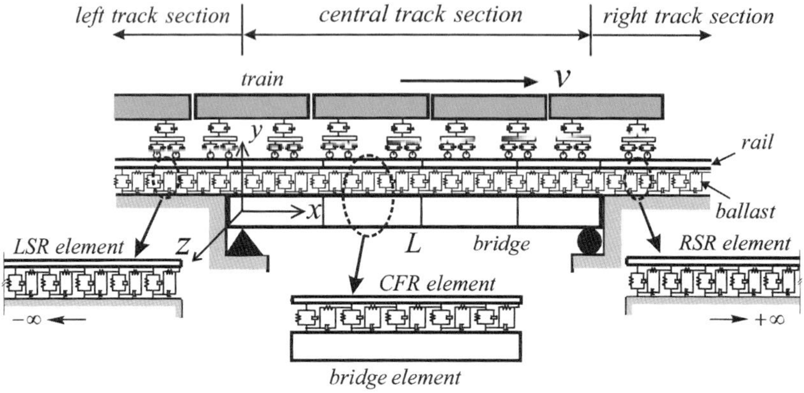


Fig. 10.1. Simply-supported railway bridge traveled by a moving train.

ballast reaction is provided by the cross-sectional areas of the two rails rigidly-connected by sleepers. Furthermore, the four wheelsets of each vehicle are supposed to be in full contact with the rails at all times (i.e., no jumps occur) and move with the two rails in the vertical and lateral directions. It should be added that the mass and mass moment of inertia of the sleepers are considered as part of the two rails of each track.

The bridge has constant sections and uniform properties, which is idealized as a three-dimensional Bernoulli–Euler beam, as shown in Fig. 10.3(a). The mass and mass moment of inertia of the ballast layer on the bridge are included as part of the bridge. The physical parameters of the vehicle, track and bridge have been shown in the figure, with their definitions summarized in Tables 10.1 and 10.2. In addition, the deviations in geometry of the track, i.e., track irregularities, are also taken into account.

## 10.3. Vehicle Equations and Contact Forces

All the degrees of freedom (DOFs) permitted for the vehicle body, bogies, and wheelsets have been shown in Fig. 10.2. The vehicle body has a total of five DOFs with respect to its center of gravity $G$,

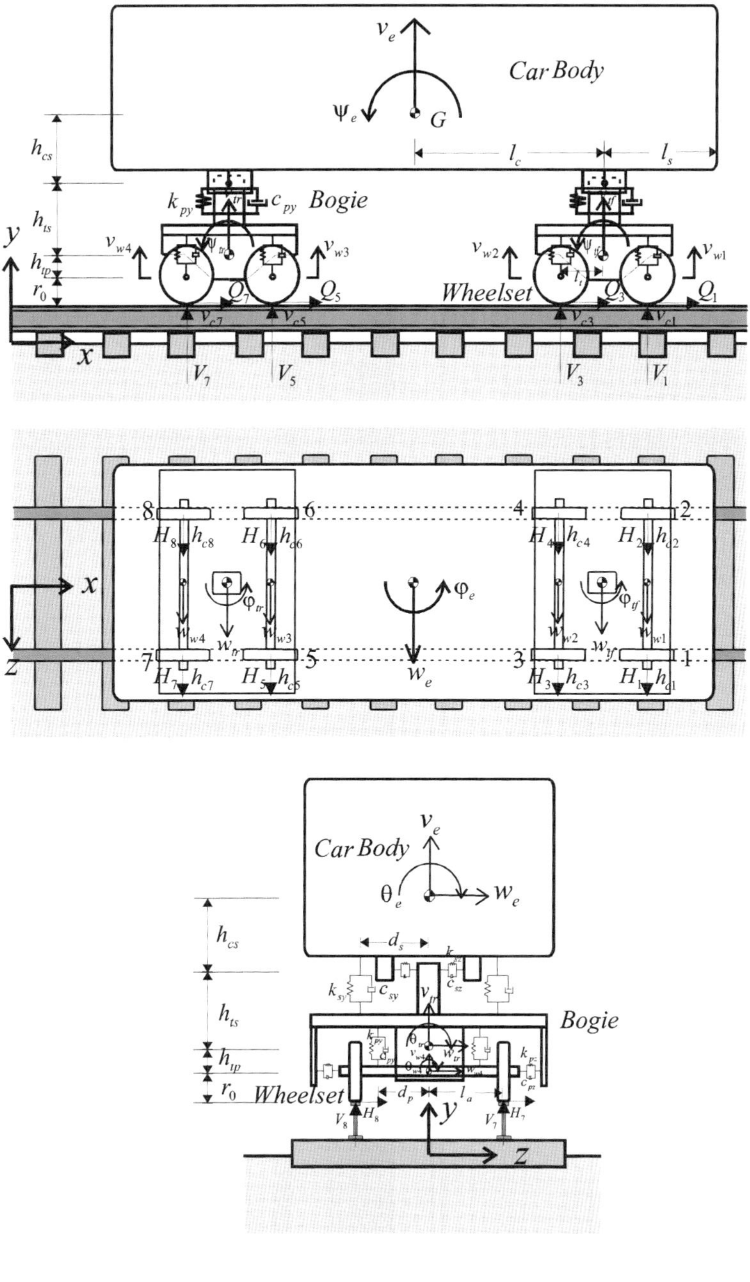


Fig. 10.2. Three dimensional vehicle–rails model.

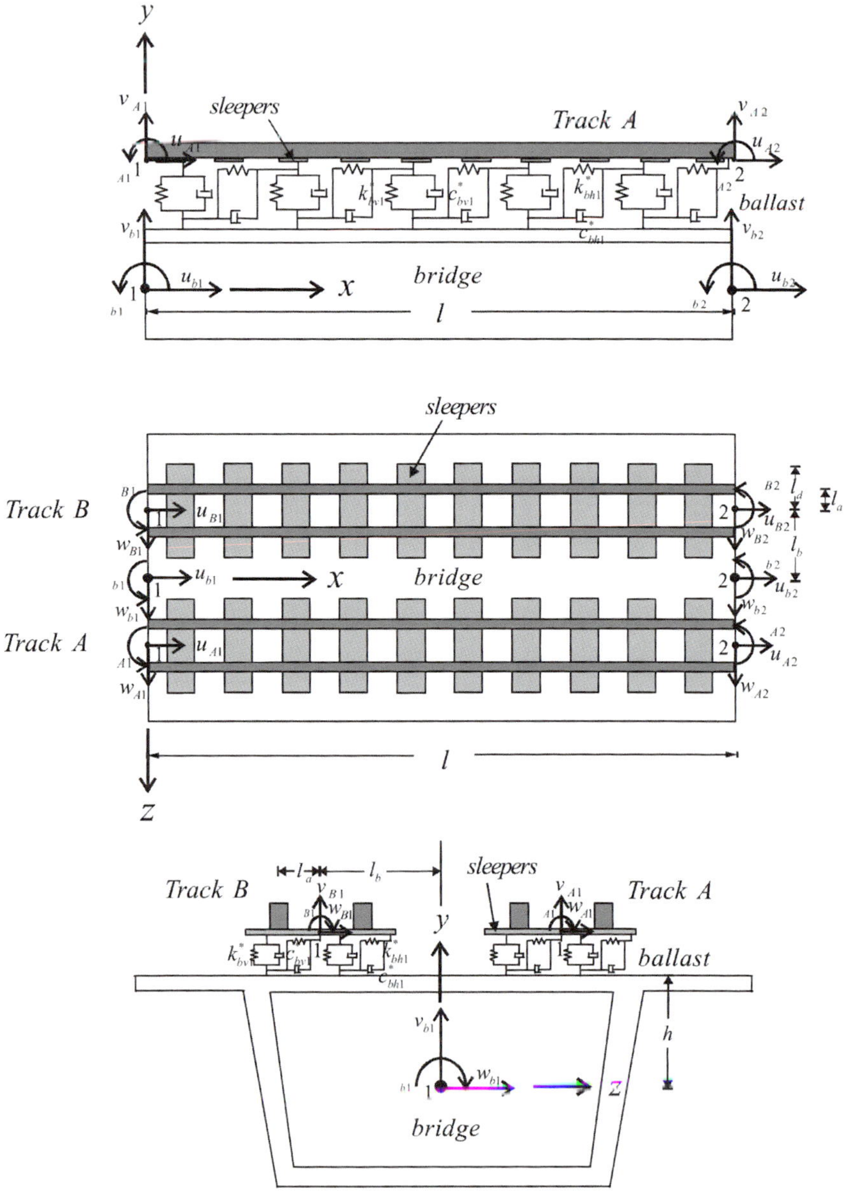


(a)

Fig. 10.3. Rail elements: (a) CFR, (b) LSR, and (c) RSR.

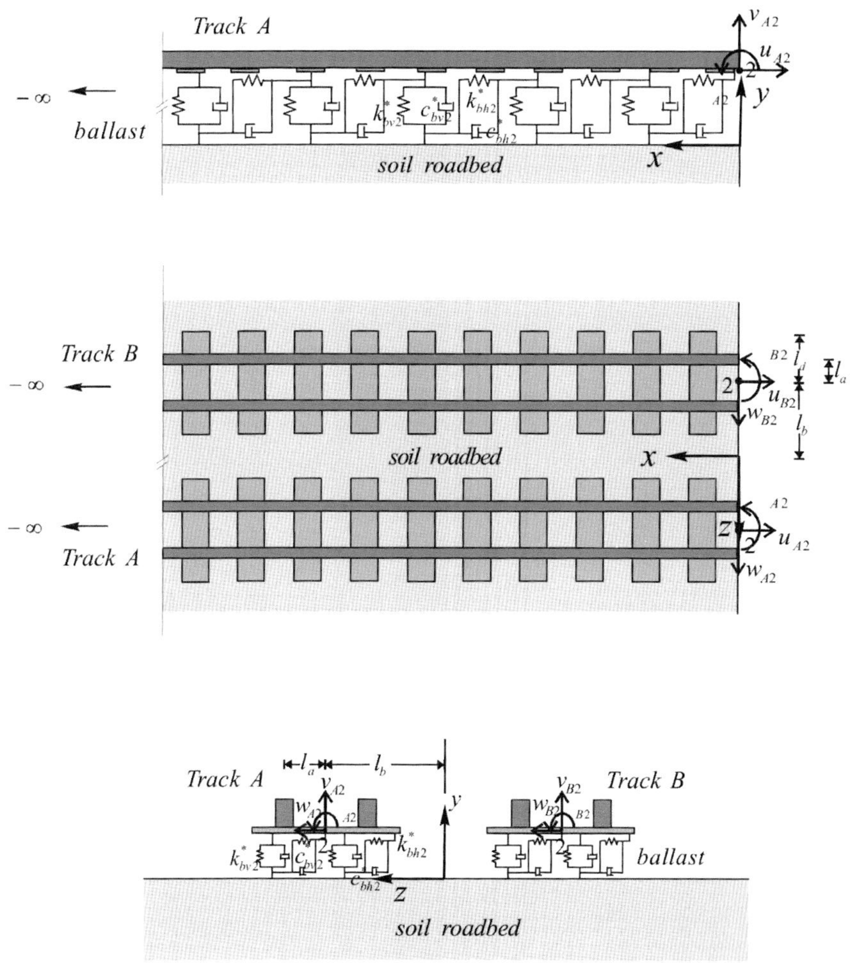


(b)

Fig. 10.3. (*Continued*).

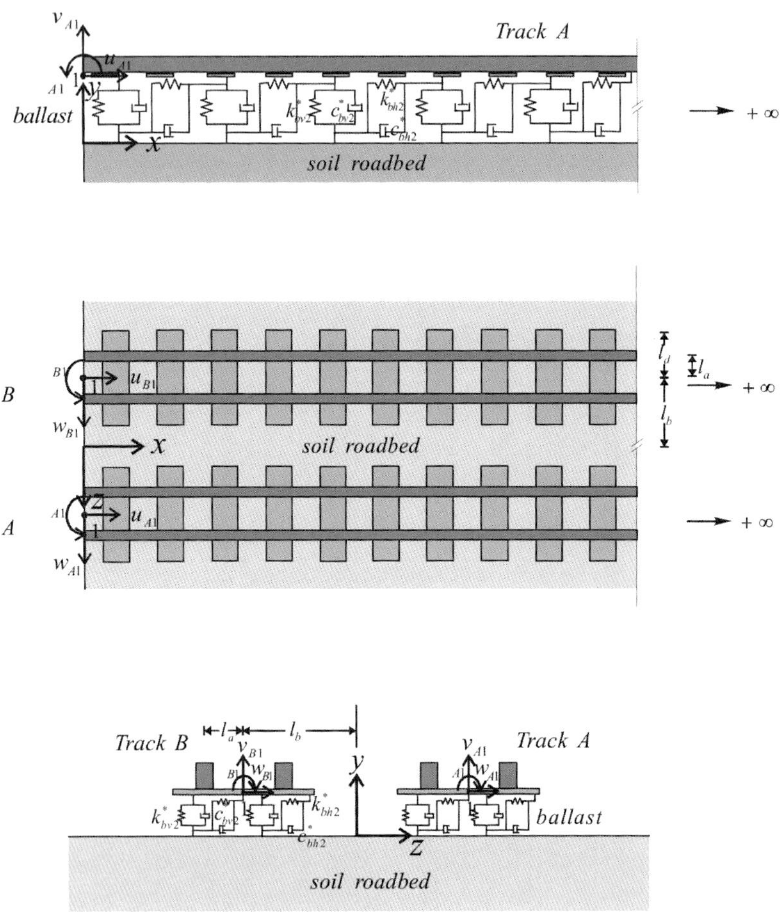


(c)

Fig. 10.3. (*Continued*).

Table 10.1. The properties of vehicle model.

| Item | Notation | Value* |
|---|---|---|
| Mass of car body | $M_c$ | 41.75 t |
| Mass moment of inertia of car body around $x$ axis | $I^*_{cx}$ | 23.2 t-m$^2$ |
| Mass moment of inertia of car body around $y$ axis | $I^*_{cy}$ | 2100 t-m$^2$ |
| Mass moment of inertia of car body around $z$ axis | $I^*_{cz}$ | 2080 t-m$^2$ |
| Mass of bogie | $M_t$ | 3.04 t |
| Mass moment of inertia of bogie around $x$ axis | $I^*_{tx}$ | 1.58 t-m$^2$ |
| Mass moment of inertia of bogie around $y$ axis | $I^*_{ty}$ | 2.34 t-m$^2$ |
| Mass moment of inertia of bogie around $z$ axis | $I^*_{tz}$ | 3.93 t-m$^2$ |
| Mass of wheelset | $M_w$ | 1.78 t |
| Mass moment of inertia of wheelset around $x$ axis | $I^*_w$ | 1.14 t-m$^2$ |
| Half of longitudinal distance between centers of gravity of front and rear bogies | $l_c$ | 8.75 m |
| Half of wheelbase | $l_t$ | 1.25 m |
| Longitudinal distance between center of gravity of bogie and nearest side of car body | $l_s$ | 3.75 m |
| Half of transverse distance between contact points of wheel and rail | $l_a$ | 0.75 m |
| Half of transverse distance between vertical primary suspension systems | $d_p$ | 1.00 m |
| Half of transverse distance between vertical secondary suspension systems | $d_s$ | 1.23 m |
| Vertical distance between center of gravity of car body and lateral secondary suspension system | $h_{cs}$ | 0.75 m |
| Vertical distance between lateral secondary suspension system and center of gravity of bogie | $h_{ts}$ | 0.42 m |
| Vertical distance between center of gravity of bogie and lateral primary suspension system | $h_{tp}$ | 0.20 m |
| Stiffness of vertical primary suspension system | $k_{py}$ | 590 kN/m |
| Damping of vertical primary suspension system | $c_{py}$ | 19.6 kN-s/m |
| Stiffness of lateral primary suspension system | $k_{pz}$ | 2350 kN/m |
| Damping of lateral primary suspension system | $c_{pz}$ | 0 kN/m |
| Stiffness of vertical secondary suspension system | $k_{sy}$ | 265 kN/m |
| Damping of vertical secondary suspension system | $c_{sy}$ | 45.1 kN-s/m |
| Stiffness of lateral secondary suspension system | $k_{sz}$ | 176 kN/m |
| Damping of lateral secondary suspension system | $c_{sz}$ | 39.2 kN-s/m |
| Nominal radius of wheel | $r_0$ | 0.455 m |

*For SKS series 300 vehicle model, extracted from Wakui *et al.* (1995) with some modifications.

Table 10.2. The properties of track and bridge

| Item | Notation | Value |
|---|---|---|
| *Track* | | |
| Young's modulus | $E_t$ | 210 GPa |
| Poisson's ratio | $v_t$ | 0.3 |
| Per-unit-length mass[1] | $m_t$ | 0.587 t |
| Per-unit-length mass moment of inertia about $x$ axis[1] | $I_t^*$ | 0.383 t-m$^2$ |
| Sectional area[2] | $A_t$ | $1.54 \times 10^{-5}$ m$^2$ |
| Flexural moment of inertia about $y$ axis[2] | $I_{ty}$ | $1.03 \times 10^{-5}$ m$^4$ |
| Flexural moment of inertia about $z$ axis[2] | $I_{tz}$ | $6.12 \times 10^{-5}$ m$^4$ |
| Half of gauge of rails | $l_a$ | 0.75 m |
| Half of length of sleeper | $l_d$ | 1.3 m |
| Transverse distance between center lines of track and bridge | $l_b$ | 2.35 m |
| Per-unit-area vertical stiffness of ballast on bridge | $k_{bv1}^*$ | 92.3 MN/m$^2$ |
| Per-unit-area vertical damping of ballast on bridge | $c_{bv1}^*$ | 22.6 MN-s/m$^2$ |
| Per-unit-area lateral stiffness of ballast on bridge | $k_{bh1}^*$ | 3.85 MN/m$^2$ |
| Per-unit-area lateral damping of ballast on bridge | $c_{bh1}^*$ | 22.6 MN-s/m$^2$ |
| Per-unit-area longitudinal stiffness of ballast on bridge | $k_{bh1}^*$ | 3.85 MN/m$^2$ |
| Per-unit-area longitudinal damping of ballast on bridge | $c_{bh1}^*$ | 22.6 MN-s/m$^2$ |
| Per-unit-area vertical stiffness of ballast on approach | $k_{bv2}^*$ | 92.3 MN/m$^2$ |
| Per-unit-area vertical damping of ballast on approach | $c_{bv2}^*$ | 22.6 MN-s/m$^2$ |
| Per-unit-area lateral stiffness of ballast on approach | $k_{bh2}^*$ | 3.85 MN/m$^2$ |
| Per-unit-area lateral damping of ballast on approach | $c_{bh2}^*$ | 22.6 MN-s/m$^2$ |
| Per-unit-area longitudinal stiffness of ballast on approach | $k_{bh2}^*$ | 3.85 MN/m$^2$ |
| Per-unit-area longitudinal damping of ballast on approach | $c_{bh2}^*$ | 22.6 MN-s/m$^2$ |
| Sleeper space | D | 0.6 m |
| *Bridge* | | |
| Young's modulus | $E_b$ | 28.25 GPa |
| Poison's ratio | $v_b$ | 0.2 |
| Per-unit-length mass[3] | $m_b$ | 41.74 t |
| Per-unit-length mass moment of inertia about $x$ axis[3] | $I_b^*$ | 495 t-m$^2$ |
| Sectional area | $A_b$ | 7.73 m$^2$ |
| Torsional moment of inertia about $x$ axis | $I_{bx}$ | 15.65 m$^4$ |
| Flexural moment of inertia about $y$ axis | $I_{by}$ | 74.42 m$^4$ |
| Flexural moment of inertia about $z$ axis | $I_{bz}$ | 7.84 m$^4$ |
| Bridge length | $L$ | 30 m |
| Vertical distance between bridge deck and center of torsion | $h$ | 1.2 m |

1. Including the masses of the rails and sleepers.
2. For two rails.
3. Including the masses of the bridge and ballast.

i.e., the vertical, lateral, rolling, yawing and pitching DOFs, which are denoted as $\{d_e\} = \langle v_e \; w_e \; \theta_e \; \varphi_e \; \psi_e\rangle^T$. Similarly, both the front and rear bogies of the vehicle are assigned five DOFs each with respect to their center points, denoted as $\{d_f\} = \langle v_{tf} \; w_{tf} \; \theta_{tf} \; \varphi_{tf} \; \psi_{tf}\rangle^T$ and $\{d_r\} = \langle v_{tr} \; w_{tr} \; \theta_{tr} \; \varphi_{tr} \; \psi_{tr}\rangle^T$, respectively. Each wheelset is assigned three DOFs, i.e., the vertical, lateral and rolling DOFs, at the center of the axle, denoted as $\{d_{wi}\} = \langle v_{wi} \; w_{wi} \; \theta_{wi}\rangle^T$, where $i = 1 \sim 4$. Thus, the *upper part* of the 3D vehicle model, that is, the part *not* in direct contact with the rails, has a total of 15 DOFs, as indicated by the vector $\{d_u\} = \langle\{d_e\} \; \{d_f\} \; \{d_r\}\rangle^T$, and the *wheel part* has a total of 12 DOFs, which is denoted as $\{d_w\} = \langle\{d_{w1}\} \; \{d_{w2}\} \; \{d_{w3}\} \; \{d_{w4}\}\rangle^T$. In addition, there exists a total of eight contact points with two rails (see also Fig. 10.2), each of which has a vertical and a lateral DOF. Let $v_{ci}$ and $h_{ci}$ respectively denote the vertical and lateral displacements of the $i$th contact point. The total *contact-point* DOFs for one vehicle can be denoted as

$$\begin{aligned}\{d_c\} = \langle\, & v_{c1} \quad h_{c1} \quad v_{c2} \quad h_{c2} \quad v_{c3} \quad h_{c3} \quad v_{c4} \quad h_{c4} \\ & v_{c5} \quad h_{c5} \quad v_{c6} \quad h_{c6} \quad v_{c7} \quad h_{c7} \quad v_{c8} \quad h_{c8}\,\rangle^T . \end{aligned} \tag{10.1}$$

Correspondingly, the *contact forces* are as follows:

$$\begin{aligned}\{f_c^*\} = \langle\, & V_1 \quad H_1 \quad V_2 \quad H_2 \quad V_3 \quad H_3 \quad V_4 \quad H_4 \\ & V_5 \quad H_5 \quad V_6 \quad H_6 \quad V_7 \quad H_7 \quad V_8 \quad H_8\,\rangle^T , \end{aligned} \tag{10.2}$$

where $V_i$ and $H_i$ respectively denote the vertical and lateral contact forces of the $i$th contact point. Because the lateral contact forces acting on the two wheels of each wheelset are the same (i.e., $H_1 = H_2$, $H_3 = H_4, \ldots$), the contact forces $\{f_c^*\}$ can be rewritten in a *compact form* as

$$\begin{aligned}\{f_c\} = \langle\, & V_1 \quad H_1 \quad V_2 \quad V_3 \quad H_3 \quad V_4 \quad V_5 \quad H_5 \quad V_6 \\ & V_7 \quad H_7 \quad V_8\,\rangle^T . \end{aligned} \tag{10.3}$$

In this connection, the vector $\{f_c^*\}$ is referred to as a *complete form*.

In a time-history analysis, one is interested in the behavior of a structure during the time step from $t$ to $t + \Delta t$. The equations of motion for the *vehicle* at time $t + \Delta t$ with partitions for the *upper* (noncontact) and *wheel* (contact) parts can be written as

$$\begin{bmatrix} [m_{uu}] & [m_{uw}] \\ [m_{wu}] & [m_{ww}] \end{bmatrix} \begin{Bmatrix} \{\ddot{d}_u\} \\ \{\ddot{d}_w\} \end{Bmatrix}_{t+\Delta t} + \begin{bmatrix} [c_{uu}] & [c_{uw}] \\ [c_{wu}] & [c_{ww}] \end{bmatrix} \begin{Bmatrix} \{\dot{d}_u\} \\ \{\dot{d}_w\} \end{Bmatrix}_{t+\Delta t} + \begin{bmatrix} [k_{uu}] & [k_{uw}] \\ [k_{wu}] & [k_{ww}] \end{bmatrix} \begin{Bmatrix} \{d_u\} \\ \{d_w\} \end{Bmatrix}_{t+\Delta t} = \begin{Bmatrix} \{f_{ue}\} \\ \{f_{we}\} \end{Bmatrix}_{t+\Delta t} + \begin{bmatrix} [l_u] \\ [l_w] \end{bmatrix} \{f_c\}_{t+\Delta t}\,, \tag{10.4}$$

where $[m_{uu}]$, $[m_{uw}]$, $[m_{wu}]$ and $[m_{ww}]$ denote the components of the mass matrices, $[c_{uu}]$, $[c_{uw}]$, $[c_{wu}]$ and $[c_{ww}]$ the components of the damping matrices, and $[k_{uu}]$, $[k_{uw}]$, $[k_{wu}]$ and $[k_{ww}]$ the components of the stiffness matrices of the vehicle, with the subscripts $u$ and $w$ denoting the upper and wheel parts, respectively; $\{f_{ue}\}$ and $\{f_{we}\}$ denote the force vectors acting on the upper and wheel parts; $[l_u]$ and $[l_w]$ the associated transformation matrices; and $\{f_c\}$ denotes the contact forces acting on the four wheelsets as stated above. Since the contact forces are acting only on the wheel part, i.e., the wheelsets of the vehicle, the transformation matrix for the upper part $[l_u]$ should be set to $[0]$. The transformation matrix for the wheel part $[l_w]$ is given as

$$[l_w] = \begin{bmatrix} [l] & & & \\ & [l] & & \\ & & [l] & \\ & & & [l] \end{bmatrix}_{12\times 12} \tag{10.5}$$

with

$$[l] = \begin{bmatrix} 1 & 0 & 1 \\ 0 & 2 & 0 \\ -l_a & 2r_0 & l_a \end{bmatrix}_{3\times 3}\,, \tag{10.6}$$

where $l_a$ denotes half of the axle length of the wheelset and $r_0$ the nominal radius of the wheel (also see Table 10.1). All of the other partitioned matrices and vectors involved have been listed in Appendix I, where $W$ is the static axle load for each wheelset and $g$ is the acceleration of gravity. In the above derivation, the vehicle is assumed to vibrate from the static equilibrium position.

The wheelset displacements $\{d_w\}$ can be related to the contact-point displacements $\{d_c\}$ of the two rails with surface irregularities by the *constraint equation*,

$$\{d_w\} = [\Gamma]\{d_c\} + \{r\}\,. \tag{10.7}$$

Here, $[\Gamma]$ is a constraint matrix and $\{r\}$ is a vector used to represent the *track irregularities*. For the vehicle and rail models considered in this study, the constraint matrix $[\Gamma]$ is

$$[\Gamma] = \begin{bmatrix} [\gamma] & & & \\ & [\gamma] & & \\ & & [\gamma] & \\ & & & [\gamma] \end{bmatrix}_{12\times 16} \tag{10.8}$$

with

$$[\gamma] = \begin{bmatrix} \dfrac{1}{2} & 0 & \dfrac{1}{2} & 0 \\ 0 & 1 & 0 & 0 \\ -\dfrac{1}{2l_a} & 0 & \dfrac{1}{2l_a} & 0 \end{bmatrix}_{3\times 4} \tag{10.9}$$

and the vector $\{r\}$ represents the rail irregularities at four wheelset positions,

$$\{r\} = \langle\, \{r_1\} \quad \{r_2\} \quad \{r_3\} \quad \{r_4\}\,\rangle^T\,, \tag{10.10}$$

where

$$\{r_i\} = \left\langle \frac{1}{2}[r_{v1}(x_i) + r_{v2}(x_i)] \quad r_h(x_i) \quad \frac{1}{2l_a}[-r_{v1}(x_i) + r_{v2}(x_i)] \right\rangle^T, \quad i = 1 \sim 4\,. \tag{10.11}$$

In Eq. (10.8), $r_{v1}(x)$ and $r_{v2}(x)$ denote the vertical deviations of the two rails, $r_h(x)$ denotes the deviations in lateral alignment of the two rails, and $x_i$ is the position of the $i$th wheelset of the vehicle at time $t + \Delta t$.

One key step in analysis of the present interaction problem is to solve for the contact forces $\{f_c\}$ existing between the *moving* (i.e., the vehicle) and the *nonmoving* (i.e., the rails) parts of the train–bridge system. This can be done by following the procedure outlined in Chapter 8 from Eqs. (8.5) to (8.12), namely, by solving the upper-part vehicle displacement vector $\{d_u\}$ from the first row of Eq. (10.4) using the Newmark finite difference scheme, and then substituting the displacement vector $\{d_u\}$ and derivatives into the second row of Eq. (10.4). This will result in the following expression for the contact forces $\{f_c\}$ at time $t + \Delta t$:

$$\begin{aligned}\{f_c\}_{t+\Delta t} &= \underline{[m_c][\Gamma]}\{\ddot{d}_c\}_{t+\Delta t} + \underline{[c_c][\Gamma]}\{\dot{d}_c\}_{t+\Delta t} + \underline{[k_c][\Gamma]}\{d_c\}_{t+\Delta t} \\ &\quad + \underline{[k_c]\{r\}_{t+\Delta t} + \{p_c\}_{t+\Delta t}} + \{q_c\}_t \\ &= [\bar{m}_c]\{\ddot{d}_c\}_{t+\Delta t} + [\bar{c}_c]\{\dot{d}_c\}_{t+\Delta t} + [\bar{k}_c]\{d_c\}_{t+\Delta t} \\ &\quad + \{\bar{p}_c\}_{t+\Delta t} + \{q_c\}_t\,, \end{aligned} \tag{10.12}$$

where the contact matrices $[m_c]$, $[c_c]$, $[k_c]$ and load vectors $\{p_c\}_{t+\Delta t}$, $\{q_c\}_t$ have been given in Eq. (8.15). Since the procedure for deriving the contact forces $\{f_c\}$ follows exactly the same lines as those presented as Eqs. (8.5)–(8.16) in Chapter 8, no attempt will be made herein to recapitulate the related details.

The contact force vector $\{f_c\}$ as presented in Eq. (10.12) relates only to the contact-point displacements $\{d_c\}$, which can be augmented through introduction of a transformation matrix $[\lambda]$ to yield the contact forces $\{f_c^*\}$ in *complete form* as

$$\begin{aligned}\{f_c^*\}_{t+\Delta t} &= [\lambda]\{f_c\}_{t+\Delta t} \\ &= [\tilde{m}_c]\{\ddot{d}_c\}_{t+\Delta t} + [\tilde{c}_c]\{\dot{d}_c\}_{t+\Delta t} \\ &\quad + [\tilde{k}_c]\{d_c\}_{t+\Delta t} + \{\tilde{p}_c\}_{t+\Delta t} + \{\tilde{q}_c\}_t\,, \end{aligned} \tag{10.13}$$

where

$$[\lambda] = \begin{bmatrix} [\delta] & & & \\ & [\delta] & & \\ & & [\delta] & \\ & & & [\delta] \end{bmatrix}_{16\times 12} \tag{10.14}$$

with

$$[\delta] = \begin{bmatrix} 1 & 0 & 0 \\ 0 & 1 & 0 \\ 0 & 0 & 1 \\ 0 & 1 & 0 \end{bmatrix}_{4\times 3}, \tag{10.15}$$

the contact matrices are

$$\begin{aligned} [\tilde{m}_c] &= [\lambda][\bar{m}_c]\,, \\ [\tilde{c}_c] &= [\lambda][\bar{c}_c]\,, \\ [\tilde{k}_c] &= [\lambda][\bar{k}_c]\,, \end{aligned} \tag{10.16}$$

and the load vectors associated with the contact forces are

$$\begin{aligned} \{\tilde{p}_c\}_{t+\Delta t} &= [\lambda]\{\bar{p}_c\}_{t+\Delta t}\,, \\ \{\tilde{q}_c\}_t &= [\lambda]\{q_c\}_t\,. \end{aligned} \tag{10.17}$$

Through expansion of Eq. (10.13), one can obtain the vertical and lateral contact forces for each contact point as

$$\begin{aligned} V_{i,t+\Delta t} &= \tilde{p}_{c(2i-1),t+\Delta t} + \tilde{q}_{c(2i-1),t} \\ &\quad + \sum_{j=1}^{16} (\tilde{m}_{c(2i-1)j}\ddot{d}_{cj,t+\Delta t} + \tilde{c}_{c(2i-1)j}\dot{d}_{cj,t+\Delta t} + \tilde{k}_{c(2i-1)j}d_{cj,t+\Delta t})\,, \end{aligned} \tag{10.18}$$

$$H_{i,t+\Delta t} = \tilde{p}_{c(2i),t+\Delta t} + \tilde{q}_{c(2i),t} + \sum_{j=1}^{16} (\tilde{m}_{c(2i)j} \ddot{d}_{cj,t+\Delta t} + \tilde{c}_{c(2i)j} \dot{d}_{cj,t+\Delta t} + \tilde{k}_{c(2i)j} d_{cj,t+\Delta t}) , \tag{10.19}$$

where $i = 1 \sim 8$, looping over all the wheels.

## 10.4. Equations for the Rail and Bridge Elements

Each of the two tracks can be divided into three parts, i.e., the central, left, and right track sections, as shown in Fig. 10.1. Furthermore, with reference to the top view in Fig. 10.3, the track located on the right side (when viewed along the positive $x$ direction) of the bridge or approach will be referred to as *Track A*, and the track on the left side as *Track B*. Each of the two track structures in the central track section is modeled as a rail element supported by the uniformly-distributed spring-dashpot units and in turn by the bridge elements. The rail element used to represent each track (which consists of two rails) is of length $l$ and will be referred to as the *central finite rail* (CFR) element. Similarly, each of the two track structures on the two side approaches will be modeled as a semi-infinite rail element supported by the uniformly-distributed spring-dashpot units, which will be referred to either as the *left semi-infinite rail* (LSR) element or the *right semi-infinite rail* (RSR) element. As for the central section, since the two tracks (i.e., Tracks $A$ and $B$) each lie on one side of the cross section of the bridge, the CFR element associated with one track is somewhat different from the other. However, the rail elements for the two tracks on the two side approaches are exactly the same, because they sit on the same stationary roadbeds. It follows that no distinction need to be made for the LSR elements used to represent the two tracks on the left approach. The same is also true for the RSR elements on the right approach. Because the CFR elements for the two tracks (i.e., Tracks $A$ and $B$) are different, they will be derived separately.

### 10.4.1. *Central Finite Rail (CFR) Element for Track A*

As shown in Fig. 10.3(a), there are two tracks of which each consists of two rails connected by the sleepers. In this study, a central finite rail (CFR) element of length $l$ and modeled as conventional beam will be used to represent the combination effect of the two rails of each track, which are treated as two beams of negligible torsional rigidity connected by the sleepers, which are modeled as rigid bars. With reference to the top view in Fig. 10.3(a), we shall first discuss the CFR element associated with Track $A$, namely, the one connected to the right side of the three-dimensional bridge element (when viewed along the positive $x$ direction) by a uniformly-distributed spring-dashpot-unit (ballast) layer. The nodal DOFs of the CFR element used to represent the combined effect of the two rails on Track $A$ are specified at the two ends as

$$\begin{aligned}\{d_A\} = \langle\, & u_{A1} \quad v_{A1} \quad w_{A1} \quad \theta_{A1} \quad \varphi_{A1} \quad \psi_{A1} \\ & u_{A2} \quad v_{A2} \quad w_{A2} \quad \theta_{A2} \quad \varphi_{A2} \quad \psi_{A2}\,\rangle^T .\end{aligned} \tag{10.20}$$

From here on, the subscript $A$ of each quantity will be used to denote that the quantity is associated with *Track A*. Correspondingly, the nodal DOFs of the bridge element are

$$\begin{aligned}\{d_b\} = \langle\, & u_{b1} \quad v_{b1} \quad w_{b1} \quad \theta_{b1} \quad \varphi_{b1} \quad \psi_{b1} \\ & u_{b2} \quad v_{b2} \quad w_{b2} \quad \theta_{b2} \quad \varphi_{b2} \quad \psi_{b2}\,\rangle^T .\end{aligned} \tag{10.21}$$

By the principle of virtual work, the equation of equilibrium for the CFR element can be written as

$$\begin{aligned}&\int_0^l E_t A_t u'_A \delta u'_A dx + \int_0^l E_t I_{tz} v''_A \delta v''_A dx + \int_0^l E_t I_{ty} w''_A \delta w''_A dx \\ &\quad = -\int_0^l m_t(\ddot{u}_A \delta u_A + \ddot{v}_A \delta v_A + \ddot{w}_A \delta w_A)dx \\ &\qquad \int_0^l I_t^* \ddot{\theta}_A \delta\theta_A dx - \int_0^l c_t(\dot{u}_A \delta u_A + \dot{v}_A \delta v_A + \dot{w}_A \delta w_A)dx\end{aligned}$$

$$
\begin{aligned}
& - \int_0^l c_t^* \dot{\theta}_A \delta\theta_A dx + 2l_d \int_0^l [p_c(x) + p_k(x)] \delta u_A dx \\
& + \int_0^l \int_{l_b - l_d}^{l_b + l_d} [q_c(x,z) + q_k(x,z)][\delta v_{TA} - (z - l_b)\delta\theta_A] dz dx \\
& + 2l_d \int_0^l [r_c(x) + r_k(x)] \delta w_A dx + \langle \delta d_A \rangle \{f_A\} \,, \qquad (10.22)
\end{aligned}
$$

where all the physical parameters involved have been defined in Table 10.2, in particular, all the quantities with subscript $t$ denote quantities associated with the *track* or *rails*, $\delta$ denotes a variational quantity, $c_t$ and $c_t^*$ are the material damping coefficients for the translational and torsional motions of the rails, $x$ is the local coordinate, $0 \leqq x \leqq l$, $(u_A, v_A, w_A)$ the axial (longitudinal), vertical and lateral displacements, $\{f_A\}$ denotes the external forces directly acting on the nodal points of the rail element of Track $A$, i.e.,

$$
\begin{aligned}
\{f_A\} = \langle\, & F_{Ax1} \quad F_{Ay1} \quad F_{Az1} \quad M_{Ax1} \quad M_{Ay1} \quad M_{Az1} \\
& F_{Ax2} \quad F_{Ay2} \quad F_{Az2} \quad M_{Ax2} \quad M_{Ay2} \quad M_{Az2} \,\rangle^T \,, \qquad (10.23)
\end{aligned}
$$

and $(p_c, p_k)$, $(q_c, q_k)$ and $(r_c, r_k)$ the unit axial, vertical and lateral interaction forces arising from the relative motion of the rail and bridge elements, which can be defined as

$$
\begin{aligned}
p_c(x) &= c_{bh1}^* [\dot{u}_b(x) - \dot{u}_A(x)] \,, \\
p_k(x) &= k_{bh1}^* [u_b(x) - u_A(x)] \,,
\end{aligned} \qquad (10.24)
$$

$$
\begin{aligned}
q_c(x,z) &= c_{bv1}^* [\dot{v}_b(x) - z\dot{\theta}_b(x) - \dot{v}_A(x) + (z - l_b)\dot{\theta}_A(x)] \,, \\
q_k(x,z) &= k_{bv1}^* [v_b(x) - z\theta_b(x) - v_A(x) + (z - l_b)\theta_A(x)] \,,
\end{aligned} \qquad (10.25)
$$

$$
\begin{aligned}
r_c(x) &= c_{bh1}^* [\dot{w}_b(x) + h\dot{\theta}_b(x) - \dot{w}_A(x)] \,, \\
r_k(x) &= k_{bh1}^* [w_b(\xi) + h\theta_b(x) - w_A(x)] \,.
\end{aligned} \qquad (10.26)
$$

The subscripts $c$ and $k$ in the preceding three equations indicate that the forces are damping or stiffness-related. The displacements $(u_A, v_A, w_A, \theta_A)$ of the rail element associated with Track $A$

(consisting of two rails) can be related to the nodal DOFs as

$$\begin{aligned} u_A(x) &= \langle N_u \rangle \{d_A\}\,, \\ v_A(x) &= \langle N_v \rangle \{d_A\}\,, \\ w_A(x) &= \langle N_w \rangle \{d_A\}\,, \\ \theta_A(x) &= \langle N_\theta \rangle \{d_A\}\,. \end{aligned} \tag{10.27}$$

Similarly, the displacements $(u_b, v_b, w_b, \theta_b)$ of the bridge element can be expressed as

$$\begin{aligned} u_b(x) &= \langle N_u \rangle \{d_b\}\,, \\ v_b(x) &= \langle N_v \rangle \{d_b\}\,, \\ w_b(x) &= \langle N_w \rangle \{d_b\}\,, \\ \theta_b(x) &= \langle N_\theta \rangle \{d_b\}\,, \end{aligned} \tag{10.28}$$

where the interpolation vectors for the axial, vertical, lateral, and torsional displacements are given as follows:

$$\begin{aligned} \langle N_u \rangle &= \langle\, N_1 \quad 0 \quad 0 \quad 0 \quad 0 \quad 0 \quad N_2 \quad 0 \quad 0 \quad 0 \quad 0 \quad 0 \,\rangle^T\,, \\ \langle N_v \rangle &= \langle\, 0 \quad N_3 \quad 0 \quad 0 \quad 0 \quad N_4 \quad 0 \quad N_5 \quad 0 \quad 0 \quad 0 \quad N_6 \,\rangle^T\,, \\ \langle N_w \rangle &= \langle\, 0 \quad 0 \quad N_3 \quad 0 \quad -N_4 \quad 0 \quad 0 \quad 0 \quad N_5 \quad 0 \quad -N_6 \quad 0 \,\rangle^T\,, \\ \langle N_\theta \rangle &= \langle\, 0 \quad 0 \quad 0 \quad N_1 \quad 0 \quad 0 \quad 0 \quad 0 \quad 0 \quad N_2 \quad 0 \quad 0 \,\rangle^T\,. \end{aligned} \tag{10.29}$$

Here, $N_1$ and $N_4$ are linear, and $N_2$, $N_3$, $N_5$ and $N_6$ are cubic Hermitian functions. It should be noted that the torsional strain energy was not included in Eq. (10.22) due to the assumption of negligible torsional deformation in each rail.

Substituting the preceding displacement fields, i.e., Eqs. (10.27) and (10.28), into the virtual work equation in Eq. (10.22) yields the equation of motion for the CFR element on Track $A$ as follows:

$$[m_A]\{\ddot{d}_A\} + [c_A]\{\dot{d}_A\} + [k_A]\{d_A\} = \{f_A\} + [c_{Ab}]\{\dot{d}_b\} + [k_{Ab}]\{d_b\}\,. \tag{10.30}$$

Here, the mass, damping and stiffness matrices $[m_A]$, $[c_A]$ and $[k_A]$ are

$$\begin{aligned}[m_A] &= [m_{A0}]\,,\\ [c_A] &= [c_{A0}] + [c_{A1}] + [c_{A2}] + [c_{A3}] + [c_{A4}]\,,\\ [k_A] &= [k_{A0}] + [k_{A1}] + [k_{A2}] + [k_{A3}] + [k_{A4}]\,,\end{aligned} \tag{10.31}$$

and the interaction matrices $[c_{Ab}]$ and $[k_{Ab}]$ are

$$\begin{aligned}[c_{Ab}] &= [c_{Ab1}] + [c_{Ab2}] + [c_{Ab3}] + [c_{Ab4}] + [c_{Ab5}] + [c_{Ab6}]\,,\\ [k_{Ab}] &= [k_{Ab1}] + [k_{Ab2}] + [k_{Ab3}] + [k_{Ab4}] + [k_{Ab5}] + [k_{Ab6}]\,,\end{aligned} \tag{10.32}$$

where $[m_{A0}]$, $[c_{A0}]$ and $[k_{A0}]$ are due to the inertial, material damping and stiffness effects of the rail element. The matrices $[m_{A0}]$ and $[k_{A0}]$ have been listed in Appendix J. The material damping matrix $[c_{A0}]$ will be included in the system equation, along with those for the other elements, based on the Rayleigh damping assumption, which will just be skipped herein. The matrices $[c_{A1}] \sim [c_{A4}]$, $[c_{Ab1}] \sim [c_{Ab6}]$, $[k_{A1}] \sim [k_{A4}]$ and $[k_{Ab1}] \sim [k_{Ab6}]$ related to the interactions between the bridge and CFR elements on Track $A$, can be given as

$$\begin{aligned}[c_{A1}] &= 2l_d c_{bh1}^*[\psi_u]\,,\\ [c_{A2}] &= 2l_d c_{bv1}^*[\psi_v]\,,\\ [c_{A3}] &= l_1 c_{bv1}^*[\psi_\theta]\,,\\ [c_{A4}] &= 2l_d c_{bh1}^*[\psi_w]\,,\end{aligned} \tag{10.33}$$

$$\begin{aligned}[c_{Ab1}] &= 2l_d c_{bh1}^*[\psi_u]\,,\\ [c_{Ab2}] &= 2l_d c_{bv1}^*[\psi_v]\,,\\ [c_{Ab3}] &= -2l_d l_b c_{bv1}^*[\psi_{v\theta}]\,,\\ [c_{Ab4}] &= l_1 c_{bv1}^*[\psi_\theta]\,,\\ [c_{Ab5}] &= 2l_d c_{bh1}^*[\psi_w]\,,\\ [c_{Ab6}] &= 2hl_d c_{bh1}^*[\psi_{w\theta}]\,,\end{aligned} \tag{10.34}$$

$$\begin{aligned}
[k_{A1}] &= 2l_d k_{bh1}^*[\psi_u]\,,\\
[k_{A2}] &= 2l_d k_{bv1}^*[\psi_v]\,,\\
[k_{A3}] &= l_1 k_{bv1}^*[\psi_\theta]\,,\\
[k_{A4}] &= 2l_d k_{bh1}^*[\psi_w]\,,
\end{aligned} \tag{10.35}$$

$$\begin{aligned}
[k_{Ab1}] &= 2l_d k_{bh1}^*[\psi_u]\,,\\
[k_{Ab2}] &= 2l_d k_{bv1}^*[\psi_v]\,,\\
[k_{Ab3}] &= -2l_d l_b k_{bv1}^*[\psi_{v\theta}]\,,\\
[k_{Ab4}] &= l_1 k_{bv1}^*[\psi_\theta]\,,\\
[k_{Ab5}] &= 2l_d k_{bh1}^*[\psi_w]\,,\\
[k_{Ab6}] &= 2hl_d k_{bh1}^*[\psi_{w\theta}]\,,
\end{aligned} \tag{10.36}$$

where $l_1 = (l_{b2}^3 - l_{b1}^3 - 6l_s \cdot l_b^2)/3$, $l_{b2} = l_b + l_d$, $l_{b1} = l_b - l_d$ and the matrices $[\psi_u]$, $[\psi_v]$, $[\psi_w]$, $[\psi_\theta]$, $[\psi_{v\theta}]$ and $[\psi_{w\theta}]$ are defined as follows:

$$\begin{aligned}
[\psi_u] &= \int_0^l \{N_u\}\langle N_u\rangle dx\,,\\
[\psi_v] &= \int_0^l \{N_v\}\langle N_v\rangle dx\,,\\
[\psi_w] &= \int_0^l \{N_w\}\langle N_w\rangle dx\,,\\
[\psi_\theta] &= \int_0^l \{N_\theta\}\langle N_\theta\rangle dx\,,\\
[\psi_{v\theta}] &= \int_0^l \{N_v\}\langle N_\theta\rangle dx\,,\\
[\psi_{w\theta}] &= \int_0^l \{N_w\}\langle N_\theta\rangle dx\,.
\end{aligned} \tag{10.37}$$

The results for all the matrices in Eq. (10.37) can be found in Appendix J.

### 10.4.2. *Central Finite Rail (CFR) Element for Track B*

The CFR element for *Track B*, that is, the one interacting with the left side of the cross sections of the bridge element (when viewed along the positive $x$ direction), is also shown in Fig. 10.3(a). The nodal DOFs of the CFR element are denoted as follows:

$$\begin{aligned}\{d_B\} = \langle\, & u_{B1} \quad v_{B1} \quad w_{B1} \quad \theta_{B1} \quad \varphi_{B1} \quad \psi_{B1} \\ & u_{B2} \quad v_{B2} \quad w_{B2} \quad \theta_{B2} \quad \varphi_{B2} \quad \psi_{B2}\,\rangle^T ,\end{aligned} \tag{10.38}$$

in which the subscript $B$ denotes *Track B*. Similarly, by the principle of virtual work, the equation of equilibrium for the CFR element on Track $B$ can be written as

$$\begin{aligned}&\int_0^l E_t A_t u_B' \delta u_B' dx + \int_0^l E_t I_{tz} v_B'' \delta v_B'' dx + \int_0^l E_t I_{ty} w_B'' \delta w_B'' dx \\ &\quad = -\int_0^l m_t(\ddot{u}_B \delta u_B + \ddot{v}_B \delta v_B + \ddot{w}_B \delta w_B) dx \\ &\qquad - \int_0^l I_t^* \ddot{\theta}_B \delta\theta_B dx - \int_0^l c_t(\dot{u}_B \delta u_B + \dot{v}_B \delta v_B + \dot{w}_B \delta w_B) dx \\ &\qquad - \int_0^l c_t^* \dot{\theta}_B \delta\theta_B dx + 2l_d \int_0^l [p_c(x) + p_k(x)] \delta u_B dx \\ &\qquad + \int_0^l \int_{-l_d-l_b}^{l_d-l_b} [q_c(x,z) + q_k(x,z)][\delta v_{TB} - (z+l_b)\delta\theta_B] dz dx \\ &\qquad + 2l_d \int_0^l [r_c(x) + r_k(x)] \delta w_B dx + \langle \delta d_B \rangle \{f_B\} ,\end{aligned} \tag{10.39}$$

where $\{f_B\}$ denotes the external nodal forces of the rail element, i.e.,

$$\begin{aligned}\{f_B\} = \langle\, & F_{Bx1} \quad F_{By1} \quad F_{Bz1} \quad M_{Bx1} \quad M_{By1} \quad M_{Bz1} \\ & F_{Bx2} \quad F_{By2} \quad F_{Bz2} \quad M_{Bx2} \quad M_{By2} \quad M_{Bz2}\,\rangle^T .\end{aligned} \tag{10.40}$$

The unit axial, vertical and lateral interaction forces $(p_c, p_k)$, $(q_c, q_k)$ and $(r_c, r_k)$ between the rail and bridge elements are as follows:

$$\begin{aligned} p_c(x) &= c^*_{bh1}[\dot{u}_b(x) - \dot{u}_B(x)]\,, \\ p_k(x) &= k^*_{bh1}[u_b(x) - u_B(x)]\,, \end{aligned} \tag{10.41}$$

$$\begin{aligned} q_c(x,z) &= c^*_{bv1}[\dot{v}_b(x) + z\dot{\theta}_b(x) - \dot{v}_B(x) + (z + l_b)\dot{\theta}_B(x)]\,, \\ q_k(x,z) &= k^*_{bv1}[v_b(x) - z\theta_b(x) - v_B(x) + (z + l_b)\theta_B(x)]\,, \end{aligned} \tag{10.42}$$

$$\begin{aligned} r_c(x) &= c^*_{bh1}[\dot{w}_b(x) + h\dot{\theta}_b(x) - \dot{w}_B(x)]\,, \\ r_k(x) &= k^*_{bh1}[w_b(x) + h\theta_b(x) - w_B(x)]\,. \end{aligned} \tag{10.43}$$

The displacements $(u_B, v_B, w_B, \theta_B)$ of the rail element on Track $B$ can be related to the element nodal DOFs as

$$\begin{aligned} u_B &= \langle N_u \rangle \{d_B\}\,, \\ v_B &= \langle N_v \rangle \{d_B\}\,, \\ w_B &= \langle N_w \rangle \{d_B\}\,, \\ \theta_B &= \langle N_\theta \rangle \{d_B\}\,. \end{aligned} \tag{10.44}$$

Here, the interpolation functions $\langle N_u \rangle$, $\langle N_v \rangle$, $\langle N_w \rangle$ and $\langle N_\theta \rangle$ are the same as those given previously for the CFR element for Track $A$. Using Eqs. (10.41)–(10.44), one obtains from Eq. (10.39) after some manipulations the equation of motion for the CFR element of Track $B$ as follows:

$$[m_B]\{\ddot{d}_B\} + [c_B]\{\dot{d}_B\} + [k_B]\{d_B\} = \{f_B\} + [c_{Bb}]\{\dot{d}_b\} + [k_{Bb}]\{d_b\}\,. \tag{10.45}$$

Clearly, Eq. (10.45) is identical in form to Eq. (10.30) for the CFR element of Track $A$. The system matrices in Eq. (10.45) can be given as

$$\begin{aligned}[m_B] &= [m_{B0}]\,,\\ [c_B] &= [c_{B0}] + [c_{B1}] + [c_{B2}] + [c_{B3}] + [c_{B4}]\,,\\ [k_B] &= [k_{B0}] + [k_{B1}] + [k_{B2}] + [k_{B3}] + [k_{B4}]\,,\end{aligned} \tag{10.46}$$

and the interaction matrices as

$$\begin{aligned}[c_{Bb}] &= [c_{Bb1}] + [c_{Bb2}] + [c_{Bb3}] + [c_{Bb4}] + [c_{Bb5}] + [c_{Bb6}]\,,\\ [k_{Bb}] &= [k_{Bb1}] + [k_{Bb2}] + [k_{Bb3}] + [k_{Bb4}] + [k_{Bb5}] + [k_{Bb6}]\,,\end{aligned} \tag{10.47}$$

where all the matrices on the right side are exactly the same as the corresponding terms given in Eqs. (10.33)–(10.36), except for the matrices $[c_{Bb3}]$ and $[k_{Bb3}]$, which are given as

$$\begin{aligned}[c_{Bb3}] &= -[c_{Ab3}]\,,\\ [k_{Bb3}] &= -[k_{Ab3}]\,.\end{aligned} \tag{10.48}$$

As can be seen, the difference between the CFR elements for Tracks $A$ and $B$ originates mainly from interaction with the torsional deformation of the bridge element.

### 10.4.3. *The Bridge Element*

By the principle of virtual work, along with the displacement fields defined previously in Eqs. (10.27) and (10.28), the equation of motion for the bridge element can be derived as

$$\begin{aligned}[m_b]\{\ddot{d}_b\} + [c_b]\{\dot{d}_b\} + [k_b]\{d_b\} &= \{f_b\} + [c_{bA}]\{\dot{d}_A\} + [k_{bA}]\{d_A\}\\ &\quad + [c_{bB}]\{\dot{d}_B\} + [k_{bB}]\{d_B\}\,,\end{aligned} \tag{10.49}$$

where $\{f_b\}$ denotes the external nodal forces,

$$\{f_b\} = \langle\, F_{bx1} \quad F_{by1} \quad F_{bz1} \quad M_{bx1} \quad M_{by1} \quad M_{bz1} \quad F_{bx2} \quad F_{by2} \quad F_{bz2} \quad M_{bx2} \quad M_{by2} \quad M_{bz2} \,\rangle^T , \qquad (10.50)$$

and the mass, damping and stiffness matrices $[m_b]$, $[c_b]$ and $[k_b]$ are defined as follows:

$$\begin{aligned}
[m_b] &= [m_{b0}] , \\
[c_b] &= [c_{b0}] + [c_{b1}] + [c_{b2}] + [c_{b3}] + [c_{b4}] + [c_{b5}] + [c_{b6}] \\
&\quad + [c_{b7}] + [c_{b8}] + [c_{b9}] + [c'_{b1}] + [c'_{b2}] + [c'_{b3}] + [c'_{b4}] \\
&\quad + [c'_{b5}] + [c'_{b6}] + [c'_{b7}] + [c'_{b8}] + [c'_{b9}] , \qquad (10.51) \\
[k_b] &= [k_{b0}] + [k_{b1}] + [k_{b2}] + [k_{b3}] + [k_{b4}] + [k_{b5}] + [k_{b6}] \\
&\quad + [k_{b7}] + [k_{b8}] + [k_{b9}] + [k'_{b1}] + [k'_{b2}] + [k'_{b3}] + [k'_{b4}] \\
&\quad + [k'_{b5}] + [k'_{b6}] + [k'_{b7}] + [k'_{b8}] + [k'_{b9}] ,
\end{aligned}$$

where $[m_{b0}]$ and $[k_{b0}]$ can be found in Appendix J, and the matrices $[c_{b1}] \sim [c_{b9}]$, $[k_{b1}] \sim [k_{b9}]$ and $[c'_{b1}] \sim [c'_{b9}]$, $[k'_{b1}] \sim [k'_{b9}]$ arise from the interaction of Tracks $A$ and $B$ with the bridge, which can be given as

$$\begin{aligned}
[c_{b1}] &= [c'_{b1}] = 2l_d c^*_{bh1} [\psi_u] , \\
[c_{b2}] &= [c'_{b2}] = 2l_d c^*_{bv1} [\psi_v] , \\
[c_{b3}] &= [c'_{b3}] = l_2 c^*_{bv1} [\psi_\theta] , \\
[c_{b4}] &= [c'_{b4}] = 2l_d c^*_{bh1} [\psi_w] , \\
[c_{b5}] &= [c'_{b5}] = 2h^2 l_d c^*_{bh1} [\psi_\theta] , \qquad (10.52) \\
[c_{b6}] &= -[c'_{b6}] = -2l_d l_b c^*_{bv1} [\psi_{v\theta}] , \\
[c_{b7}] &= -[c'_{b7}] = -2l_d l_b c^*_{bv1} [\psi_{\theta v}] , \\
[c_{b8}] &= [c'_{b8}] = 2hl_d c^*_{bh1} [\psi_{w\theta}] , \\
[c_{b9}] &= [c'_{b9}] = 2hl_d c^*_{bh1} [\psi_{\theta w}] ,
\end{aligned}$$

and

$$\begin{aligned}
[k_{b1}] &= [k'_{b1}] = 2l_d k^*_{bh1}[\psi_u]\,,\\
[k_{b2}] &= [k'_{b2}] = 2l_d k^*_{bv1}[\psi_v]\,,\\
[k_{b3}] &= [k'_{b3}] = l_2 k^*_{bv1}[\psi_\theta]\,,\\
[k_{b4}] &= [k'_{b4}] = 2l_d k^*_{bh1}[\psi_w]\,,\\
[k_{b5}] &= [k'_{b5}] = 2h^2 l_d k^*_{bh1}[\psi_\theta]\,,\\
[k_{b6}] &= -[k'_{b6}] = -2l_d l_b k^*_{bv1}[\psi_{v\theta}]\,,\\
[k_{b7}] &= -[k'_{b7}] = -2l_d l_b k^*_{bv1}[\psi_{\theta v}]\,,\\
[k_{b8}] &= [k'_{b8}] = 2hl_d k^*_{bh1}[\psi_{w\theta}]\,,\\
[k_{b9}] &= [k'_{b9}] = 2hl_d k^*_{bh1}[\psi_{\theta w}]\,,
\end{aligned} \tag{10.53}$$

where $l_2 = (l_{b2}^3 - l_{b1}^3)/3$, $[\psi_{\theta v}] = [\psi_{v\theta}]^T$ and $[\psi_{\theta w}] = [\psi_{w\theta}]^T$, and $[\psi_u]$, $[\psi_v]$, $[\psi_w]$, $[\psi_\theta]$, $[\psi_{v\theta}]$ and $[\psi_{w\theta}]$ are the same as those defined in Eq. (10.37). The matrices $[c_{bA}]$, $[k_{bA}]$, $[c_{bB}]$ and $[k_{bB}]$ originate from the interactions between the bridge and the two tracks, which are identical to the matrices $[c_{Ab}]$, $[k_{Ab}]$, $[c_{Bb}]$ and $[k_{Bb}]$, respectively, i.e.,

$$\begin{aligned}
[c_{bA}] &= [c_{Ab}]\,,\\
[k_{bA}] &= [k_{Ab}]\,,\\
[c_{bB}] &= [c_{Bb}]\,,\\
[k_{bB}] &= [k_{Bb}]\,.
\end{aligned} \tag{10.54}$$

From the preceding equations, it is clear that the lateral and torsional vibrations of the bridge are coupled, as implied by the terms involving the matrices $[\psi_{\theta w}]$ and $[\psi_{w\theta}]$ in Eqs. (10.52) and (10.53), which can be attributed to the fact that the torsional center of the box-girder bridge is located at a distance below the bridge deck on which the two tracks are lying. However, the vertical and torsional vibrations of the bridge do not interact each other, although there also exist terms involving the matrices $[\psi_{v\theta}]$ and $[\psi_{\theta v}]$ in Eqs. (10.52) and (10.53), due to the fact that all these terms will cancel each other, as can be observed from the following relations: $[c_{b6}] = -[c'_{b6}]$, $[c_{b7}] = -[c'_{b7}]$, $[k_{b6}] = -[k'_{b6}]$, and $[k_{b7}] = -[k'_{b7}]$.

### 10.4.4. *Left Semi-Infinite Rail (LSR) Element for Track A*

As shown in Fig. 10.3(b), the two tracks (each consisting of two rails) extend to infinity on the left side. In this study, each of the two tracks will be represented by a left semi-infinite rail (LSR) element, which is a semi-infinite beam element used to represent the combination effect of the two rails assumed to be of negligible torsional rigidity but connected by rigid bars (i.e., sleepers). The LSR element has only a single node situated at the middle point between the start ends of the two beams. The nodal displacements of the *LSR element* associated with *Track A* may be denoted as

$$\{d_{Al}\} = \langle\, u_{A2} \quad v_{A2} \quad w_{A2} \quad \theta_{A2} \quad \varphi_{A2} \quad \psi_{A2}\,\rangle^T . \tag{10.55}$$

Correspondingly, the nodal forces of the element are

$$\{f_{Al}\} = \langle\, F_{Ax2} \quad F_{Ay2} \quad F_{Az2} \quad M_{Ax2} \quad M_{Ay2} \quad M_{Az2}\,\rangle^T . \tag{10.56}$$

By the principle of virtual work, one can write

$$\begin{aligned}
&\int_{-\infty}^{0} E_t A_t u'_A \delta u'_A dx + \int_{-\infty}^{0} E_t I_{tz} v''_A \delta v''_A dx + \int_{-\infty}^{0} E_t I_{ty} w''_A \delta w''_A dx \\
&\quad = -\int_{-\infty}^{0} m_t (\ddot{u}_A \delta u_A + \ddot{v}_A \delta v_A + \ddot{w}_A \delta w_A) dx \\
&\qquad - \int_{-\infty}^{0} I_t^* \ddot{\theta}_A \delta\theta_A dx - \int_{-\infty}^{0} c_t (\dot{u}_A \delta u_A + \dot{v}_A \delta v_A + \dot{w}_A \delta w_A) dx \\
&\qquad - \int_{-\infty}^{0} c_t^* \dot{\theta}_A \delta\theta_A dx - 2l_d \int_{-\infty}^{0} (c^*_{bh2} \dot{u}_A + k^*_{bh2} u_A) \delta u_A dx
\end{aligned}$$

$$
\begin{aligned}
&- \int_{-\infty}^{0} \int_{l_b - l_d}^{l_b + l_d} [q_c(x, z) + q_k(x, z)][\delta v_A - (z - l_b)\delta\theta_A] dz dx \\
&- 2l_d \int_{-\infty}^{0} (c_{bh2}^{*} \dot{w}_A + k_{bh2}^{*} w_A)\delta w_A dx + \langle \delta d_{Al} \rangle \{ f_{Al} \} , \qquad (10.57)
\end{aligned}
$$

where a quantity preceded by $\delta$ denotes a virtual displacement, $x$ is the local coordinate, $-\infty < x \leqq 0$, and the interaction forces $q_c(x, z)$, $q_k(x, z)$ are defined as

$$
\begin{aligned}
q_c(x, z) &= c_{bv2}^{*}[\dot{v}_A(x) - (z - l_b)\dot{\theta}_A(x)] , \\
q_k(x, z) &= k_{bv2}^{*}[v_A(x) - (z - l_b)\theta_A(x)] .
\end{aligned} \qquad (10.58)
$$

The axial, vertical, lateral and torsional displacements of the element can also be related to the nodal displacements as

$$
\begin{aligned}
u_A &= \langle N_u \rangle \{ d_{Al} \} , \\
v_A &= \langle N_v \rangle \{ d_{Al} \} , \\
w_A &= \langle N_w \rangle \{ d_{Al} \} , \\
\theta_A &= \langle N_\theta \rangle \{ d_{Al} \} ,
\end{aligned} \qquad (10.59)
$$

where the interpolation vectors are

$$
\begin{aligned}
\langle N_u \rangle &= \langle N_1 \quad 0 \quad 0 \quad 0 \quad 0 \quad 0 \rangle^T , \\
\langle N_v \rangle &= \langle 0 \quad N_3 \quad 0 \quad 0 \quad 0 \quad N_4 \rangle^T , \\
\langle N_w \rangle &= \langle 0 \quad 0 \quad N_5 \quad 0 \quad -N_6 \quad 0 \rangle^T , \\
\langle N_\theta \rangle &= \langle 0 \quad 0 \quad 0 \quad N_2 \quad 0 \quad 0 \rangle^T .
\end{aligned} \qquad (10.60)
$$

The interpolation functions, i.e., $N_1 \sim N_6$, can be obtained from the static solution to the problem of two rigidly-connected beams resting on the *Winkler foundation* subjected to a unit force or moment, that is,

$$\begin{aligned}
N_1 &= e^{\lambda_u x}\,, \\
N_2 &= e^{\lambda_\theta x}\,, \\
N_3 &= e^{\lambda_v x}(\cos\lambda_v x - \sin\lambda_v x)\,, \\
N_4 &= \frac{1}{\lambda_v} e^{\lambda_v x}\sin\lambda_v x\,, \\
N_5 &= e^{\lambda_w x}(\cos\lambda_w x - \sin\lambda_w x)\,, \\
N_6 &= \frac{1}{\lambda_w} e^{\lambda_w x}\sin\lambda_w x\,,
\end{aligned} \tag{10.61}$$

where $\lambda_u, \lambda_v, \lambda_w$ and $\lambda_\theta$ denote the longitudinal, vertical, lateral, and torsional characteristic numbers of the twin beam-Winkler foundation system,

$$\begin{aligned}
\lambda_u &= \sqrt{\frac{2l_d k^*_{bh2}}{E_t A_t}}\,, \qquad & \lambda_v &= \sqrt[4]{\frac{l_d k^*_{bv2}}{2E_t I_{tz}}}\,, \\
\lambda_w &= \sqrt[4]{\frac{l_d k^*_{bh2}}{2E_t I_{ty}}}\,, \qquad & \lambda_\theta &= \sqrt{\frac{2l_d^3 k^*_{bv2}}{3E_t I_{tz}}}\,.
\end{aligned} \tag{10.62}$$

By using Eq. (10.61) and the definition of the displacement fields, Eq. (10.57) can be manipulated to yield the equation of motion for the *LSR element* on *Track A*,

$$[m_l]\{\ddot{d}_{Al}\} + [c_l]\{\dot{d}_{Al}\} + [k_l]\{d_{Al}\} = \{f_{Al}\}\,, \tag{10.63}$$

where the mass, damping and stiffness matrices $[m_l]$, $[c_l]$ and $[k_l]$ of the LSR element can be computed as

$$\begin{aligned}
[m_l] &= [m_{l0}]\,, \\
[c_l] &= [c_{l0}] + [c_{l1}] + [c_{l2}] + [c_{l3}] + [c_{l4}]\,, \\
[k_l] &= [k_{l0}] + [k_{l1}] + [k_{l2}] + [k_{l3}] + [k_{l4}]\,.
\end{aligned} \tag{10.64}$$

Here, the matrices $[m_{l0}]$ and $[k_{l0}]$ have been listed in Appendix J, and the other matrices are given as follows:

$$\begin{aligned}
[c_{l1}] &= 2l_d c^*_{bh2}[\psi_u]\,, \\
[c_{l2}] &= 2l_d c^*_{bv2}[\psi_v]\,, \\
[c_{l3}] &= l_1 c^*_{bv2}[\psi_\theta]\,, \\
[c_{l4}] &= 2l_d c^*_{bh2}[\psi_w]\,,
\end{aligned} \tag{10.65}$$

$$\begin{aligned}
[k_{l1}] &= 2l_d k^*_{bh2}[\psi_u]\,, \\
[k_{l2}] &= 2l_d k^*_{bv2}[\psi_v]\,, \\
[k_{l3}] &= l_1 k^*_{bv2}[\psi_\theta]\,, \\
[k_{l4}] &= 2l_d k^*_{bh2}[\psi_w]\,.
\end{aligned} \tag{10.66}$$

The matrices $[\psi_u]$, $[\psi_v]$, $[\psi_\theta]$ and $[\psi_w]$ in the preceding equation have also been listed in Appendix J.

### 10.4.5. *Right Semi-Infinite Rail (RSR) Element for Track A*

Similar to the LSR element for Track $A$ discussed in Sec. 10.4.4, the right semi-infinite rail (RSR) element is a semi-infinite beam element used to simulate the combination effect of two rails with negligible torsional rigidity connected by rigid bars (i.e., sleepers), having one node situated at the middle point between the start ends of the two rails, as shown in Fig. 10.3(c). The nodal displacements for the *RSR element* on *Track A* are

$$\{d_{Ar}\} = \langle\, u_{A1} \quad v_{A1} \quad w_{A1} \quad \theta_{A1} \quad \varphi_{A1} \quad \psi_{A1}\,\rangle^T\,. \tag{10.67}$$

Correspondingly, the nodal forces of the element are

$$\{f_{Ar}\} = \langle\, F_{Ax1} \quad F_{Ay1} \quad F_{Az1} \quad M_{Ax1} \quad M_{Ay1} \quad M_{Az1}\,\rangle^T\,. \tag{10.68}$$

The axial, vertical, lateral and torsional displacements of the element can be related to the nodal DOFs as:

$$\begin{aligned} u_A &= \langle N_u \rangle \{d_{Ar}\}, \\ v_A &= \langle N_v \rangle \{d_{Ar}\}, \\ w_A &= \langle N_w \rangle \{d_{Ar}\}, \\ \theta_A &= \langle N_\theta \rangle \{d_{Ar}\}, \end{aligned} \tag{10.69}$$

where the interpolation vectors $\langle N_u \rangle$, $\langle N_v \rangle$, $\langle N_w \rangle$ and $\langle N_\theta \rangle$ are identical in form to those for the LSR element. For the present case, the interpolation functions can also be determined from the static solution as

$$\begin{aligned} N_1 &= e^{-\lambda_u x}, \\ N_2 &= e^{-\lambda_\theta x}, \\ N_3 &= e^{-\lambda_v x}(\cos \lambda_v x + \sin \lambda_v x), \\ N_4 &= \frac{1}{\lambda_v} e^{-\lambda_v x} \sin \lambda_v x, \\ N_5 &= e^{-\lambda_w x}(\cos \lambda_w x + \sin \lambda_w x), \\ N_6 &= \frac{1}{\lambda_w} e^{-\lambda_w x} \sin \lambda_w x, \end{aligned} \tag{10.70}$$

where $\lambda_u$, $\lambda_v$, $\lambda_w$ and $\lambda_\theta$ are the same as those given in Eq. (10.62). Following the same procedure as that for the LSR element, one can derive the equation of motion for the RSR element as:

$$[m_r]\{\ddot{d}_{Ar}\} + [c_r]\{\dot{d}_{Ar}\} + [k_r]\{d_{Ar}\} = \{f_{Ar}\}, \tag{10.71}$$

where the mass, damping and stiffness matrices $[m_r]$, $[c_r]$ and $[k_r]$ are

$$\begin{aligned} [m_r] &= [m_{r0}], \\ [c_r] &= [c_{r0}] + [c_{r1}] + [c_{r2}] + [c_{r3}] + [c_{r4}], \\ [k_r] &= [k_{r0}] + [k_{r1}] + [k_{r2}] + [k_{r3}] + [k_{r4}]. \end{aligned} \tag{10.72}$$

Here, $[m_{r0}]$ and $[k_{r0}]$ have been given in Appendix J, and the matrices $[c_{r1}] \sim [c_{r4}]$ and $[k_{r1}] \sim [k_{r4}]$ are the same as those for the LSR element, that is,

$$\begin{aligned} [c_{r1}] &= [c_{l1}]\,, \\ [c_{r2}] &= [c_{l2}]\,, \\ [c_{r3}] &= [c_{l3}]\,, \\ [c_{r4}] &= [c_{l4}]\,, \end{aligned} \tag{10.73}$$

$$\begin{aligned} [k_{r1}] &= [k_{l1}]\,, \\ [k_{r2}] &= [k_{l2}]\,, \\ [k_{r3}] &= [k_{l3}]\,, \\ [k_{r4}] &= [k_{l4}]\,. \end{aligned} \tag{10.74}$$

### 10.4.6. *Left Semi-Infinite Rail (LSR) Element for Track B*

The LSR element for Track $B$ shown in Fig. 10.3(b) is exactly the same as that for Track $A$, except that the DOFs should be replaced by

$$\{d_{Bl}\} = \langle\, u_{B2} \quad v_{B2} \quad w_{B2} \quad \theta_{B2} \quad \varphi_{B2} \quad \psi_{B2}\,\rangle^T\,. \tag{10.75}$$

Correspondingly, the nodal forces of the element are

$$\{f_{Bl}\} = \langle\, F_{Bx2} \quad F_{By2} \quad F_{Bz2} \quad M_{Bx2} \quad M_{By2} \quad M_{Bz2}\,\rangle^T\,, \tag{10.76}$$

The equation of motion for the LSR element on Track $B$ is

$$[m_l]\{\ddot{d}_{Bl}\} + [c_l]\{\dot{d}_{Bl}\} + [k_l]\{d_{Bl}\} = \{f_{Bl}\}\,, \tag{10.77}$$

where the matrices $[m_l]$, $[c_l]$ and $[k_l]$ have been given in Eq. (10.64).

### 10.4.7. *Right Semi-Infinite Rail (RSR) Element for Track B*

The RSR element for Track $B$ is also the same as that for Track $A$, as shown in Fig. 10.3(c), except that the nodal DOFs should be replaced by

$$\{d_{Br}\} = \langle\, u_{B1} \quad v_{B1} \quad w_{B1} \quad \theta_{B1} \quad \varphi_{B1} \quad \psi_{B1}\,\rangle^T . \tag{10.78}$$

Correspondingly, the nodal forces of the element are

$$\{f_{Br}\} = \langle\, F_{Bx1} \quad F_{By1} \quad F_{Bz1} \quad M_{Bx1} \quad M_{By1} \quad M_{Bz1}\,\rangle^T . \tag{10.79}$$

The equation of motion for the present element is

$$[m_r]\{\ddot{d}_{Br}\} + [c_r]\{\dot{d}_{Br}\} + [k_r]\{d_{Br}\} = \{f_{Br}\}, \tag{10.80}$$

where $[m_r]$, $[c_r]$ and $[k_r]$ are the same as those given in Eq. (10.72).

Note that in a step-by-step time-history analysis, the equations of motion for the CFR, LSR, RSR and bridge elements formulated above should be interpreted as those established for the deformed position of the system at time $t + \Delta t$.

## 10.5. VRI Element Considering Vertical and Lateral Contact Forces

Assume that at time $t+\Delta t$, the four wheelsets of a vehicle are acting simultaneously at the $e_1$, $e_2$, $e_3$ and $e_4$th rail elements of Track $A$ or $B$. The rail elements and the wheelsets acting over them will be collectively referred to as the *vehicle–rails interaction* (VRI) elements, as they are directly interacting with each other. Consider the $e_i$th element that is acted upon by the vertical and lateral components of the $(2i-1)$th and $(2i)$th contact forces, i.e., $V_{(2i-1),t+\Delta t}$, $H_{(2i-1),t+\Delta t}$, $V_{2i,t+\Delta t}$, and $H_{2i,t+\Delta t}$. The equation of motion for the $e_i$th rail element of *Track A* at time $t+\Delta t$ may be written as follows:

$$[m_{Ai}]\{\ddot{d}_{Ai}\}_{t+\Delta t} + [c_{Ai}]\{\dot{d}_{Ai}\}_{t+\Delta t} + [k_{Ai}]\{d_{Ai}\}_{t+\Delta t}$$
$$= \{f_{Ai}\}_{t+\Delta t} + \varepsilon_i([c_{Ab}]\{\dot{d}_{bi}\} + [k_{Ab}]\{d_{bi}\}) - \{f_{Aci}\}_{t+\Delta t}\,, \tag{10.81}$$

where $[m_{Ai}] = [m_A]$, $[c_{Ai}] = [c_A]$, $[k_{Ai}] = [k_A]$, $\{d_{Ai}\} = \{d_A\}$, and $\varepsilon_i = 1$ for the case with the contact forces acting on the CFR element; $[m_{Ai}] = [m_l]$, $[c_{Ai}] = [c_l]$, $[k_{Ai}] = [k_l]$, $\{d_{Ai}\} = \{d_{Al}\}$, and $\varepsilon_i = 0$ for the LSR element; $[m_{Ai}] = [m_r]$, $[c_{Ai}] = [c_r]$, $[k_{Ai}] = [k_r]$, $\{d_{Ai}\} = \{d_{Ar}\}$, and $\varepsilon_i = 0$ for the RSR element; and $\{f_{Aci}\}$ denotes the vector of equivalent nodal forces resulting from the action of the $(2i-1)$th and $(2i)$th vertical and lateral contact forces $V_{(2i-1),t+\Delta t}$, $H_{(2i-1),t+\Delta t}$, $V_{2i,t+\Delta t}$, and $H_{2i,t+\Delta t}$:

$$\{f_{Aci}\}_{t+\Delta t} = \{N_{vi}^R\}V_{(2i-1),t+\Delta t} + \{N_{wi}^R\}H_{(2i-1),t+\Delta t}$$
$$+ \{N_{vi}^L\}V_{2i,t+\Delta t} + \{N_{wi}^L\}H_{2i,t+\Delta t}\,, \tag{10.82}$$

where $\{N_v^R\}$ and $\{N_w^R\}$ denote the interpolation vectors for the vertical and lateral displacements of the *right rail* of the $e_i$th rail element (see Fig. 10.3, when viewed along the positive $x$ direction), which actually consists of two rails. These interpolation vectors vary according to the type of elements, i.e., CFR, RSR or LSR elements, to which the contact forces are acting. Similarly, $\{N_v^L\}$ and $\{N_w^L\}$ denote the interpolation vectors for the vertical and lateral displacements of the left rail of the $e_i$th element, which are also dependent on the type of elements. All the interpolation vectors mentioned here are somewhat different from those for the rail element. For instance, the interpolation vectors $\{N_v^R\}$, $\{N_w^R\}$, $\{N_v^L\}$ and $\{N_w^L\}$ for the CFR element are:

$$\begin{aligned}
\{N_v^R\} &= \langle 0 \quad N_3 \quad 0 \quad -l_aN_1 \quad 0 \quad N_4 \quad 0 \quad N_5 \quad 0 \quad -l_aN_2 \quad 0 \quad N_6\rangle^T\,,\\
\{N_w^R\} &= \langle 0 \quad 0 \quad N_3 \quad 0 \quad -N_4 \quad 0 \quad 0 \quad 0 \quad N_5 \quad 0 \quad -N_6 \quad 0\rangle^T\,,\\
\{N_v^L\} &= \langle 0 \quad N_3 \quad 0 \quad l_aN_1 \quad 0 \quad N_4 \quad 0 \quad N_5 \quad 0 \quad l_aN_2 \quad 0 \quad N_6\rangle^T\,,\\
\{N_w^L\} &= \{N_w^R\}\,,
\end{aligned} \tag{10.83}$$

and those for the LSR or RSR elements are

$$
\begin{aligned}
\{N_v^R\} &= \langle 0 \quad N_3 \quad 0 \quad -l_a N_2 \quad 0 \quad N_4 \rangle^T ,\\
\{N_w^R\} &= \langle 0 \quad 0 \quad N_5 \quad 0 \quad -N_6 \quad 0 \rangle^T ,\\
\{N_v^L\} &= \langle 0 \quad N_3 \quad 0 \quad l_a N_2 \quad 0 \quad N_4 \rangle^T ,\\
\{N_w^L\} &= \{N_w^R\} .
\end{aligned}
\tag{10.84}
$$

In Eqs. (10.83) and (10.84), the interpolation functions $N_1 \sim N_6$ involved are identical to those defined for the CFR, LSR or RSR elements.

In Eq. (10.82), the subscript $i$ associated with the vectors $\{N_{vi}^R\}$, $\{N_{wi}^R\}$, $\{N_{vi}^L\}$ and $\{N_{wi}^L\}$ is used to indicate that they should be evaluated at the contact position of the $i$th wheelset, i.e.,

$$
\begin{aligned}
\{N_{vi}^R\} &= \{N_v^R(x_i)\} ,\\
\{N_{wi}^R\} &= \{N_w^R(x_i)\} ,\\
\{N_{vi}^L\} &= \{N_v^L(x_i)\} ,\\
\{N_{wi}^L\} &= \{N_w^L(x_i)\} ,
\end{aligned}
\tag{10.85}
$$

where $x_i$ is the local coordinate of the position of the $i$th wheelset on the $e_i$th element. By using Eqs. (10.82), (10.18) and (10.19), Eq. (10.81) for the $e_i$th rail element of Track $A$ can be rewritten as follows:

$$
\begin{aligned}
&[m_{Ai}]\{\ddot{d}_{Ai}\}_{t+\Delta t} + [c_{Ai}]\{\dot{d}_{Ai}\}_{t+\Delta t} + [k_{Ai}]\{d_{Ai}\}_{t+\Delta t}\\
&\quad = \{f_{Ai}\}_{t+\Delta t} + \varepsilon_i([c_{Ab}]\{\dot{d}_{bi}\}_{t+\Delta t} + [k_{Ab}]\{d_{bi}\}_{t+\Delta t})\\
&\qquad - \sum_{j=1}^{4}([m_{cij}^*]\{\ddot{d}_{rj}\}_{t+\Delta t} + [c_{cij}^*]\{\dot{d}_{rj}\}_{t+\Delta t} + [k_{cij}^*]\{d_{rj}\}_{t+\Delta t})\\
&\qquad - \{p_{ci}^*\}_{t+\Delta t} - \{q_{ci}^*\}_t ,
\end{aligned}
\tag{10.86}
$$

where the asterisked matrices represent the *linking action* transmitted through the car body by the $e_j$th element (under the $j$th wheel load) on the $e_i$th element (under the $i$th wheel load) and the interaction between the right and left rails of the same track through the four wheelsets of the vehicle considered,

$$
\begin{aligned}
[m^*_{cij}] &= \sum_{k=1}^{4}\left[\{N^k_i\}\left(\sum_{l=1}^{4}\tilde{m}_{c[4(i-1)+k][4(j-1)+l]}\langle N^l_j\rangle\right)\right],\\
[c^*_{cij}] &= \sum_{k=1}^{4}\left[\{N^k_i\}\left(\sum_{l=1}^{4}\tilde{c}_{c[4(i-1)+k][4(j-1)+l]}\langle N^l_j\rangle\right)\right],\\
[k^*_{cij}] &= \sum_{k=1}^{4}\left[\{N^k_i\}\left(\sum_{l=1}^{4}\tilde{k}_{c[4(i-1)+k][4(j-1)+l]}\langle N^l_j\rangle\right)\right],
\end{aligned}
\tag{10.87}
$$

and the equivalent nodal loads are

$$
\begin{aligned}
\{p^*_{ci}\}_{t+\Delta t} &= \sum_{k=1}^{4}(\{N^k_i\}\tilde{p}_{c[4(i-1)+k],t+\Delta t}),\\
\{q^*_{ci}\}_{t} &= \sum_{k=1}^{4}(\{N^k_i\}\tilde{q}_{c[4(i-1)+k],t}),
\end{aligned}
\tag{10.88}
$$

where $\{N^\beta_\alpha\}$ is defined as

$$
\{N^\beta_\alpha\} = \begin{cases}
\{N^R_{v\alpha}\} = \{N^R_v(\xi_\alpha)\}, & \text{for} \quad \beta = 1,\\
\{N^R_{w\alpha}\} = \{N^R_w(\xi_\alpha)\}, & \text{for} \quad \beta = 2,\\
\{N^L_{v\alpha}\} = \{N^L_v(\xi_\alpha)\}, & \text{for} \quad \beta = 3,\\
\{N^L_{w\alpha}\} = \{N^L_w(\xi_\alpha)\}, & \text{for} \quad \beta = 4.
\end{cases}
\tag{10.89}
$$

The equation of motion as given in Eq. (10.86) is the *condensed equation of motion* for the VRI element of Track $A$, as all the relevant vehicle DOFs have been eliminated. Similarly, the condensed equation of motion for the VRI element of Track $B$ can be written as follows:

$$\begin{aligned}
&[m_{Bi}]\{\ddot{d}_{Bi}\}_{t+\Delta t} + [c_{Bi}]\{\dot{d}_{Bi}\}_{t+\Delta t} + [k_{Bi}]\{d_{Bi}\}_{t+\Delta t} \\
&\quad = \{f_{Bi}\}_{t+\Delta t} + \varepsilon_i([c_{Bb}]\{\dot{d}_{bi}\}_{t+\Delta t} + [k_{Bb}]\{d_{bi}\}_{t+\Delta t}) \\
&\qquad - \sum_{j=1}^{4}([m^*_{cij}]\{\ddot{d}_{Bj}\}_{t+\Delta t} + [c^*_{cij}]\{\dot{d}_{Bj}\}_{t+\Delta t} \\
&\qquad + [k^*_{cij}]\{d_{Bj}\}_{t+\Delta t}) - \{p^*_{ci}\}_{t+\Delta t} - \{q^*_{ci}\}_t \,, \qquad (10.90)
\end{aligned}$$

where $[m_{Bi}] = [m_B]$, $[c_{Bi}] = [c_B]$, $[k_{Bi}] = [k_B]$, $\{d_{Bi}\} = \{d_B\}$, and $\varepsilon_i = 1$ for the case with the contact forces acting on the CFR element; $[m_{Bi}] = [m_l]$, $[c_{Bi}] = [c_l]$, $[k_{Bi}] = [k_l]$, $\{d_{Bi}\} = \{d_{Bl}\}$, and $\varepsilon_i = 0$ for the LSR element; $[m_{Bi}] = [m_r]$, $[c_{Bi}] = [c_r]$, $[k_{Bi}] = [k_r]$, $\{d_{Bi}\} = \{d_{Br}\}$, and $\varepsilon_i = 0$ for the the RSR element, and $[m^*_{cij}]$, $[c^*_{cij}]$, $[k^*_{cij}]$, $\{p^*_{ci}\}_{t+\Delta t}$, and $\{q^*_{ci}\}_t$ are the same as those defined in Eqs. (10.87) and (10.88). All the asterisked matrices and load vectors involved in Eqs. (10.86) and (10.90) are *time-dependent* and should be updated at each time step in an incremental analysis. The condensation process described above should be repeated until the *interaction effects* of all the vehicles moving on Track $A$ or $B$ (or both) have been included.

## 10.6. VRI Element Considering General Contact Forces

For the case where the longitudinal ($x$-direction) contact forces originate from acceleration or deceleration of the vehicles are considered, the contact force $\{f_{Aci}\}_{t+\Delta t}$ in Eq. (10.82) for Track $A$ should be modified as

$$\begin{aligned}
&\{f_{Aci}\}_{t+\Delta t} \\
&\quad = \{N^R_{vi}\}V_{(2i-1),t+\Delta t} + \{N^R_{wi}\}H_{(2i-1),t+\Delta t} + \{N^R_{ui}\}Q_{(2i-1),t+\Delta t} \\
&\qquad + \{N^L_{vi}\}V_{(2i),t+\Delta t} + \{N^L_{wi}\}H_{(2i),t+\Delta t} + \{N^L_{ui}\}Q_{(2i),t+\Delta t} \,,
\end{aligned} \qquad (10.91)$$

where $Q_{(2i-1),t+\Delta t}$ and $Q_{(2i),t+\Delta t}$ denote the longitudinal contact forces exerted on the two rails by the $i$th wheelset, which can be related to the vertical contact forces $V_{(2i-1),t+\Delta t}$ and $V_{(2i),t+\Delta t}$ in terms of the *frictional coefficient* $\mu_i$ as

$$\begin{aligned} Q_{(2i-1),t+\Delta t} &= \mu_i V_{(2i-1),t+\Delta t}\,, \\ Q_{(2i),t+\Delta t} &= \mu_i V_{(2i),t+\Delta t}\,. \end{aligned} \tag{10.92}$$

In Eq. (10.91), $\{N_u\}$ denotes the interpolation vector for the axial displacement of the $e_i$th element, which varies according to the type of rail elements, i.e., CFR, RSR or LSR elements, and $\{N_{ui}^R\} = \{N_u^R(x_i)\}$ and $\{N_{ui}^L\} = \{N_u^L(x_i)\}$ are evaluated at the position of the $i$th wheelset.

Considering the more general expression in Eq. (10.91) for the contact forces, along with Eq. (10.92), one can derive a condensed equation of motion for the $e_i$th rail element of Track $A$ or $B$ that is identical in form to Eq. (10.86) or (10.90), but with the asterisked matrices given as

$$\begin{aligned} [m_{cij}^*] &= \sum_{k=1}^{4}\left[\{N_i^k\}\left(\sum_{l=1}^{4}\tilde{m}_{c[4(i-1)+k][4(j-1)+l]}\langle N_j^l\rangle\right)\right] \\ &\quad + \sum_{m=1}^{2}\left[\mu_i\{N_i^m\}'\left(\sum_{l=1}^{4}\tilde{m}_{c[4(i-1)+m][4(j-1)+l]}\langle N_j^l\rangle\right)\right], \\ [c_{cij}^*] &= \sum_{k=1}^{4}\left[\{N_i^k\}\left(\sum_{l=1}^{4}\tilde{c}_{c[4(i-1)+k][4(j-1)+l]}\langle N_j^l\rangle\right)\right] \\ &\quad + \sum_{m=1}^{2}\left[\mu_i\{N_i^m\}'\left(\sum_{l=1}^{4}\tilde{c}_{c[4(i-1)+m][4(j-1)+l]}\langle N_j^l\rangle\right)\right], \\ [k_{cij}^*] &= \sum_{k=1}^{4}\left[\{N_i^k\}\left(\sum_{l=1}^{4}\tilde{k}_{c[4(i-1)+k][4(j-1)+l]}\langle N_j^l\rangle\right)\right] \\ &\quad + \sum_{m=1}^{2}\left[\mu_i\{N_i^m\}'\left(\sum_{l=1}^{4}\tilde{k}_{c[4(i-1)+m][4(j-1)+l]}\langle N_j^l\rangle\right)\right], \end{aligned} \tag{10.93}$$

and the following for the nodal loads,

$$\begin{aligned}\{p^*_{ci}\}_{t+\Delta t} &= \sum_{k=1}^{4}(\{N^k_i\}\tilde{p}_{c[4(i-1)+k],t+\Delta t}) \\ &\quad + \sum_{m=1}^{2}(\mu_i\{N^m_i\}'\tilde{p}_{c[4(i-1)+m],t+\Delta t})\,, \\ \{q^*_{ci}\}_t &= \sum_{k=1}^{4}(\{N^k_i\}\tilde{q}_{c[4(i-1)+k],t}) \\ &\quad + \sum_{m=1}^{2}(\mu_i\{N^m_i\}'\tilde{q}_{c[4(i-1)+m],t})\,,\end{aligned} \tag{10.94}$$

where $\{N^\beta_\alpha\}$ has been defined in Eq. (10.89) and $\{N^\beta_\alpha\}'$ can be given as follows:

$$\{N^\beta_\alpha\}' = \begin{cases}\{N^R_{u\alpha}\} = \{N^R_u(\xi_\alpha)\}\,, & \text{for} \quad \beta = 1\,, \\ \{N^L_{u\alpha}\} = \{N^L_u(\xi_\alpha)\}\,, & \text{for} \quad \beta = 2\,.\end{cases} \tag{10.95}$$

## 10.7. System Equations and Structural Damping

In simulating the vehicle–rails–bridge interaction system, the parts of the rails of each track that are directly acted upon by the wheel loads should be modeled by the VRI elements. However, for the remaining parts of the rails that are not directly under the action of wheel loads, they should be modeled by the ordinary rail elements. By assembling all the VRI elements, rail elements and bridge elements for the system considered, the structural equations at time $t + \Delta t$ can be established:

$$\begin{aligned}&[M]\{\ddot{D}\}_{t+\Delta t} + [C]\{\dot{D}\}_{t+\Delta t} + [K]\{D\}_{t+\Delta t} \\ &\quad = \{F\}_{t+\Delta t} - \{P^*_c\}_{t+\Delta t} - \{Q^*_c\}_t\,,\end{aligned} \tag{10.96}$$

where $[M]$, $[C]$ and $[K]$ are the mass, damping and stiffness matrices; $\{D\}$ denotes the nodal DOF's of the entire system, $\{D\} =$

$\langle\{D_A\}\ \{D_B\}\ \{D_b\}\rangle^T$, in particular, $\{D_A\}$, $\{D_B\}$ and $\{D_b\}$ denote the DOFs associated with Track $A$, Track $B$, and the bridge; $\{F\}$ denotes the corresponding external loads, $\{F\} = \langle\{F_A\}\ \{F_B\}\ \{F_b\}\rangle^T$, with $\{F_A\}$, $\{F_B\}$, $\{F_b\}$ indicating the forces pertaining to Track $A$, Track $B$, and the bridge; and $\{P_c^*\}_{t+\Delta t}$ and $\{Q_c^*\}_t$ are the equivalent contact forces in global coordinates.

In establishing the system matrices, one first constructs the matrices $[M_0]$, $[C_0]$ and $[K_0]$ for the railway bridge that is free of any vehicle actions, that is,

$$[M_0] = \begin{bmatrix} [M_A] & 0 & 0 \\ 0 & [M_B] & 0 \\ 0 & 0 & [M_b] \end{bmatrix} = \begin{bmatrix} [m_l] + [m_r] + \sum[m_A] & 0 & 0 \\ 0 & [m_l] + [m_r] + \sum[m_B] & 0 \\ 0 & 0 & \sum[m_b] \end{bmatrix}, \tag{10.97}$$

$$[C_0] = [C_{00}] + [C_{0b}] = [C_{00}] + \begin{bmatrix} [C_A] & 0 & -[C_{Ab}] \\ 0 & [C_B] & -[C_{Bb}] \\ -[C_{bA}] & -[C_{bB}] & [C_b] \end{bmatrix}$$

$$= [C_{00}] + \begin{bmatrix} \left(\sum_{i=1}^{4}[c_{li}]\right) + \left(\sum_{i=1}^{4}[c_{ri}]\right) + \left(\sum\sum_{i=1}^{4}[c_{Ai}]\right) & 0 & -\sum\sum_{i=1}^{6}[c_{Abi}] \\ 0 & \left(\sum_{i=1}^{4}[c_{li}]\right) + \left(\sum_{i=1}^{4}[c_{ri}]\right) + \left(\sum\sum_{i=1}^{4}[c_{Bi}]\right) & -\sum\sum_{i=1}^{6}[c_{Bbi}] \\ -\sum\sum_{i=1}^{6}[c_{bAi}] & -\sum\sum_{i=1}^{6}[c_{bBi}] & \sum\sum_{i=1}^{9}([c_{bi}] + [c'_{bi}]) \end{bmatrix}, \tag{10.98}$$

$$
[K_0] = \begin{bmatrix}
\left(\sum_{i=1}^{4}[k_{li}]\right) + \left(\sum_{i=1}^{4}[k_{ri}]\right) + \left(\sum\sum_{i=1}^{4}[k_{Ai}]\right) & 0 & -\sum\sum_{i=1}^{6}[k_{Abi}] \\
0 & \left(\sum_{i=1}^{4}[k_{li}]\right) + \left(\sum_{i=1}^{4}[k_{ri}]\right) + \left(\sum\sum_{i=1}^{4}[k_{Bi}]\right) & -\sum\sum_{i=1}^{6}[k_{Bbi}] \\
-\sum\sum_{i=1}^{6}[k_{bAi}] & -\sum\sum_{i=1}^{6}[k_{bBi}] & \sum\sum_{i=1}^{9}([k_{bi}]+[k'_{bi}])
\end{bmatrix},
\tag{10.99}
$$

and then add to them the interaction effects resulting from the moving vehicles via the VRI elements, as represented by the asterisked terms given in Eqs. (10.87) and (10.93), to form the system matrices $[M]$, $[C]$ and $[K]$, as given below:

$$
\begin{aligned}
[M] &= \begin{bmatrix}
[M_A]+[M_c^*] & 0 & 0 \\
0 & [M_B]+[M_c^*] & 0 \\
0 & 0 & [M_b]
\end{bmatrix} \\
&= \begin{bmatrix}
[M_A] + \sum_{k=1}^{n_A}\left(\sum_{i=1}^{4}\sum_{j=1}^{4}[m_{cij}^*]\right)_k & 0 & 0 \\
0 & [M_B] + \sum_{k=1}^{n_B}\left(\sum_{i=1}^{4}\sum_{j=1}^{4}[m_{cij}^*]\right)_k & 0 \\
0 & 0 & [M_b]
\end{bmatrix},
\end{aligned}
\tag{10.100}
$$

$$[C] = [C_{00}] + \begin{bmatrix} [C_A] + [C_c^*] & 0 & -[C_{Ab}] \\ 0 & [C_B] + [C_c^*] & -[C_{Bb}] \\ -[C_{bA}] & -[C_{bB}] & [C_b] \end{bmatrix}$$

$$= [C_{00}] + \begin{bmatrix} [C_A] + \sum_{k=1}^{n_A} \left( \sum_{i=1}^{4} \sum_{j=1}^{4} [c_{cij}^*] \right)_k & 0 & -[C_{Ab}] \\ 0 & [C_B] + \sum_{k=1}^{n_B} \left( \sum_{i=1}^{4} \sum_{j=1}^{4} [c_{cij}^*] \right)_k & -[C_{Bb}] \\ -[C_{bA}] & -[C_{bB}] & [C_b] \end{bmatrix}, \tag{10.101}$$

$$[K] = \begin{bmatrix} [K_A] + [K_c^*] & 0 & -[K_{Ab}] \\ 0 & [K_B] + [K_c^*] & -[K_{Bb}] \\ -[K_{bA}] & -[K_{bB}] & [K_b] \end{bmatrix}$$

$$= \begin{bmatrix} [K_A] + \sum_{k=1}^{n_A} \left( \sum_{i=1}^{4} \sum_{j=1}^{4} [k_{cij}^*] \right)_k & 0 & -[K_{Ab}] \\ 0 & [K_B] + \sum_{k=1}^{n_B} \left( \sum_{i=1}^{4} \sum_{j=1}^{4} [k_{cij}^*] \right)_k & -[K_{Bb}] \\ -[K_{bA}] & -[K_{bB}] & [K_b] \end{bmatrix}, \tag{10.102}$$

where $n_A$ and $n_B$ denote the number of the vehicles comprising the trains moving on Tracks $A$ and $B$, respectively. Similarly, the equivalent contact forces $\{P_c^*\}_{t+\Delta t}$ and $\{Q_c^*\}_t$ are

$$\{P_c^*\}_{t+\Delta t} = \begin{Bmatrix} \sum_{k=1}^{n_A} \left( \sum_{i=1}^{4} \{p_{ci}^*\}_{t+\Delta t} \right)_k \\ \sum_{k=1}^{n_B} \left( \sum_{i=1}^{4} \{p_{ci}^*\}_{t+\Delta t} \right)_k \\ 0 \end{Bmatrix}, \tag{10.103}$$

$$\{Q_c^*\}_t = \begin{Bmatrix} \sum_{k=1}^{n_A} \left( \sum_{i=1}^{4} \{q_{ci}^*\}_t \right)_k \\ \sum_{k=1}^{n_B} \left( \sum_{i=1}^{4} \{q_{ci}^*\}_t \right)_k \\ 0 \end{Bmatrix}. \tag{10.104}$$

Note that the subscript $k$, which indicates the $k$th vehicle of the train moving on Track $A$ (or $B$), should be looped over from 1 to $n_A$ (or $n_B$) in order to account for the vehicles effects.

At each time step, it is necessary to check whether a rail element becomes a VRI element and vice versa, and to update the entries of the system matrices and vectors, concerning the contribution of the asterisked terms or components in Eqs. (10.100)–(10.104), for the DOFs that are directly affected by the vehicle actions, according to the change in position of the contact points. One advantage of the present approach is that the total number of DOFs of the VRBI system remains identical to that of the original railway bridge, while the *symmetry* property of the original system is fully preserved. In addition, the damping matrix $[C_{00}]$ of the railway bridge in Eq. (10.101) can be determined based on the *Rayleigh damping* assumption as

$$[C_{00}] = \alpha_0[M_0] + \alpha_1[K_0], \tag{10.105}$$

where $[M_0]$ and $[K_0]$ are the mass and stiffness matrices of the railway bridge, respectively, and the two coefficients $\alpha_0$ and $\alpha_1$ are

$$\begin{aligned} \alpha_0 &= \frac{2\xi\omega_1\omega_2}{\omega_1+\omega_2}, \\ \alpha_1 &= \frac{2\xi}{\omega_1+\omega_2}, \end{aligned} \tag{10.106}$$

where $\xi$ is the damping ratio, $\omega_1$ and $\omega_2$ are the first and second frequencies of natural vibration of the railway bridge. The Newmark $\beta$ method with $\beta = 0.25$ and $\gamma = 0.5$ will be employed for

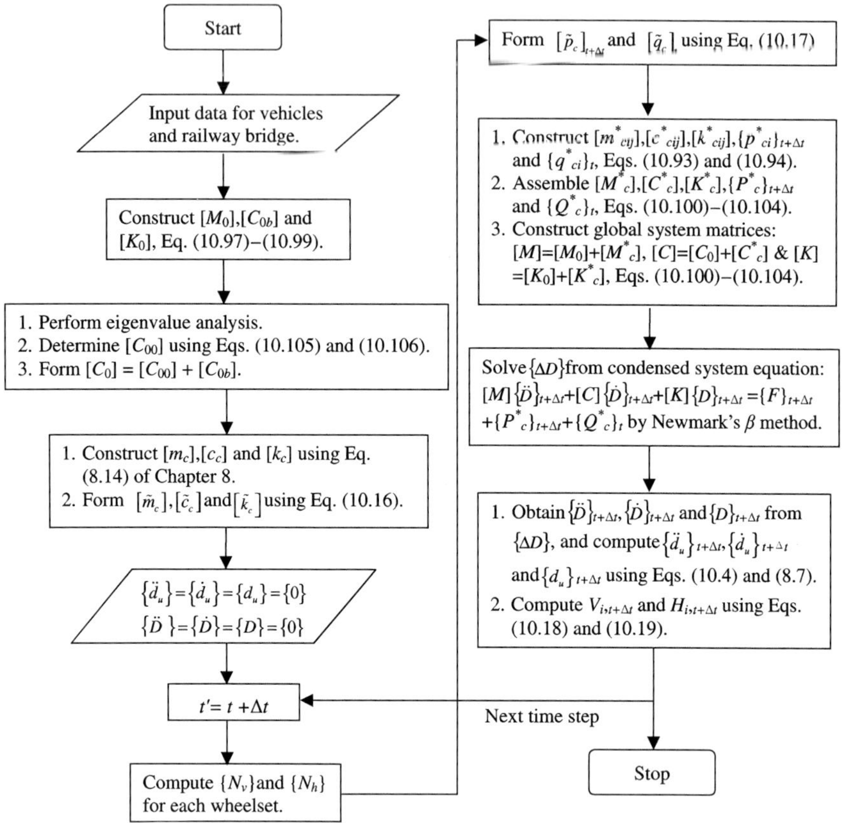


Fig. 10.4. Procedure for time-history analysis.

solving the system equations presented in Eq. (10.96). The procedure for incremental analysis of the three-dimensional vehicle–rails–bridge interaction system is summarized in the flow chart given in Fig. 10.4.

## 10.8. Simulation of Track Irregularities

Track irregularities consist of deviations of rails from ideal geometry of track layout. As shown in Fig. 10.5, four types of *track irregularity*

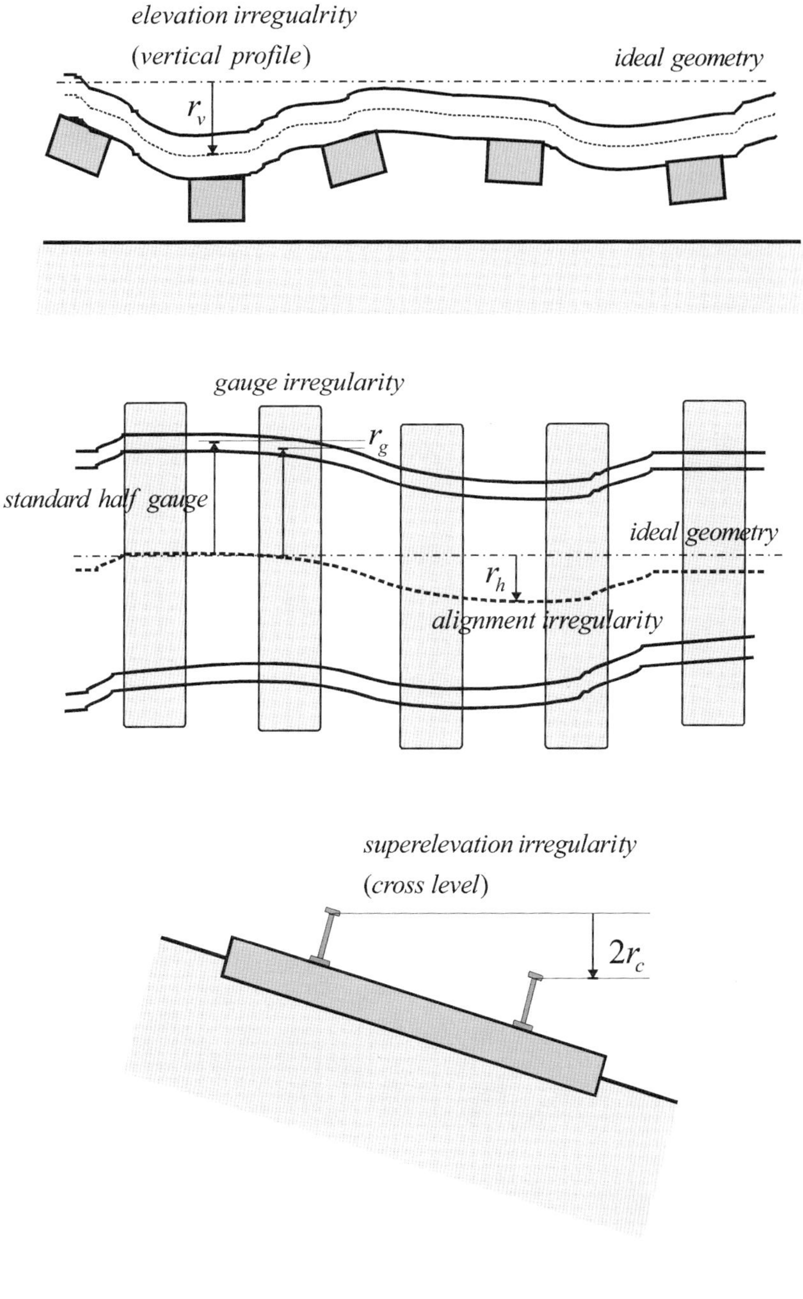


Fig. 10.5 Four types of track irregularity.

can be distinguished, i.e., elevation, alignment, superelevation and gauge irregularities (Frýba, 1996), which are caused mainly by wear, initial installation errors, degradation of support materials, improper clearances, bridge support or pier settlement and their combinations. Only the deviations in elevation, alignment and superelevation are considered in this study, which can be expressed as stationary processes in space, i.e., as random functions in terms of the longitudinal coordinate $x$. In railway engineering practice, the track irregularity is frequently characterized by the one-sided *power spectral density* (PSD) function of the track geometry. The PSD functions used in the study for the elevation irregularity (vertical profile), alignment irregularity and superelevation irregularity (cross level) are given as follows (Fries and Coffey, 1990):

$$
\begin{aligned}
S_{v,a}(\Omega) &= \frac{A_v \Omega_c^2}{(\Omega^2 + \Omega_r^2)(\Omega^2 + \Omega_c^2)} && \text{for elevation and alignment irregularities}\,,\\
S_c(\Omega) &= \frac{(A_v \Omega_c^2 / l_a)\Omega^2}{(\Omega^2 + \Omega_r^2)(\Omega^2 + \Omega_c^2)(\Omega^2 + \Omega_s^2)} && \text{for superelevation irregularity}\,,
\end{aligned}
\tag{10.107}
$$

where $\Omega = 1/L_r$ denotes the spatial frequency (Hz) and $L_r$ is the length of the irregularity (m). Table 10.3 contains the values for the coefficients involved in Eqs. (10.107), which are equivalent to Classes 4, 5 and 6 of track classification used by the

Table 10.3. Track PSD model parameters.

| Quality (FRA class) | Very poor (4) | Poor (5) | Moderate (6) |
|---|---|---|---|
| $A_v$(m) | $2.39 \times 10^{-5}$ | $9.35 \times 10^{-6}$ | $1.50 \times 10^{-6}$ |
| $\Omega_s$(rad/m) | 1.130 | 0.821 | 0.438 |
| $\Omega_r$(rad/m) | $2.06 \times 10^{-2}$ | $2.06 \times 10^{-2}$ | $2.06 \times 10^{-2}$ |
| $\Omega_c$(rad/m) | 0.825 | 0.825 | 0.825 |

*Federal Railroad Administration* (FRA) (Fries and Coffey, 1990). The track classes refer to track designations that range from 1 to 6, with class 6 indicating the best and class 1 the worst. However, the PSD function cannot be directly used in time-domain analysis because of its frequency-based nature. To overcome this problem, the spectral representation method was implemented to generate the vertical profile and alignment irregularity of the rails from the PSD functions as described in Eqs. (10.107). The rail profiles generated will be included in the time-history analysis in this chapter.

By applying the *spectral representation method*, the deviations in the vertical profile, horizontal alignment, and cross level (or superelevation), i.e., $r_v(x)$, $r_h(x)$ and $r_c(x)$, respectively, of the two rails of each track can be written along the longitudinal axis $x$ as (Claus and Schiehlen, 1998)

$$\begin{aligned} r_v(x) &= \sqrt{2}\sum_{n=0}^{N-1} A_n \cos(\Omega_n x + \alpha_n)\,, \\ r_h(x) &= \sqrt{2}\sum_{n=0}^{N-1} B_n \cos(\Omega_n x + \beta_n)\,, \\ r_c(x) &= \sqrt{2}\sum_{n=0}^{N-1} C_n \cos(\Omega_n x + \gamma_n)\,, \end{aligned} \tag{10.108}$$

where $N$ denotes the total number of discrete spatial frequencies considered, and $\Omega_n$ is the $n$th discrete frequency, which is computed as

$$\Omega_n = n\Delta\Omega = n\frac{(\Omega_u - \Omega_l)}{N}\,, \tag{10.109}$$

where $\Omega_u$ and $\Omega_l$ respectively denote the uppermost and lowest frequencies considered, and $n = 1, 2, \ldots, N-1$. The coefficients $A_n$,

$B_n$ and $C_n$ in Eq. (10.108) are defined as

$$
\begin{aligned}
A_0 &= B_0 = C_0 = 0\,,\\
A_1 = B_1 &= \sqrt{\left(\frac{1}{\pi}S_{v,a}(\Delta\Omega) + \frac{4}{6\pi}S_{v,a}(0)\right)\Delta\Omega}\,,\\
C_1 &= \sqrt{\left(\frac{1}{\pi}S_c(\Delta\Omega) + \frac{4}{6\pi}S_c(0)\right)\Delta\Omega}\,,\\
A_2 = B_2 &= \sqrt{\left(\frac{1}{\pi}S_{v,a}(2\Delta\Omega) + \frac{1}{6\pi}S_{v,a}(0)\right)\Delta\Omega}\,,\\
C_2 &= \sqrt{\left(\frac{1}{\pi}S_c(2\Delta\Omega) + \frac{1}{6\pi}S_c(0)\right)\Delta\Omega}\,,\\
A_n = B_n &= \sqrt{\left(\frac{1}{\pi}S_{v,a}(n\Delta\Omega)\right)\Delta\Omega}\,,\\
C_n &= \sqrt{\left(\frac{1}{\pi}S_c(n\Delta\Omega)\right)\Delta\Omega}\,,
\end{aligned}
\tag{10.110}
$$

for $n = 3, 4, \ldots, N-1$. The independent random phase angles $\alpha_n$, $\beta_n$ and $\gamma_n$ ($n = 1, 2, \ldots, N-1$) are uniformly-distributed in the range $[0, 2\pi]$. Note that the results generated in this way may not satisfy the requirements for the specified deviations in track geometry, and thus normalization on the final results is necessary. In this study, the results computed for $r_v(x)$, $r_h(x)$ and $r_c(x)$ for FRA track Classes 4, 5 and 6 are normalized so that their maximum deviations equal the maximum tolerable deviations specified in Table 10.4 for very poor, poor and moderate tracks according to the standards for high-speed tracks (Esveld, 1989). After the normalization, the vertical profile and alignment irregularities for the right and left rails of the track

Table 10.4. Maximum tolerable deviations for various rail irregularities.

| Quality (FRA class) | Very poor (4) | Poor (5) | Moderate (6) |
|---|---|---|---|
| $r_{v,\max}$(mm) | 4.05 | 3.38 | 2.70 |
| $r_{h,\max}$(mm) | 5.10 | 4.25 | 3.40 |
| $r_{c,\max}$(mm) | 1.50 | 1.25 | 1.00 |

can be computed as

$$\begin{aligned} r_{vr}(x) &= r_v(x) - r_c(x)\,, \\ r_{vl}(x) &= r_v(x) + r_c(x)\,, \\ r_{hr}(x) &= r_{hl}(x) = r_h(x)\,. \end{aligned} \tag{10.111}$$

In the simulation, the following are assumed: $\Omega_l = 0.0209$ rad/m, $\Omega_u = 12.5664$ rad/m and $N = 3540$. The normalized vertical profile and alignment irregularities for the right and left rails of the track for FRA Classes 4–6 were shown in Figs. 10.6(a)–10.6(c), respectively.

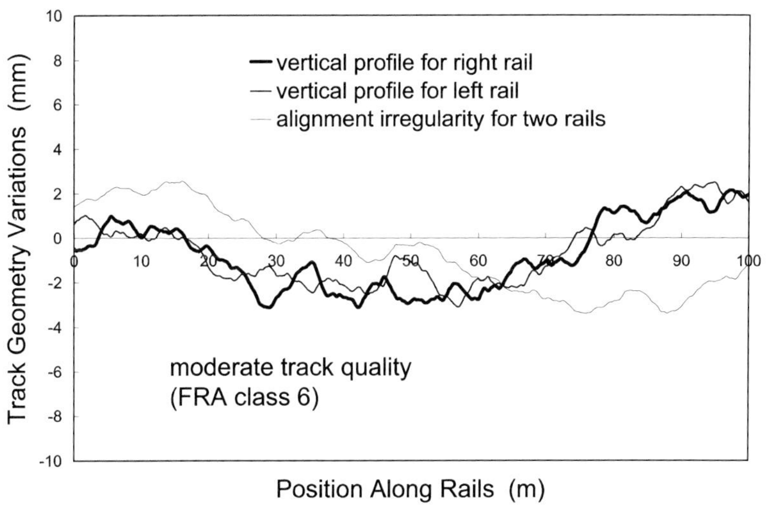


(a)

Fig. 10.6. Track irregularity for: (a) class 6, (b) class 5, and (c) class 4.

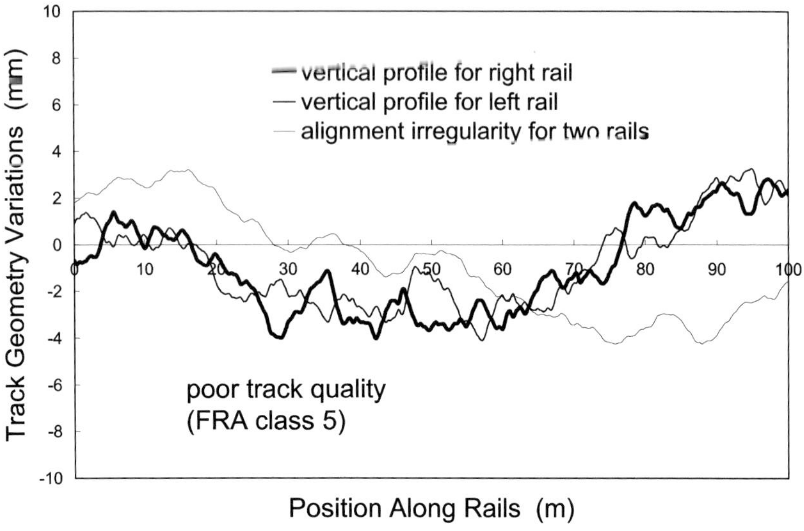


(b)

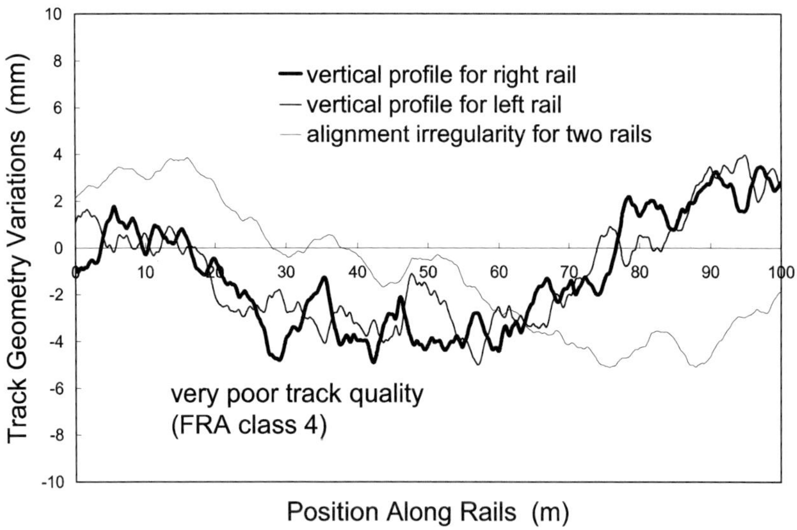


(c)

Fig. 10.6. (*Continued*).

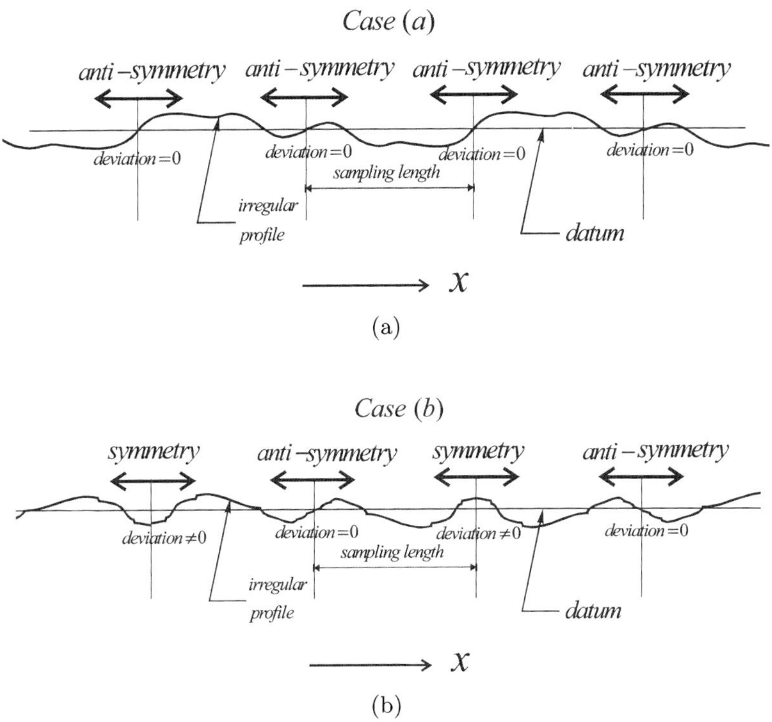


Fig. 10.7. Generation of continuous irregular profile from finite length with: (a) zero deviations and (b) nonzero deviations.

Note that if the length for the irregular track needed in analysis exceeds the sampling length, which is selected as 100 m in this study, the same irregularities should be used repeatedly in certain manner until the entire length of the track is fully covered, as illustrated in Fig. 10.7.

## 10.9. Verification of the Proposed Theory and Procedure

In this section, a typical vehicle–rails–bridge interaction problem will be analyzed both by the present 3D procedure and the 2D procedure presented in Chapter 8. The results obtained by the former will

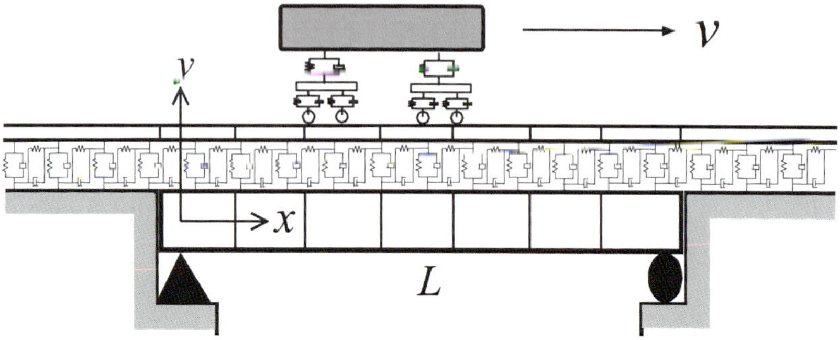


Fig. 10.8. Simply-supported bridge subjected to a moving vehicle.

then be compared with those by the latter to validate the model and formulation developed in this chapter. Figure 10.8 shows a simply-supported single-track bridge of length $L = 30$ m traveled by a moving vehicle. For comparison, only a single track located along the centerline of the bridge is considered in this example. The data adopted for the vehicle are extracted from those for the Shinkansen (SKS) train (Wakui *et al.*, 1995), as listed in Table 10.1. We choose to use this car model since it is the only one of which the fundamental properties are readily available to us. The rails are assumed to be of the UIC-60 type and the sleepers to be made of concrete. With the properties of the rails and sleepers given, the data specified for the track have been listed in Table 10.2. The bridge is made of prestressed concrete and has constant cross sections of uniform properties. The data adopted for the bridge were also listed in Table 10.2.

In the 2D analysis, the following data, equivalent to those for the 3D model, are used for the vehicle: $M_c = 41.75$ t, $I_c = 2080$ t-m$^4$, $M_t = 3.04$ t, $I_t = 3.93$ t-m$^4$, $M_w = 1.78$ t, $k_p = 1180$ kN/m, $c_p = 39.2$ kN-s/m, $k_s = 530$ kN/m, $c_s = 90.2$ kN-s/m, $l_c = 8.75$ m, $l_t = 1.25$ m, and $l_s = 3.75$ m; the following for the track (i.e., rails) and ballast: $E_t = 210$ GPa, $v_t = 0.3$, $I_t = 6.12 \times 10^{-5}$ m$^4$, $m_t = 0.587$ t/m, $A_t = 1.5374 \times 10^{-5}$ m$^2$, $k_{bv1} = k_{bv2} = 240$ MPa and $c_{bv1} = c_{bv2} = 58.8$ kN-s/m$^2$; and the following for the bridge: $E_b = 28.25$ GPa, $v_b = 0.2$, $I_b = 7.839$ m$^4$, $A_b = 8.730$ m$^2$, $m_b = 41.74$ t/m and $L = 30$ m. In each analysis, the vehicle is assumed to pass through the bridge from $x = 0$ to 100 m with a constant speed $v = 100$ m/s and the bridge is modeled as 10 beam elements.

The first and second frequencies of the railway bridge simulated by the 3D model have been obtained respectively as 25.09 and 100.34 rad/s through an independent eigenvalue analysis, which are identical to those obtained by the 2D model. The results for the midspan vertical response of the bridge obtained by the two models have been plotted in Figs. 10.9 and 10.10, and those of the track in Figs. 10.11 and 10.12. The acceleration response of the vertical and pitching motions of the vehicle was also plotted in Figs. 10.13 and 10.14, respectively. From these figures, it is observed that the responses obtained by the two models agree very well. As a result, the reliability of the present formulation and procedure in dealing with the 3D vehicle–rails–bridge interaction problems is confirmed.

In these figures, the responses identified as Track $A$ for a double-track bridge, which were obtained using the present procedure, assuming the same properties and $l_b = 2.35$ m (see Fig. 10.3 for definition) for each track are also shown. As can be seen, the results obtained for the case of a double-track bridge are very close to those

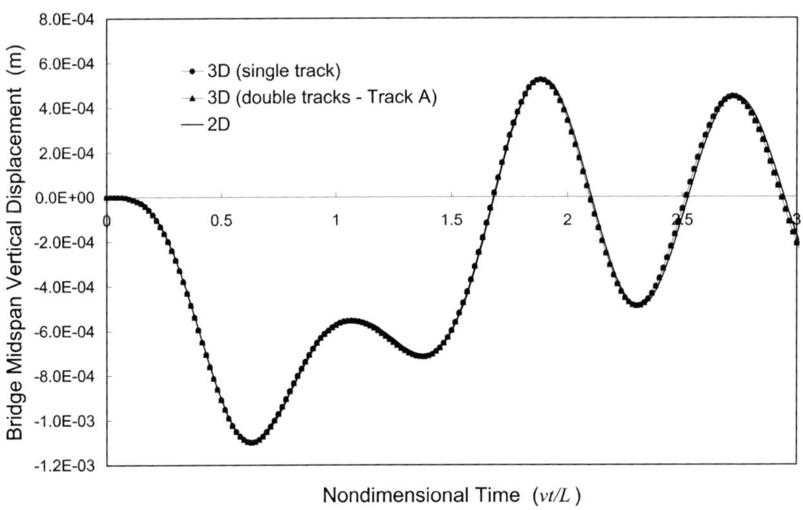


Fig. 10.9. Midspan vertical displacement of bridge.

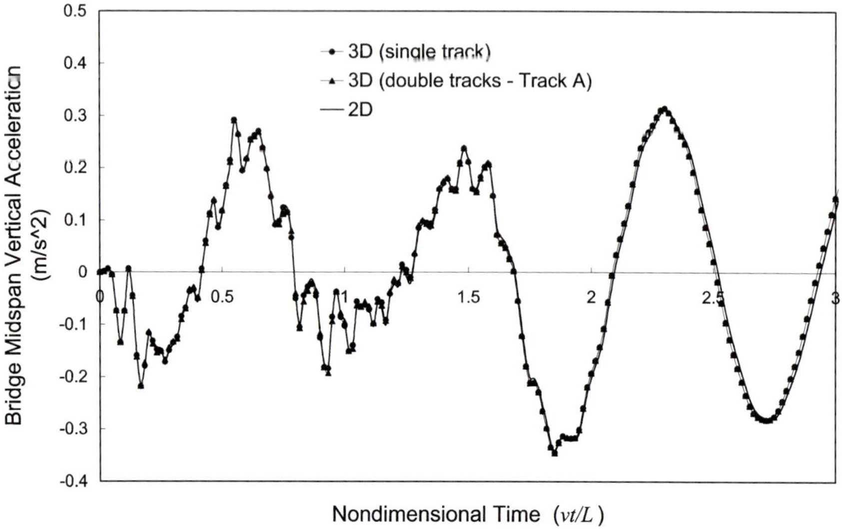


Fig. 10.10. Midspan vertical acceleration of bridge.

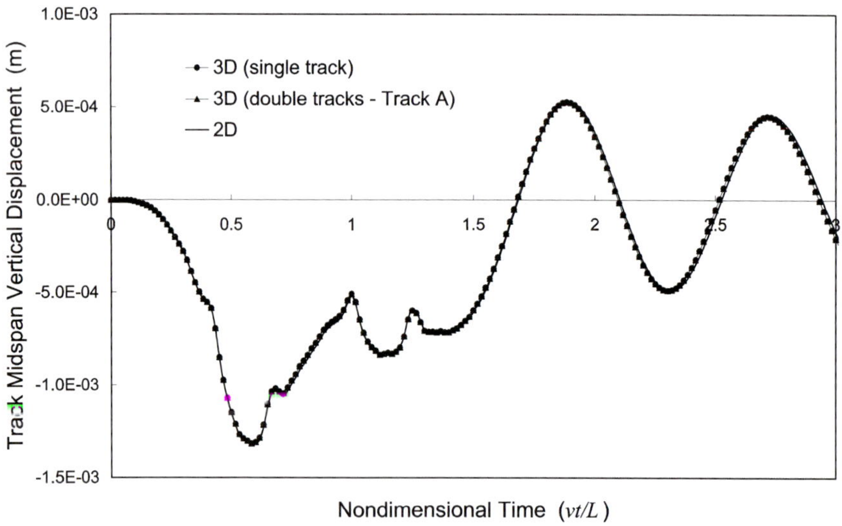


Fig. 10.11. Midspan vertical displacement of track.

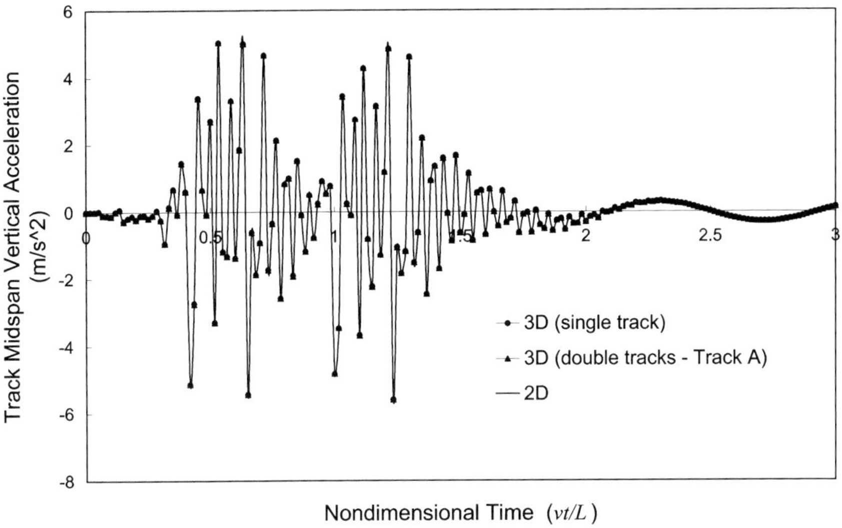


Fig. 10.12. Midspan vertical acceleration of track.

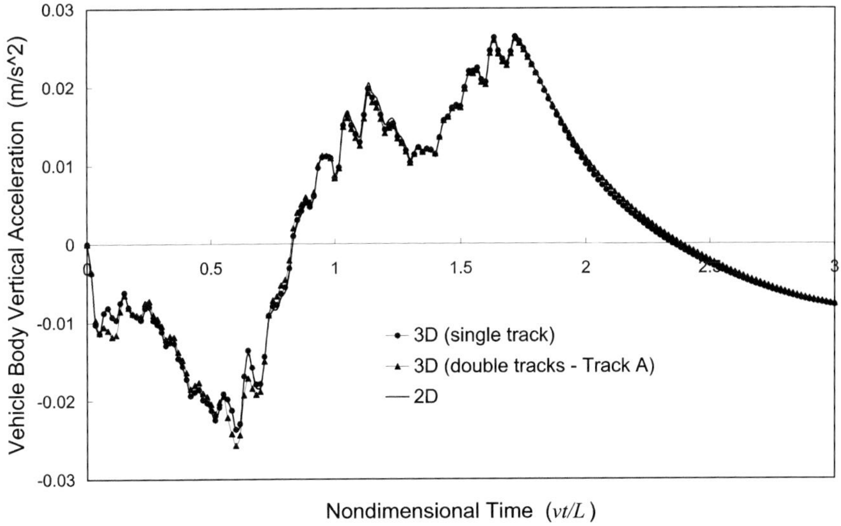


Fig. 10.13. Vertical acceleration of vehicle.

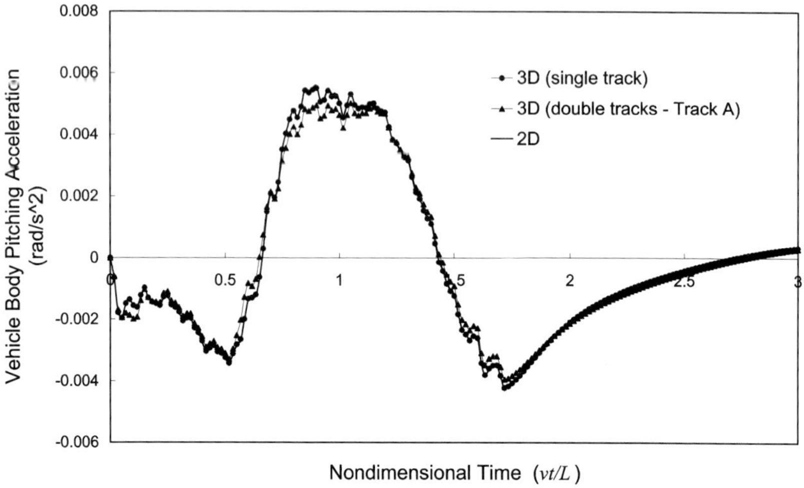


Fig. 10.14. Rotational acceleration (pitching) of vehicle.

for the case of a single-track bridge, an indication of the reliability of the present model and procedure in dealing with the interactions between the double-track bridge and running vehicles.

## 10.10. Dynamic Characteristics of Train–Rails–Bridge Systems

In this section, the three-dimensional modeling scheme developed in previous sections for the vehicle–rails–bridge interaction system will be employed to study the dynamic interactions between the railway bridge and moving trains under various conditions.

### 10.10.1. *Properties of the Railway Vehicles and Bridge*

The data adopted for the train car have been listed in Table 10.1, which are modified from those for the Series 300 rail cars used in Japan's Shinkansen system (Wakui *et al.*, 1995). The rails are assumed to be of UIC-60 type and the sleepers made of monolithic concrete blocks. The bridge is a prestressed concrete bridge of constant cross sections and uniform properties. The properties of the rails, sleepers, ballast layer and bridge have been made available in Table 10.2.

### 10.10.2. *Natural Frequencies of the Railway Vehicles and Bridge*

The natural frequencies and vibration modes of the vehicle obtained by performing an eigenvalue analysis were listed in Table 10.5. It can be seen from the first and fifth modes that the lateral and rolling motions of the vehicle are coupled with each other, as implied in the modeling of vehicle in Sec. 10.3. In addition, the first five modes are mainly related to the motion of the vehicle body, indicating that the vehicle body will be more sensitive to *low-frequency excitations* than to high-frequency excitations, which can be attributed primarily to the isolation effect of the suspension systems installed on the vehicle. The first five natural frequencies of the railway bridge were also shown in Table 10.5, along with the attributes of the corresponding vibration modes. As can be seen, the modes corresponding to lower frequencies (i.e., the first and second frequencies) are typical of vertical vibrations, due to the relatively small vertical rigidity of the bridge, compared with the lateral and torsional rigidities. As a result, it can be expected that the vertical vibration of the bridge plays a role more important than the lateral and torsional vibrations under the action of the moving trains.

### 10.10.3. *Dynamic Interactions Between the Train and Bridge*

In this subsection, the dynamic interactions between the train and the railway bridge will be investigated. The train is assumed to consist of 15 identical vehicles and is allowed to pass the bridge through Track $A$ (see Fig. 10.3). Two running cases with train speeds of $v = 60$ m/s (216 km/h) and 100 m/s (360 km/h) are considered. In each case, the train starts at position $x_0 = -50$ m relative to the bridge and stops at position $x_f = 130$ m. Track irregularity is not

Table 10.5. Natural frequencies and vibration modes for vehicle and bridge.

| Vehicle | | | | | |
|---|---|---|---|---|---|
| Mode | 1 | 2 | 3 | 4 | 5 |
| $f$ (Hz) | 0.504 | 0.723 | 0.775 | 0.895 | 1.423 |
| $v_e^*$ | 0 | 0.9666 | 0 | 0 | 0 |
| $w_e$ | −0.2090 | 0 | 0 | 0 | −0.2090 |
| $\theta_e$ | 0.9448 | 0 | 0 | 0 | 0.9448 |
| $\varphi_e$ | 0 | 0 | 0.7079 | 0 | 0 |
| $\psi_e$ | 0 | 0 | 0 | 0.3919 | 0 |
| $v_{tf}$ | 0 | 0.1812 | 0 | 0.6505 | 0 |
| $w_{tf}$ | −0.0015 | 0 | −0.2971 | 0 | −0.0015 |
| $\theta_{tf}$ | 0.1783 | 0 | −0.4015 | 0 | 0.1783 |
| $\varphi_{tf}$ | 0 | 0 | 0 | 0 | 0 |
| $\psi_{tf}$ | 0 | 0 | 0 | 0 | 0 |
| $v_{tr}$ | 0 | 0.1812 | 0 | −0.6505 | 0 |
| $w_{tr}$ | −0.0015 | 0 | 0.2971 | 0 | −0.0015 |
| $\theta_{tr}$ | 0.1783 | 0 | 0.4015 | 0 | 0.1783 |
| $\varphi_{tr}$ | 0 | 0 | 0 | 0 | 0 |
| $\psi_{tr}$ | 0 | 0 | 0 | 0 | 0 |
| **Railway bridge** | | | | | |
| Mode | 1 | 2 | 3 | 4 | 5 |
| $f$ (Hz) | 3.97 | 15.86 | 20.16 | 20.31 | 20.75 |
| $\omega$ (rad/s$^2$) | (24.93) | (99.64) | (126.69) | (127.62) | (130.37) |
| Attribute | Vertical | vertical | longitudinal | lateral-torsional | lateral-torsional |

*The symbols denote the degrees of freedom of the car body and bogies of the vehicle (see Sec. 10.3).

considered here, namely, smooth and straight track geometry is assumed. The railway bridge is modeled as 10 elements and the time increment is selected to be $\Delta t = 0.005$ s.

The midspan vertical, lateral and torsional displacements of the centerline of the bridge were plotted in Figs. 10.15(a)–10.15(c). The midspan vertical, lateral and torsional displacements of Track $A$ were plotted in Figs. 10.16(a)–10.16(c). As can be seen, the bridge and the track vibrate periodically as the vehicles pass sequentially through the bridge. The displacement responses of the track are nearly the

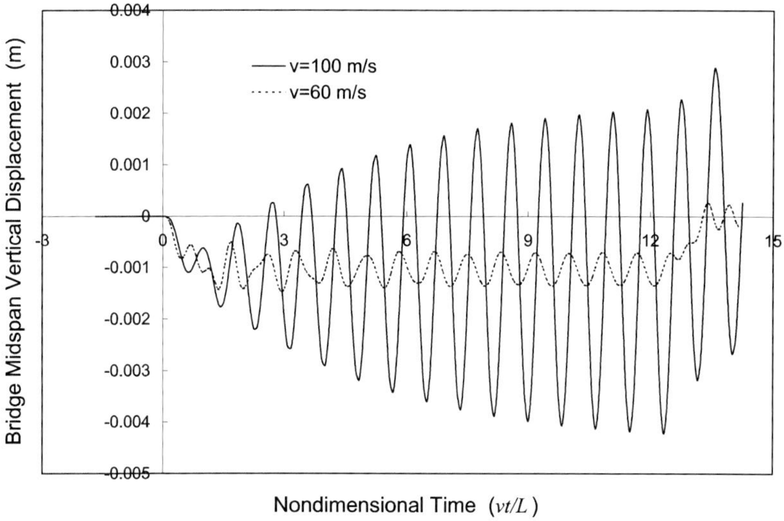


(a)

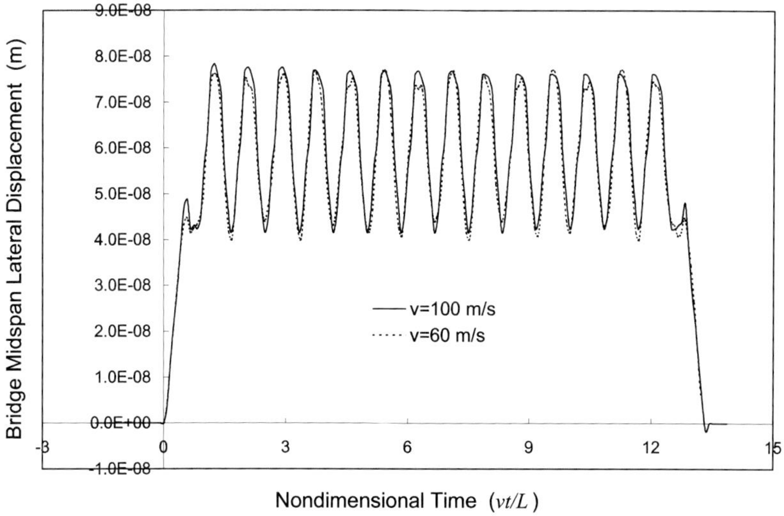


(b)

Fig. 10.15. Midspan displacements of the bridge: (a) vertical, (b) lateral, and (c) torsional.

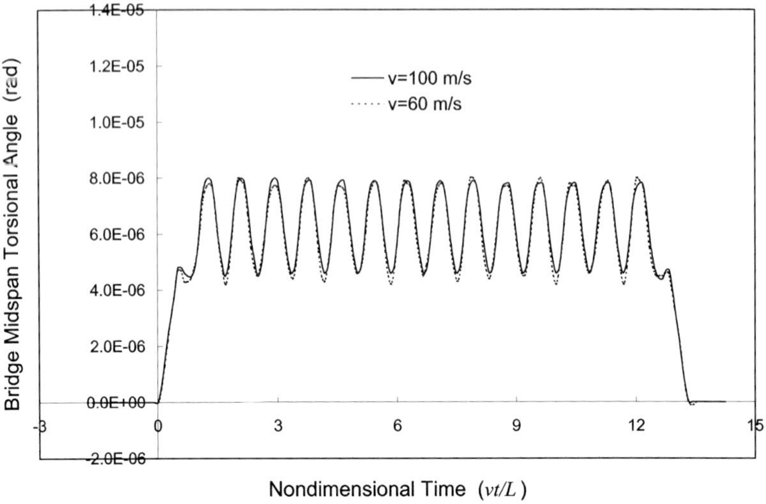


(c)

Fig. 10.15. (*Continued*).

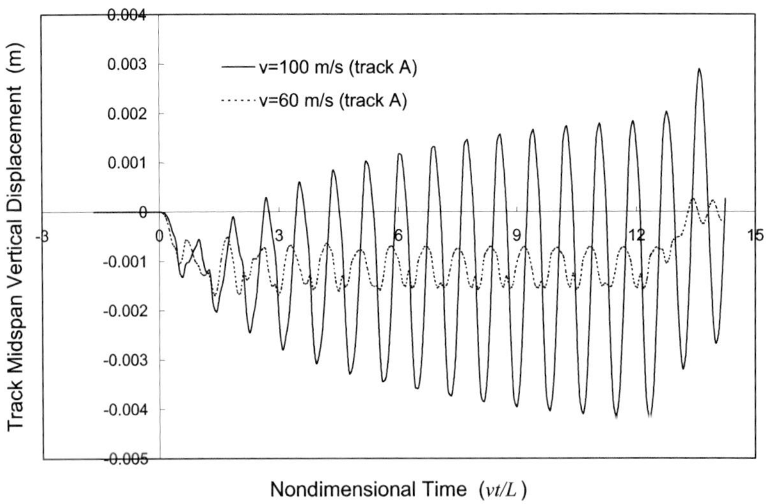


(a)

Fig. 10.16. Midspan displacements of the track: (a) vertical, (b) lateral, and (c) torsional.

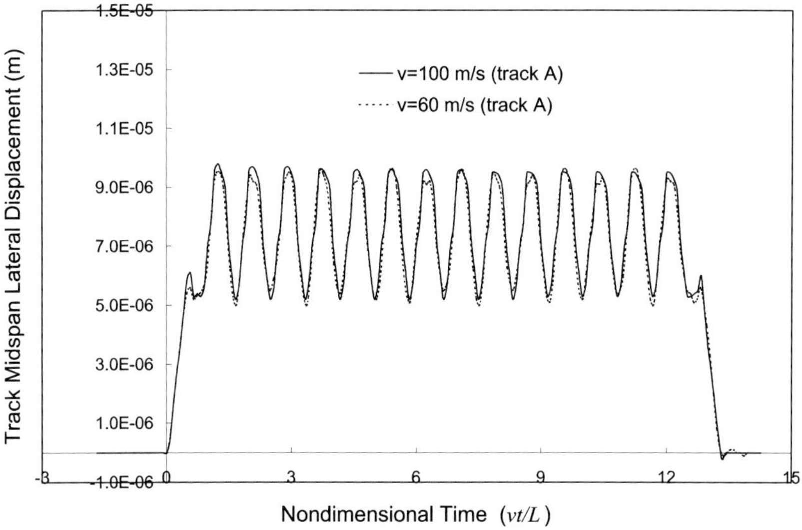


(b)

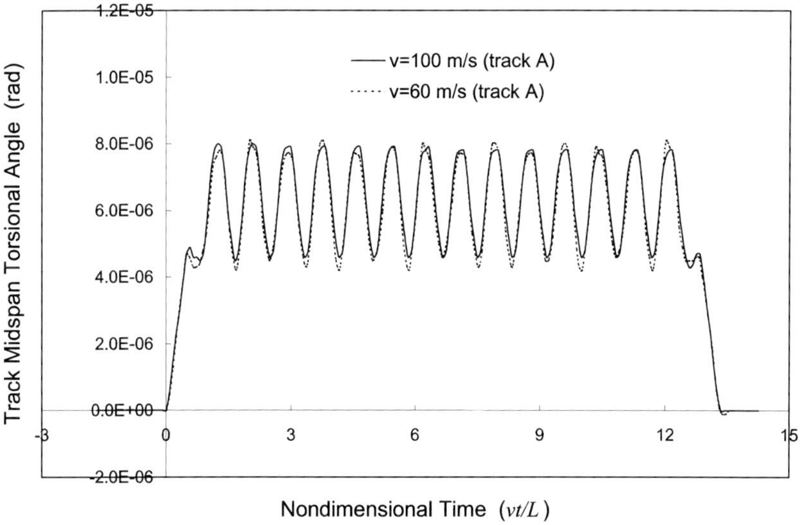


(c)

Fig. 10.16. (*Continued*).

same as that of the bridge, except for the lateral displacement, which is much larger than that of the bridge (the latter is virtually zero). This can be attributed mainly to the relatively larger vertical, but smaller lateral stiffnesses of the ballast. In contrast, the acceleration of Track $A$, in particular, the vertical acceleration (not shown here), is much larger than that of the bridge due to the direct action of the vehicular loads. In Figs. 10.17(a)–10.17(f), the acceleration responses of the first and the fifteenth (last) vehicles were presented, revealing that the last vehicle tends to vibrate more violently than the first one, particularly for the vertical and pitching motions. The implication is that the dynamic response of the last car of a train should be given much more attention concerning the safety and riding comfort of the train moving over a bridge.

In general, the vertical responses of the bridge and track for $v = 100$ m/s are much larger than those for $v = 60$ m/s, mainly owing to the occurrence of *resonance* between the train and the railway bridge, as was discussed in Chapter 9. The resonance phenomenon is further confirmed by the fact that the amplitude of the vertical vibration for $v = 100$ m/s increases with the number of vehicles passing through the bridge. The lateral and torsional responses for the same train speed, however, do not display similar resonance phenomenon. It should be noted that the lateral and torsional responses of the bridge for $v = 60$ m/s are slightly larger than those for $v = 100$ m/s due to the occurrence of secondary lateral and torsional resonances at that speed. In general, the acceleration responses (especially the vertical and pitching responses) of the first and the fifteenth vehicles for $v = 100$ m/s are larger than those for $v = 60$ m/s, also due to the vertical train–railway bridge resonance at $v = 100$ m/s. The dynamic characteristics of the train and bridge under different train speeds will be studied later.

### 10.10.4. *Train–Rails–Bridge Interaction Considering Track Irregularities*

It is desirable to study the influence of track irregularities, i.e., vertical profile, alignment irregularity, and cross level, on the

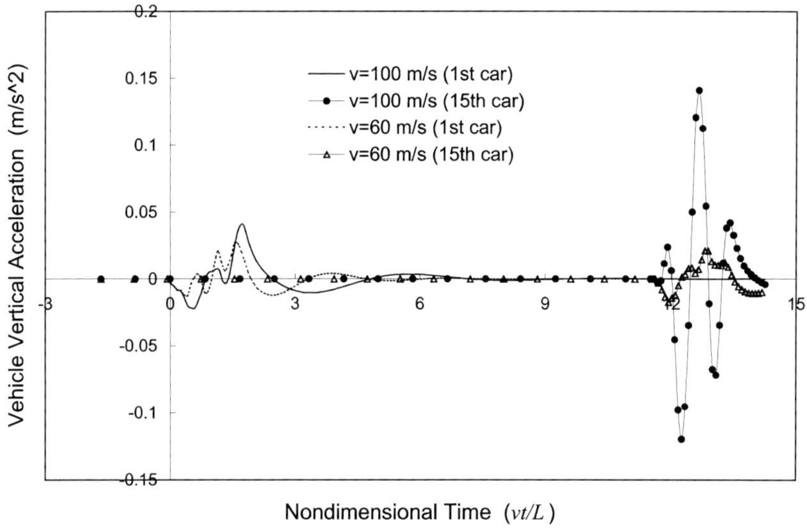


(a)

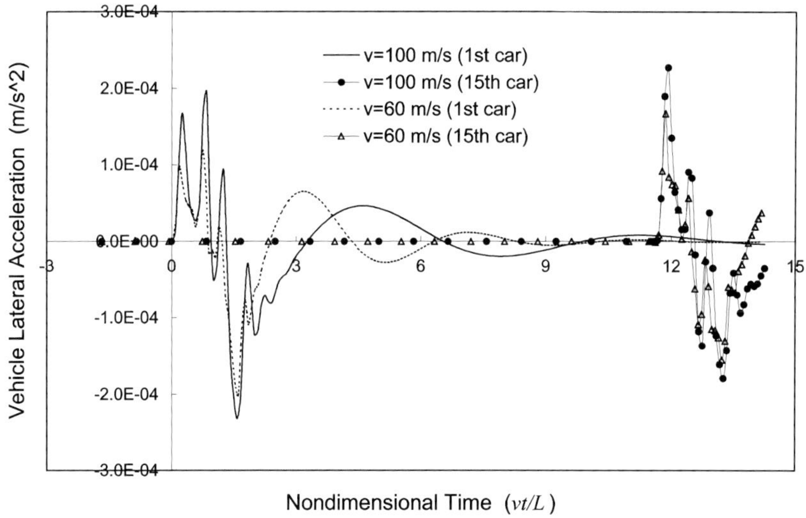


(b)

Fig. 10.17. Acceleration responses of the vehicles: (a) vertical, (b) lateral, (c) rolling, (d) yawing, and (e) pitching.

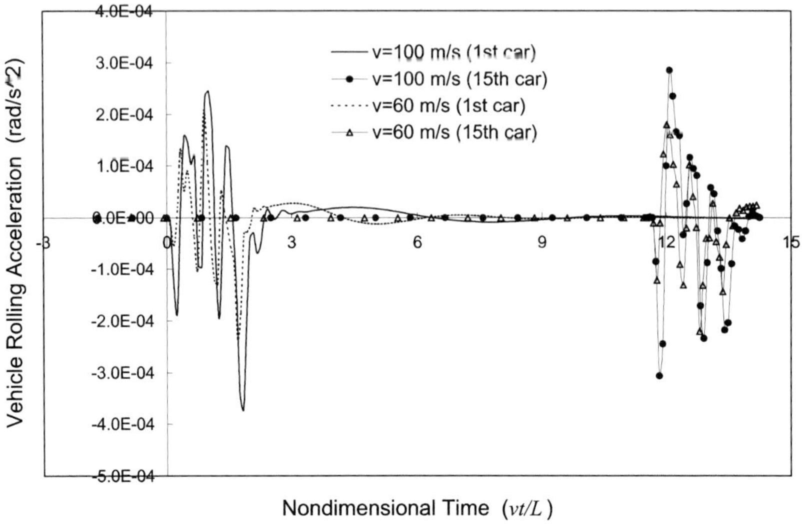


(c)

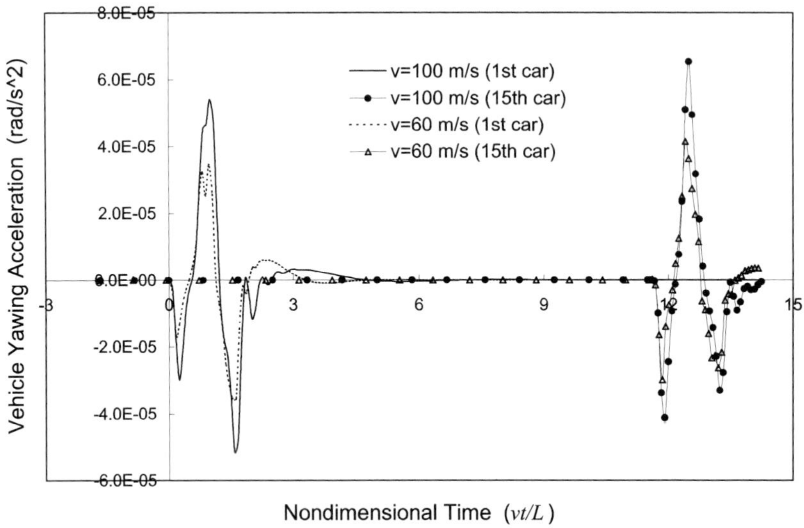


(d)

Fig. 10.17. (*Continued*).

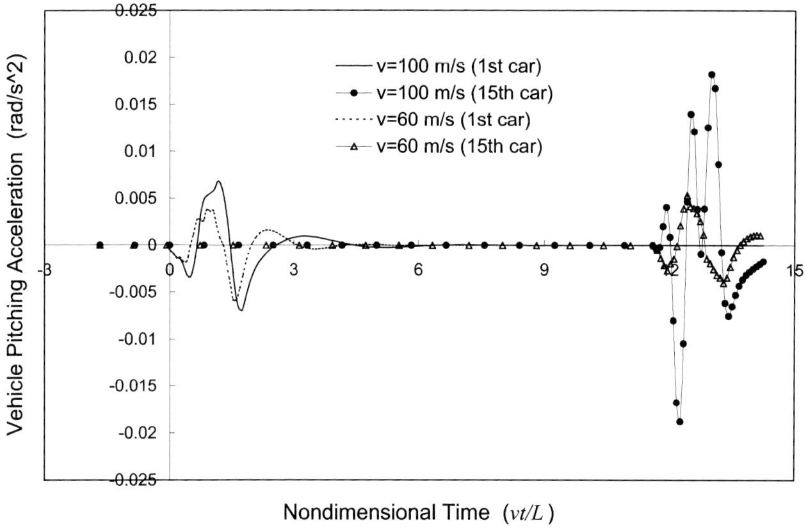


(e)

Fig. 10.17. (*Continued*).

train–rails–bridge interactions. Furthermore, for the purpose of safe operation and track maintenance, it is necessary to investigate the dynamic behaviors of trains and railway bridges under various track conditions. The deviations in ideal geometry of the two rails of the track that have been generated in Figs. 10.6(a)–10.6(c) considering simultaneously the three types of irregularity (i.e., vertical profile, alignment irregularity, and cross level) for three different track qualities (FRA Classes 4, 5 and 6) will be adopted. The conditions for the time-history analysis in this subsection are the same as those of the preceding subsection.

The time-history responses of the bridge, Track $A$, and the fifteenth vehicle of the train for the three classes of track were shown in Figs. 10.18–10.20, respectively, along with those for case with smooth and straight track geometry. From these figures, it is observed that the vertical *profile irregularity* has only marginal influence on the responses of the bridge and the track. On the other hand, the *alignment irregularity* can significantly aggravate the bridge and track

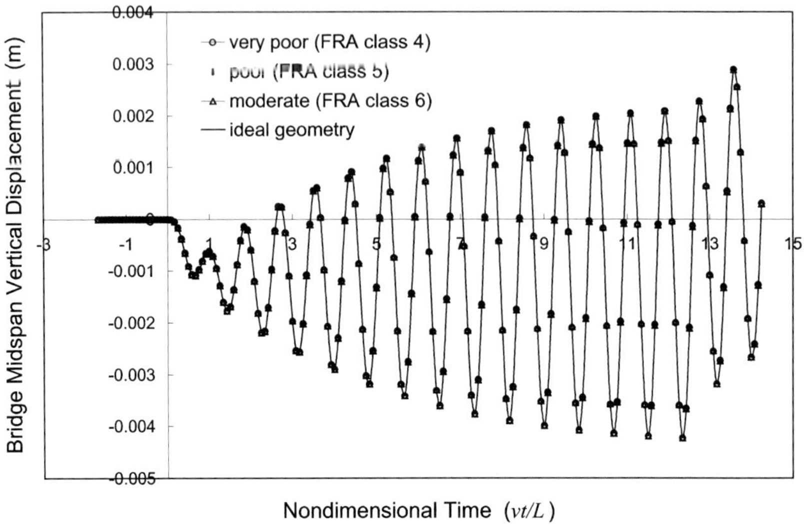


(a)

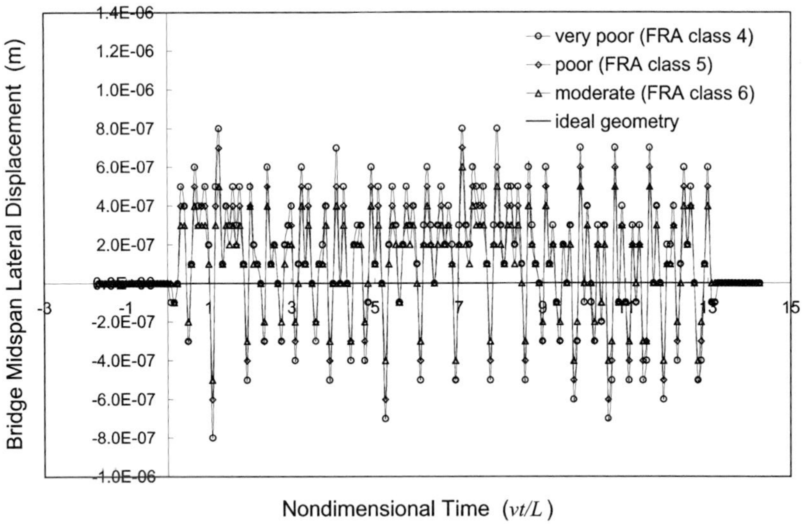


(b)

Fig. 10.18. Bridge displacements for different track qualities: (a) vertical, (b) lateral, and (c) torsional.

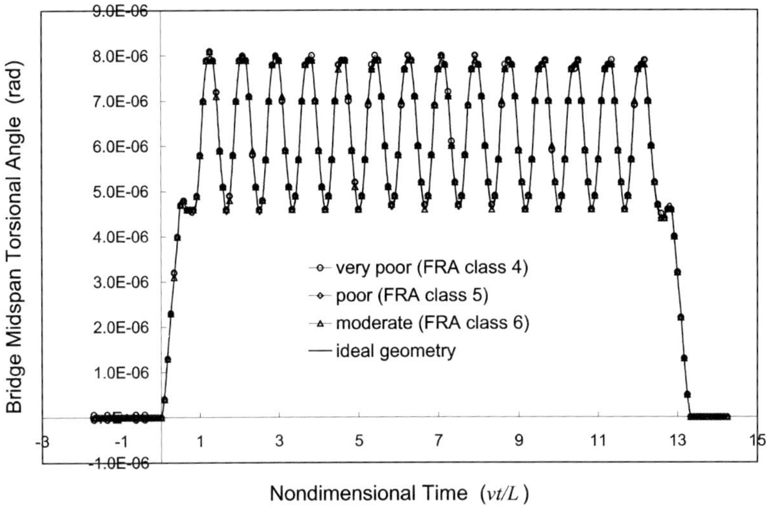


(c)

Fig. 10.18. (*Continued*).

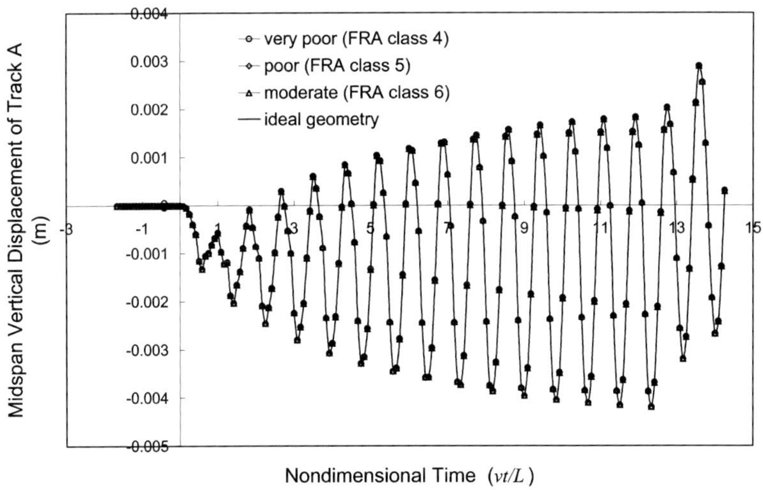


(a)

Fig. 10.19. Track displacements for different track qualities: (a) vertical, (b) lateral, and (c) torsional.

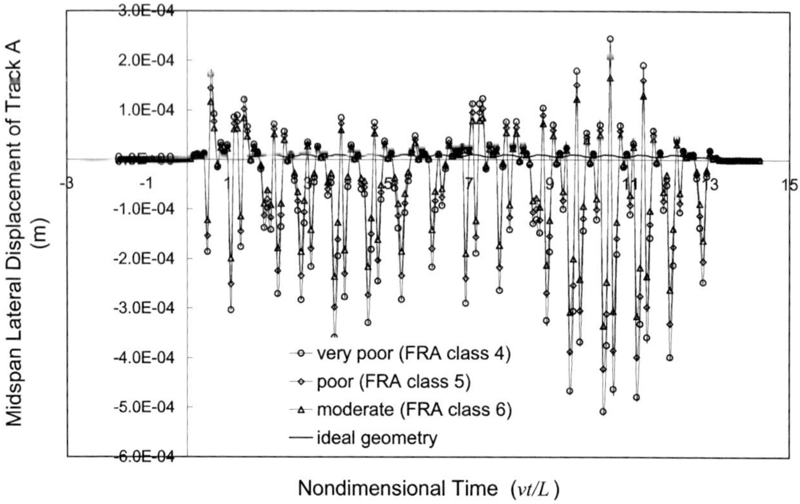


(b)

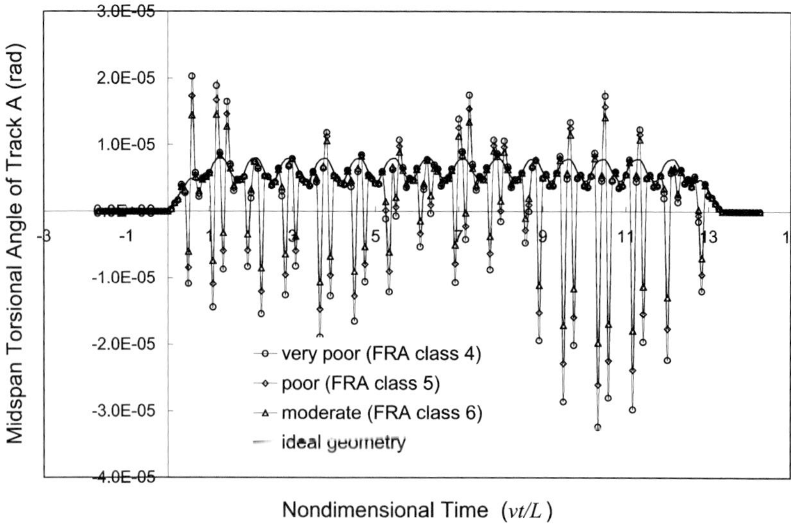


(c)

Fig. 10.19. (*Continued*).

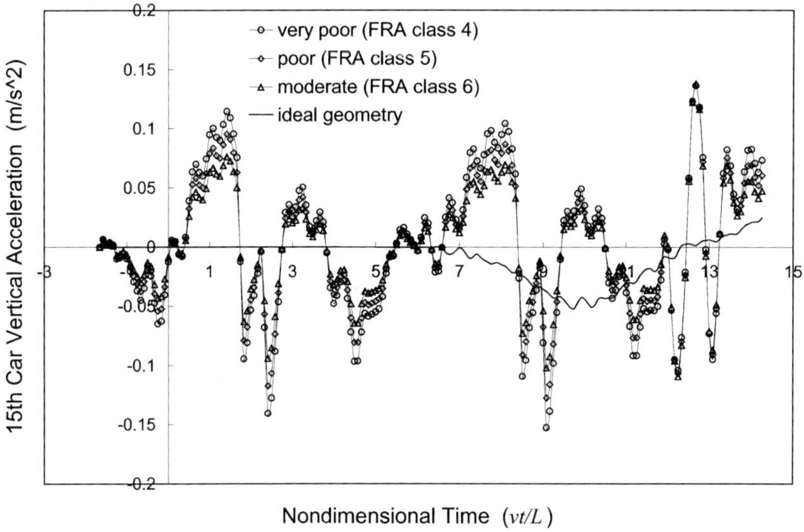


(a)

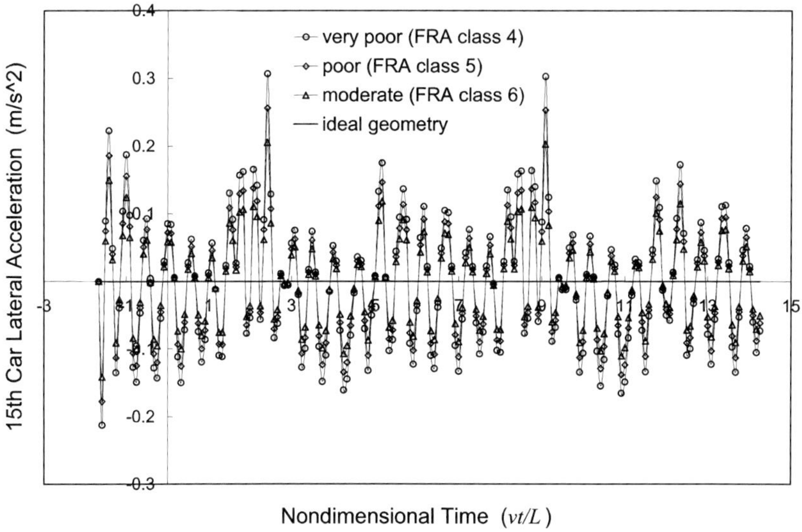


(b)

Fig. 10.20. Vehicle responses for different track qualities: (a) vertical, (b) lateral, and (c) rolling.

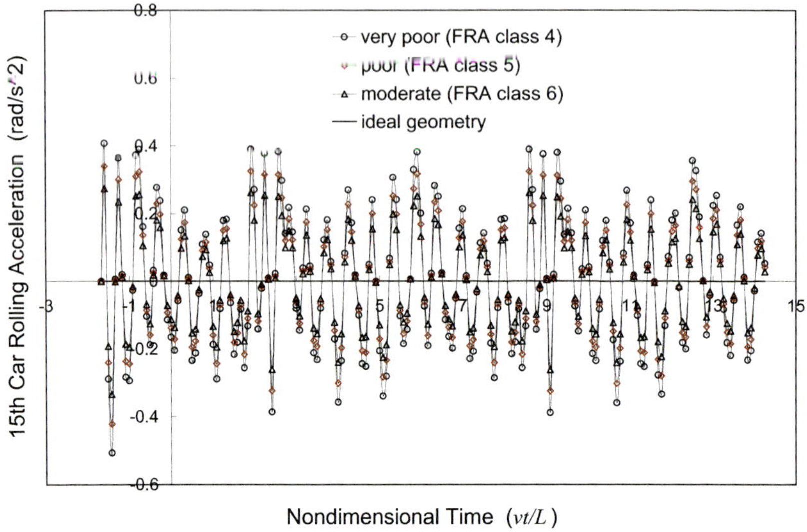


(c)

Fig. 10.20. (*Continued*).

lateral vibrations, and the *cross-level irregularity* can increase significantly the torsional response of the track. As for the vehicle, the irregular vertical profiles can drastically increase the level of vertical and pitching vibrations, and the alignment irregularity can greatly amplify the lateral, rolling and yawing motions. The cross level can also influence the lateral, rolling and yawing vibrations of the vehicle, but the degree of influence is less than that of the alignment irregularity. In general, the vehicles appear to be much more sensitive to track irregularities than the track and bridge.

In addition, the vibrations of the train–rails–bridge system for a track with better quality are much smaller than those for a track with poorer quality. The dynamic characteristics of vibrations for the three track classes are similar in trend, but different in magnitude, due to the fact that the same property of spatial frequency is implied by the three classes of irregularity. Figure 10.21 shows the variation in the contact forces between the rails and the two front and two rear wheels of the 15th car (see Fig. 10.2 for numbering of

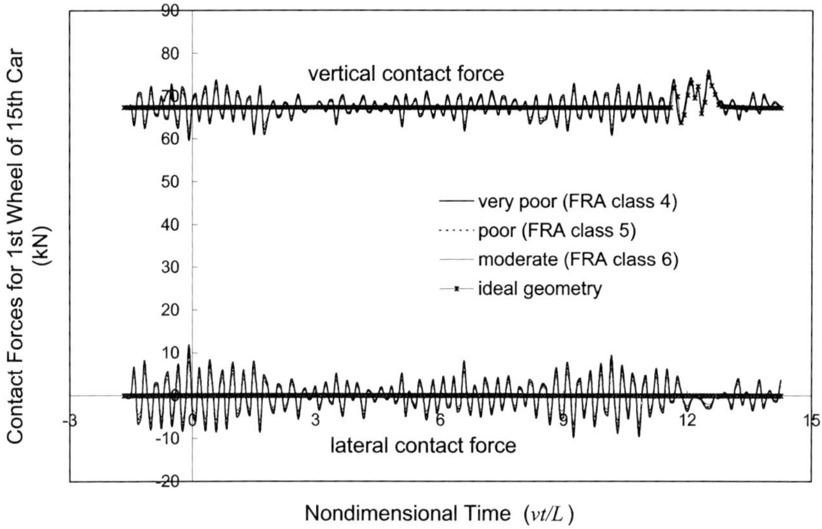


(a)

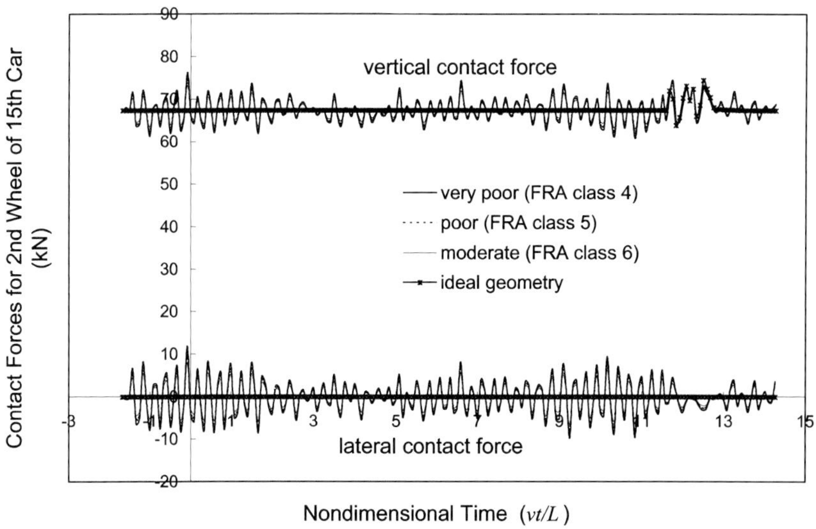


(b)

Fig. 10.21. Contact forces for the front and rear wheels of 15th car due to various track irregularities. (a) 1st wheel, (b) 2nd wheel, (c) 7th wheel, and (d) 8th wheel.

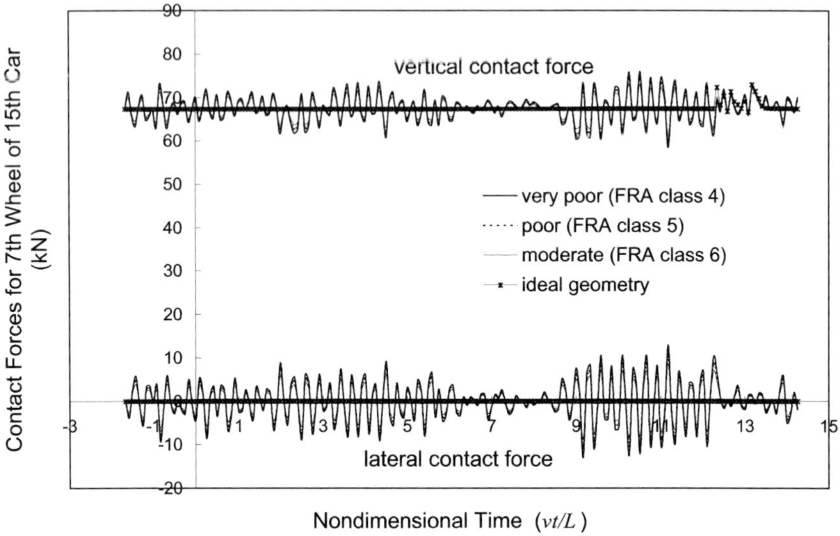


(c)

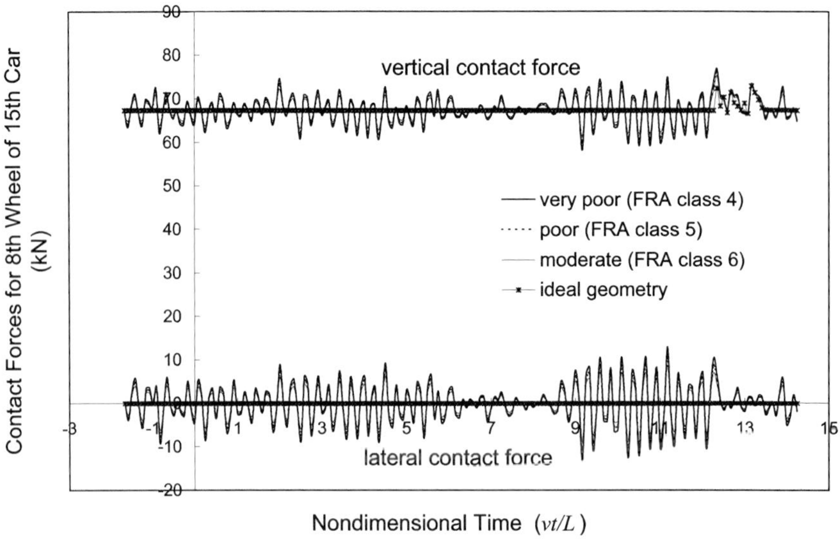


(d)

Fig. 10.21. (*Continued*).

the wheels). As can be seen, the vertical contact forces vary less rapidly compared with the lateral contact forces. Both the vertical and lateral contact forces vary more drastically as the vehicle is acting on the bridge. In Table 10.6, the maximum, mean, and standard deviation of the vertical and lateral contact forces for the four wheels of the vehicle under the three track conditions were listed. In the table the measured values for the lateral force (Ma and Zhu, 1998) are also shown. It can be seen that the maximum values obtained by the present analysis for the axle lateral force agree well with the measured ones.

Table 10.6. Maximum, mean, and standard deviation for vertical and lateral contact forces.

| | | | | | | | | | |
|---|---|---|---|---|---|---|---|---|---|
| | | | | Vertical contact force* (kN) | | | | | |
| Quality (FRA class) | Very poor (4) | | | Poor (5) | | | Moderate (6) | | |
| Item** | $Z$ | $\mu$ | $\sigma$ | $\zeta$ | $\mu$ | $\sigma$ | $\zeta$ | $\mu$ | $\sigma$ |
| 1st Wheelset | 148.6 | 135.2 | 1.2 | 148.6 | 135.2 | 1.1 | 148.7 | 135.2 | 1.0 |
| 2nd Wheelset | 144.1 | 135.2 | 1.2 | 144.2 | 135.2 | 1.0 | 144.3 | 135.1 | 0.9 |
| 3rd Wheelset | 150.7 | 135.2 | 1.2 | 150.5 | 135.2 | 1.1 | 150.4 | 135.2 | 1.0 |
| 4th Wheelset | 145.8 | 135.1 | 1.1 | 145.7 | 135.1 | 1.0 | 145.8 | 135.1 | 0.9 |
| | | | | Lateral contact force* (kN) | | | | | |
| Quality (FRA class) | Very poor (4) | | | Poor (5) | | | Moderate (6) | | |
| Item** | $Z$ | $\mu$ | $\sigma$ | $\zeta$ | $\mu$ | $\sigma$ | $\zeta$ | $\mu$ | $\sigma$ |
| 1st Wheelset | 23.8 | 6.7 | 4.1 | 19.8 | 5.6 | 3.4 | 15.9 | 4.5 | 2.8 |
| 2nd Wheelset | 21.6 | 7.2 | 4.4 | 18.0 | 6.0 | 3.7 | 14.4 | 4.8 | 2.9 |
| 3rd Wheelset | 26.3 | 7.1 | 4.7 | 21.9 | 5.9 | 3.9 | 17.5 | 4.7 | 3.1 |
| 4th Wheelset | 26.0 | 7.8 | 5.0 | 21.7 | 6.5 | 4.1 | 17.3 | 5.2 | 3.3 |
| Measured values | SKS ($v$ = 260 km/h) = 0.2P = 0.2 × 134.8 = 27 kN.<br>TGV ($v$ = 515.3 km/h) = 32 kN for front wheelset<br>= 35 kN for rear wheelset | | | | | | | | |

$^*v = 360$ km/h.
$^{**}\zeta$ = maximum, $\mu$ = mean, $\sigma$ = standard deviation

## 10.11. Dynamic Effects Induced by Trains at Different Speeds

In reality, the train can move over the approach and bridge at different speeds. There exists a need to investigate the dynamic responses of the train–rails–bridge systems under different train speeds, as they can be quite different, especially when the resonance phenomenon is to be considered. The dynamic effects to be investigated in this section include the impact factor for the midspan displacement of the bridge and the maximum acceleration of the train. The impact factor $I$ used here is the same as the one defined in Eq. (1.1), i.e.,

$$I = \frac{R_d(x) - R_s(x)}{R_s(x)}, \tag{10.112}$$

where $R_d(x)$ and $R_s(x)$ denote the maximum dynamic and static responses, respectively, of the bridge at cross-section $x$ caused by the moving train. Because of the lack of a lateral static deflection (no lateral static axle loads acting on the bridge), the impact effect for the lateral vibration of the bridge cannot be investigated using the definition of Eq. (10.112). For this reason, only the maximum response will be investigated for the lateral displacement. In addition, the same nondimensional speed parameter $S$ as the one used in previous chapters will be used to denote the speed of train, namely,

$$S = \frac{\pi v}{\omega_1 L}, \tag{10.113}$$

where $\omega_1$ denotes the first natural frequency, $L$ the length of the bridge and $v$ the train speed. For a specific bridge, the speed parameter $S$ is proportional to the train speed. The value of $S$ considered in the present study varies from $S = 0$ ($v = 0$ km/h) to 0.7 ($v = 600$ km/h). The other assumptions are identical to those of Sec. 10.10.

The impact responses for the midspan of the bridge were plotted in Fig. 10.22 with respect to the speed parameter $S$ for four classes of track quality, namely, the FRA Classes 4, 5, 6 and ideal track geometry. As can be seen, the impact factor for the vertical displacement reaches a peak value at $S = 0.425$ ($v = 364$ km/h), indicating the

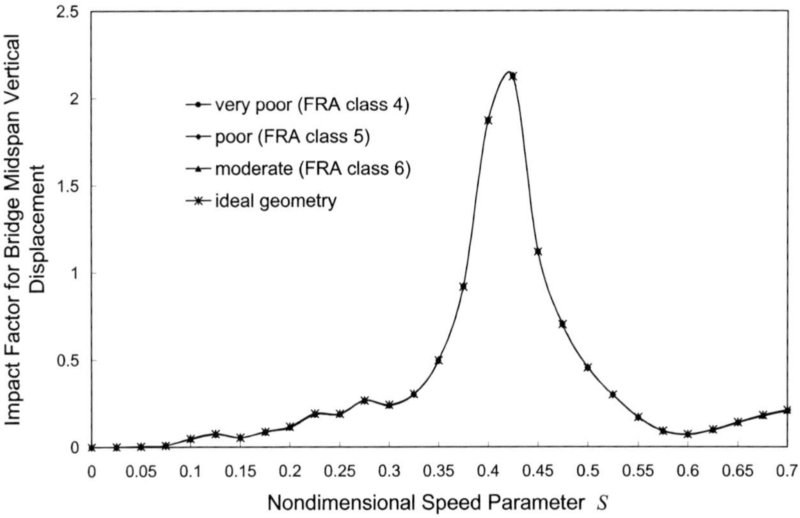


(a)

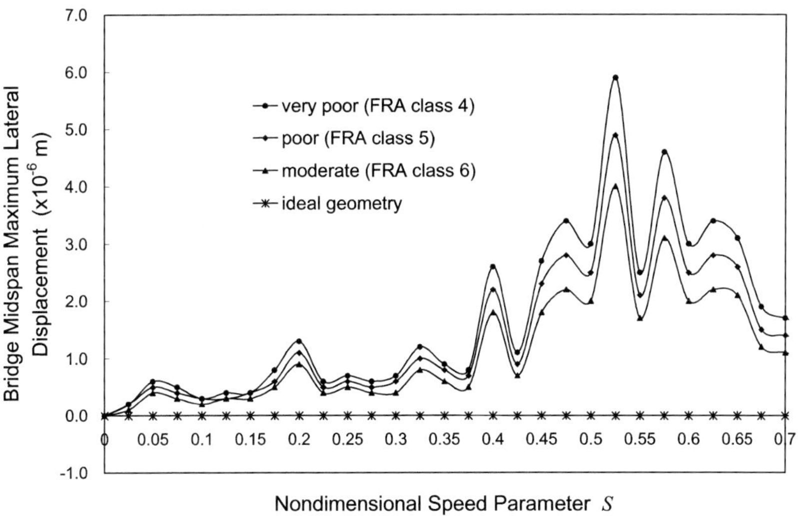


(b)

Fig. 10.22. Impact factors for bridge midspan displacement due to different track qualitites: (a) vertical, (b) lateral, and (c) torsional.

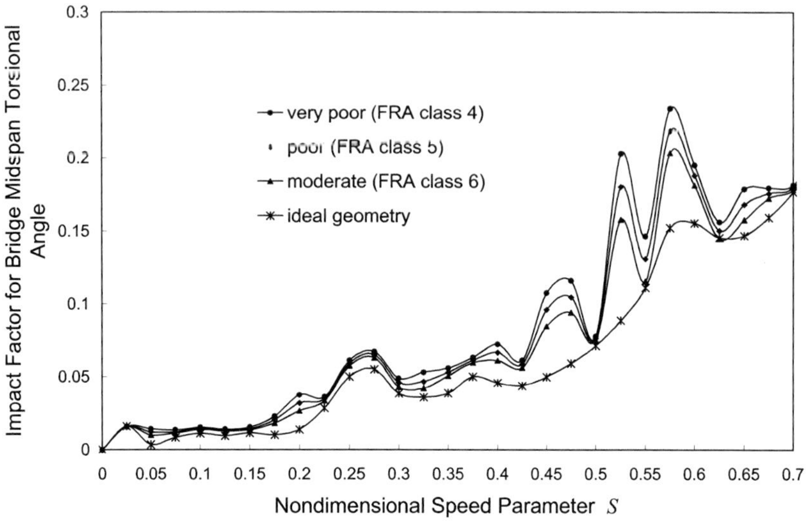


(c)

Fig. 10.22. (*Continued*).

occurrence of *resonance* between the train and the bridge. Note that the impact factor is nearly independent of the track quality. The maximum lateral displacement of the bridge has a peak value at $S = 0.525$ ($v = 450$ km/h) for all the three FRA classes, which vanishes for the case of ideal track. This is mainly due to resonance in the lateral vibration, caused by the coincidence of any of the driving frequencies of the train with any of the frequencies implied by the track irregularity. Unlike the vertical displacement, substantial difference exists between the lateral displacements for the four classes of track quality considered. In particular, nearly zero lateral displacement is observed for the case with ideal track geometry.

The impact factor for the midspan torsional angle of the bridge increases generally with the increase in $S$, with rather small difference existing between the four track classes. In addition, two local maxima can be observed for the impact factor at $S = 0.525$ and 0.575, which can be attributed to the occurrence of resonance in torsion between the train and the bridge. The maximum impact factors for the midspan vertical and torsional displacements of the bridge are 2.13 and 0.23, respectively, while the maximum midspan lateral displacement is 0.006 mm.

The maximum acceleration of the train with respect to the speed parameter $S$ was plotted in Fig. 10.23 for the four classes of track quality. As can be seen, the maximum acceleration of the train for the track with irregularities appears to be much larger than that for the ideal track. Moreover, much larger difference can be observed for the maximum lateral, rolling and yawing accelerations of the train between the irregular cases and the ideal case than for the maximum vertical and pitching accelerations, due to the fact that nearly no lateral, rolling and yawing vibrations are induced on the train as it moves over a smooth and straight track.

Note that owing to the relatively large responses induced by the irregularities, the *resonance* effect on the vertical and pitching responses becomes almost invisible for the irregular tracks, except for

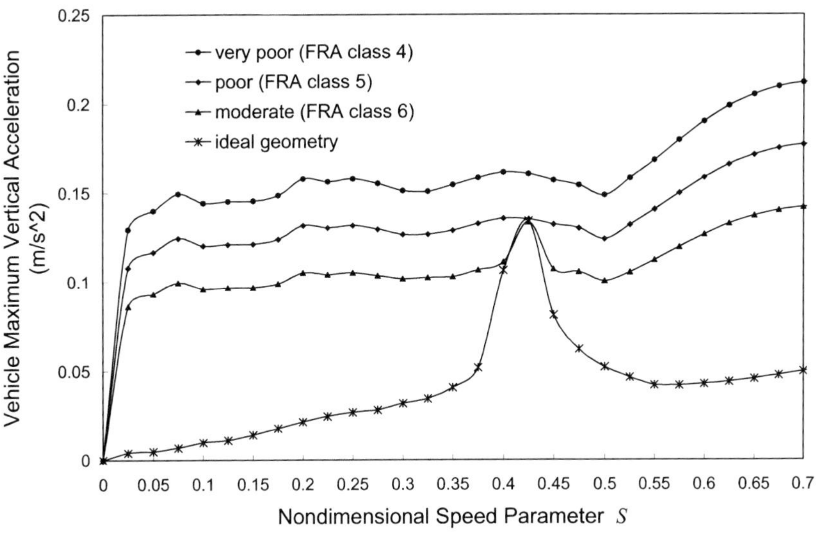


(a)

Fig. 10.23. Maximum vehicle acceleration for different train speeds: (a) vertical, (b) lateral, (c) rolling, (d) yawing, and (e) pitching.

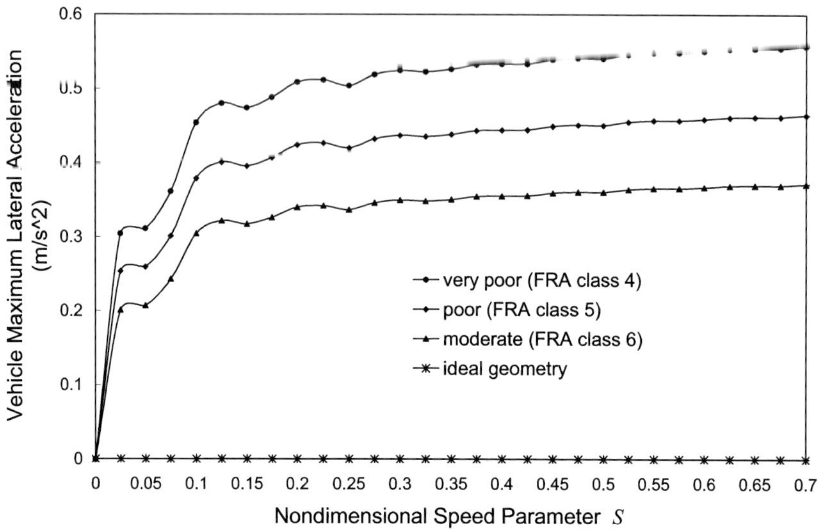


(b)

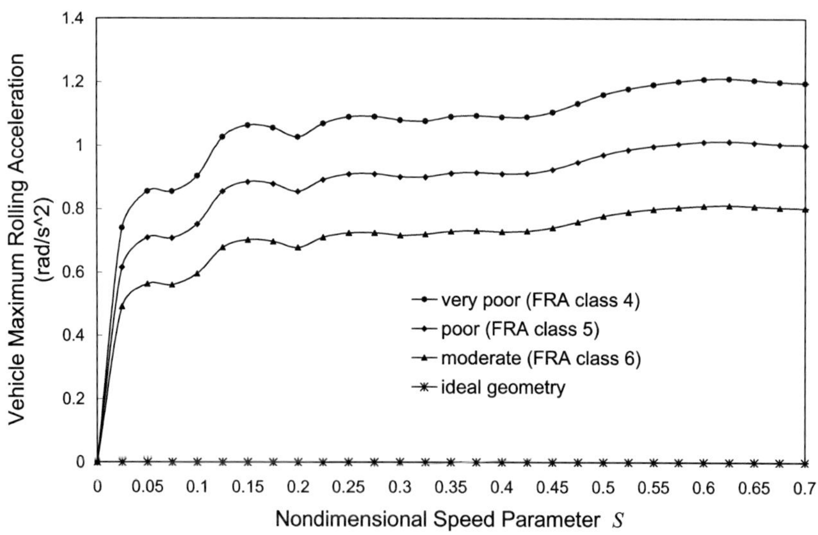


(c)

Fig. 10.23. (*Continued*).

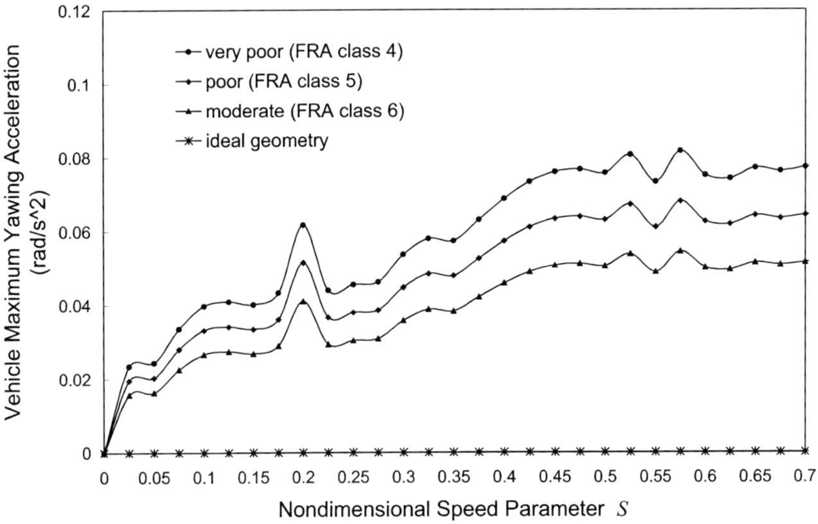


(d)

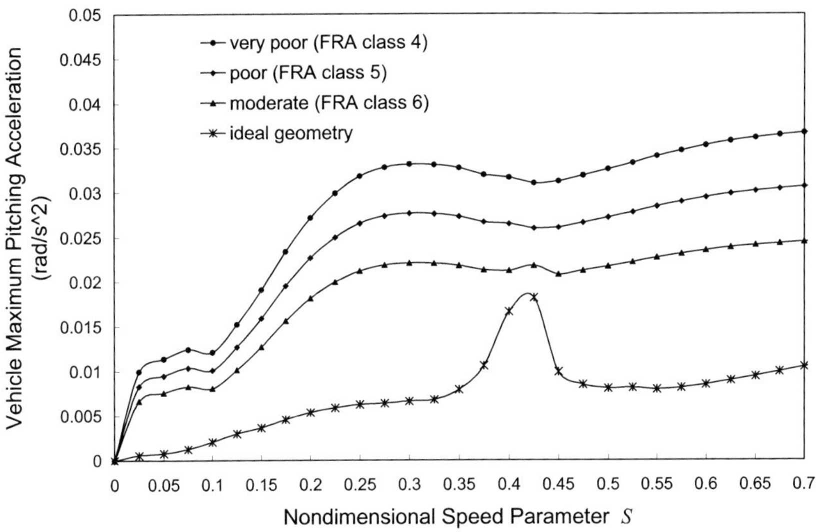


(e)

Fig. 10.23. (*Continued*).

Table 10.7. Asymptotic limits and maximum values of vehicle acceleration.

| Vehicle acceleration | Vertical ($m/s^2$) | Lateral ($m/s^2$) | Rolling ($rad/s^2$) | Yawing ($rad/s^2$) | Pitching ($rad/s^2$) |
|---|---|---|---|---|---|
| Asymptotic Limit[a] | 0.22 | 0.57 | 1.22 | — | 0.04 |
| Maximum Value[b] | 0.16 | 0.53 | 1.09 | 0.07 | 0.03 |
| Tolerable Value[c] | 0.98 | 0.49 | — | — | — |

[a] $v = 0 \sim 600$ km/h (very poor track).
[b] $v = 0 \sim 360$ km/h (very poor track).
[c] Adopted by Taiwan HSR. For vertical acceleration, the following values are used by UIC, SKS, and ICE: 1, 1.96, 0.49 $m/s^2$, respectively.

the yawing motion, the maximum accelerations of the train appear to have an *asymptotic limit* over the whole range of speeds considered. The asymptotic limits have been listed in Table 10.7, along with the maximum accelerations of the train in the speed range $S < 0.42$ (= 360 km/h) considered by the Taiwan High Speed Railways. It can be seen that the maximum lateral acceleration of the train for the very poor track (FRA Class 4) exceeds the limit of 0.49 $m/s^2$, but that for the poor track (FRA Class 5) is below the limit, indicating that the track should be maintained frequently to ensure that the maximum deviations are less than those for the poor quality, i.e., $r_{v,\max} = 3.38$ mm, $r_{h,\max} = 4.25$ mm and $r_{c,\max} = 1.25$ mm.

## 10.12. Response Induced by Two Trains in Crossing

For a two-way railroad bridge, two trains on two different tracks may cross each other on the bridge with the same or different speeds. The crossing movement of the trains can result in drastically larger vertical vibrations for the trains and bridge than those by a single train, which may be harmful to the riding quality of the trains themselves, as well as to the maintenance of the track structures. Three cases will be analyzed to investigate the effects induced by the crossing of two trains: (a) $v_A = 100$ m/s, $v_B = 100$ m/s, $x_{A0} = 0$ m, $x_{Af} = 430$ m, $x_{B0} = 30$ m, $x_{Bf} = -400$ m, (b) $v_A = 100$ m/s, $v_B = 100$ m/s, $x_{A0} = 0$ m, $x_{Af} = 430$ m, $x_{B0} = 220$ m, $x_{Bf} = -400$ m, and (c) $v_A = 100$ m/s, $x_{A0} = 0$ m, $x_{Af} = 430$ m (single train), where $v_A$, $x_{A0}$ and $x_{Af}$ respectively denote the speed, starting and stopping positions of the train on Track $A$, and $v_B$, $x_{B0}$ and $x_{Bf}$ denote the corresponding quantities for the train on Track $B$. The train on Track $A$ is assumed to move along the positive $x$ direction, which will be referred to as train $A$ and the train on Track $B$ along the negative $x$ direction, which will be referred to as train $B$. Each of the two trains consists of 15 identical cars. Case (a) is conceived for studying the train and bridge responses caused by two trains crossing each other on the midspan of the bridge ($v = 100$ m/s). Case (b) is used to study the system responses induced when the mid-portion (i.e., the 8th car) of train $A$ moves to the center of the bridge, while train $B$ starts to enter the bridge ($v = 100$ m/s). Case (c) is conceived only for the purpose of comparison ($v = 100$ m/s). In the following, the crossing movement of Case (a) will be referred to as *symmetric crossing movement* (in the view point of the bridge), and those encountered in Cases (b) as *asymmetric crossing movement.*

The results computed for the midpoint response of the bridge were shown in Fig. 10.24. As can be seen, the vertical response of the bridge to the symmetric crossing of the two trains (Case (a)) is larger than that to the passage of train $A$ alone (Case (c)). The maximum displacements of the bridge for Case (a) and Case (c) are 7.8 and 4.2 mm, respectively. Clearly, the maximum bridge response to the passage of two trains need not be two times larger as that to the passage of a single train. Notice that the vertical response of the bridge for Case (b) becomes smaller than that for Case (c) after train $B$ enters the bridge, i.e., for $vt/L \geqq 6.3$. This can be attributed primarily to the cancellation of the vehicular loads under the asymmetric crossing of the trains in Case (b). Meanwhile, the lateral displacements of the bridge for the three cases are negligibly small.

The midspan torsional response of the bridge for Case (a) is rather small due to cancellation of the torsional moments induced by the

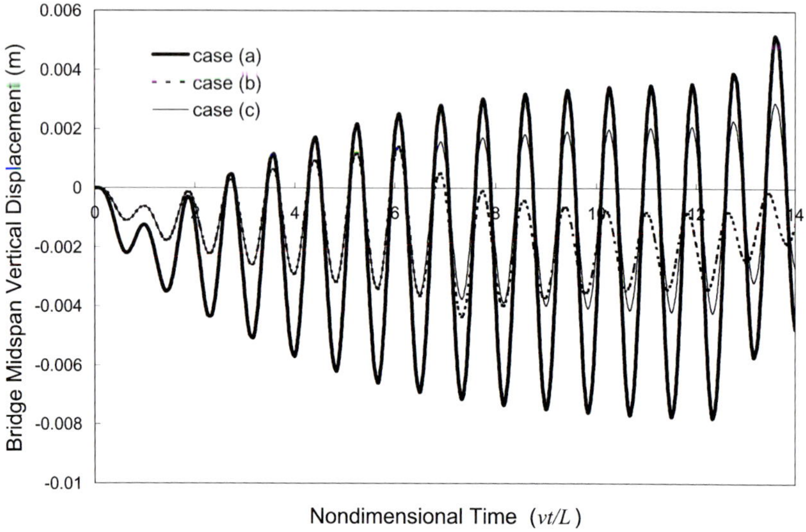


(a)

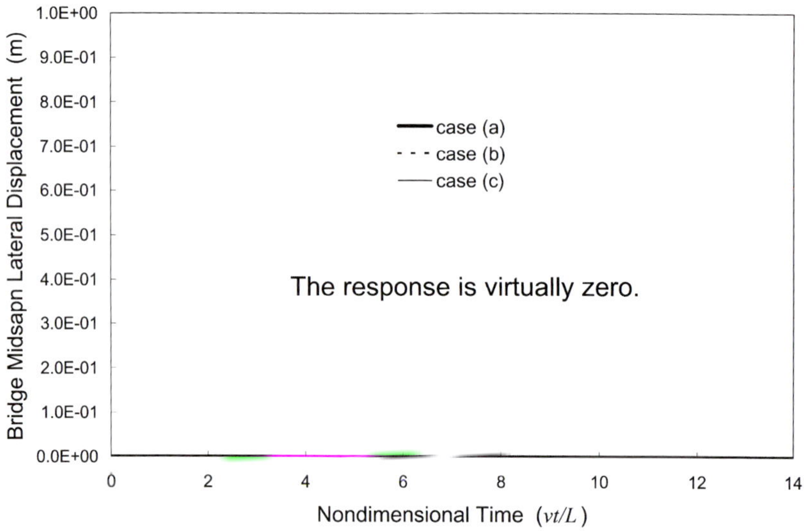


(b)

Fig. 10.24. Bridge responses due to crossing of two trains: (a) vertical, (b) lateral, and (c) torsional.

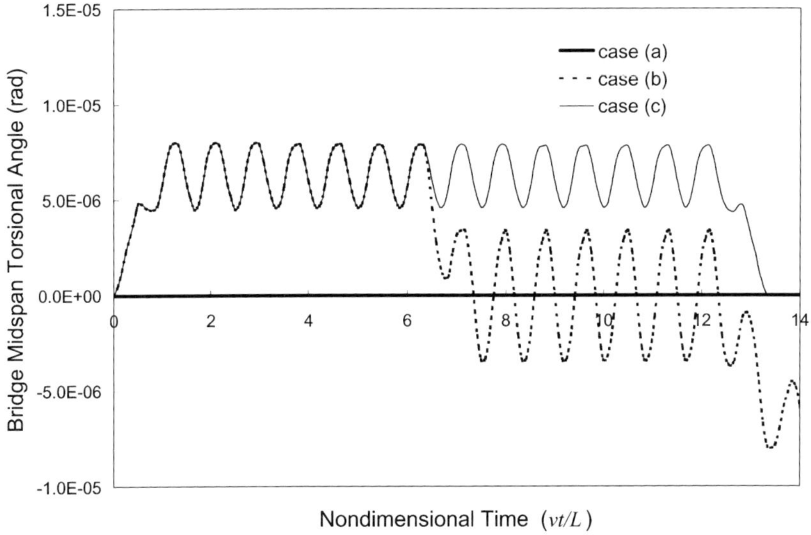


(c)

Fig. 10.24. (*Continued*).

two series of loads, as can be seen from Fig. 10.24(c). The midspan torsional vibrations of the bridge in Cases (b) and (c) are, however, rather large compared with that of Case (a). Furthermore, the torsional vibration in Case (b) changes its equilibrium position from the deformed one by a single train ($\theta_b = 6.510^{-6}$ rad) to the undeformed one ($\theta_b = 0$ rad), and oscillates more drastically than that in Case (c), after the entrance of train $B$ into the bridge, i.e., for $vt/L \geqq 6.3$.

The response of the 8th car of train $A$ and that of the 1st car of train $B$ are selected for investigation, as they are the typical cars of the two trains crossing on the bridge. From the vehicle responses given in Fig. 10.25(a), one observes that the vertical acceleration of a train under symmetric crossing is larger than that for a single train. Moreover, a train that first travels over the bridge under the condition of asymmetric crossing (e.g., train $A$ in Case (b)) vibrates less severely than it does when traveling alone over the bridge (see Case (c)). On the contrary, the maximum response of a train

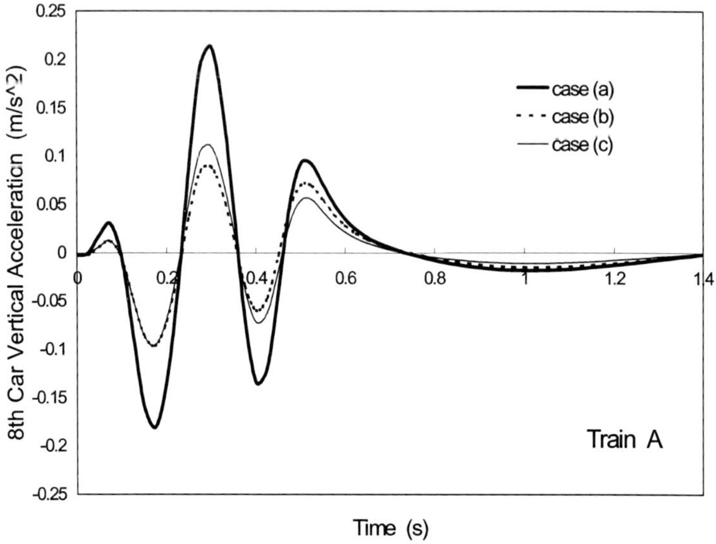


(a)

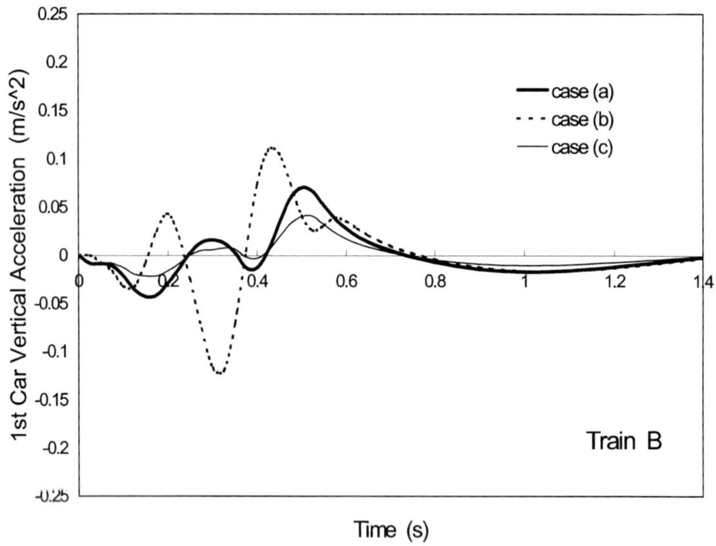


(b)

Fig. 10.25. Vehicle accelerations due to crossing of two trains: (a) and (b) vertical, (c) and (d) lateral, (e) and (f) rolling, (g) and (h) yawing, (i) and (j) pitching.

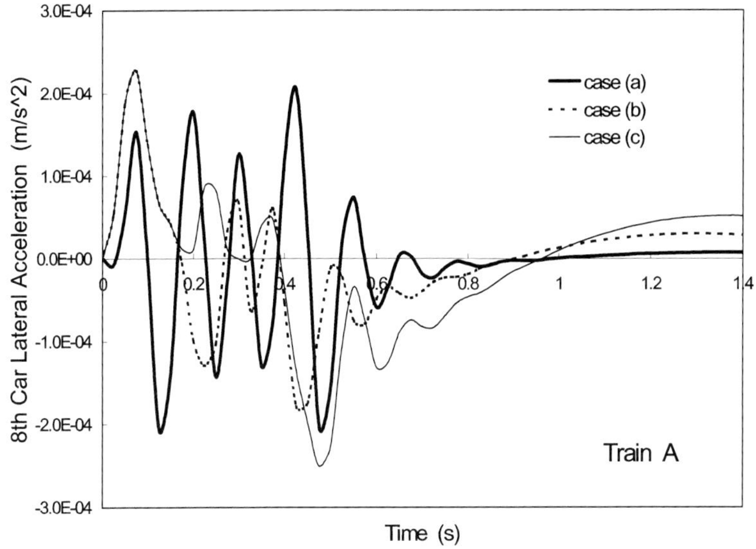


(c)

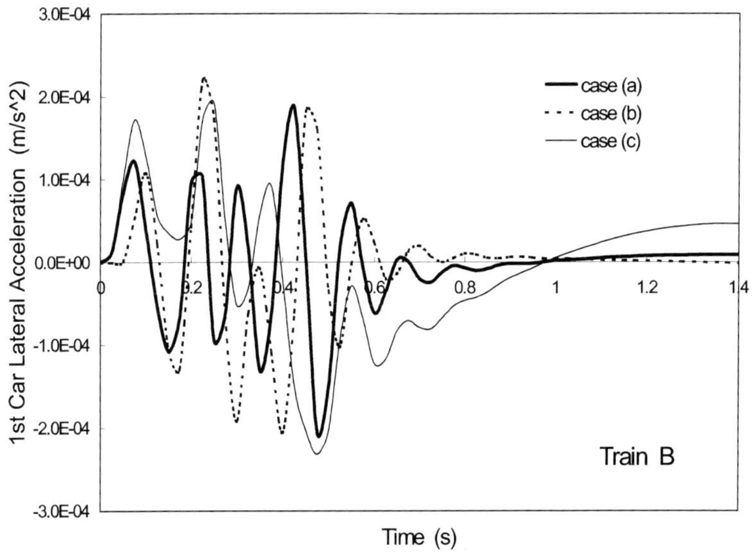


(d)

Fig. 10.25. (*Continued*).

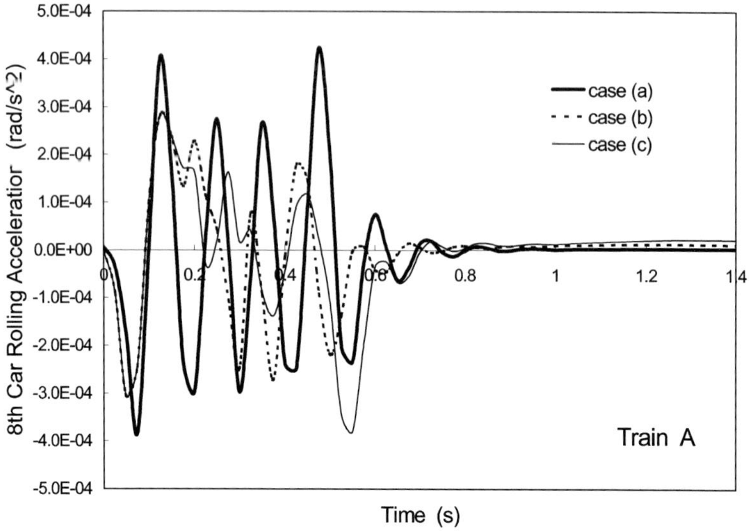


(e)

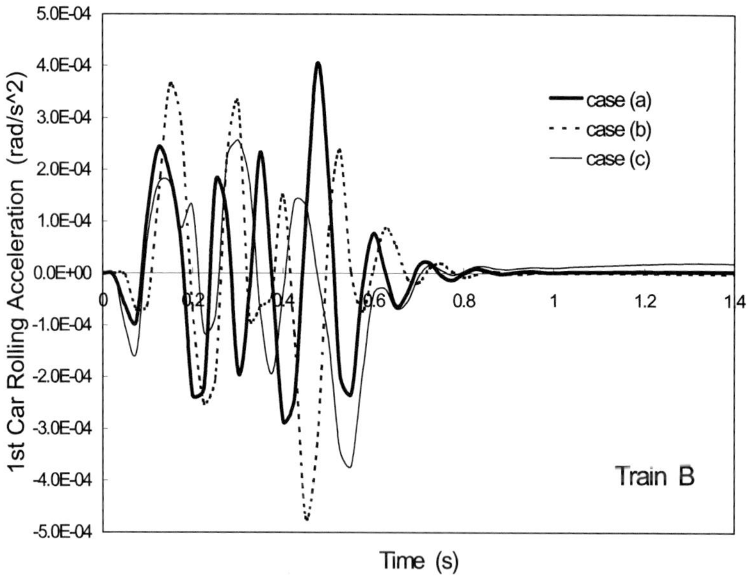


(f)

Fig. 10.25. (*Continued*).

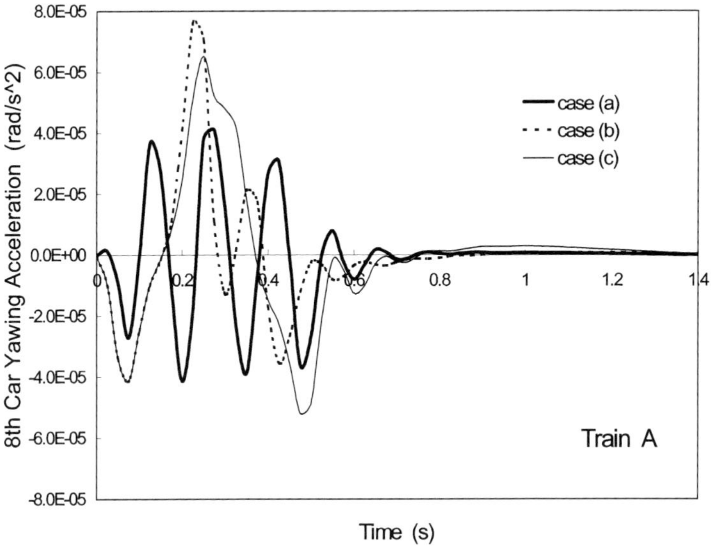


(g)

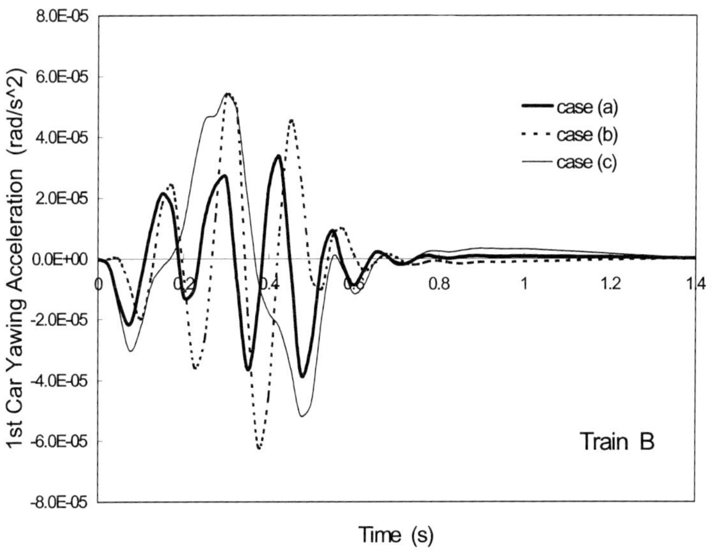


(h)

Fig. 10.25. (*Continued*).

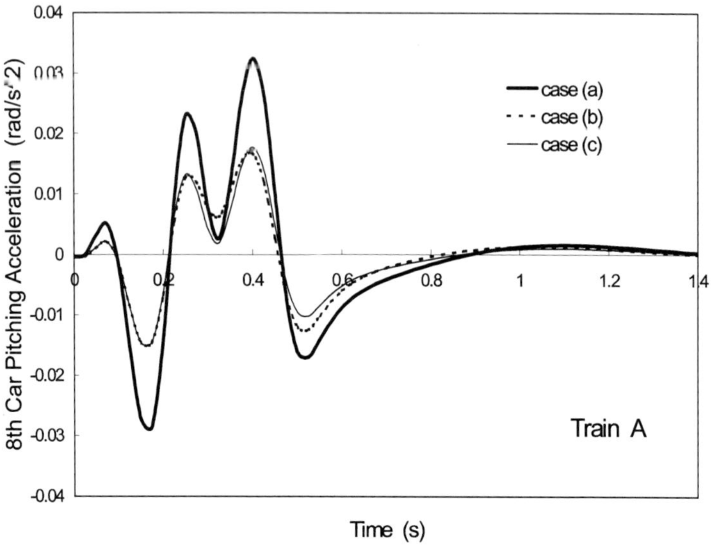


(i)

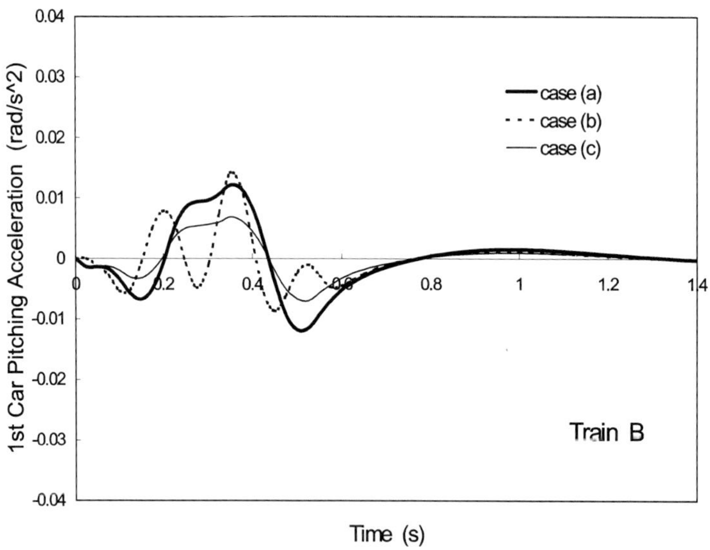


(j)

Fig. 10.25. (*Continued*).

that travels over the bridge lately under asymmetric crossing (e.g., train $B$ in Case (b)) is larger than the case when it passes alone over the bridge (see Fig. 10.25(b)). The maximum lateral acceleration of a train induced by symmetric or asymmetric crossing does not differ markedly from that by the passage of a single train (see Figs. 10.25(c) and 10.25(d)). Similar phenomenon can also be observed for the *rolling* acceleration of the trains (see Figs. 10.25(e) and 10.25(f)). However, the maximum response of the *yawing* acceleration of the train under the condition of asymmetric crossing is larger than those caused by the other two types of train movement (see Figs. 10.25(g) and 10.25(h)). Least yawing response will be induced on the trains under symmetric crossing. As for the *pitching* acceleration, the behaviors of the train are similar to those of the vertical acceleration, namely, trains $A$ and $B$ have the largest responses when the two trains cross each other symmetrically and asymmetrically, respectively, over the bridge (see Figs. 10.25(i) and 10.25(j)).

## 10.13. Criteria for Derailment and Safety Assessment of Trains

The running safety of trains has been of great concern in railway engineering for a long time, particularly due to the development of high-speed railways and the need to upgrade existing railways. Several mechanisms that can result in the derailment of a running train have been identified through analytical and experimental investigations, based on which some indices have been proposed for evaluating the possibility or *risk of derailment* of trains. In this section, five simple and practical criteria will be employed for assessing the running safety of trains traveling over bridges.

The first index is the *axle load decrement ratio* (PD). This index is defined as the ratio of the decrement in vertical axle load of a wheelset to the vertical static axle load of the same wheelset, i.e.,

$$\mathrm{PD} = \frac{\Delta Q}{Q_s} = \frac{Q_s - Q}{Q_s}, \tag{10.114}$$

where $Q_s$ denotes the axle static load of a wheelset and $Q$ the dynamic vertical axle load ($0 < Q < Q_s$). Large PD values indicate that the axle load or the vertical contact force acting on the wheelset is substantially reduced, which is detrimental to the lateral stability of the wheelset. Therefore, a limit should be placed on the value of the PD index to prevent the wheelset from the occurrence of derailment. An upper limit of 0.25 on the PD value was used in Japan's specifications for the design of high-speed railways (Ma and Zhu, 1998), which will also be adopted herein. Note that in this study the axle static load $Q_s$ is set equal to $W$ (see Sec. 10.3) and $Q = [V_{(2i-1)} + V_{(2i)}]$, where $[V_{(2i-1)} + V_{(2i)}] > 0$ and $i = 1,2, 3$ or 4 (for 4 wheelsets of a railway car). If $[V_{(2i-1)} + V_{(2i)}] < 0$, which indicates that the $i$th wheelset will jump and the train has the potential for derailment.

The second index adopted is the *single wheel lateral to vertical force ratio* (SYQ). This index is used as a measure of proximity to a flange climb derailment situation for a single wheel (not wheelset), which is defined as

$$\mathrm{SYQ} = \frac{Y}{Q}, \tag{10.115}$$

where $Y$ and $Q$ respectively denote the lateral and vertical contact forces acting through the right or left wheel of a wheelset. A value of 1.2 will be adopted as the upper bound of the SYQ ratio (Elkins and Carter, 1993), over which the train is said to have the potential of derailment. Here, $V_i$ and $H_i$ ($i = 1, 8$) should be used in computing the SYQ value for each wheel (or contact point), that is, $Q = V_i$ and $Y = |H_i|$, for $V_i > 0$. If $V_i < 0$ and $H_i = 0$, the wheel will jump; however, if $V_i < 0$ and $H_i \neq 0$, the wheel has the potential to exhibit flange climb derailment.

The third index is the *wheelset lateral to vertical force ratio* (YQ). This index is similar to the SYQ index, but is considered for the entire wheelset (which consists of two wheels), rather than a single wheel. The YQ ratio is, in general, more realistic than the SYQ ratio, and has been used by many authorities and high-speed rail lines in their specifications, for example, the UIC, ICE, SKS and Mainland China.

The definition of the YQ index is

$$\text{YQ} = \frac{Y}{Q}, \tag{10.116}$$

where $Y$ and $Q$ respectively denote the lateral and vertical contact forces acting on a wheelset: $Q = [V_{(2i-1)} + V_{(2i)}] > 0$ and $Y = |H_{(2i-1)} + H_{(2i)}|$, $i = 1, 2, 3$ or 4. If $Q = [V_{(2i-1)} + V_{(2i)}] < 0$, one or two wheels of the $i$th wheelset encounter the jump condition, and the train is at a high degree of risk of derailment. According to the existing specifications, the value of the YQ ratio must not exceed 0.8 to ensure the safety against derailment (Ma and Zhu, 1998).

Another index adopted herein is the *bogie-side lateral to vertical force ratio* (BYQ). This index is associated with gauge spreading or rail roll over derailment. Derailments of this type can occur when the gauge faces of the two rails are spread apart sufficiently for the rim face of one wheel to drop inside of the gauge face of the rail with which it is in contact (Elkins and Carter, 1993). Moreover, this type of derailment is closely related to the combined effect of all wheels on one side of a bogie. Accordingly, the BYQ criterion is defined as the ratio of the sum of the lateral forces on all the wheels on one side of a bogie to the sum of all the vertical loads on the same wheels, namely,

$$\text{BYQ} = \frac{Y}{Q}, \tag{10.117}$$

where $Y$ and $Q$ represent the lateral and vertical contact forces acting on one rail (right or left) by the two wheels on one side of a bogie, respectively, that is, $Q = [V_{(4j-3)} + V_{(4j-1)}]$, $Y = |H_{(4j-3)} + H_{(4j-1)}|$ for right rail and $Q = [V_{(4j-2)} + V_{(4j)}]$, $Y = |H_{(4j-2)} + H_{(4j)}|$ for left rail, where $j = 1$ (for front bogie) and 2 (for rear bogie). It should be noted that (1) $Q = [V_{(4j-3)} + V_{(4j-1)}]$ (or $Q = [V_{(4j-2)} + V_{(4j)}]$) for $V_{(4j-3)}, V_{(4j-1)} > 0$ (or $V_{(4j-2)}, V_{(4j)} > 0$) and $Y = |H_{(4j-3)} + H_{(4j-1)}|$ (or $Y = |H_{(4j-2)} + H_{(4j)}|$), (2) $Q = V_{(4j-3)}$ for $V_{(4j-3)} > 0$, $V_{(4j-1)} \leqq 0$ and $Y = |H_{(4j-3)}|$, and (3) $Q = V_{(4j-1)}$ for $V_{(4j-1)} > 0$, $V_{(4j-3)} \leqq 0$ and $Y = |H_{(4j-1)}|$. The BYQ ratio should not exceed 0.6 in order to be safe from a rail roll over derailment (Elkins and Carter, 1993).

The last index adopted herein is the *lateral track force* ($Y$). A limit is placed on the maximum lateral force exerted by an axle on the track in order to minimize the risk of track panel shift. It should be noted that track panel shift has become increasingly important with increased train speeds and popular use of continuously welded rails. The maximum allowable lateral axle force $Y_{lim}$ (kN) specified by Prud'homme can be given as follows (Elkins and Carter, 1993):

$$Y_{lim} = \alpha \left( 10 + \frac{Q_s}{3} \right) \text{ (kN)}, \tag{10.118}$$

where $Q_s$ is the static axle force (kN) $= W$, and $\alpha$ is a modification factor; $\alpha = 1.0$ in general and $\alpha = 0.85$ for poorer quality track. The maximum allowable lateral axle force for the present study is obtained as 55 kN by Eq. (10.118) for $W = 135$ kN and $\alpha = 1.0$.

The relations between the maximum values of the above five indices and the speed parameter $S$ under different track conditions have been plotted in Figs. 10.26(a)–10.26(e). As can be seen, larger values of the indices have been predicated for tracks with poorer quality, implying that a bad-maintained or deteriorating track structure has a higher possibility of derailment, except for the PD ratio, the indices exhibit some local peaks at certain speeds, primarily due to the lateral train–rails–bridge resonance occurring at these speeds, as evidenced by Fig. 10.26(e) for the lateral force $Y$. Moreover, larger difference exists in the values of the indices between different track qualities at the resonant speeds.

The maximum SYQ ratio for the train exceeds the limit of 1.2 and the maximum YQ ratio approaches the allowable value of 0.8 at some resonant speeds for the poorest track quality considered (FRA Class 4). Such an observation indicates that the resonance occurring between the train, track and bridge can greatly aggravate the running instability of the train, particularly when the train moves on a poor track structure.

In general, the index values for higher resonant speeds ($S \geqq 0.5$) are larger than those for lower resonant speeds ($S < 0.5$). Although the values for the SYQ ratio at some speeds are larger than the allowable one, it does not necessarily mean that derailment will

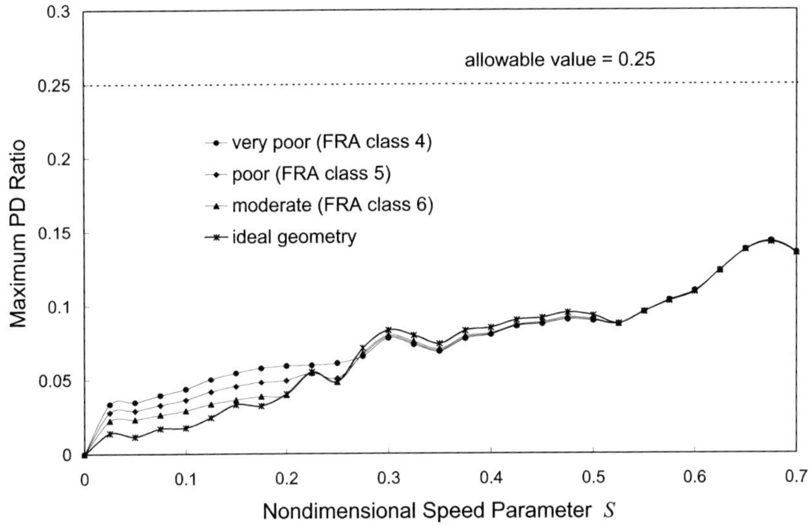


(a)

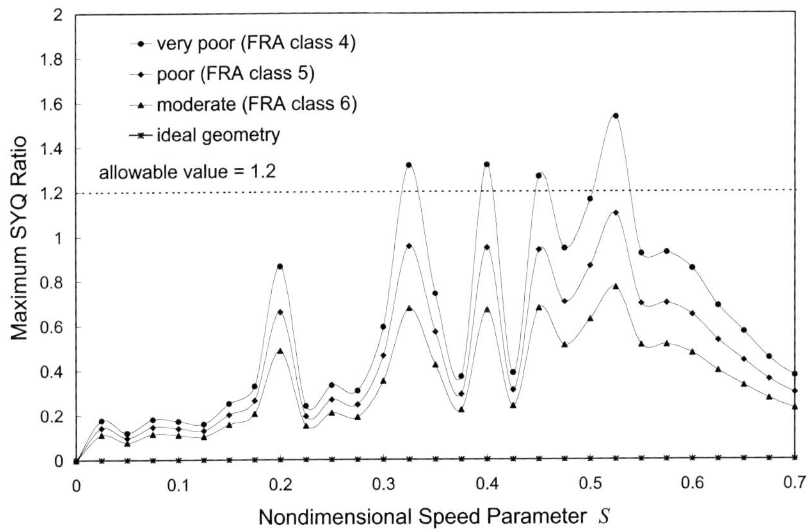


(b)

Fig. 10.26. Maximum index values with relation to the speed parameter: (a) PD ratio, (b) SYQ ratio, (c) YQ ratio, (d) BYQ ratio, and (c) lateral force Y.

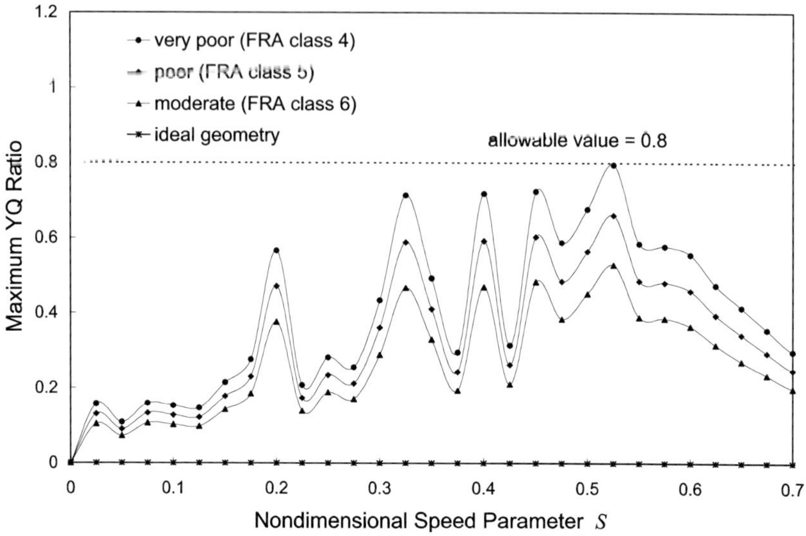


(c)

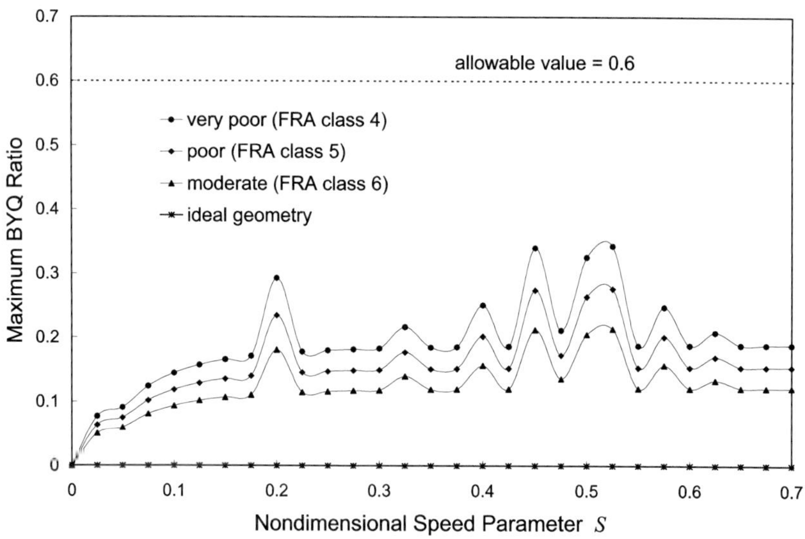


(d)

Fig. 10.26. (*Continued*).

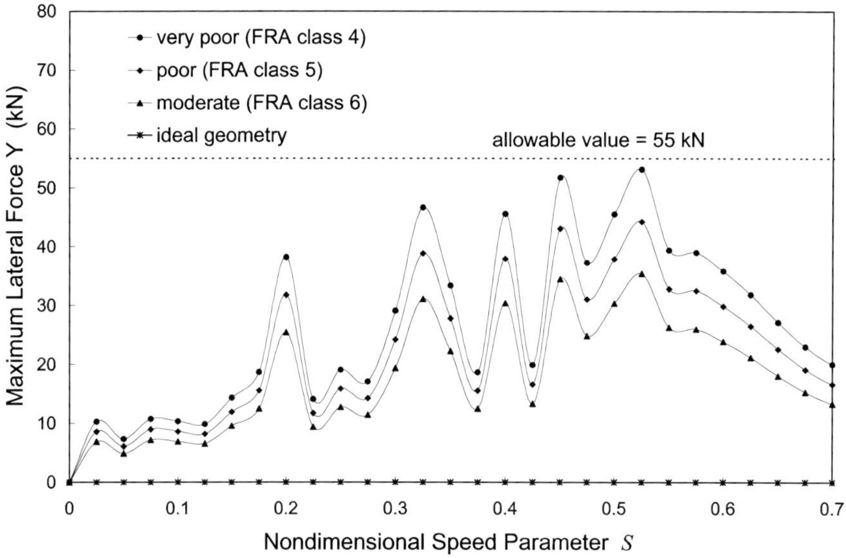


(e)

Fig. 10.26. (*Continued*).

immediately occur on the train at these speeds. One reason for this is that the allowable value of 1.2 has been proved to be a conservative estimate by relevant investigations. Another reason is that even though a very large lateral force is exerted on a wheel of the train, it does require some duration of time for the wheel to derail, which therefore may not occur in practice. However, it is true that the train would be at a higher risk of derailment when traveling on a poor-quality track at the resonant speeds. Based on the results obtained for the five criteria, it is concluded that the safety (or stability) of the train passing over the bridge with various speeds for the four track qualities considered is acceptable. However, the train deserves special attention when it moves over tracks of bad qualities. Besides, the train considered herein shows higher potential for the wheel climb derailment than for other types of derailment, as implied by comparatively large SYQ values.

## 10.14. Concluding Remarks

A three-dimensional VRBI model for analyzing the train–rails–bridge interactions was established. The vehicle is modeled as an assembly of a car body, two bogies and four wheelsets, which has a total of 27 DOFs. Such a model allows us to consider the coupling effect between the lateral and rolling vibrations, which may be caused by the difference in elevation of the center of gravity of the car body, bogie and wheelsets, and the linking action of any two wheels connected by a rigid axle. The track structure is idealized as a continuous twin rails system of infinite length supported by spring-dashpot units. The use of the twin rails model enables us to minimize the number of DOFs used in modeling of the track, while taking into account the constraint imposed by the sleepers on the two rails. In addition to the vertical vibration, the lateral and torsional responses of the train–rails–bridge system can be obtained simultaneously using the present VRBI model, which are useful for evaluation of the running safety of trains over the bridge. The procedure developed for analyzing the 3D VRBI system was verified in the numerical study of a typical example.

The procedure developed above was applied to analysis of the VRBI systems in the three-dimensional sense, with due account taken of the random track irregularities. The results indicate that resonance can occur in the lateral and torsional vibrations of the bridge at some specific speeds, similar to that encountered in vertical vibrations. Besides, the presence of track irregularity can greatly increase the response of the train, track and bridge. Moreover, the extent of influence is much larger for the train than for the track and bridge. There is no doubt that track irregularity should be taken into account in the design and maintenance of railway bridges concerning the running safety of the train.

For a two-way railroad bridge, it is necessary to consider the impact effects caused by the crossing movement of two trains on the bridge, in addition to those by a single moving train. In general, the vertical vibration of the bridge appears to be more violent under the crossing of two trains, but the lateral and torsional responses may be increased or reduced, depending on the way the two trains cross each other. The vertical and pitching vibrations of a train will be aggravated when it crosses with another train on the bridge. Such a phenomenon is more severe for trains moving at high speeds.

Finally, the possibility of derailment of the train was assessed through the use of five common indices. The results of computation for these five indices show that the train will be at a higher risk of derailment as it travels on a bad-conditioned track, mainly due to the relatively large lateral forces induced between the wheelset of the train and the rails. By comparison of the maximum index values computed with the tolerable limits, it is concluded that the train can safely pass through the bridge under the conditions specified in the present study, i.e., those assumed for the track irregularity, train model, track and bridge properties, and so on, over a wide range of speeds.

## Chapter 11

# Stability of Trains Moving over Bridges Shaken by Earthquakes

The dynamic stability of trains moving over bridges shaken by earthquakes is studied in this chapter. Unlike the dynamic analysis of structures subjected only to an earthquake motion, the dynamic analysis of a bridge subjected to a moving train and earthquake motion requires not only information on the acceleration, but also velocity and displacement of the ground motion. Four typical earthquakes, including the 1999 Chi–Chi Earthquake, were adopted as the input excitations, each of which was normalized to have a moderate intensity. The results indicate that a train initially resting on the bridge can stay safely when subjected to any of the four earthquakes, on the condition that no inelastic deformations occur on the bridge and track structures. The characteristics of the vertical component of ground motions can affect significantly the stability of the train–rails–bridge system. As a preliminary attempt, safety, possible instability and instability regions will be established for a train running over a bridge for each of the four earthquakes considered using a three-phase plot, from which the maximum allowable speed for the train to run safely under the specific ground acceleration can be obtained.

### 11.1. Introduction

Seismic resistance of bridge structures is an issue of great concern in earthquake-prone regions. As for railway bridges, it is possible that the bridge itself may remain safe during an earthquake, but may not be safe enough for the trains to move over it due to excessive vibrations of the sustaining bridge. Evidently, the safety of the moving trains over a bridge shaken by earthquakes is a subject of great concern for railway engineers, especially in those countries that are earthquake prone. This problem is becoming more important due to the increasing use of elevated bridges as the supporting structures for railways in metropolitan areas, and the advancement in locomotive technologies for enhancing the driving power of railway trains. To the knowledge of the authors, however, rather limited research works have been conducted on such a subject.

In the study by Miura (1996), emphasis was placed on the earthquake-induced displacement of tracks and structures, as well as the damage of trains caused by earthquake excitations, rather than on the dynamic stability of trains during an earthquake. Miyamoto *et al.* (1997) investigated analytically the running safety of railway vehicles under the action of earthquakes using a three-dimensional simplified vehicle model, where sine waves are used as the input excitation and the vehicle is assumed to remain stationary on the track. Ma and Zhu (1998) studied the response of high-speed trains and continuous rigid-frame bridges under three different ground motions by the random vibration theory. In their work, only the lateral forces were obtained, while the track system was not taken into account. It follows that the index values computed for evaluation of safety (or derailment) of the train may not be accurate enough. Moreover, the assessment of running safety of the train was made only for ground motions with specific intensities, which may not be sufficient for devising any safety guidelines that are useful for the safe operation of trains, as a train may be attacked by earthquakes of various intensities.

In this chapter, the stability of a train initially resting on or traveling over a bridge under different seismic excitations will be investigated using the train and bridge models developed in Chapter 10 with inclusion of the earthquake-induced effect. As a preliminary effort toward assessment of the safety of moving trains shaken by earthquakes, the bridge is assumed to remain fully elastic during the earthquake, while some simple criteria, such as the *derailment index*, are adopted for evaluating the safety of the moving train. It is realized that margin does exist for improving the train and bridge models, say, to include the effect of inelastic deformations, while the derailment of a moving train may have to be evaluated on a statistics basis, rather than on some simplified threshold values. For this reason, the present formulation should be regarded as the one that serves to lay out the basic framework of the problem considered. Even though the results have been computed for several different cases, they should be regarded as a qualitative description, rather than as a quantitative assessment. The materials presented in this chapter are based mainly on the paper by Yang and Wu (2002), but with significant revisions and supplements.

## 11.2. Analysis Model for Train–Rails–Bridge System

Figure 11.1 shows a train traveling with speed $v$ over a simply-supported bridge shaken by an earthquake. The train is modeled in the same way as the one described in Chapter 10, namely, we assume the train to consist of a series of separate identical cars, of which

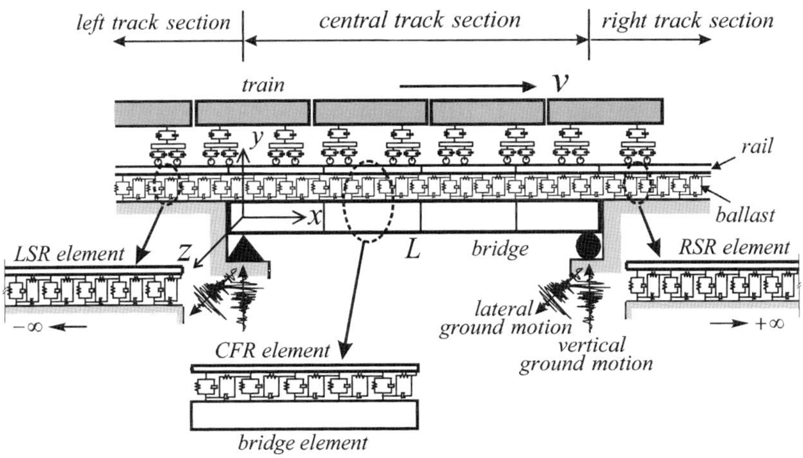


Fig. 11.1. A train traveling over a simply-supported bridge shaken by earthquakes.

each is composed of one car body, two bogies and four wheelsets, as shown in Fig. 11.2. The total number of degrees of freedom (DOFs) implied by each vehicle is 27. Such a model enables us to simulate the vertical, lateral, rolling, yawing and pitching motions of the car body, as well as the vertical and lateral contact forces between the rails and wheels.

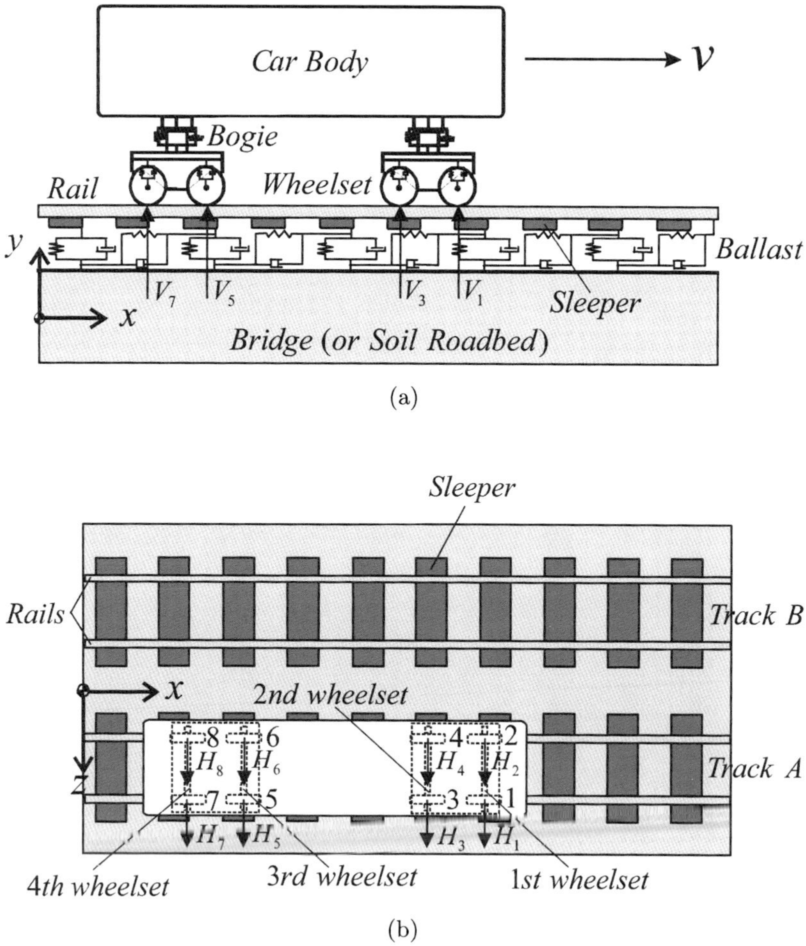


Fig. 11.2. Train car, track and bridge models: (a) side view, (b) top view, and (c) rear view.

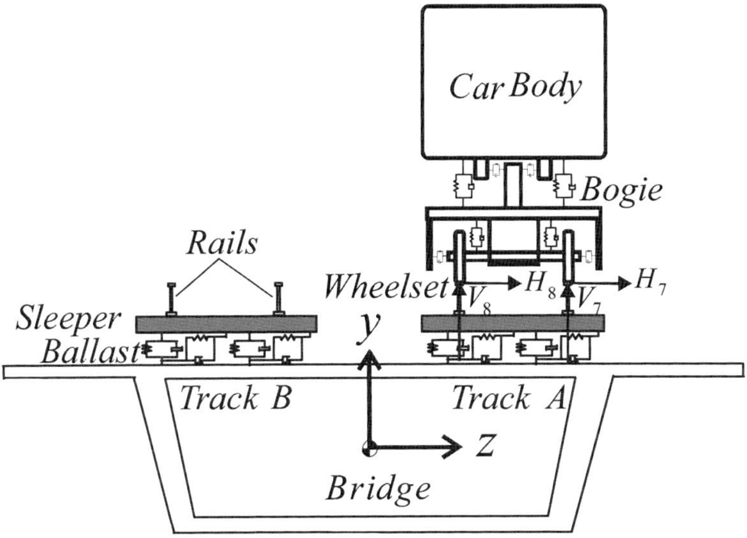


(c)

Fig. 11.2. (*Continued*).

The bridge is assumed to be made of a box girder of uniform cross sections that can carry two parallel tracks, i.e., Tracks $A$ and $B$. Each of the two tracks is simulated as a set of infinite, continuous twin rails lying on a single-layer ballast foundation. The bridge girder, by itself, is modeled as a three-dimensional Bernoulli–Euler beam (see Fig. 11.2). As can be seen, the train car, rails and bridge form a *train–bridge interaction* system, or more specifically, a *vehicle–rails–bridge interaction* (VRBI) system. For the sake of simplicity, each of the subsystems, i.e., the vehicle, track and bridge, is assumed to remain fully elastic during the earthquake excitation.

Again, the fundamental data provided by Wakui *et al.* (1995) for the Series 300 car model of Japan's Shinkansen (SKS) high-speed train will be adopted in this chapter. We choose to use this model, simply because this is the only vehicle model of which the mechanical data were made available to us. The rails are assumed to be of the UIC-60 type. The fundamental data of the vehicle, track and bridge models that are required in analysis have been listed in Tables 10.1.

and 10.2, which are also available in Wu *et al.* (2001). It should be noted that all the conclusions made in this chapter remain strictly valid only for the particular bridge and vehicle models, as well as the assumptions, adopted in the study.

## 11.3. Railway–Bridge System with Ground Motions

In this section, focus will be placed on inclusion of the ground motion effects in the equations of motion for each component of the railway–bridge system. The methodology is similar to the one presented in Chapter 10. In particular, let us consider the railway–bridge system shown in Fig. 11.1, where three types of elements are identified for the rails, i.e., the *central finite rail* (CFR), *left semi-infinite rail* (LSR) and *right semi-infinite rail* (RSR) *elements*. The bridge girder is modeled by the conventional 3D beam elements, each of which has a total of 12 DOFs at the two nodes, with three translations and three rotations at each node. The equations of motion for the CFR, LSR, RSR and bridge elements that have been presented in Chapter 10 will be extended to include the earthquake-induced effects. All the quantities associated with Track $A$, Track $B$, and the bridge will be denoted by symbols with subscripts "$A$", "$B$" and "$b$", respectively.

### 11.3.1. *Central Finite Rail (CFR) Element for Track A*

The two rails of each track are assumed to be identical and connected by rigid sleepers that are uniformly-distributed. Just as in Sec. 10.4.1, the two rails will be considered together and represented by a single line of rail elements, also known as the *central finite rail* (CFR) *elements*, lying on the ballast layer that is modeled by uniformly-distributed spring-dashpot units. The CFR element is nothing but a conventional 3D beam element, of which the nodal displacement vector can be given as follows:

$$\{d_A\} = \langle u_{A1} \quad v_{A1} \quad w_{A1} \quad \theta_{A1} \quad \varphi_{A1} \quad \psi_{A1} \quad u_{A2} \quad v_{A2} \quad w_{A2} \quad \theta_{A2} \quad \varphi_{A2} \quad \psi_{A2} \rangle^T, \tag{11.1}$$

where the subscript $A$ indicates that the quantity is associated with *Track A*. Correspondingly, the nodal force vector $\{f_A\}$ is

$$\begin{aligned}\{f_A\} = \langle\, & F_{Ax1} \quad F_{Ay1} \quad F_{Az1} \quad M_{Ax1} \quad M_{Ay1} \quad M_{Az1} \\ & F_{Ax2} \quad F_{Ay2} \quad F_{Az2} \quad M_{Ax2} \quad M_{Ay2} \quad M_{Az2}\,\rangle^T . \end{aligned} \tag{11.2}$$

By the principle of virtual work, the equation of equilibrium for the CFR element located on Track $A$ can be written as

$$\begin{aligned}&\int_0^l E_t A_t u'_A \delta u'_A dx + \int_0^l E_t I_{tz} v''_A \delta v''_A dx + \int_0^l E_t I_{ty} w''_A \delta w''_A dx \\ &\quad = -\int_0^l m_t(\ddot{u}_A \delta u_A + \ddot{v}_A \delta v_A + \ddot{w}_A \delta w_A) dx - \int_0^l I_t^* \ddot{\theta}_A \delta\theta_A dx \\ &\quad - \int_0^l c_t(\dot{u}_A \delta u_A + \dot{v}_A \delta v_A + \dot{w}_A \delta w_A) dx \\ &\quad - \int_0^l c_t^* \dot{\theta}_A \delta\theta_A dx + 2l_d \int_0^l [p_c(x) + p_k(x)] \delta u_A dx \\ &\quad + \int_0^l \int_{l_b - l_d}^{l_b + l_d} [q_c(x,z) + q_k(x,z)][\delta v_A - (z - l_b)\delta\theta_A] dz dx \\ &\quad + 2l_d \int_0^l [r_c(x) + r_k(x)] \delta w_A dx + \langle \delta d_A \rangle \{f_A\} ,\end{aligned} \tag{11.3}$$

where all the quantities with subscript $t$ indicates that they are associated with the *track* or *rails*, $E_t$ is Young's modulus, $A_t$ is the cross-sectional area, $I_{ty}$, $I_{tz}$ are the moment of inertia about $y$, $z$ axes, $m_t$ is the per-unit-length mass (including the mass of sleepers), $I_t^*$ the per-unit-length mass moment of inertia about $x$ axis (including the contribution of sleepers), $c_t$, $c_t^*$ the visco-damping coefficients, $l_b$ the distance between the center line of Track $A$ and the bridge, $l_d$ the half-length of sleepers, $l$ the length of the rail element; $\{d_A\}$ denotes the nodal displacement of the rail element, $\{f_A\}$ the corresponding external loads; $u_A$, $v_A$, $w_A$ the displacements along $x$, $y$, $z$-axes; and $\theta_A$ the rotation about $x$ axis of the rail element.

The interaction forces arising from the relative motion of the rail and bridge elements, i.e., $(p_c, p_k)$, $(q_c, q_k)$ and $(r_c, r_k)$, can be defined as

$$\begin{aligned}
p_c(x) &= c^*_{bh1}(\dot{u}_b(x) - \dot{u}_A(x)) \,, \\
p_k(x) &= k^*_{bh1}(u_b(x) - u_A(x)) \,, \\
q_c(x,z) &= c^*_{bv1}(\dot{v}_b(x) - z\dot{\theta}_b(x) - \dot{v}_A(x) + (z - l_b)\dot{\theta}_A(x)) \,, \\
q_k(x,z) &= k^*_{bv1}(v_b(x) - z\theta_b(x) - v_A(x) + (z - l_b)\theta_A(x)) \,, \\
r_c(x) &= c^*_{bh1}(\dot{w}_b(x) + h\dot{\theta}_b(x) - \dot{w}_A(x)) \,, \\
r_k(x) &= k^*_{bh1}(w_b(x) + h\theta_b(\xi) - w_A(x)) \,,
\end{aligned} \tag{11.4}$$

where $c^*_{bh1}$, $k^*_{bh1}$ and $c^*_{bv1}$, $k^*_{bv1}$, respectively, denote the unit-area horizontal and vertical damping and stiffness coefficients of the ballast on the bridge; $u_b$, $v_b$, $w_b$ the displacements along the three axes; and $\theta_b$ the rotation about $x$ axis of the bridge element; $h$ is the vertical distance between the deck and the center of torsion of the bridge section. The longitudinal and lateral damping and stiffness coefficients of the ballast are assumed to be the same, denoted as $c^*_{bh1}$ and $k^*_{bh1}$. The displacement fields of the rail and bridge elements can be expressed in terms of the nodal DOFs using linear or cubic Hermitian interpolation functions (Paz, 1985). Substituting these displacement fields into Eq. (11.3) yields the equation of motion for the CFR element composed of twin rails for Track $A$ that is free of any ground motions as

$$\begin{aligned}
&[m_A]\{\ddot{d}_A\} + [c_A]\{\dot{d}_A\} + [k_A]\{d_A\} \\
&\quad = \{f_A\} + [c_{Ab}]\{\dot{d}_b\} + [k_{Ab}]\{d_b\} \,,
\end{aligned} \tag{11.5}$$

where $[m_A]$, $[c_A]$ and $[k_A]$ denote the mass, damping and stiffness matrices of the rail element, $\{d_A\}$ is the nodal displacement of the rail element, as given in Eq. (11.1), and $\{d_b\}$ is the nodal displacement vector of the underlying bridge element,

$$\begin{aligned}
\{d_b\} = \langle\, & u_{b1} \quad v_{b1} \quad w_{b1} \quad \theta_{b1} \quad \varphi_{b1} \quad \psi_{b1} \\
& u_{b2} \quad v_{b2} \quad w_{b2} \quad \theta_{b2} \quad \varphi_{b2} \quad \psi_{b2} \,\rangle^T \,,
\end{aligned} \tag{11.6}$$

The damping and stiffness matrices on the right-hand side of Eq. (11.5), i.e., $[c_{Ab}]$ and $[k_{Ab}]$, represent the interaction effects caused by the relative motion of the two rails and bridge elements through the ballast layer simulated as uniformly-distributed spring-dashpot units. Details for the matrices and vectors involved in Eq. (11.5) are available in Sec. 10.4.1.

For a railway bridge subjected to a ground motion, both the displacement vectors $\{d_A\}$ and $\{d_b\}$ appearing in Eq. (11.5) should be interpreted as the *total* or *absolute* displacements of the rail and bridge elements, respectively. The total displacements of the bridge element $\{d_b\}$ can be divided into two parts:

$$\{d_b\} = \{d_b^n\} + \{d_b^r\}, \tag{11.7}$$

where $\{d_b^n\}$ denotes the *natural deformations* and $\{d_b^r\}$ the *rigid displacements* of the bridge element due to the ground motion. The latter can be determined as

$$\{d_b^r\} = [R]\{u_g\}, \tag{11.8}$$

where $[R]$ denotes the transformation matrix and $\{u_g\}$ the support displacements of the bridge due to the ground motion, which are prescribed in general. Based on the assumption that the support motions occur *synchronously* and that no rotations are induced by the ground motions on the bridge supports, the transformation matrix $[R]$ can be given as

$$\begin{aligned} [R] &= \begin{bmatrix} [r] \\ [r] \end{bmatrix}, \\ [r] &= \begin{bmatrix} [I]_{3\times3} \\ [0]_{3\times3} \end{bmatrix}, \end{aligned} \tag{11.9}$$

where $[I]$ is a $3 \times 3$ unit matrix and $[0]$ a zero matrix. The support displacements $\{u_g\}$ are assumed to be three-dimensional,

$$\{u_g\} = \langle\, u_{gx} \quad u_{gy} \quad u_{gz} \,\rangle^T, \tag{11.10}$$

where $u_{gx}$, $u_{gy}$ and$u_{gz}$ denote the displacement components along the three axes.

Substituting Eqs. (11.7)–(11.10) into Eq. (11.5), one obtains the equation of motion for the CFR element with due account taken of the ground motion as

$$[m_A]\{\ddot{d}_A\} + [c_A]\{\dot{d}_A\} + [k_A]\{d_A\} = \{f_A^t\} + [c_{Ab}]\{\dot{d}_b^n\} + [k_{Ab}]\{d_b^n\}\,, \qquad (11.11)$$

where $\{f_A^t\}$ denotes the total equivalent nodal forces of the element under the ground motion,

$$\{f_A^t\} = \{f_A\} + \{f_A^c\} + \{f_A^k\} = \{f_A\} + \underline{[c_{Ab}][R]\{\dot{u}_g\}} + \underline{[k_{Ab}][R]\{u_g\}}\,. \qquad (11.12)$$

As can be seen, the parts of the equivalent nodal forces induced by ground motions, i.e., $\{f_A^c\}$ and $\{f_A^k\}$, relate to the displacement and velocity, but not acceleration, of the ground or supports of the bridge.

### 11.3.2. *Central Finite Rail (CFR) Element for Track B*

Let the nodal displacement vector of the central finite rail (CFR) element on Track $B$ be denoted as

$$\{d_B\} = \langle\, u_{B1} \quad v_{B1} \quad w_{B1} \quad \theta_{B1} \quad \varphi_{B1} \quad \psi_{B1} \quad u_{B2} \quad v_{B2} \quad w_{B2} \quad \theta_{B2} \quad \varphi_{B2} \quad \psi_{B2}\,\rangle^T\,, \qquad (11.13)$$

and the associated nodal force vector be denoted as

$$\{f_B\} = \langle\, F_{Bx1} \quad F_{By1} \quad F_{Bz1} \quad M_{Bx1} \quad M_{By1} \quad M_{Bz1} \quad F_{Bx2} \quad F_{By2} \quad F_{Bz2} \quad M_{Bx2} \quad M_{By2} \quad M_{Bz2}\,\rangle^T\,. \qquad (11.14)$$

By following the procedure presented in Sec. 11.3.1, the equation of motion of the CFR element located on Track $B$ with due account for the ground motions can be given as

$$[m_B]\{\ddot{d}_B\} + [c_B]\{\dot{d}_B\} + [k_B]\{d_B\} = \{f_B^t\} + [c_{Bb}]\{\dot{d}_b^n\} + [k_{Bb}]\{d_b^n\}\,, \qquad (11.15)$$

where

$$\{f_B^t\} = \{f_B\} + \{f_B^c\} + \{f_B^k\}$$
$$= \{f_B\} + \underline{[c_{Bb}][R]\{\dot{u}_g\}} + \underline{[k_{Bb}][R]\{u_g\}}. \qquad (11.16)$$

Based on the assumption that *no rocking motion* are induced by the earthquake on the bridge supports, the extra equivalent nodal forces induced by the earthquake on the CFR element for both tracks should be the same, namely,

$$\{f_B^c\} = \{f_A^c\},$$
$$\{f_B^k\} = \{f_A^k\}. \qquad (11.17)$$

The same results can be obtained for the two tracks through manipulation of the triple products underlined in Eq. (11.12) involving the matrices $[c_{Ab}]$ and $[k_{Ab}]$ and in Eq. (11.16) involving the matrices $[c_{Bb}]$ and $[k_{Bb}]$.

### 11.3.3. *Bridge Element*

The bridge element is also regarded as a three-dimensional solid beam element, which has a total of 12 DOFs, as indicated in Eq. (11.6). Following the same procedure as that for the CFR element above, one can derive the equation of motion for the bridge element, with due account taken of the interaction with the rail elements of the two tracks through the ballast layer, as follows:

$$[m_b]\{\ddot{d}_b^n\} + [c_b]\{\dot{d}_b^n\} + [k_b]\{d_b^n\}$$
$$= \{f_b^t\} + [c_{bA}]\{\dot{d}_A\} + [k_{bA}]\{d_A\}$$
$$+ [c_{bB}]\{\dot{d}_B\} + [k_{bB}]\{d_B\}, \qquad (11.18)$$

where $\{d_b^n\}$ denotes the natural deformations of the bridge element, $\{d_A\}$ and $\{d_B\}$ the nodal displacements of the associated rail elements of Tracks $A$ and $B$, respectively, $[m_b]$, $[c_b]$ and $[k_b]$ the mass, damping and stiffness matrices of the beam element, and $[c_{bA}]$, $[k_{bA}]$ and $[c_{bB}]$, $[k_{bB}]$ the damping and stiffness effects resulting from the interaction with the rail elements of Tracks $A$ and $B$, respectively, through the ballast layer. The system matrices $[m_b]$, $[c_b]$ and $[k_b]$ are identical to those given in Eq. (10.51). Due to the interaction between the rail and bridge elements, the matrices $[c_{bA}]$ and $[k_{bA}]$ are identical to $[c_{Ab}]$ and $[k_{Ab}]$, respectively, that is, $[c_{bA}] = [c_{Ab}]$ and $[k_{bA}] = [k_{Ab}]$. Similarly, $[c_{bB}]$ and $[k_{bB}]$ are identical to $[c_{Bb}]$ and $[k_{Bb}]$, respectively. All the interaction matrices $[c_{Ab}]$, $[k_{Ab}]$ and $[c_{Bb}]$, $[k_{Bb}]$ have been made available in Eqs. (10.32) and (10.47). The total forces $\{f_b^t\}$ of the bridge element are

$$\begin{aligned}\{f_b^t\} &= \{f_b\} + \{f_b^m\} + \{f_b^c\} + \{f_b^k\} \\ &= \{f_b\} - \underline{[m_b][R]\{\ddot{u}_g\}} - \underline{([c_{bA}] + [c_{bB}])[R]\{\dot{u}_g\}} \\ &\quad - \underline{([k_{bA}] + [k_{bB}])[R]\{u_g\}}\,. \end{aligned} \tag{11.19}$$

Here, $\{f_b^m\}$ represents the equivalent inertia forces due to the rigid motions of the bridge element caused by the ground motion, and $\{f_b^c\}$ and $\{f_b^k\}$ the equivalent damping and restoring forces due to the restraint effect of the ballast layer beneath Tracks $A$ and $B$ relative to the bridge element under the ground motion. Evidently, the parts of nodal forces induced by the ground motion on the bridge element relate not only to the acceleration, but also to the displacement and velocity of the ground.

### 11.3.4. *Left Semi-Infinite Rail (LSR) Element for Tracks A and B*

Both the left semi-infinite rail (LSR) and right semi-infinite rail (RSR) elements have one side with infinite boundary, which therefore has only a single node at one end. As was described in Sec. 10.4.4, the infinite boundary of both the LSR and RSR elements is represented by functions of decaying nature. Unlike the CFR element that is lying on the bridge girder, both the LSR and RSR elements are lying on the embankments and therefore are directly affected by the ground motion. The equation of motion for the LSR element used to simulate Track $A$ under the ground motion can be written as follows:

$$[m_l]\{\ddot{d}_{Al}\} + [c_l]\{\dot{d}_{Al}\} + [k_l]\{d_{Al}\} = \{f_{Al}^t\}, \tag{11.20}$$

where $\{d_{Al}\}$ denotes the nodal displacements of the element, which consists of three translations and three rotations at the starting node,

$$\{d_{Al}\} = \langle\, u_{A2} \quad v_{A2} \quad w_{A2} \quad \theta_{A2} \quad \varphi_{A2} \quad \psi_{A2}\,\rangle^T , \tag{11.21}$$

and $[m_l]$, $[c_l]$ and $[k_l]$ denote the mass, damping and stiffness matrices of the LSR element, which are identical to those given in Eq. (10.64). The total nodal forces $\{f^t_{Al}\}$ of the element, which consist of the loads directly acting on the node and the equivalent nodal forces due to ground motion, can be given as

$$\{f^t_{Al}\} = \{f_{Al}\} + \{f^c_{Al}\} + \{f^k_{Al}\} , \tag{11.22}$$

where $\{f_{Al}\}$ denotes the loads directly acting on the nodes,

$$\{f_{Al}\} = \langle\, F_{Ax2} \quad F_{Ay2} \quad F_{Az2} \quad M_{Ax2} \quad M_{Ay2} \quad M_{Az2}\,\rangle^T \tag{11.23}$$

and $\{f^c_{Al}\}$ and $\{f^k_{Al}\}$ the forces induced by the ground motion

$$\begin{aligned}
\{f^c_{Al}\} &= 2l_d c^*_{bh2} \int_{-\infty}^{0} \dot{u}_{gx}\{N_u\}dx + 2l_d c^*_{bv2} \int_{-\infty}^{0} \dot{u}_{gy}\{N_v\}dx \\
&\quad + 2l_d c^*_{bh2} \int_{-\infty}^{0} \dot{u}_{gz}\{N_w\}dx , \\
\{f^k_{Al}\} &= 2l_d k^*_{bh2} \int_{-\infty}^{0} u_{gx}\{N_u\}dx + 2l_d k^*_{bv2} \int_{-\infty}^{0} u_{gy}\{N_v\}dx \\
&\quad + 2l_d k^*_{bh2} \int_{-\infty}^{0} u_{gz}\{N_w\}dx ,
\end{aligned} \tag{11.24}$$

where $\{N_u\}$, $\{N_v\}$ and $\{N_w\}$ denote the interpolation vectors for the LSR element of Track $A$ as defined in Sec. 10.4.4 and $c^*_{bh2}$, $k^*_{bh2}$ and $c^*_{bv2}$, $k^*_{bv2}$ the damping and stiffness coefficients of the ballast per-unit-area on the soil roadbed along the horizontal ($h$) and vertical ($v$) directions. Supposing that the ground displacement and velocity do not vary along the track, Eq. (11.24) can be rewritten,

$$\begin{aligned}
\{f^c_{Al}\} &= [c_{gl}]\{\dot{u}_g\} = [\,\{c_{glx}\} \quad \{c_{gly}\} \quad \{c_{glz}\}\,]\{\dot{u}_g\} , \\
\{f^k_{Al}\} &= [k_{gl}]\{u_g\} = [\,\{k_{glx}\} \quad \{k_{gly}\} \quad \{k_{glz}\}\,]\{u_g\} ,
\end{aligned} \tag{11.25}$$

where

$$\{c_{glx}\} = \left\langle \frac{2l_d c^*_{bh2}}{\lambda_u} \quad 0 \quad 0 \quad 0 \quad 0 \quad 0 \right\rangle^T,$$

$$\{c_{gly}\} = \left\langle 0 \quad \frac{2l_d c^*_{bv2}}{\lambda_v} \quad 0 \quad 0 \quad 0 \quad \frac{-2l_d c^*_{bv2}}{2\lambda_v^2} \right\rangle^T, \tag{11.26}$$

$$\{c_{glz}\} = \left\langle 0 \quad 0 \quad \frac{2l_d c^*_{bh2}}{\lambda_w} \quad 0 \quad \frac{2l_d c^*_{bh2}}{2\lambda_w^2} \quad 0 \right\rangle^T,$$

$$\{k_{glx}\} = \left\langle \frac{2l_d k^*_{bh2}}{\lambda_u} \quad 0 \quad 0 \quad 0 \quad 0 \quad 0 \right\rangle^T,$$

$$\{k_{gly}\} = \left\langle 0 \quad \frac{2l_d k^*_{bv2}}{\lambda_v} \quad 0 \quad 0 \quad 0 \quad \frac{-2l_d k^*_{bv2}}{2\lambda_v^2} \right\rangle^T, \tag{11.27}$$

$$\{k_{glz}\} = \left\langle 0 \quad 0 \quad \frac{2l_d k^*_{bh2}}{\lambda_w} \quad 0 \quad \frac{2l_d k^*_{bh2}}{2\lambda_w^2} \quad 0 \right\rangle^T,$$

where $\lambda_u$, $\lambda_v$, $\lambda_w$ denote the longitudinal, vertical and lateral characteristic numbers of the beam-Winkler foundation system, which have been given in Eqs. (10.62).

Similarly, the equation of motion for the *LSR element* used to represent *Track B* under the ground motion can be derived as

$$[m_l]\{\ddot{d}_{Bl}\} + [c_l]\{\dot{d}_{Bl}\} + [k_l]\{d_{Bl}\} = \{f^t_{Bl}\}, \tag{11.28}$$

where the total nodal loads acting on the element are

$$\{f^t_{Bl}\} = \{f_{Bl}\} + \{f^c_{Bl}\} + \{f^k_{Bl}\}. \tag{11.29}$$

Here, the external nodal force vector $\{f_{Bl}\}$ has been given in Eq. (10.76), i.e.,

$$\{f_{Bl}\} = \langle F_{Bx2} \quad F_{By2} \quad F_{Bz2} \quad M_{Bx2} \quad M_{By2} \quad M_{Bz2} \rangle^T \tag{11.30}$$

and the earthquake-induced forces are

$$\begin{aligned} \{f^c_{Bl}\} &= \{f^c_{Al}\}, \\ \{f^k_{Bl}\} &= \{f^k_{Al}\}. \end{aligned} \tag{11.31}$$

### 11.3.5. *Right Semi-Infinite Rail (RSR) Element for Tracks A and B*

By following the same procedure for derivation of the LSR element, the equation of motion for the *RSR element* of *Track A* considering the ground motion can be derived,

$$[m_r]\{\ddot{d}_{Ar}\} + [c_r]\{\dot{d}_{Ar}\} + [k_r]\{d_{Ar}\} = \{f^t_{Ar}\}, \tag{11.32}$$

where $\{d_{Ar}\}$ denotes the nodal displacement vector,

$$\{d_{Ar}\} = \langle\, u_{A1} \quad v_{A1} \quad w_{A1} \quad \theta_{A1} \quad \varphi_{A1} \quad \psi_{A1}\,\rangle^T \tag{11.33}$$

and $[m_r]$, $[c_r]$ and $[k_r]$ the mass, damping and stiffness matrices of the RSR element, which have been made available in Eq. (10.72). The total nodal forces $\{f^t_{Ar}\}$ of the element are

$$\{f^t_{Ar}\} = \{f_{Ar}\} + \{f^c_{Ar}\} + \{f^k_{Ar}\}, \tag{11.34}$$

where the external load vector $\{f_{Ar}\}$ is

$$\{f_{Ar}\} = \langle\, F_{Ax1} \quad F_{Ay1} \quad F_{Az1} \quad M_{Ax1} \quad M_{Ay1} \quad M_{Az1}\,\rangle^T \tag{11.35}$$

and the earthquake-induced forces are

$$\begin{aligned} \{f^c_{Ar}\} &= [c_{gr}]\{\dot{u}_g\} = [\,\{c_{grx}\} \quad \{c_{gry}\} \quad \{c_{grz}\}\,]\{\dot{u}_g\}, \\ \{f^k_{Ar}\} &= [k_{gr}]\{u_g\} = [\,\{k_{grx}\} \quad \{k_{gry}\} \quad \{k_{grz}\}\,]\{u_g\}. \end{aligned} \tag{11.36}$$

Assuming that the earthquake motion is *synchronous* for the two ends of the bridge, one therefore has the following:

$$\begin{aligned} \{c_{grx}\} &= \{c_{glx}\}, \\ \{c_{gry}\} &= \left\langle 0 \quad \frac{2l_d c^*_{bv2}}{\lambda_v} \quad 0 \quad 0 \quad 0 \quad \frac{2l_d c^*_{bv2}}{2\lambda_v^2} \right\rangle^T, \\ \{c_{grz}\} &= \left\langle 0 \quad 0 \quad \frac{2l_d c^*_{bh2}}{\lambda_w} \quad 0 \quad \frac{-2l_d c^*_{bh2}}{2\lambda_w^2} \quad 0 \right\rangle^T, \end{aligned} \tag{11.37}$$

$$\{k_{grx}\} = \{k_{glx}\},$$
$$\{k_{gry}\} = \left\langle 0 \quad \frac{2l_d k_{bv2}^*}{\lambda_v} \quad 0 \quad 0 \quad 0 \quad \frac{2l_d k_{bv2}^*}{2\lambda_v^2} \right\rangle^T, \tag{11.38}$$
$$\{k_{grz}\} = \left\langle 0 \quad 0 \quad \frac{2l_d k_{bh2}^*}{\lambda_w} \quad 0 \quad \frac{-2l_d k_{bh2}^*}{2\lambda_w^2} \quad 0 \right\rangle^T.$$

As can be observed from the derivations presented in this section for the CFR, LSR, RSR and bridge elements, the consideration of support or ground motions results only in inclusion of some earthquake-induced forces in the nodal force vectors, while the mass, damping and stiffness matrices remain exactly the same as those for the case with no ground motions. Because of this, the earthquake-induced effects can be easily included in existing vehicle–bridge interaction analysis programs with no change on the system matrices. The only thing that should be taken into account is to expand the vectors of nodal forces to include the nodal forces induced by the earthquake.

## 11.4. Method of Analysis

In the preceding section, focus has been placed on derivation of the equations of motion for each component of the railway–bridge system. By assembling the element matrices and vectors for all the components involved, the global matrices and vectors, as well as the equations of motion, for the railway–bridge or supporting system can be established. The railway–bridge system represents only the nonmoving subsystem of the train–bridge system. The other subsystem is the moving train, which consists of a number of vehicles. In this study, the equations of motion for the moving train are constructed using the procedure presented in Sec. 10.3. The supporting and moving subsystems interact with each other through the *contact points* existing between the rails and rolling wheels, as the train moves. Clearly, the equations of motion for the two subsystems are *coupled* and *time-dependent*.

As shown in Fig. 11.2, each of the train cars considered has eight wheels, i.e., $i = 1 \sim 8$. Each wheel has a vertical and a lateral contact force. Thus, there is a total of 16 contact forces, i.e., $j = 1 \sim 16$. Using the Newmark-type finite-difference formulas, together with the constraint relation for relating the wheelset displacements to the rail displacements, one can first solve the equations of motion for each vehicle to obtain the vertical and lateral contact forces, i.e., $V_i$ and $H_i$, in terms of the contact-point displacements $d_{cj}$ and vehicle forces $\tilde{p}_c$ and $\tilde{q}_c$ as in Eqs. (10.18) and (10.19), that is,

$$\begin{aligned} V_{i,t+\Delta t} &= \tilde{p}_{c(2i-1),t+\Delta t} + \tilde{q}_{c(2i-1),t} \\ &\quad + \sum_{j=1}^{16} (\tilde{m}_{c(2i-1)j} \ddot{d}_{cj,t+\Delta t} + \tilde{c}_{c(2i-1)j} \dot{d}_{cj,t+\Delta t} \\ &\quad + \tilde{k}_{c(2i-1)j} d_{cj,t+\Delta t}) \,, \end{aligned} \tag{11.39}$$

$$\begin{aligned} H_{i,t+\Delta t} &= \tilde{p}_{c(2i),t+\Delta t} + \tilde{q}_{c(2i),t} \\ &\quad + \sum_{j=1}^{16} (\tilde{m}_{c(2i)j} \ddot{d}_{cj,t+\Delta t} + \tilde{c}_{c(2i)j} \dot{d}_{cj,t+\Delta t} \\ &\quad + \tilde{k}_{c(2i)j} d_{cj,t+\Delta t}) \,, \end{aligned} \tag{11.40}$$

where the subscripts $i = 1 \sim 8$, $j = 1 \sim 16$, and the matrices $[\tilde{m}_c]$, $[\tilde{c}_c]$, $[\tilde{k}_c]$ and vectors $\{\tilde{p}_c\}$, $\{\tilde{q}_c\}$ relate to the physical properties and wheel-load effects of the vehicle, as defined in Eqs. (10.16) and (10.17).

By using the dynamic condensation technique developed in Chapter 8, the DOFs of each of the moving vehicles can be condensed into the associated rail element(s) in contact to form the *vehicle–rails interaction* (VRI) element(s). By assembling all the VRI elements, the ordinary rail elements and the bridge elements, the equations of motion for the entire railway–bridge system can be established, which appear as second-order differential equations. The Newmark $\beta$ integration method can then be called for to solve the system equations for each time step. Such an approach enables us to compute the dynamic responses of all the components involved in the train–bridge system, including the vehicle response, contact forces and bridge response. As was mentioned previously, the equations of motion for the rail elements and bridge element considering the ground motion are identical to those for the case with no ground motions, except that the earthquake-induced forces should be included in the nodal forces for the present case. Consequently, the procedure presented in Sec. 10.7 for constructing the system matrices and the one in Sec. 8.7 for performing the time-history analysis remain applicable herein for analysis of the train–bridge system shaken by earthquakes, if the earthquake-induced forces, which are functions of the acceleration, velocity and displacement of the ground motion, are duly included and updated at each incremental step.

## 11.5. Description of Input Earthquake Records

Four sets of ground accelerations induced by earthquakes are selected as the source of vibration. The first two sets are those for the 1940 *El Centro* and 1994 *Northridge Earthquakes.* The last two sets, i.e., TAP003 and TCU068, recorded at the free-field stations of Taipei and Taichung, respectively, during the 1999 *Chi–Chi Earthquake* in Taiwan, are used to simulate the *far field* and *near fault* excitations. The records for the El Centro and Northridge Earthquakes contain only a horizontal component of vibrations, while those for the Chi–Chi Earthquake contain both the EW horizontal and vertical components. In this study, the horizontal excitation of the earthquake is applied in the *lateral* ($z$) direction of the bridge for the sake of evaluating the stability of passing trains. Whenever the lateral ground motion is scaled down in terms of the PGA, the vertical ground motion is scaled down as well in a proportional manner.

Table 11.1 shows some key data for the EW and vertical motions of the TAP003 and TCU068 ground motions. As can be seen, the PGA values of the TCU068 Station are much larger than those of the TAP003 Station, since the former was recorded at a station much closer to the fault than the latter. For the TCU068 Station, the vertical PGA appears to be as large as the lateral (EW) PGA due to the near-fault effect. In contrast, for the TAP003 Station, the

Table 11.1. PGA, PGV, and PGD for TAP003 and TCU068 ground motions.

| Ground motion | TAP003 | | TCU068 | |
|---|---|---|---|---|
| Direction | EW | Vertical | EW | Vertical |
| PGA (gal) | 127.45 | 43.29 | 501.87 | 519.72 |
| PGV (cm/s) | 31.33 | 8.61 | 156.91 | 142.41 |
| PGD (cm) | 15.35 | 4.15 | 83.92 | 102.34 |

vertical PGA is only one-third of the lateral (EW) PGA, which is typical of the far-field ground motion. The EW and vertical ground accelerations of the TAP003 Station were plotted in Figs. 11.3(a) and 11.4(a), respectively, for a duration of 70 s. For comparison, the EW and vertical ground accelerations of the TCU068 Station were plotted in Figs. 11.5(a) and 11.6(a), respectively, with the same duration. The impulse-type vertical acceleration shown in Fig. 11.6(a) is typical of a near-fault earthquake.

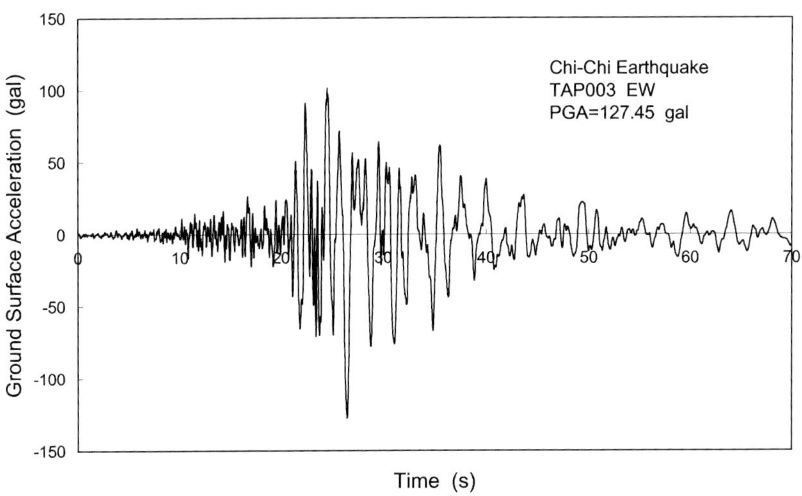


(a)

Fig. 11.3. Histogram of lateral (EW) motion of TAP003 Station: (a) acceleration, (b) velocity, and (c) displacement.

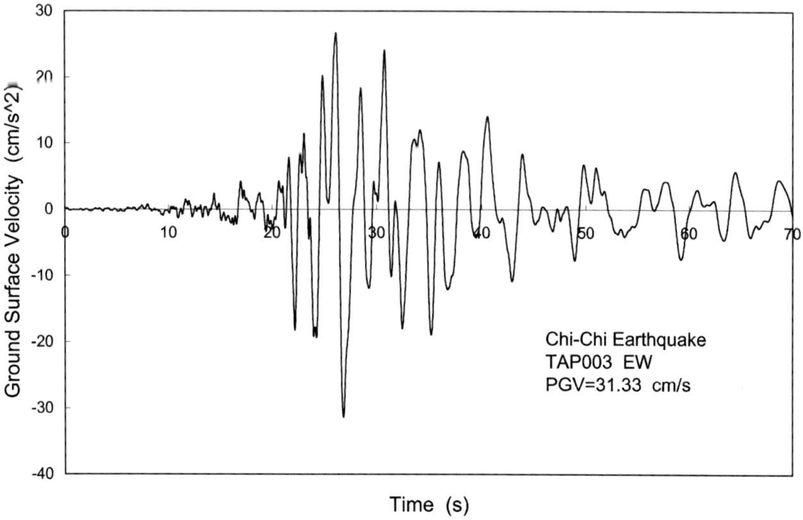


(b)

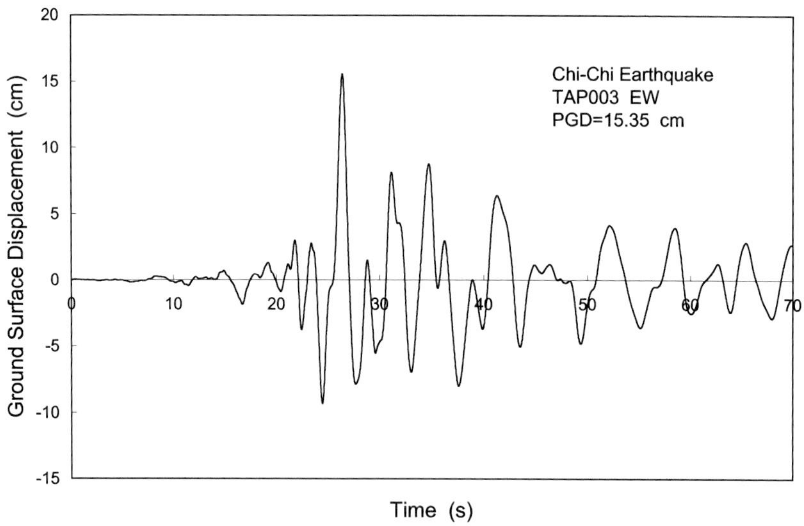


(c)

Fig. 11.3. (*Continued*).

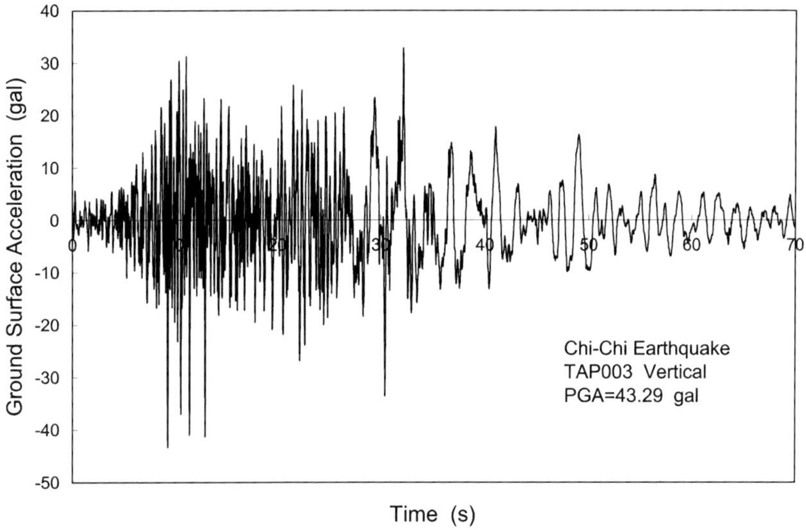


(a)

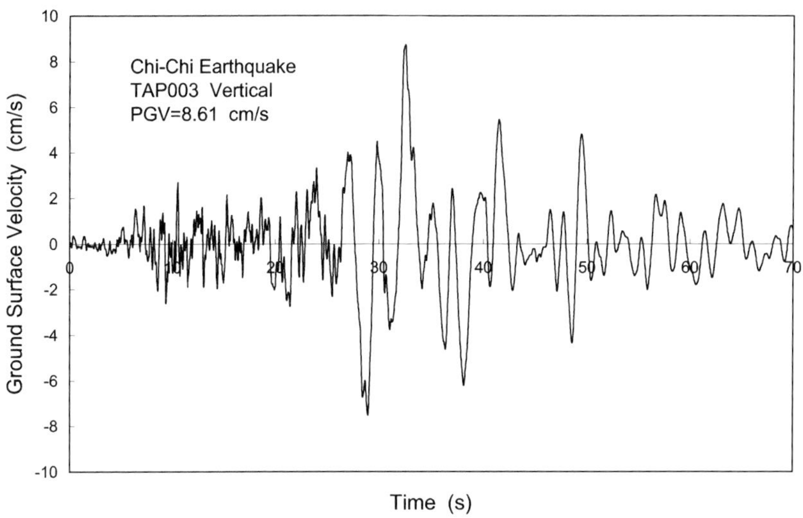


(b)

Fig. 11.4. Histogram of vertical motion of TAP003 Station: (a) acceleration, (b) velocity, and (c) displacement.

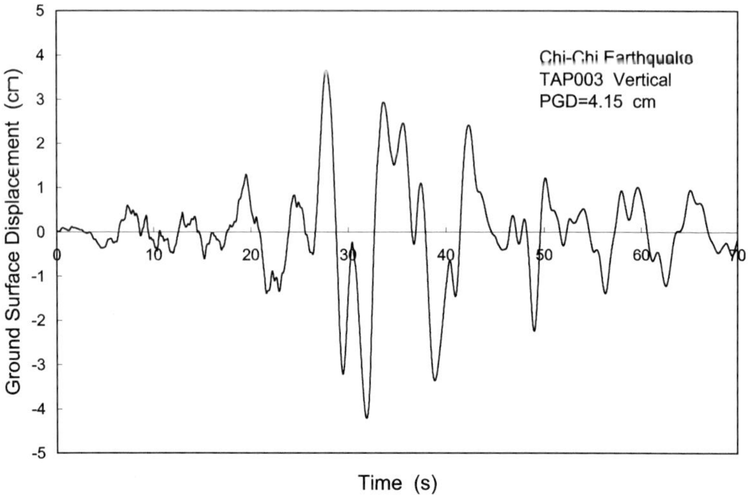


(c)

Fig. 11.4. (*Continued*).

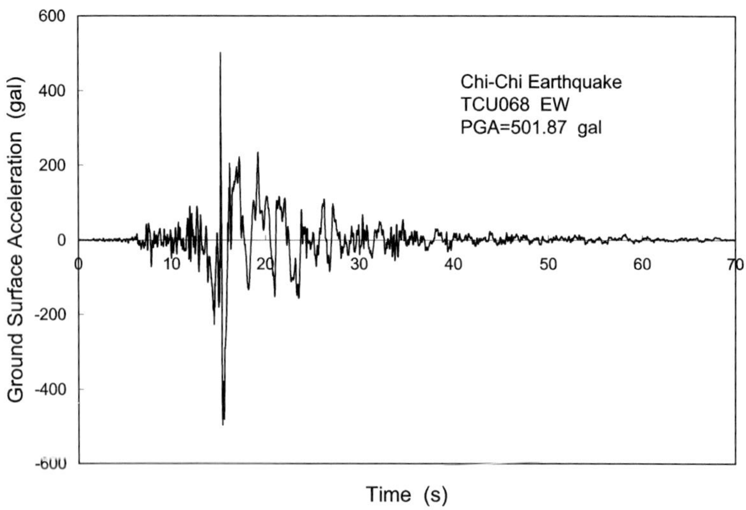


(a)

Fig. 11.5. Histogram of lateral (EW) motion of TCU068 Station: (a) acceleration, (b) velocity, and (c) displacement.

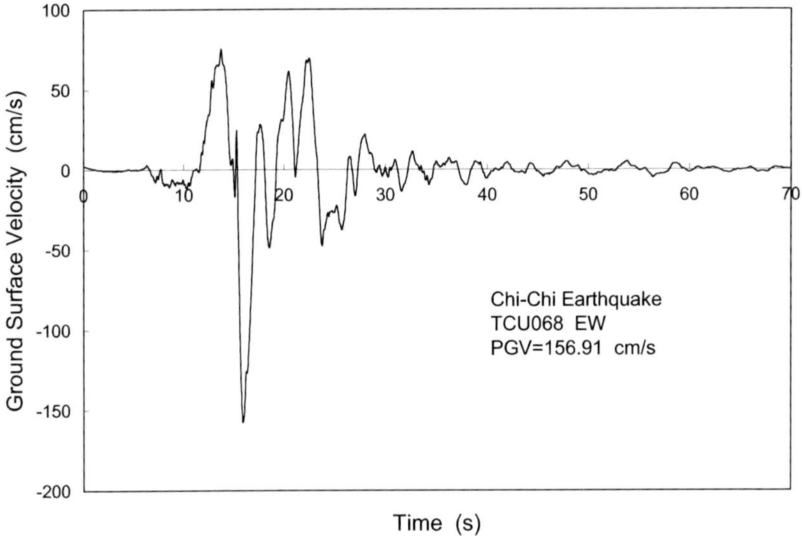


(b)

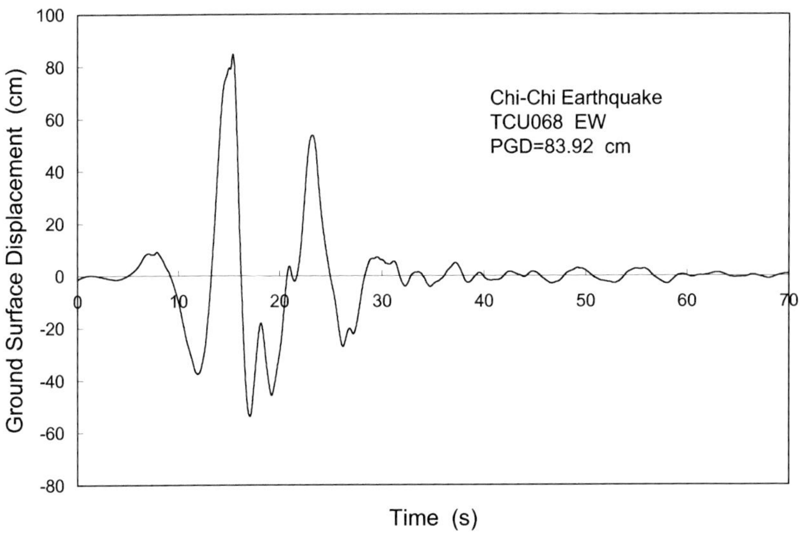


(c)

Fig. 11.5. (*Continued*).

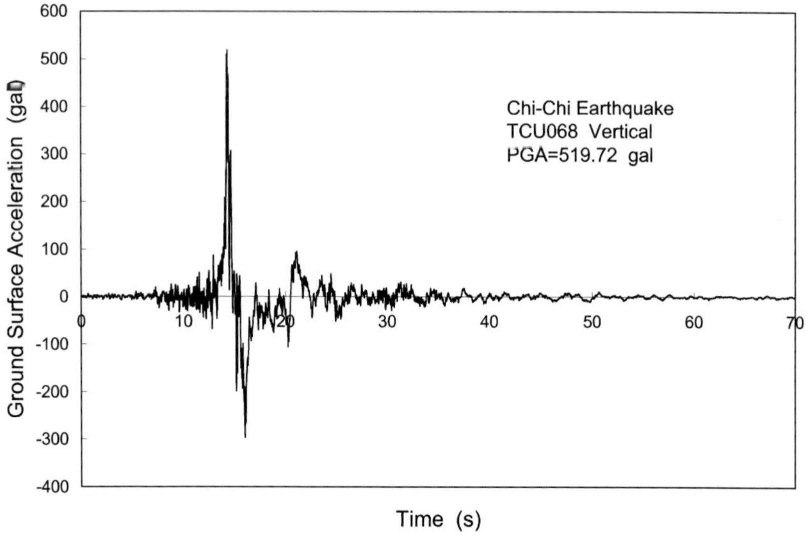


(a)

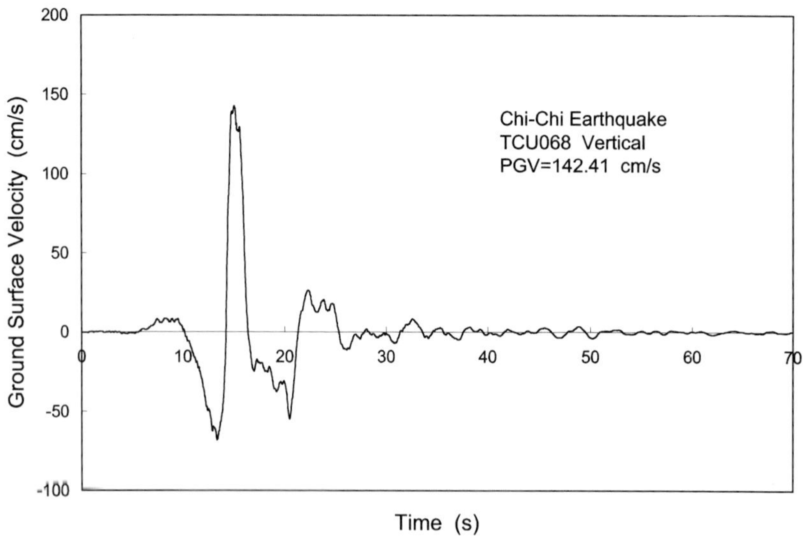


(b)

Fig. 11.6. Histogram of vertical motion of TCU068 Station: (a) acceleration, (b) velocity, and (c) displacement.

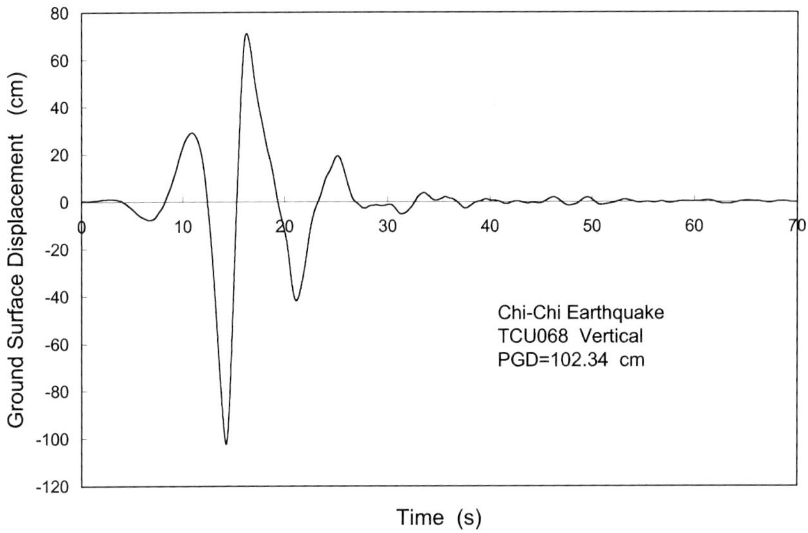


(c)

Fig. 11.6. (*Continued*).

As was stated previously, the seismic analysis of a train–railway system requires not only information on ground acceleration, but also ground displacement and velocity as part of the input source. Since only ground accelerations were recorded for most earthquakes, the ground velocity will be obtained through integration of the acceleration record with *base-line correction* and *Ormsby filtering* for eliminating the lower frequencies. Similarly, the ground displacement will be obtained from the velocity history with Ormsby filtering. The histograms of the ground motions computed in this way for the EW and vertical components of the far-field station TAP003 have been plotted as parts (b) and (c) in Figs. 11.3 and 11.4, and those of the near-fault station TCU068 in Figs. 11.5 and 11.6. The histograms for the El Centro and Northridge Earthquakes are not shown since they are well known. In these figures the peak ground acceleration (PGA), peak ground velocity (PGV) and peak ground displacement (PGD) are also indicated.

The ballast inserted between the rails and roadbed may display certain nonlinear behavior under severe ground motions, especially in the horizontal direction, which may affect the stability of moving trains, but has not been fully explored. According to *Provision DS-804* of Germany (Deutsche Bundesbahn, 1993), the horizontal resistance-relative displacement relation of the ballast can be approximately regarded as linear if the relative displacement of the ballast with respect to the bridge in the horizontal direction is less than 2 mm. For the four ground motions considered in this chapter, the maximum relative displacement between the rails and the bridge in the lateral direction caused by the ground motion is less than 2 mm if the PGA is less than 80 gal, as shown in Fig. 11.7. Furthermore, it is quite possible that a well-designed bridge remains linearly elastic when subjected to ground motions with a PGA of up to 80 gal. For the reasons stated, as well as for simplification, the lateral PGAs of all the four ground motions will be limited to 80 gal, in order not to violate the assumption of linearity for tracks and structures.

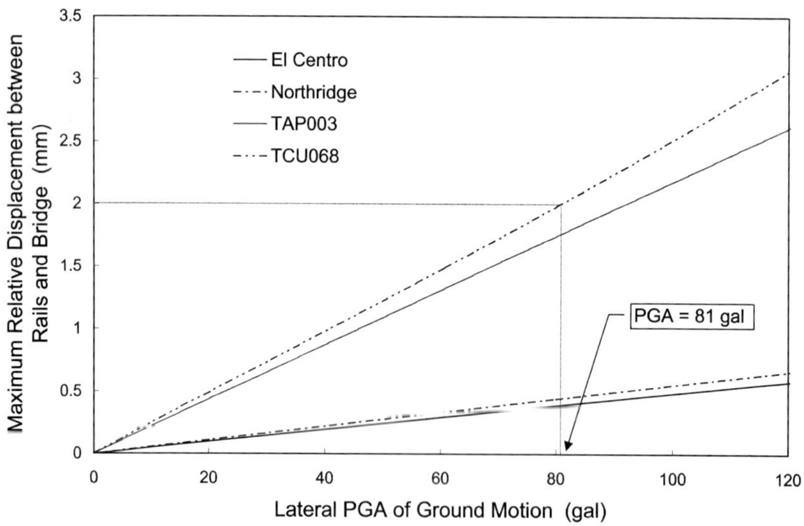


Fig. 11.7. Maximum relative displacement of ballast under the action of the four ground motions.

## 11.6. Train Resting on Railway Bridge under Earthquake

Throughout the numerical studies, the train car is assumed to be of the SKS Series 300 type and will be represented by the model briefly summarized in Sec. 11.2, or given in more details in Sec. 10.3, together with the fundamental properties listed in Table 10.1. The time increment $\Delta t$ used in the incremental analysis is 0.005 s, which is smaller than that used for recording the ground motions. It implies a frequency of $f = 1/\Delta t = 200$ Hz, higher than those implied by the wheels and rails. As a first test of the theory derived, we shall study the dynamic stability of a train car resting on a simply-supported bridge of 30 m in length shaken by the four earthquakes considered. The car is assumed to remain stationary on Track $A$ of the bridge with its first wheelset located at the position $x = 25$ m prior to the earthquake (see Fig. 11.8). As was stated previously, all the four ground motions are normalized to have a lateral PGA of 80 gal.

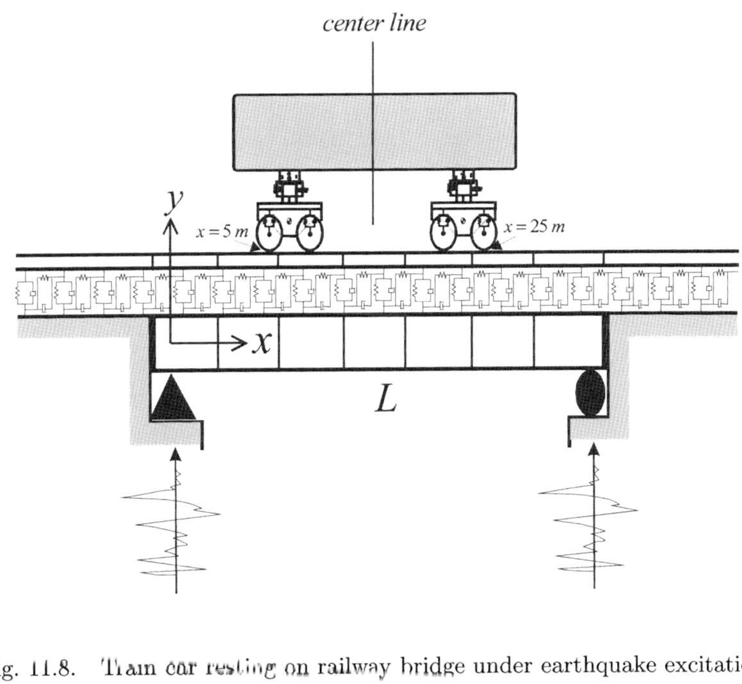


Fig. 11.8. Train car resting on railway bridge under earthquake excitation.

### 11.6.1. *Responses of Bridge and Train Car*

Both the bridge and train car are to vibrate from their static equilibrium (deformed) positions when the earthquake occurs. The responses computed from the bridge and train car subjected to the lateral (EW) and vertical excitations of the TAP003 Station were plotted in Figs. 11.9 and 11.10, respectively, along with those due to a lateral (EW) excitation of the same station only. As the car has been parked symmetrically on the bridge (see Fig. 11.8), for which no yawing or pitching vibrations will be induced on the car, the results for the yawing and pitching responses were zero and just skipped from Figs. 11.9 and 11.10.

The following observations can be made from Fig. 11.9 for the bridge: (1) The absolute vertical response of the bridge is primarily determined by the vertical component of the ground motion, but the bridge displacement is very small compared with the ground displacement shown in Fig. 11.4(c). (2) The absolute lateral response of the

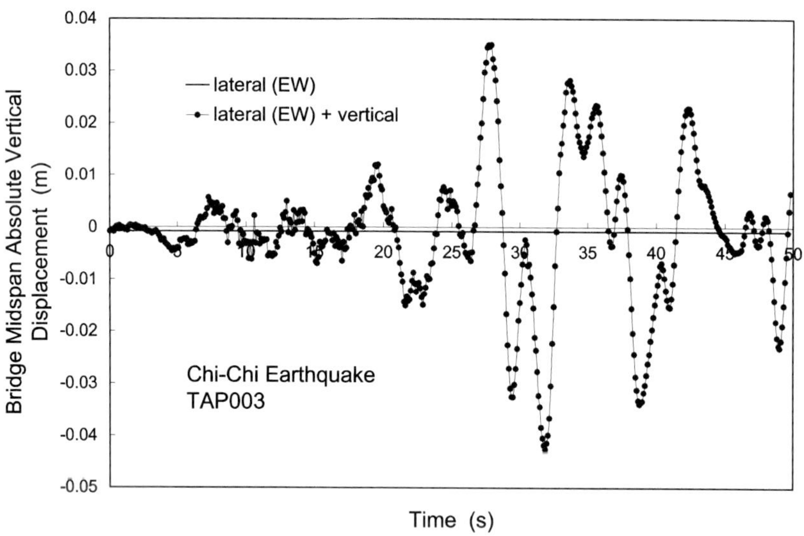


(a)

Fig. 11.9. Midspan responses of bridge carrying a train car at rest under the TAP003 excitation: (a) vertical, (b) lateral, and (c) torsional.

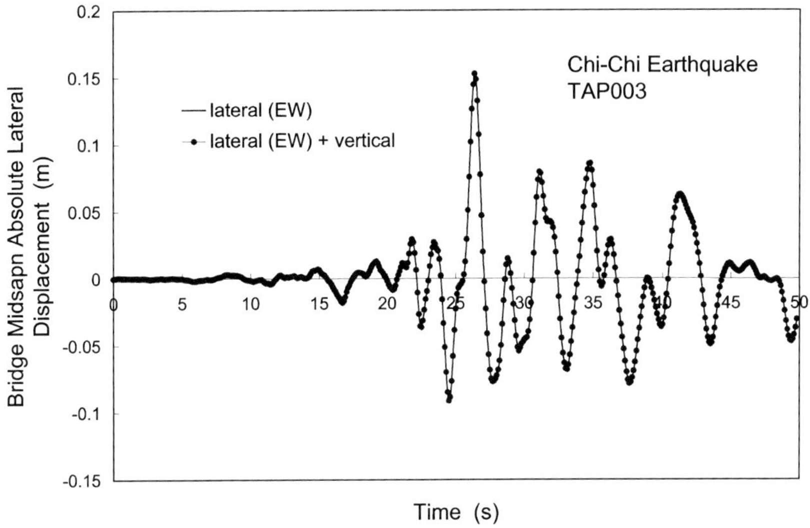


(b)

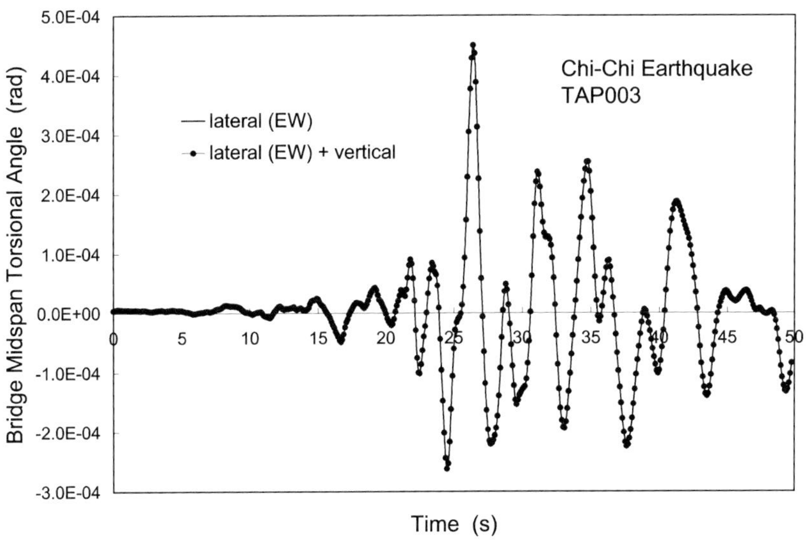


(c)

Fig. 11.9. (*Continued*).

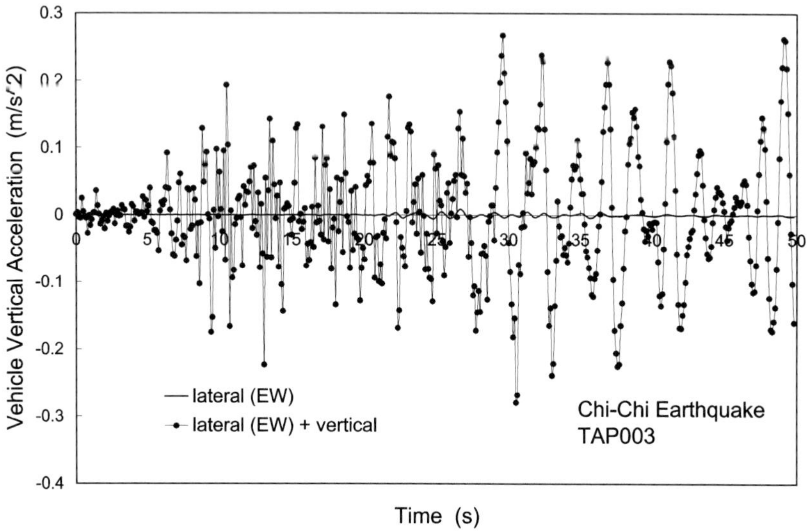


(a)

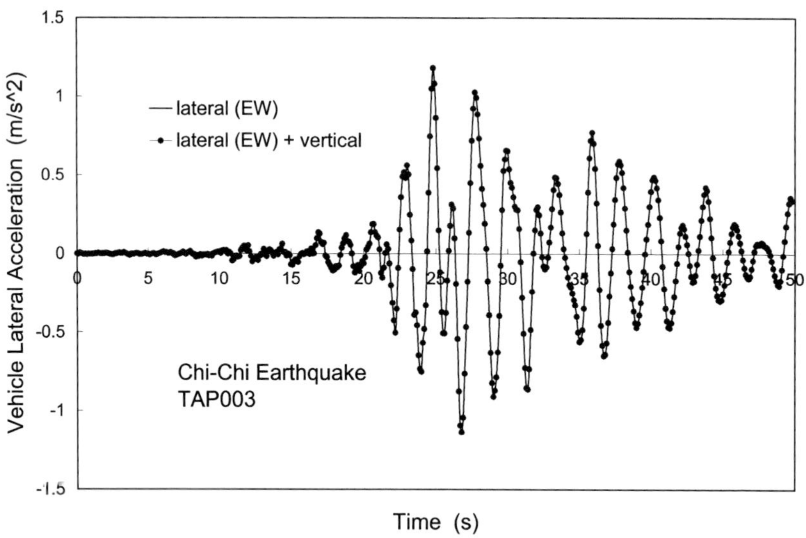


(b)

Fig. 11.10. Responses of train car resting on bridge under the TAP003 excitation: (a) vertical, (b) lateral, and (c) rolling.

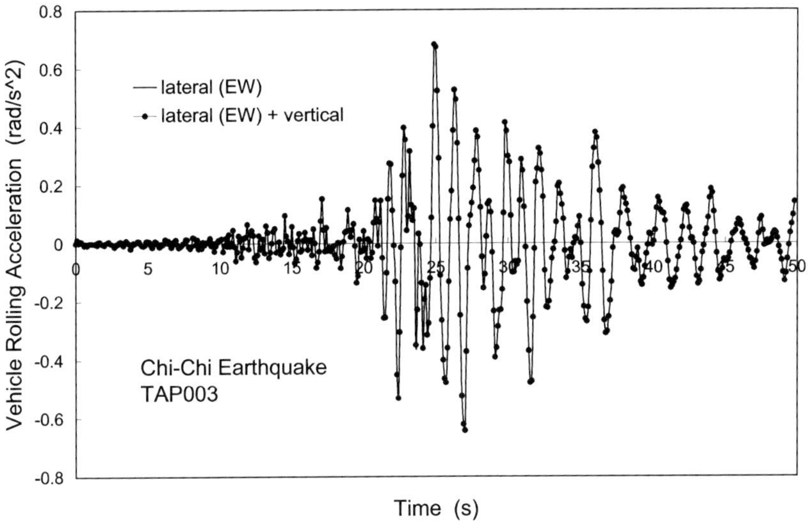


(c)

Fig. 11.10. (*Continued*).

bridge is dominated by the lateral (EW) component of the ground motion, but its magnitude is very small compared with that of the ground. (3) The torsional response of the bridge depends primarily on the lateral component of the ground motion due to the coupled lateral-torsional vibration of the bridge.

The following observations can be drawn from Fig. 11.10 for the train car: (1) The vertical acceleration of the cars is induced mainly by the vertical component of the ground motion, while the lateral and rolling accelerations of the car are induced primarily by the lateral component of the ground motion. (2) The maximum vertical acceleration of the car induced ($\approx$0.28 m/s$^2$) is less than that of the track ($\approx$1.57 m/s$^2$), which can be attributed to the isolation effect of the vertical suspension systems of the car. (3) The maximum lateral acceleration of the car induced ($\approx$1.18 m/s$^2$) is nearly as large as that of the track ($\approx$1.24 m/s$^2$), indicating that the lateral suspension systems of the car are not effective in isolating the car body from the lateral vibration of the track. The peak values for the track responses have not been shown here, which are available in Wu (2000).

The responses of the bridge and train car to the TCU068 excitations are all much larger than those to the TAP003 excitations, as shown in Figs. 11.11 and 11.12, due to greater intensity of the former. The *initial pulse phenomenon* is observed in all the acceleration responses of the bridge induced by the TCU068 excitations. It is worth noting that the lateral and rolling accelerations of the train car also display the initial pulse behavior, which were not observed in the case by the TAP003 excitations. Here, the maximum vertical and lateral accelerations of the car are equal to 3.92 m/s$^2$ and 3.21 m/s$^2$, respectively. Correspondingly, the maximum accelerations for the tracks are 7.42 m/s$^2$ and 4.8 m/s$^2$ (Wu, 2000). Again, it is indicated that the vertical suspension system of the car has a better efficiency in isolating the car body from the vibration of the track, compared with the lateral suspension system.

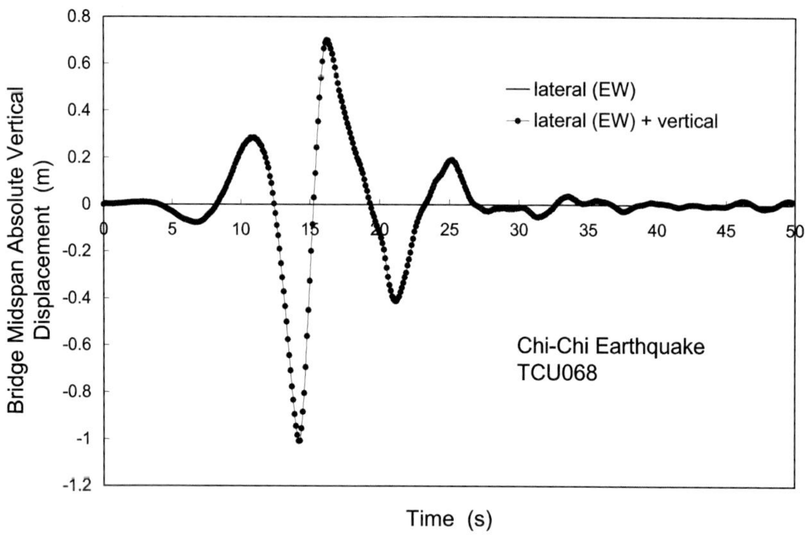


(a)

Fig. 11.11. Midspan responses of bridge carrying a train car at rest under the TCU068 excitation: (a) vertical, (b) lateral, and (c) torsional.

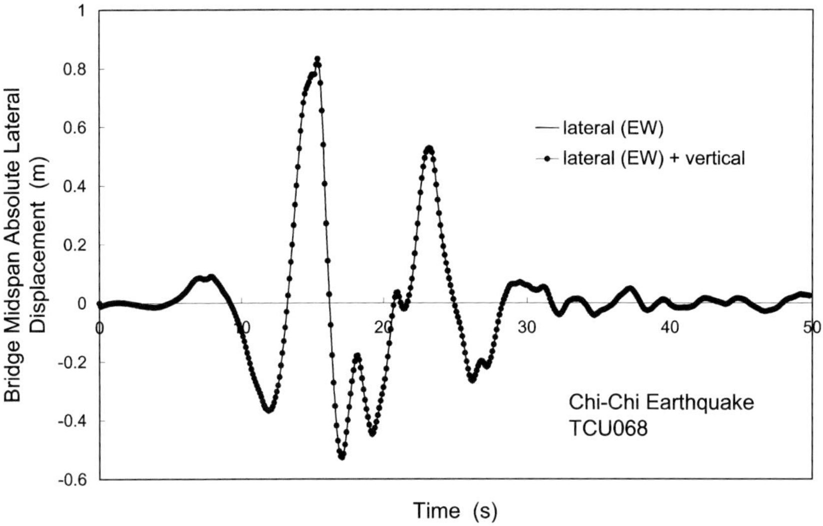


(b)

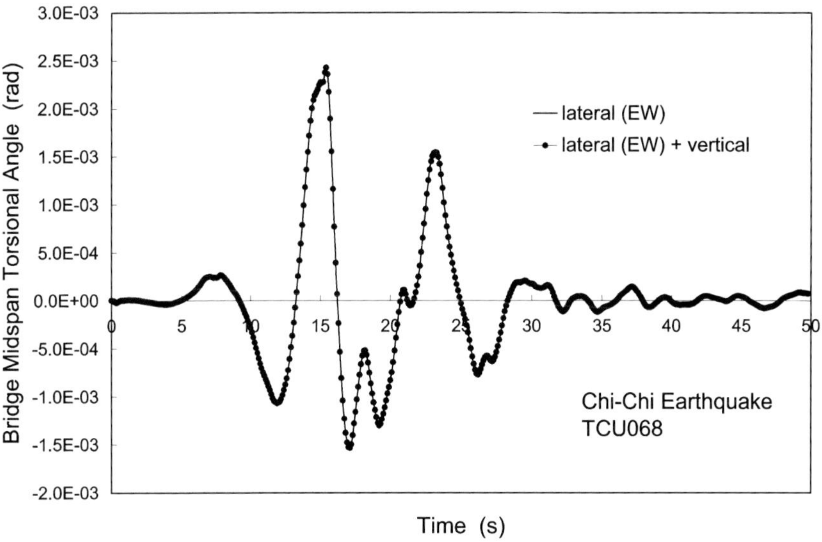


(c)

Fig. 11.11 (*Continued*).

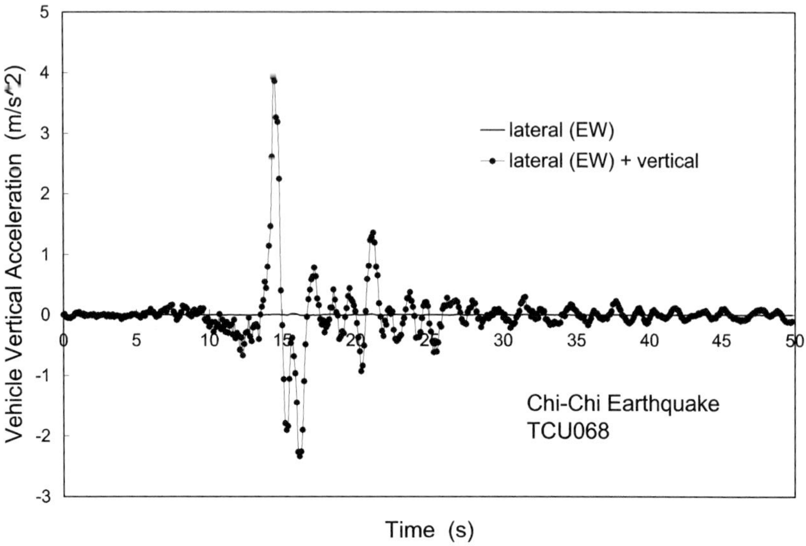


(a)

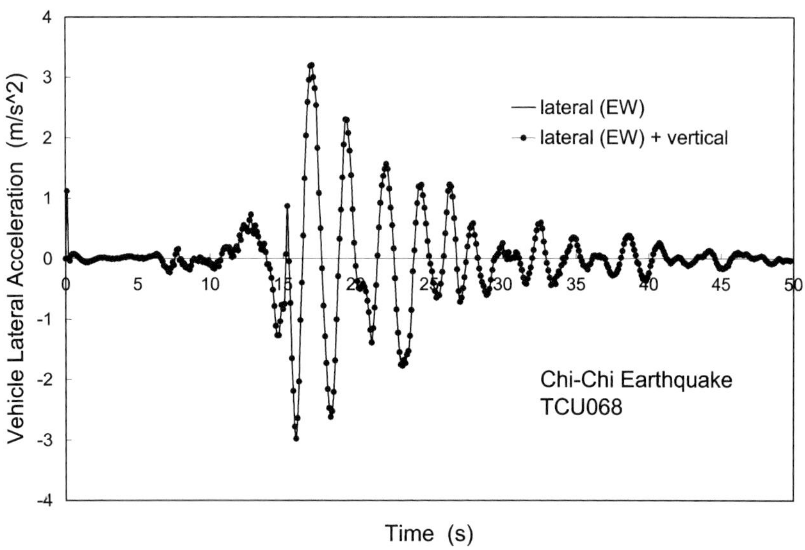


(b)

Fig. 11.12. Responses of train car resting on bridge under the TCU068 excitation: (a) vertical, (b) lateral, and (c) rolling.

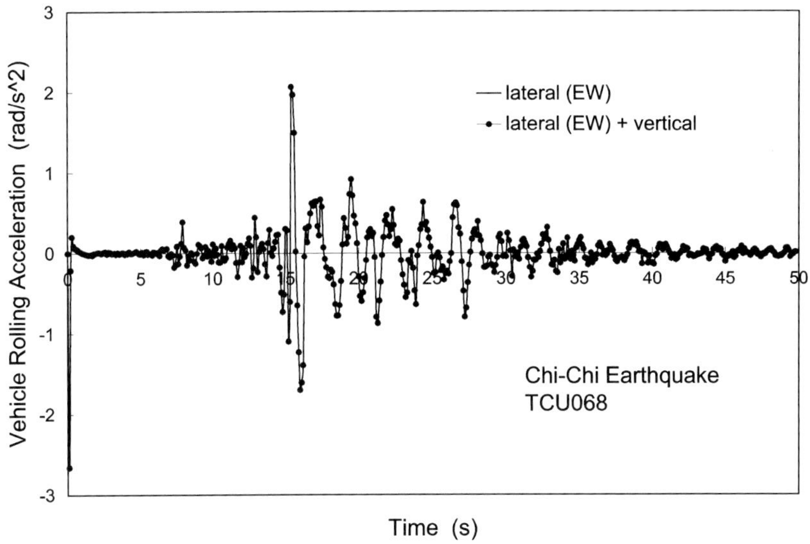


(c)

Fig. 11.12. (*Continued*).

### 11.6.2. *Contact Forces between Wheels and Rails*

The vertical and lateral contact forces induced by the four excitations between the wheelsets and rails have been shown in Figs. 11.13(a)–10.13(d), in which only the contact forces for the first and second wheels of the first wheelset of the car shown in Fig. 11.2(b) were presented. As can be seen, the vertical contact forces of the two wheels oscillate and cross each other during the earthquake. The oscillation appears to be most severe for the TAP003 excitation (see Fig. 11.13(c)), where the vertical contact force of the first wheel reaches a maximum of 77.6 kN and a minimum of 57.4 kN, and that of the second wheel reaches a maximum of 77.3 kN and a minimum of 58.0 kN, implying a fluctuation of 30% and 29%, respectively, with respect to the static axle load of 67 kN sustained by one wheel. The crossing behavior of the two wheels of the first wheelset in each figure is an indication of the occurrence of *rolling motion* during the earthquake, which is harmful to the stability of the train car. Factors

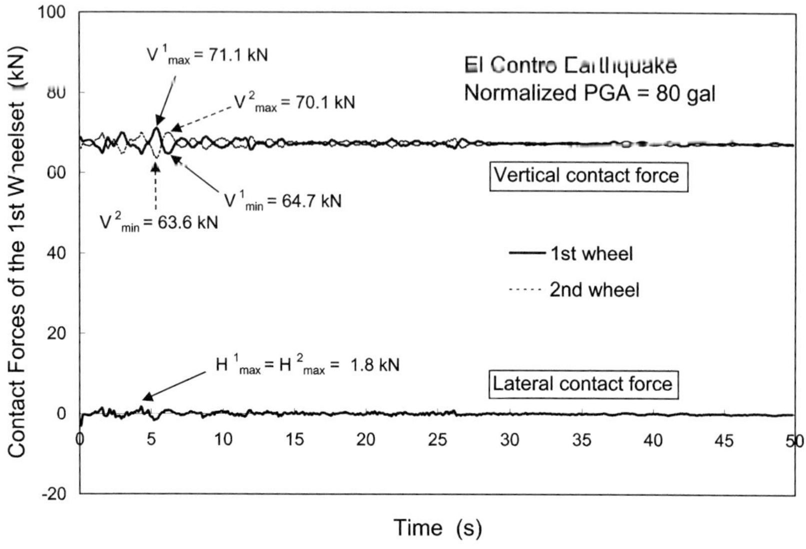


(a)

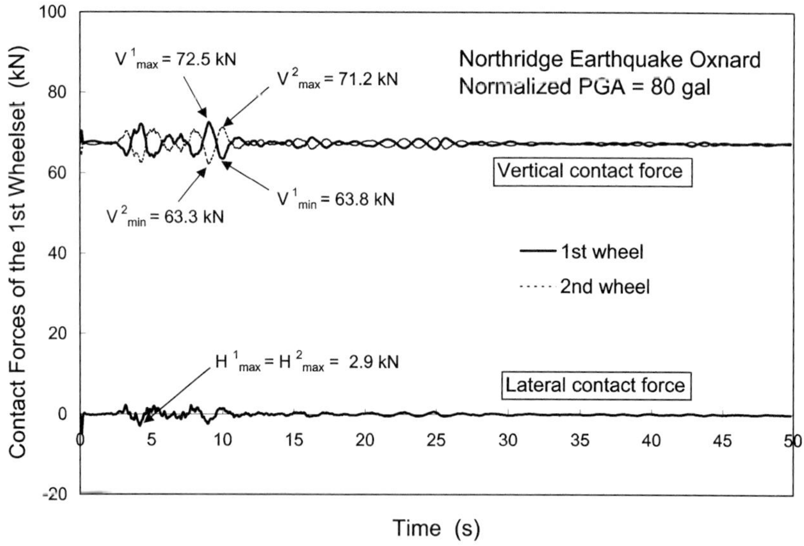


(b)

Fig. 11.13. Contact forces of the 1st wheelset of the train car under earthquake: (a) El Centro, (b) Northridge, (c) TAP003, and (d) TCU068.

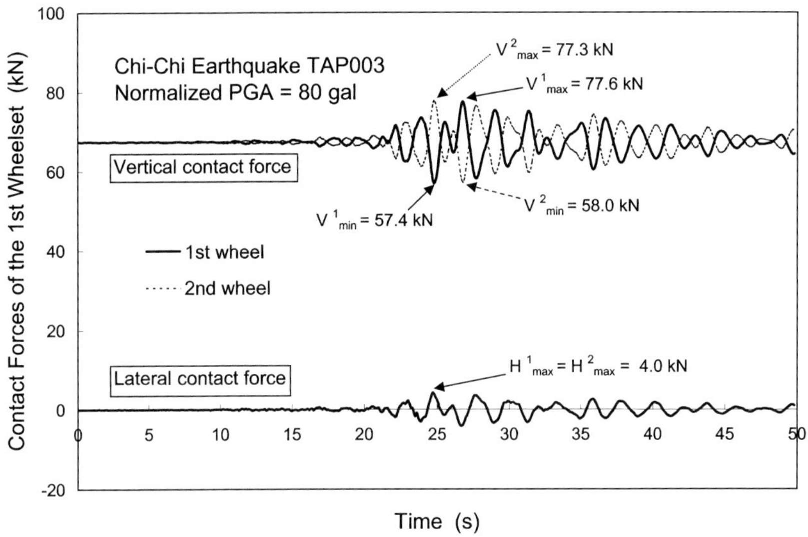


(c)

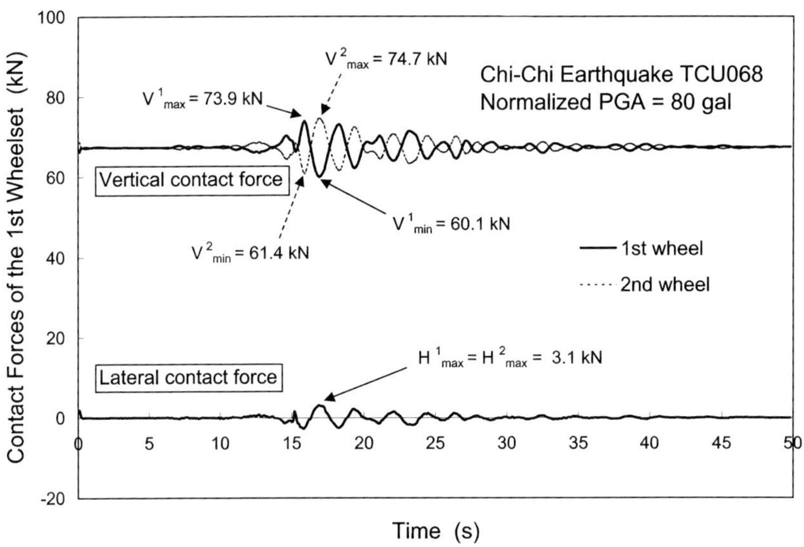


(d)

Fig. 11.13. (*Continued*).

that may affect the contact forces of the car during the earthquake include the suspension stiffness of the car, the lateral–torsional stiffness of the bridge, the intensity (PGA) and frequency intent of the earthquake. It should be added that the fourth wheelset shows the same behavior as that of the first wheelset due to symmetry, and that similar, but smaller, rolling responses exist for the second and third wheelsets, which are not shown.

### 11.6.3. *Maximum YQ Ratio for Wheelsets in Earthquake*

As shown in Fig. 11.2, a typical train car consists of four wheelsets, and each wheelset consists of two wheels, i.e., the left and right wheels. All the criteria presented in Sec. 10.13 can be used to assess the risk of derailment of each wheel, each wheelset or the entire train car. For instance, the SYQ ratio (*single wheel lateral to vertical force ratio*) is good for a single wheel, and the YQ ratio (*wheelset lateral to vertical force ratio*) can be used for a single wheelset. Both the SYQ and YQ ratios have been used as the indices by Wu (2000) in assessing the risk of derailment of train cars subjected to the same four earthquakes as in this study.

For the present purposes, we shall adopt only the YQ ratio in evaluating the stability of a train car resting on railway bridges shaken by earthquakes. This parameter can be defined as follows:

$$(YQ)_i = \left|\frac{H_{(2i-1)}}{V_{(2i-1)}}\right| + \left|\frac{H_{(2i)}}{V_{(2i)}}\right|, \tag{11.41}$$

where $i = 1$, 2, 3 or 4 for each of the four wheelsets, and $H_j$ and $V_j$, with $j = 1 \sim 8$, respectively, denote the lateral and vertical contact forces acting on a wheel (see Fig. 11.2(b)). It should be noted that the definition of the YQ ratio adopted here (Eq. (11.41)) is somewhat different from that given in Eq. (10.116). If $V_j \leqq 0$, the $j$th wheel encounters the *jump* condition, and the wheelset to which the $j$th wheel belongs is at a high risk of derailment. To prevent the wheelset from derailment, an upper limit must be set on the YQ ratio. However, the determination of the stability limit is not easy, which requires more analytical and experimental investigations. In this study, a value of YQ $= 1.5$ is set for the wheelset to remain *stable*, with no risk of derailment (Elkins and Carter, 1993). According to Elkins and Carter (1993) and Ma and Zhu (1998), a wheel will be safe and free of derailment (wheel climb), if the lateral ($H$) to vertical ($V$) force ratio (SYQ ratio) for the wheel does not exceed 1.0, namely, if SYQ $= |H/V| \leqq 1.0$.

Based on the above considerations, there exists a transition zone from YQ $= 1.5$ to 2.0, where a wheelset will be at a high risk of derailment. This transition zone will be referred to as *possible derailment zone* in this study. For the case when the YQ value computed is larger than 2.0 or when $V \leqq 0$, which implies the occurrence of derailment or jump with the wheelset, it will be regarded as the condition of *derailment*. Accordingly, the YQ value will be automatically set to 2.0, to distinguish it from the other two cases in the figures. Note that the actual derailment of a wheelset depends not only on the magnitude of the YQ ratio of the wheelset, but also on the lasting time of the YQ ratio exceeding the safety limit. Thus, it is conservative to assess the derailment risk of a wheelset using only the YQ ratio, as is done in this study, and to assess the stability of a train (car) using the maximum YQ ratios computed for all the wheelsets of the train (car).

Figure 11.14 shows the time-history plot of the maximum YQ ratio for the four wheelsets of the train car under the four ground motions. As can be seen, the peak YQ ratio computed for the TAP003 excitation is the largest among the four excitations considered, in consistence with the observation made in Sec. 11.6.2 for the contact forces. In addition, the YQ ratios computed for the TAP003 and TCU068 excitations are generally larger than those for the El Centro and Northridge Earthquakes. It can be observed that all the YQ ratios computed for all the four excitations fall well below the safety limit of 1.5, indicating that the four wheelsets of the train car do not exhibit any risk of instability under the four ground motions specified. Therefore, it is concluded that a train initially at rest on the bridge will remain stable for the four earthquakes with a PGA of 80 gal, as long as the bridge remains fully elastic. Here, one should

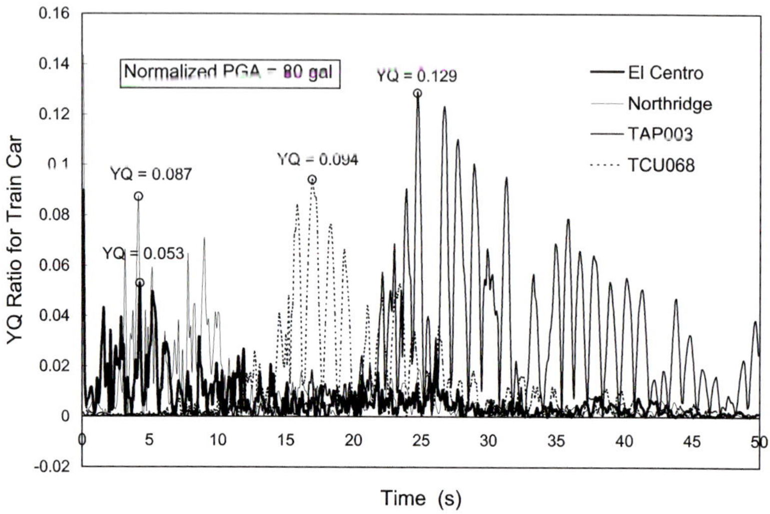


Fig. 11.14. Time history of the YQ ratio for the train car.

not forget that in reality the PGA level can be much higher for a near-fault motion than for a far-field motion.

### 11.6.4. *Stability of an Idle Train under Earthquakes of Various Intensities*

The railway engineers may be interested in under what PGA level of excitation, an initially idle train car begins to lose its stability. For the case of lateral excitations only, the maximum YQ ratios computed for the wheelsets of the train car under the four target ground motions with various PGAs have been drawn in Fig. 11.15. As can be seen, the maximum YQ ratios computed for the four excitations increase in proportion to the lateral PGA of the ground motion. Besides, the maximum YQ ratios for all range of the PGAs considered, i.e., up to 80 gal, are much less than the allowable limit of 1.5, indicating that all the wheelsets of the train car will not exhibit any instability or derailment risk under the ground motions specified if the lateral PGA does not exceed 80 gal. Of interest is the fact that the YQ ratio for the TAP003 motion is the largest among the four

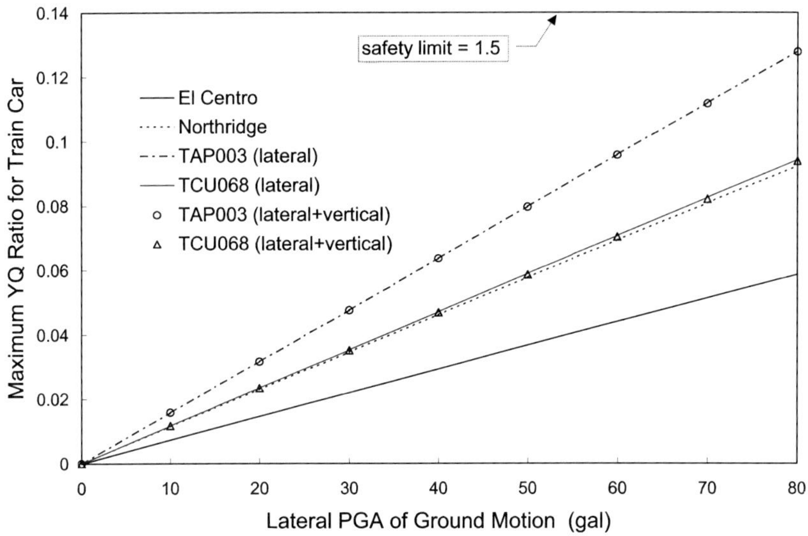


Fig. 11.15. Maximum YQ ratio for the train car versus lateral PGA of the ground motion.

ground motions of the same PGA, implying that the TAP003 motion is more detrimental to stability of the train car than the other three, given the same PGA level.

For the case of *simultaneous* lateral and vertical excitations, the maximum YQ ratios computed for the train car under the TAP003 and TCU068 ground motions have also been shown in Fig. 11.15, with the PGA of the vertical motion assumed to be proportional to that of the lateral one. As can be seen, the maximum YQ ratio computed of the train car under the action of both the lateral and vertical motions appears to be nearly the same as that under the lateral motion only, indicating that the vertical ground motion has little influence on the stability of trains initially resting on a railway bridge. An interpretation for this is that the vertical damping mechanism of the car has been given sufficient time to dissipate the vibrational energy, since the car is just "staying" on the bridge. The same is not true for cars traveling over the bridge, for which the acting time is very short.

## 11.7. Trains Moving over Railway Bridges under Earthquakes

In this section, we shall study the stability of a train traveling over a bridge shaken by earthquakes. The train is assumed to consist of 15 identical cars, which will pass through the bridge at constant speeds.

### 11.7.1. *Responses of Bridge and Train Car*

Both the TAP003 and TCU068 ground motions are used as the input excitations. For each of the two ground motions, the train is assumed to pass through the bridge at the speeds of 60 m/s (= 216 km/h) and 30 m/s (= 108 km/h). Because the time taken by the passage of the train over the bridge is much shorter than the duration of the acting time of the earthquake, the extreme state for the maximum response or instability of the train to occur cannot be easily identified. This makes the present investigation rather difficult. Here, we suppose that either the lateral or vertical PGA of the ground motion occurs at the supports of the bridge at the instant when the train begins to enter the bridge. As to whether the lateral or vertical PGA should be used as the reference for determining the ground motions depends on the characteristics of the geological conditions in the vicinity of the bridge considered. For the present purposes, we shall select the instant at which the lateral PGA occurs as the instant for the train to enter the bridge for the TAP003 motion, and the instant at which the vertical PGA occurs for the TCU068 motion.

The midspan responses of the bridge subjected simultaneously to the passage of the train and the TAP003 excitation have been plotted with respect to the nondimensional time parameter $vt/L$ in Fig. 11.16. As was expected, the response of the bridge is induced mainly by the earthquake excitation, that is, the vertical and lateral/torsional responses are caused mainly by the vertical and lateral components of the earthquake, respectively. The effect of moving vehicles on the bridge response is quite small, as can be confirmed by the fact that the results computed for the two cases $v = 30$ m/s and 60 m/s coincide generally with each other, if the time parameter

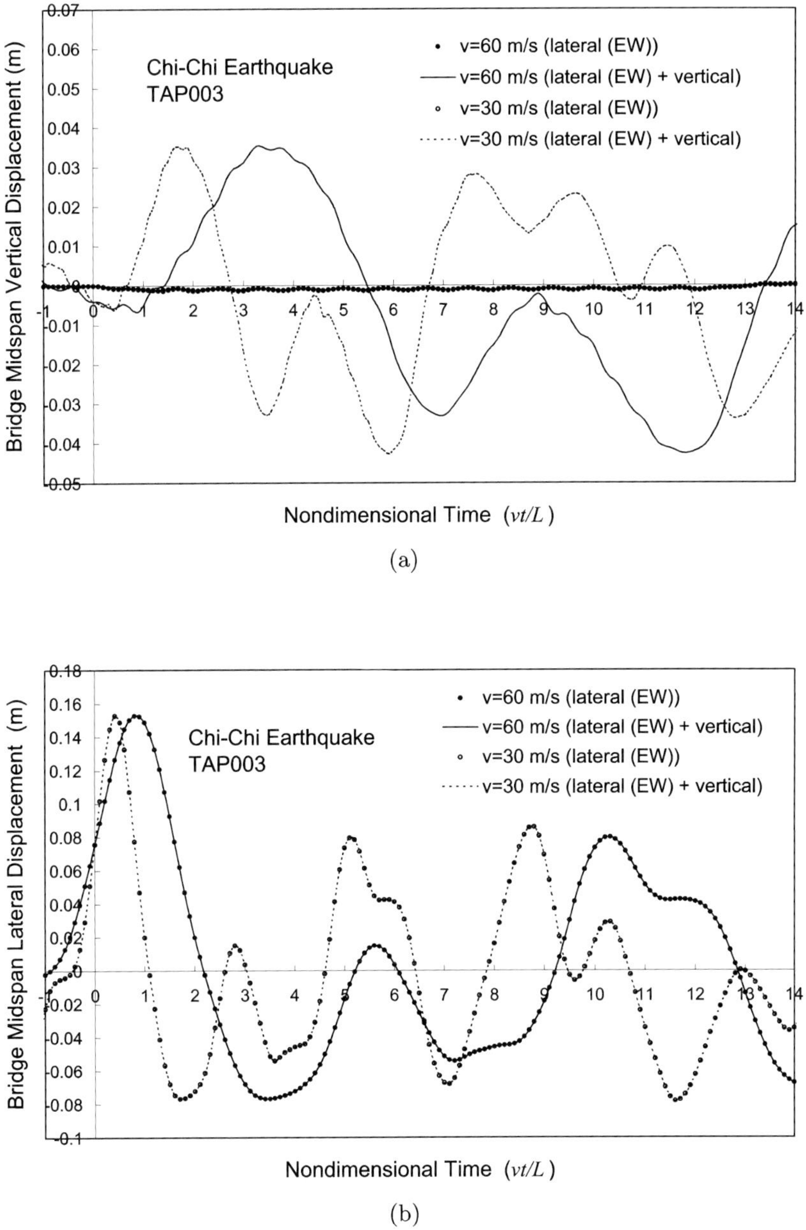


Fig. 11.16. Midspan responses of bridge to moving train under the TAP003 excitation: (a) vertical, (b) lateral, and (c) torsional.

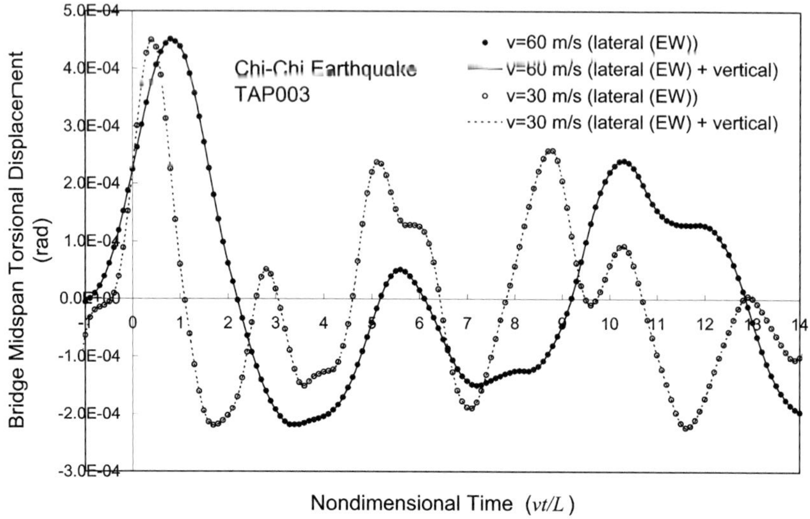


(c)

Fig. 11.16. (*Continued*).

$vt/L$ for the horizontal coordinate is transformed to real time $t$. This means that the effect of moving trains can be neglected in computing the bridge response to earthquake excitations.

In Fig. 11.17, the responses of the first (leading) car of the train under the same earthquake excitations have been plotted with respect to the nondimensional time parameter $vt/L$. By comparing the results obtained for bi-directional excitations with those for uni-directional (EW) excitations, it is observed that the vertical and pitching accelerations of the first train car are greatly enhanced by the presence of vertical excitation. In contrast, the lateral, rolling, and yawing motions of the train car are generated primarily by the lateral component of excitation and are independent of vertical excitation. Furthermore, the lateral, rolling, and yawing accelerations for the higher train speed ($v = 60$ m/s) are larger than those for the lower train speed ($v = 30$ m/s). The same is not true for the vertical and pitching accelerations, where the responses for the two accelerations for the higher speed ($v = 60$ m/s) are contrarily less than those for the lower speed ($v = 30$ m/s).

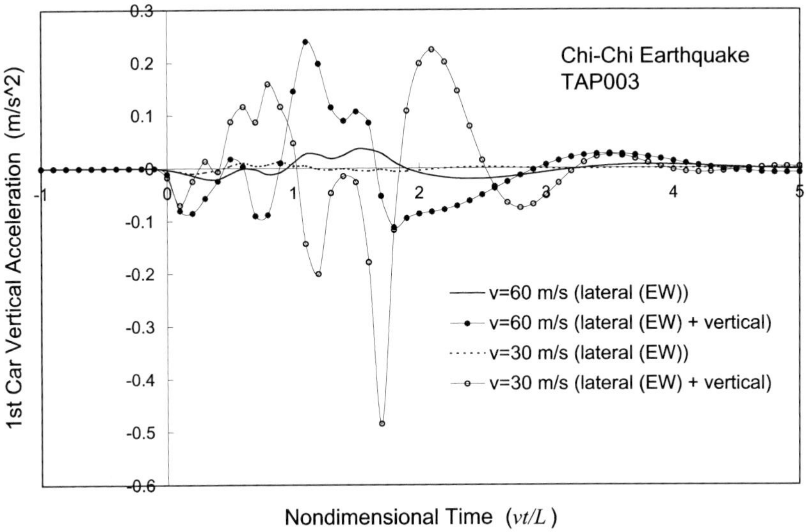


(a)

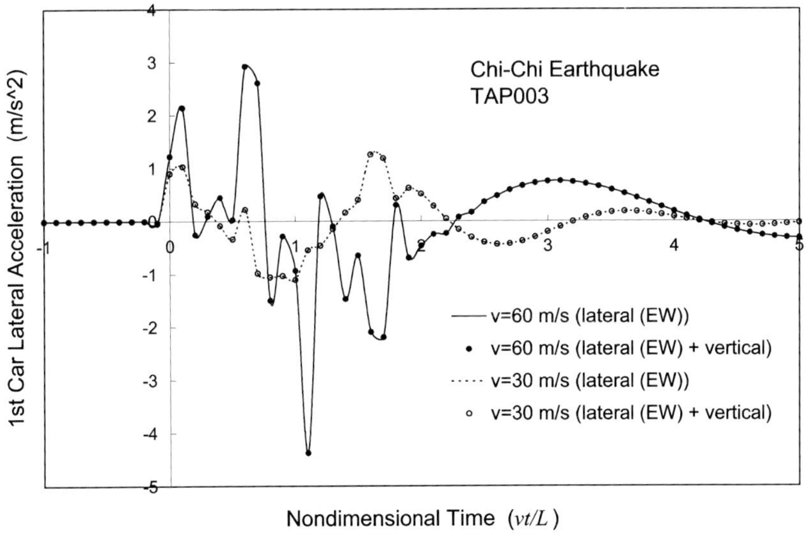


(b)

Fig. 11.17. Response of train car moving over bridge under the TAP003 excitation: (a) vertical, (b) lateral, (c) rolling, (d) yawing, and (e) pitching.

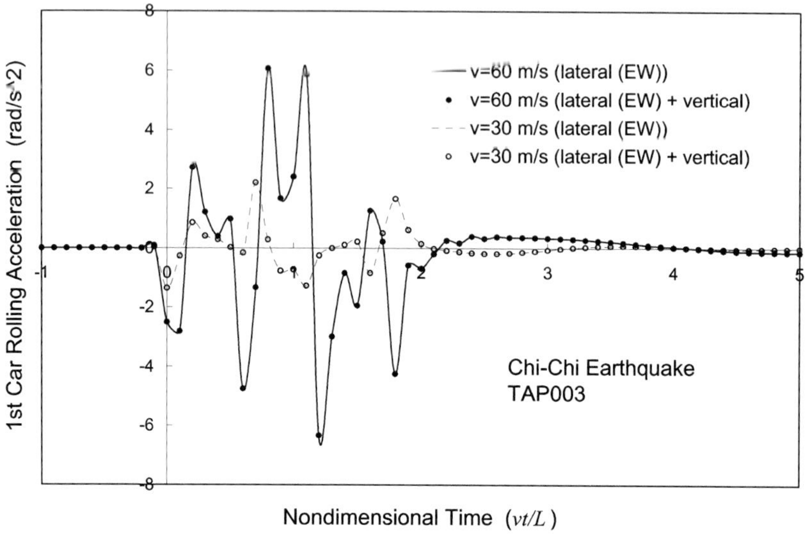


(c)

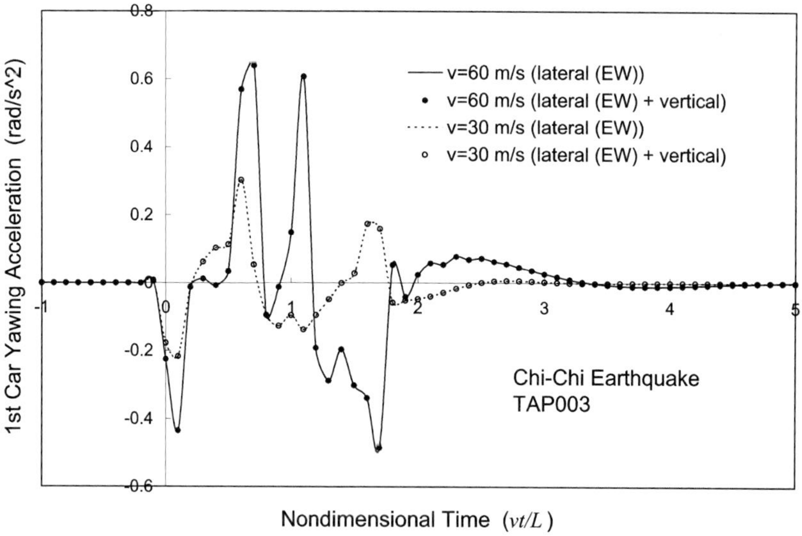


(d)

Fig. 11.17. (*Continued*).

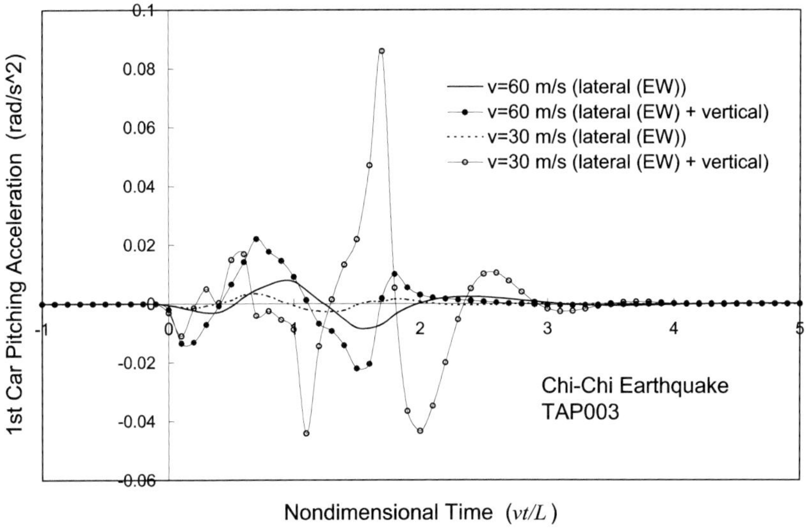


(e)

Fig. 11.17. (*Continued*).

As for the TCU068 excitation, the results computed for the bridge under a train moving at speeds $v = 60$ and 30 m/s have been plotted with respect to the nondimensional time parameter $vt/L$ in Fig. 11.18, and those for the first car of the train in Fig. 11.19. Clearly, the behaviors of the bridge and train car under the TCU068 excitation show a trend similar to those for the TAP003 excitation. There is no surprise that the responses of the bridge and vehicle to the TCU068 excitation are much larger than those to the TAP003 excitation. As can be seen from Fig. 11.19, under the TCU068 ground motion, all the acceleration responses of the vehicle for the higher speed ($v = 60$ m/s) appear to be much larger than those for the lower speed ($v = 30$ m/s), which is somewhat different from that observed for the TAP003 ground motion. It should be noted that for the present case with a rather strong earthquake, the results presented in Figs. 11.18 and 11.19 should not be regarded as a realistic representation of the actual behaviors of the train and bridge, owing to the fact that the bridge is assumed to remain as a linearly elastic

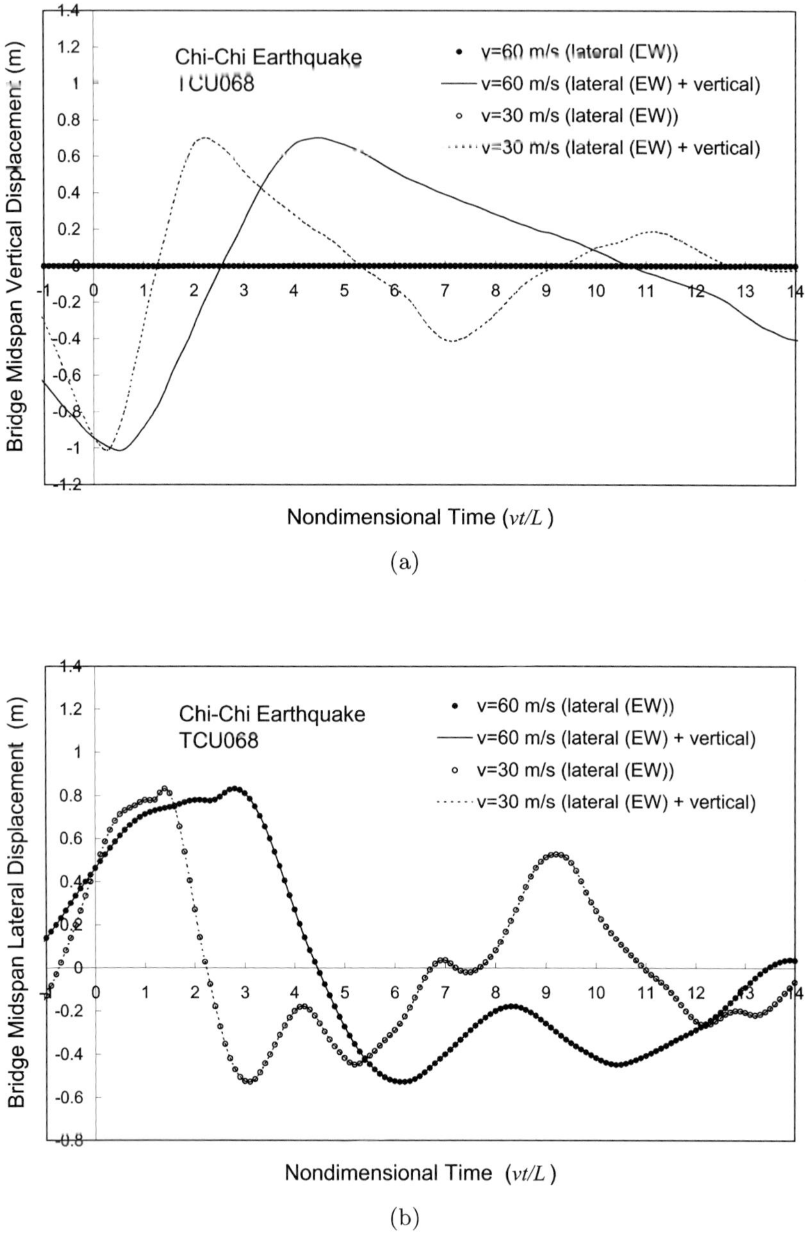


Fig. 11.18. Midspan bridge responses to the TCU068 excitation: (a) vertical, (b) lateral, and (c) torsional.

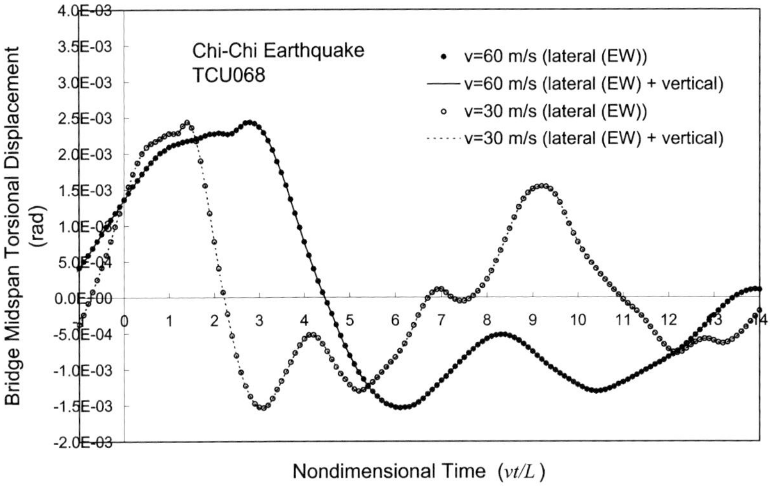


(c)

Fig. 11.18. (*Continued*).

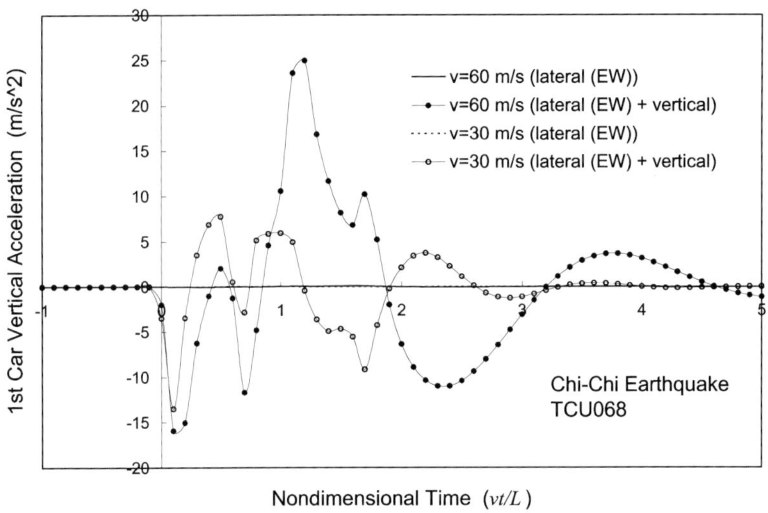


(a)

Fig. 11.19. Vehicle responses to the TCU068 excitation: (a) vertical, (b) lateral, (c) rolling, (d) yawing, and (e) pitching.

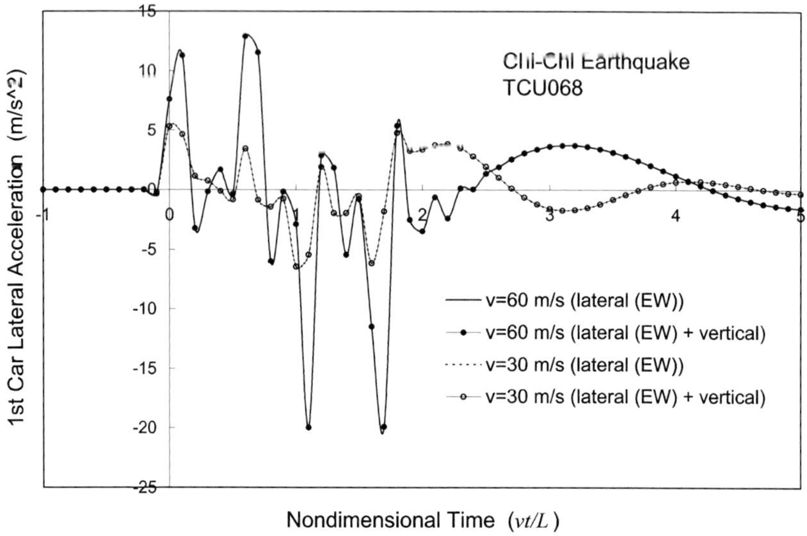


(b)

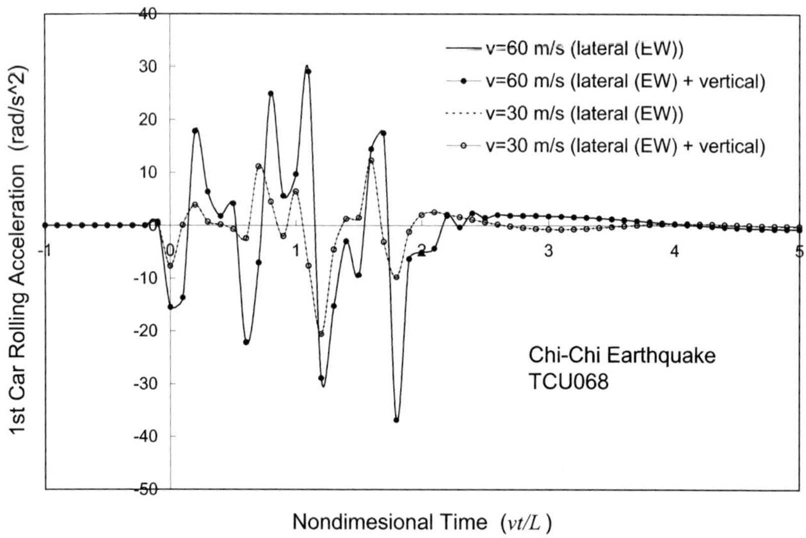


(c)

Fig. 11.19. (*Continued*).

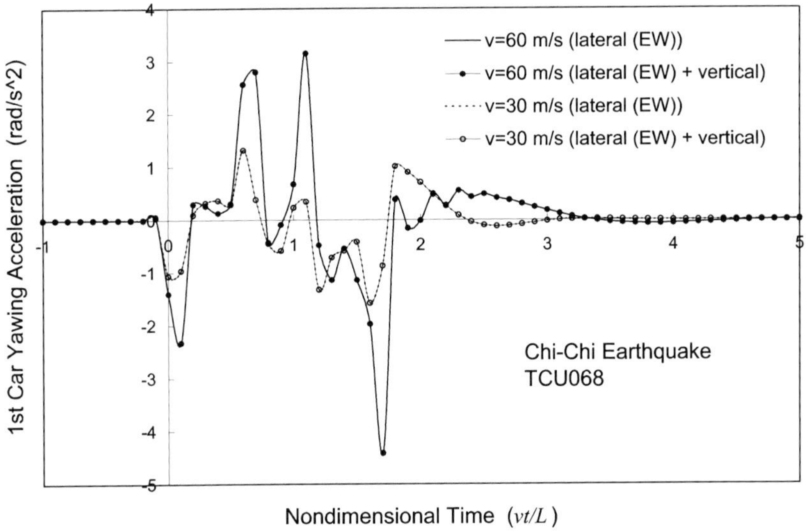


(d)

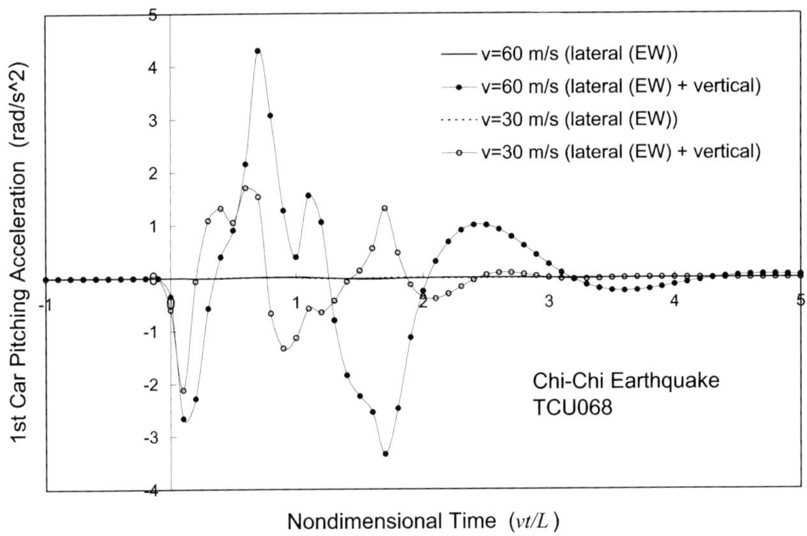


(e)

Fig. 11.19. *(Continued)*.

structure, regardless of the PGA level of the earthquake. In fact, the bridge will exhibit large inelastic deformations or even collapse under a severe earthquake, such as the one (TCU068) considered herein, or other ground motions with lateral PGA greater than 0.4 g.

### 11.7.2. *Maximum YQ Ratio for Moving Trains in Earthquake*

For the case of *uni-directional* (i.e., lateral) excitation, the train is assumed to pass through the bridge at a speed of 200 km/h (= 55.6 m/s). The bridge supports are also assumed to be excited laterally by the earthquake at the specified PGA level exactly at the instant when the train enters the bridge. The maximum YQ ratios computed for the whole train moving over the bridge subjected to the four earthquake excitations mentioned in the preceding section with PGAs of 80 and 25 gal have been plotted with respect to the nondimensional time parameter $vt/L$ in Figs. 11.20(a)–11.20(d). As can be seen from the case with PGA = 80 gal, the maximum YQ ratio for TCU068 exceeds evidently the stability limit of 1.5, indicating that derailment may occur in *near fault* areas, while the maximum YQ ratios computed for the other three excitations are far below (for El Centro and Northridge) or nearly equal to (for TAP003) the stability limit. In contrast, for the PGA of 25 gal, the maximum YQ ratios computed for the four ground motions are well below (for El Centro, Northridge and TAP003) or just slightly larger than (for TCU068) the stability limit of 1.5, indicating that generally no derailment may occur with the train under the action of the four ground motions for the PGA value specified.

### 11.7.3. *Stability Assessment of Moving Trains in Earthquake*

In earthquake-prone regions, it is important to have some feeling regarding the level of stability of a train running over bridges under some specific earthquake excitations. For *uni-directional* (i.e., lateral) excitations, the maximum YQ ratios computed for the whole train under the El Centro excitation with different PGAs (up to

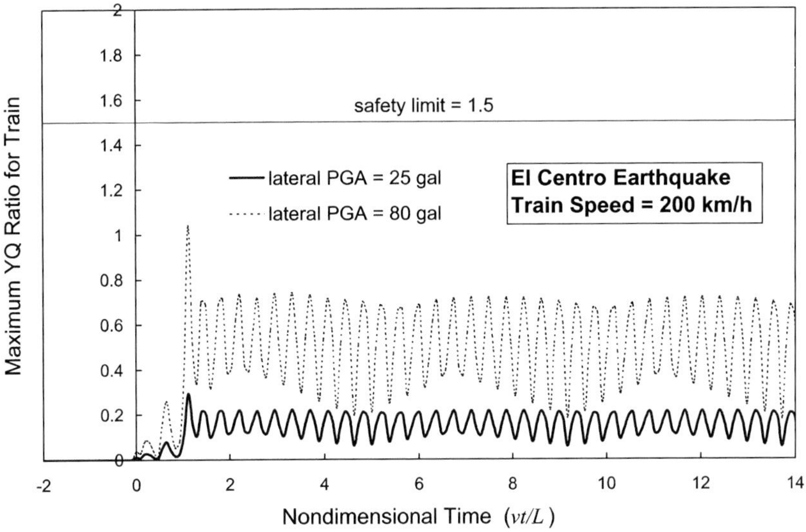


(a)

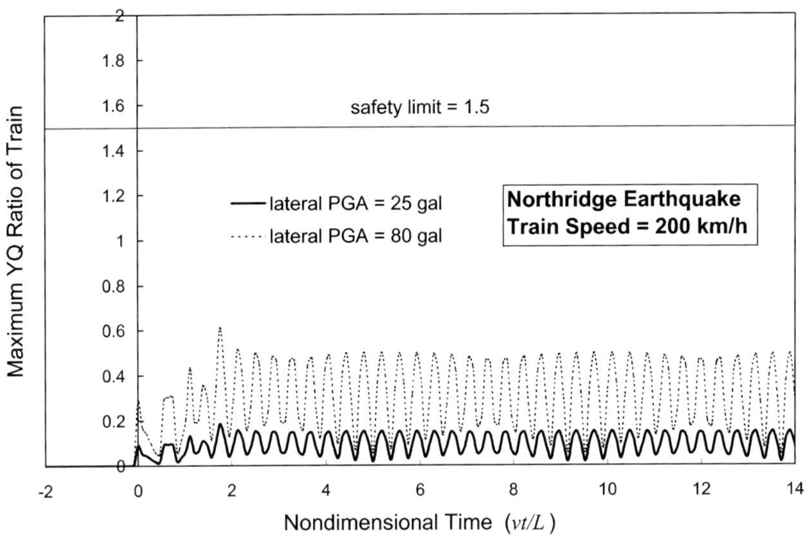


(b)

Fig. 11.20. YQ ratio for the 1st wheelset of the 1st car of the train under earthquake: (a) El Centro, (b) Northridge, (c) TAP003, and (d) TCU068.

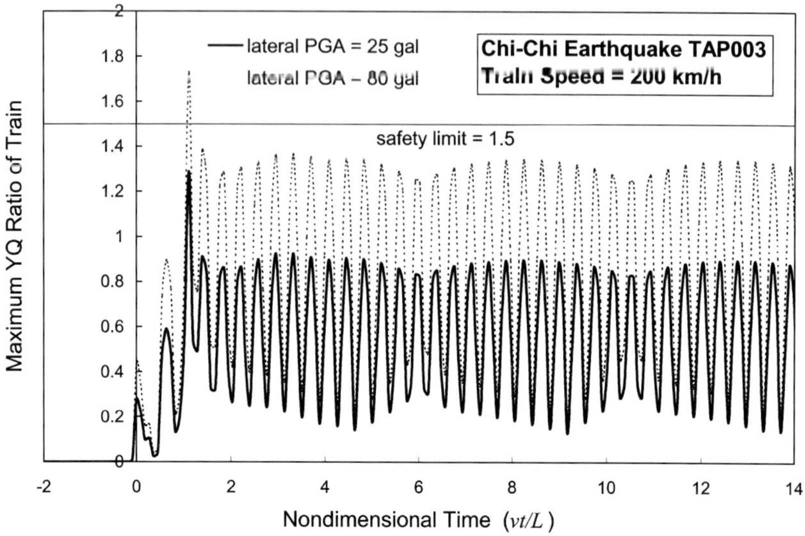


(c)

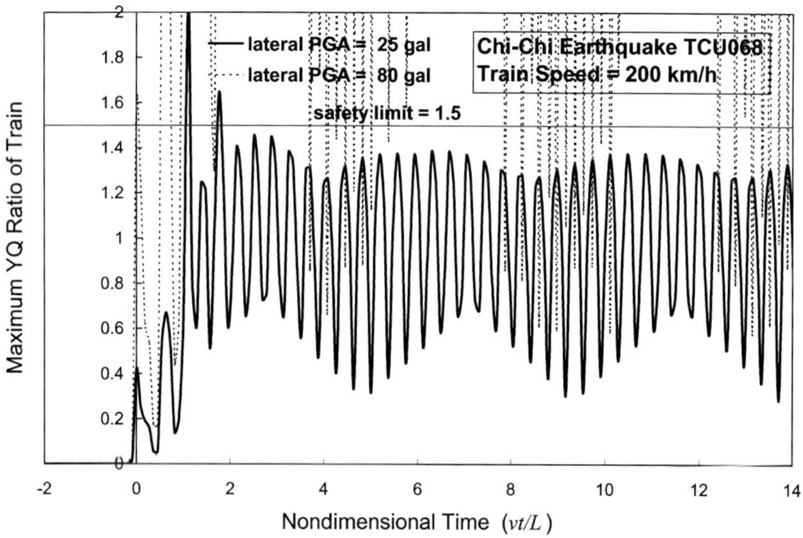


(d)

Fig. 11.20. (*Continued*).

80 gal) have been plotted with respect to the train speed (up to 200 km/h) in Fig. 11.21. As can be seen, the values of 1.5 and 2.0 were also imposed (see the vertical axis of Fig. 11.21(a)) to characterize the *safety* (with maximum YQ ratio $\leqq 1.5$), *possible instability* (with $1.5 <$ maximum YQ ratio $\leqq 2.0$) and *instability* (with maximum YQ ratio $> 2.0$) regions for the train. Similarly, the maximum YQ ratios computed for the Northridge, TAP003 and TCU068 excitations with different PGAs have been plotted with respect to the train speed in Figs. 11.21–11.24, with the different regions of stability identified in part (b) of the figures. In all these figures, the PGA values of 8 and 25 gal have been referred to as seismic zones of Levels 3 and 4, respectively, according to the classification by the *Central Weather Bureau of Taiwan.*

The following observations can be made from Figs. 11.21–11.24: (1) The maximum allowable speeds for stable running of the train are higher for smaller PGAs and lower for larger PGAs. (2) The ranges for stable running of the train under TAP003 and TCU068 are much narrower than those for the other two excitations, indicating that the train has a higher risk of derailment or instability when traveling over the bridge under the Chi–Chi Earthquake. (3) The train can move safely under El Centro and Northridge Earthquakes for all the speeds and PGAs considered. (4) If the lateral PGA of TAP003 is less than 30 gal, the train can cross the bridge safely at a speed of up to 200 km/h. The same is true for TCU068 if the PGA is less than 18 gal. (5) The train can move safely over the bridge without encountering any instability or derailment under TAP003 and TCU068 with a lateral PGA of up to 80 gal, if the train speed is kept below 122 and 85 km/h, respectively.

For *bi-directional* (i.e., lateral and vertical) excitations, the results obtained for the TAP003 and TCU068 ground motions have been plotted in Figs. 11.25 and 11.26, respectively. As can be seen, for the TAP003 excitation, the maximum allowable speeds for the train under bi-directional excitations are nearly the same as those for uni-directional excitation. In contrast, for the TCU068 motion, the maximum allowable speeds for the train under bi-directional excitations are significantly less than those under uni-directional

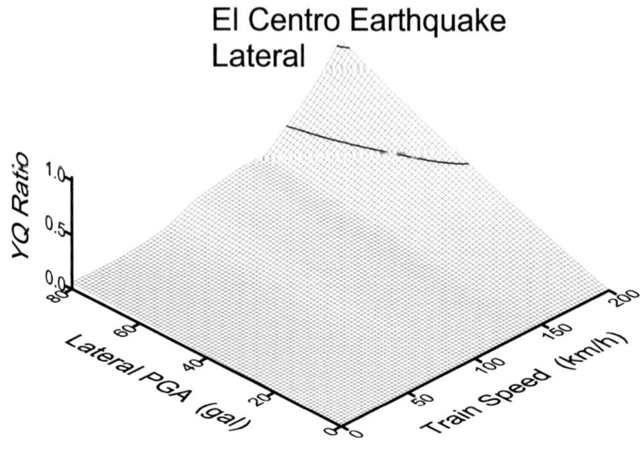


(a)

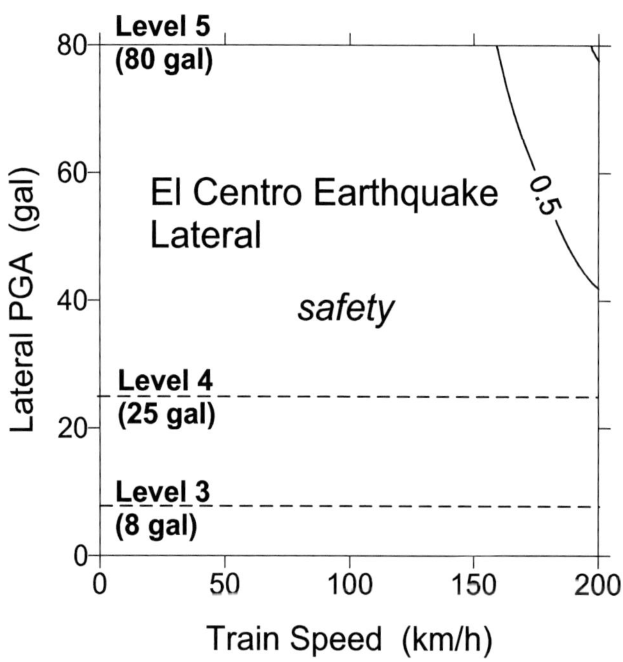


(b)

Fig. 11.21. Risk of derailment for El Centro Earthquake: (a) three-phase plot and (b) safety boundary.

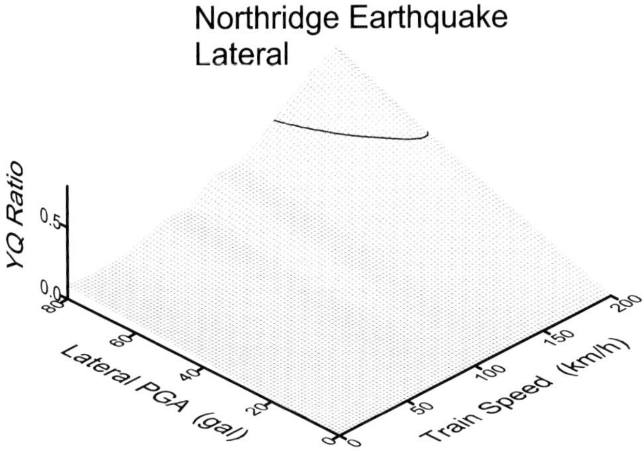


(a)

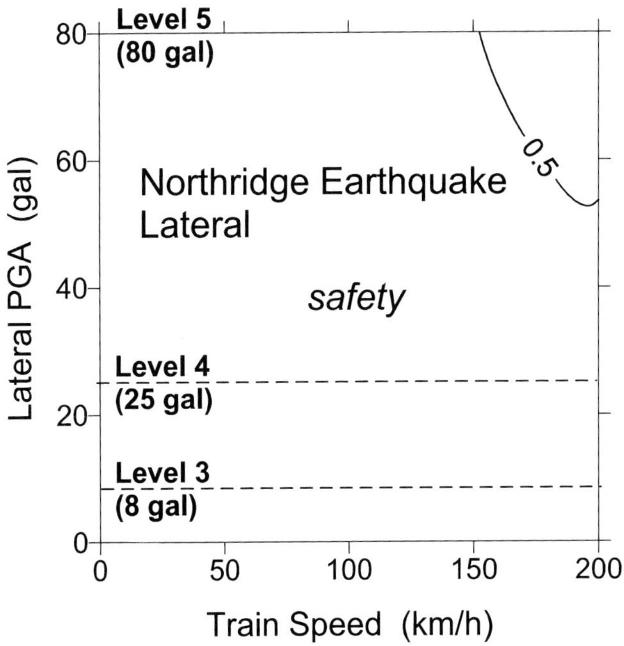


(b)

Fig. 11.22. Risk of derailment for Northridge Earthquake: (a) three-phase plot and (b) safety boundary.

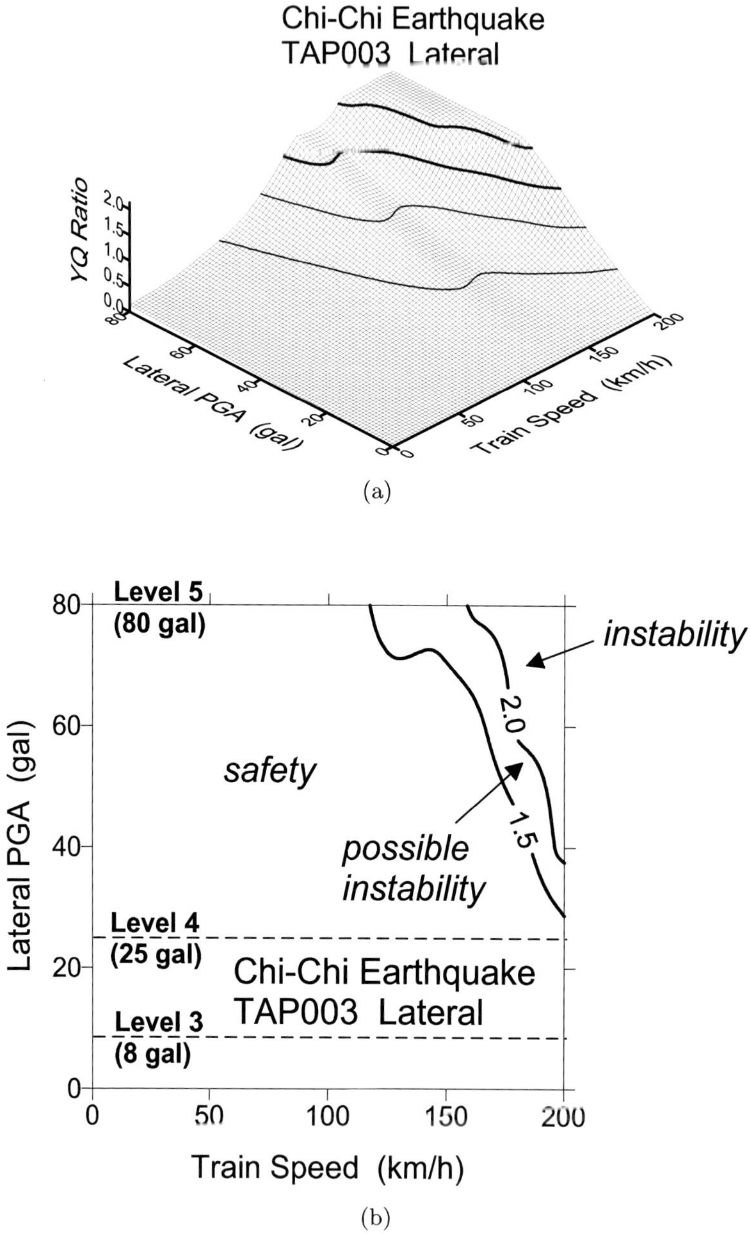


Fig. 11.23. Risk of derailment for TAP003 ground motion (lateral only): (a) three-phase plot and (b) safety boundary.

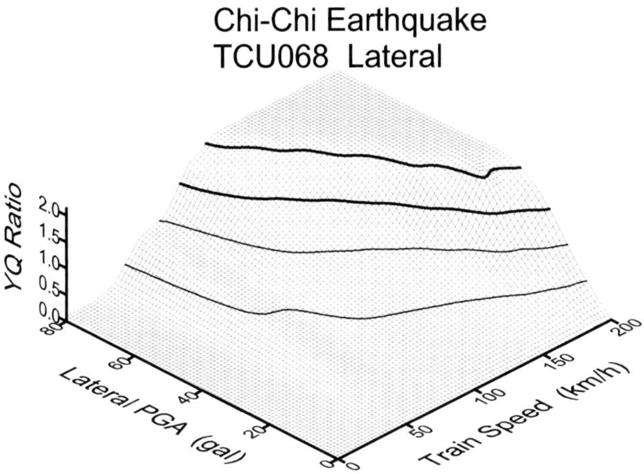


(a)

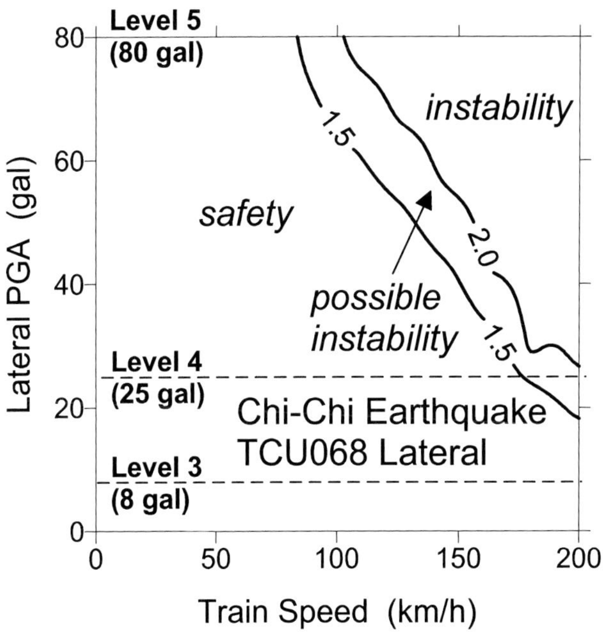


(b)

Fig. 11.24. Risk of derailment for TCU068 ground motion (lateral only): (a) three-phase plot and (b) safety boundary.

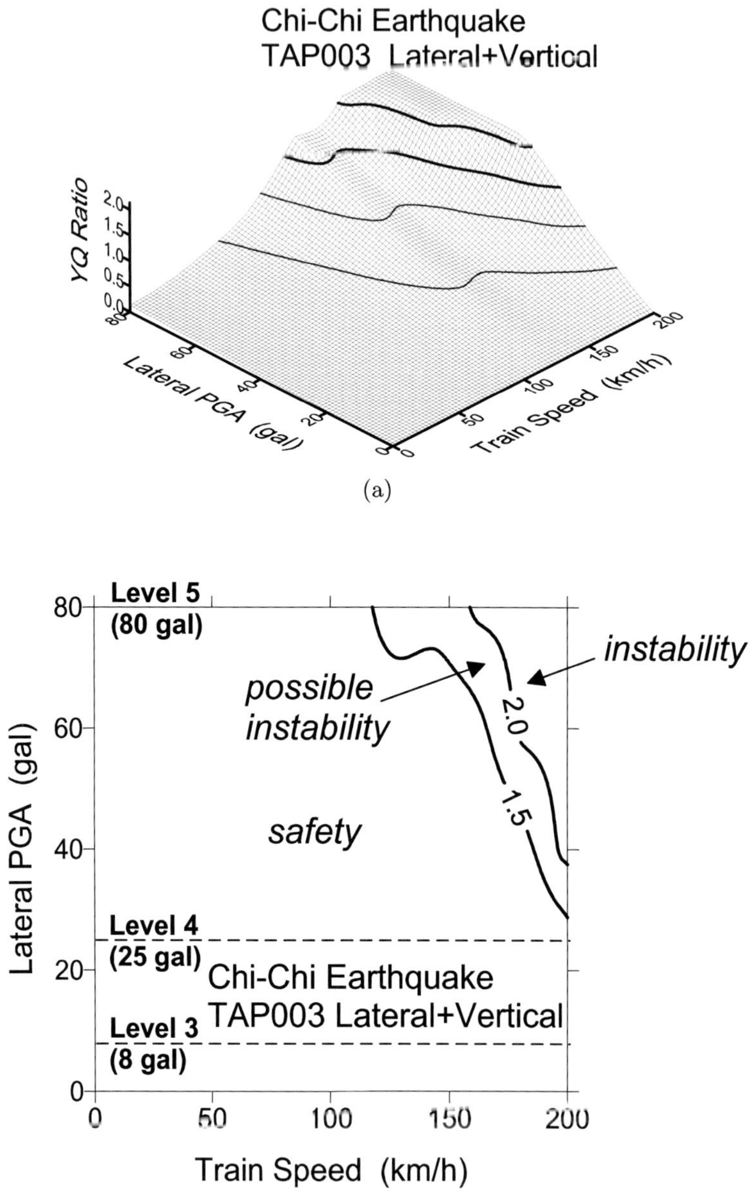


Fig. 11.25. Risk of derailment for TAP003 ground motion (lateral + vertical): (a) three-phase plot and (b) safety boundary.

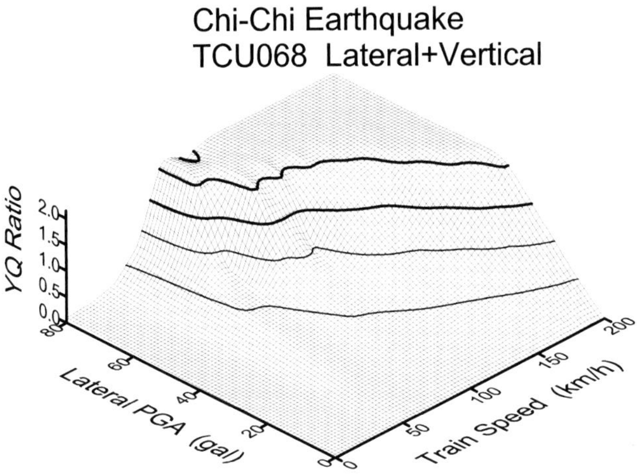


(a)

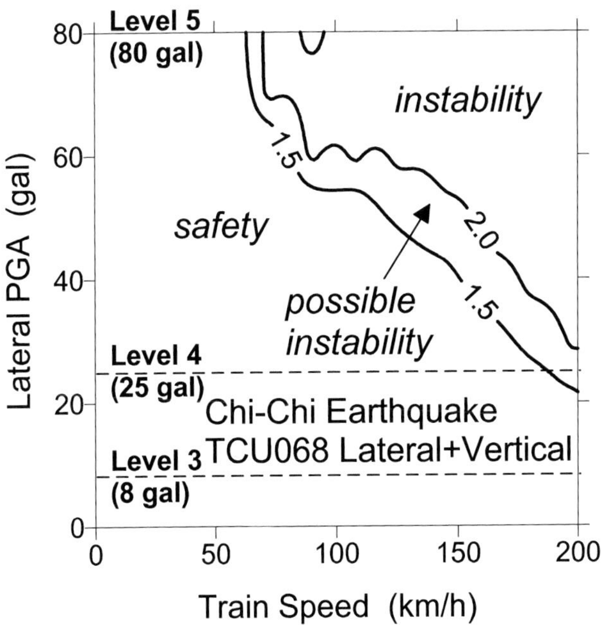


(b)

Fig. 11.26. Risk of derailment for TCU068 ground motion (lateral + vertical): (a) three-phase plot and (b) safety boundary.

excitation. This can be attributed to the fact that the vertical PGA for the far-field (TAP003) motion is very small, i.e., 27 gal, compared with the lateral PGA of 80 gal (normalized), while that for the near-fault (TCU068) motion is rather high, i.e., 83 gal, compared with the same lateral PGA. In general, the presence of vertical excitation can drastically reduce the stability region or the maximum allowable speed for the train under a specific intensity (PGA).

It should be noted that if the maximum YQ ratios for all the wheelsets of the train fall into the possible instability or instability regions, it does not necessarily mean that the train will really encounter overall derailment or instability. One reason for this is that the stability limits imposed on the YQ ratio are generally conservative due to consideration of safety. Another reason is that even though the YQ ratio of a single wheelset may exceed the safety limit, it still requires some time for derailment to develop, due to the linking action of the other parts of the car or the whole train. Further study is required in this regard to investigate the development of derailment and the mechanism involved taking into account the linking effect of all the train cars.

## 11.8. Concluding Remarks

The dynamic stability of trains moving over bridges that are simultaneously shaken by earthquakes was investigated. The equations of motion for the train–rails–bridge system presented in Chapter 10 were first generalized to include the effects of ground excitations, resulting in some nodal force terms related not only to the acceleration, but also velocity and displacement of the ground motion. Since the generalized system equations are identical in form to those with no ground motion, the analysis procedure established previously for dealing with the train–bridge interactions can directly be adopted, with modifications made only for the nodal forces to include the earthquake-induced effects.

The following conclusions remain strictly valid only for the conditions set in the numerical studies: (1) A train car initially resting on the bridge remains safe under the four ground motions specified, up to a PGA level of 80 gal, assuming that no inelastic deformation occurs on the bridge and ballast. (2) As for the train to move safely over the bridge, the speed should be kept below the allowable speed computed for each excitation based on the YQ criterion. (3) The property of ground motions and the presence of vertical excitations affect drastically the stability of the moving train, especially for near-fault excitations, as represented by the TCU068 record. (4) Various regions of stability have been established for the train under the four ground motions.

This study represents only parts of a preliminary attempt to deal with the seismic effects on train–rails–bridge interactions. Further study should be carried out to consider the effect of more severe earthquakes, say, with a PGA larger than 80 gal, and to include more representative ground motions, as well as more statistics-based criteria for derailment.

# Appendix A

# Derivation of Response Function $\bar{P}_1$ in Eq. (2.55)

We are interested in the dynamic response of a simply-supported beam of span length $L$ subjected to a sequence of $N$ identical loads of constant interval $d$ moving at speed $v$. The most severe case occurs when the former $N-1$ loads have left the beam, and the $N$th load has entered the beam, or equivalently, when $t_N < t < t_N + L/v$, where $t_N$ denotes the arriving time of the $N$th load on the beam, i.e., $t_N = (N-1)d/v$. For this particular case, one can obtain the response function $\bar{P}_1$ from Eq. (2.54) as follows:

$$
\begin{aligned}
\bar{P}_1(v,t) = {} & [\sin \Omega_1 (t - t_N) - S_1 \sin \omega_1 (t - t_N)] H(t - t_N) \\
& + \left\{ \sum_{j=1}^{N-1} \left[ \sin \Omega_1 (t - t_j) + \sin \Omega_1 \left( t - t_j - \frac{L}{v} \right) \right] \right. \\
& \left. - S_1 \sum_{j=1}^{N-1} \sin \omega_1 (t - t_j) + \sin \omega_1 \left( t - t_j - \frac{L}{v} \right) \right\} \\
& \times H \left( t - t_N - \frac{L}{v} \right),
\end{aligned} \tag{A.1}
$$

where the term containing $H(t - t_N)$ denotes the response excited by the $N$th moving load at time $t$ and the term containing $H(t - t_{N-1} - \Delta t)$, the free vibration caused by the former $N-1$ loads. By definition $\Omega_1 = \pi v/L$, it can be shown that

$$
\sin \Omega_1 (t - t_j) + \sin \Omega_1 \left( t - t_j - \frac{L}{v} \right) = 0 \,. \tag{A.2}
$$

Meanwhile,

$$\sin\omega_1(t-t_j)+\sin\omega_1\left(t-t_j-\frac{L}{v}\right)$$
$$=2\cos\frac{\omega_1 L}{2v}\sin\omega_1\left(t-t_j-\frac{L}{2v}\right). \tag{A.3}$$

Consequently, Eq. (A.1) reduces to

$$\bar{P}_1(v,t)=[\sin\Omega_1(t-t_N)-S_1\sin\omega_1(t-t_N)]H(t-t_N)$$
$$-2S_1\cos\frac{\omega_1 L}{2v}\left\{\sum_{j=1}^{N-1}\left[\sin\omega_1\left(t-t_j-\frac{L}{2v}\right)\right]\right\}$$
$$\times H\left(t-t_N-\frac{L}{v}\right). \tag{A.4}$$

Noting that $t_j=(j-1)d/v$, one may show that

$$\sum_{j=1}^{N-1}\sin\omega_1\left(t-\frac{L}{2v}-t_j\right)$$
$$=\sin\omega_1\left(t-\frac{L}{2v}\right)+\sum_{j=1}^{N-2}\sin\omega_1\left(t-\frac{L}{2v}-\frac{jd}{v}\right). \tag{A.5}$$

From Mangulis' (1965) Handbook [Eq. (9) on p. 104],

$$\sum_{n=1}^{m}\sin(n\theta+\alpha)$$
$$=\frac{-\sin\alpha+\sin(\alpha+\theta)-\sin[\alpha+(m+1)\theta]+\sin(\alpha+m\theta)}{2(1-\cos\theta)}. \tag{A.6}$$

By substituting $n = j$, $m = N - 2$, $n\theta = -j\omega_1(d/v)$, and $\alpha = \omega_1(t - L/2v)$ into the preceding equation, one obtains

$$\sum_{j=1}^{N-2} \sin\omega_1\left(t - \frac{L}{2v} - \frac{jd}{v}\right) = \frac{C}{2\left(1 - \cos\dfrac{\omega_1 d}{v}\right)}, \tag{A.7}$$

where

$$\begin{aligned} C = &-\sin\omega_1\left(t - \frac{L}{2v}\right) + \sin\omega_1\left(t - \frac{L}{2v} - \frac{d}{v}\right) \\ &- \sin\omega_1\left[t - \frac{L}{2v} - (N-1)\frac{d}{v}\right] \\ &+ \sin\omega_1\left[t - \frac{L}{2v} - (N-2)\frac{d}{v}\right]. \end{aligned} \tag{A.8}$$

By the following formula:

$$\sin a - \sin b = 2\cos\frac{a+b}{2}\sin\frac{a-b}{2}, \tag{A.9}$$

one can obtain after some operations the term $C$ as follows:

$$C = 4\sin\frac{\omega_1 d}{2v}\sin\omega_1\left(t - \frac{L}{2v} - \frac{N-1}{2v}d\right)\sin\omega_1\left(\frac{N-2}{2v}d\right). \tag{A.10}$$

In the meantime, the denominator in Eq. (A.7) can be reduced to $4\sin^2(\omega_1 d/2v)$. With these expressions, one can derive from Eq. (A.5) the following:

$$\begin{aligned} &\sum_{j=1}^{N-1} \sin\omega_1\left(t - \frac{L}{2v} - t_j\right) \\ &\quad = \sin\omega_1\left(t - \frac{L}{2v}\right) \\ &\qquad + \frac{1}{\sin(\omega_1 d/2v)}\left[\sin\omega_1\left(t - \frac{L}{2v} - \frac{t_N}{2}\right)\sin\omega_1\left(\frac{t_N}{2} - \frac{d}{2v}\right)\right]. \end{aligned} \tag{A.11}$$

By using Eqs. (A.4) and (A.11), Eq. (2.55) can be proved.

# Appendix B

# Newmark's $\beta$ Method

A great number of dynamic problems encountered in engineering appears in the form of second-order differential equations. For structures that contain only a single degree of freedom (DOF), analytical solutions in closed form can be obtained. But for structures with multi DOFs, finite difference methods are often called for to solve the second-order differential equations, which have been referred to as the *direct integration methods*, to distinguish from those based on modal superposition or others. Newmark's $\beta$ method represents a special category of finite difference methods that have frequently been used by engineers and researchers in solving the multi-DOF second-order differential equations. The following is a summary of the method proposed by Newmark (1959).

In a step-by-step nonlinear analysis, we are interested in the behavior of a structure within the incremental step from time $t$ to $t + \Delta t$, where $\Delta t$ denotes a small time increment. The following are the equations of motion for the structure at time $t + \Delta t$:

$$[M]\{\ddot{U}\}_{t+\Delta t} + [C]\{\dot{U}\}_{t+\Delta t} + [K]\{U\}_{t+\Delta t} = \{P\}_{t+\Delta t}\,, \qquad \text{(B.1)}$$

where $[M]$, $[C]$ and $[K]$ denote the mass, damping and stiffness matrices of the structure, assumed to consist of $N$ DOFs, $\{U\}$ the nodal displacements and $\{P\}$ the applied nodal forces. All quantities of the structure are assumed to be known up to time $t$.

The method proposed by Newmark is a single-step method, which requires only information of the structure at time $t$ for solution. The following are the two basic equations proposed by Newmark for determining the displacements and velocities of the structure at time $t + \Delta t$:

$$\{U\}_{t+\Delta t} = \{U\}_t + \{\dot{U}\}_t \Delta t + \left[\left(\frac{1}{2} - \beta\right)\{\ddot{U}\}_t + \beta\{\ddot{U}\}_{t+\Delta t}\right](\Delta t)^2 ,$$
$$\{\dot{U}\}_{t+\Delta t} = \{\dot{U}\}_t + [(1-\gamma)\{\ddot{U}\}_t + \gamma\{\ddot{U}\}_{t+\Delta t}]\Delta t , \tag{B.2}$$

where a dot denotes differentiation with respect to time $t$. The parameter $\beta$ denotes the variation of acceleration during the incremental step from $t$ to $t+\Delta t$. Different values of $\beta$ imply different schemes of interpolation for the acceleration over a time step. The value $\beta = 0$ indicates a scheme equivalent to the central difference method, the value $\beta = 1/4$ is a constant average acceleration method, and the value $\beta = 1/6$ is a linear acceleration method. On the other hand, the parameter $\gamma$ relates to the property of numerical or artificial damping introduced by discretization in time domain. For the case with $\gamma < 1/2$, there exists some artificial negative damping, while for $\gamma > 1/2$, artificial positive damping will occur (Weaver and Johnston, 1987). The method has been demonstrated to be unconditionally stable under the conditions when $\gamma \geq 1/2$ and $\beta \geq 1/4(1/2+\gamma)^2$. Throughout this book, the combination of $\gamma = 1/2$ and $\beta = 1/4$ will be selected.

From Eq. (B.2), the accelerations and velocities of the structure at time $t + \Delta t$ can be solved as

$$\{\ddot{U}\}_{t+\Delta t} = a_0(\{U\}_{t+\Delta t} - \{U\}_t) - a_2\{\dot{U}\}_t - a_3\{\ddot{U}\}_t ,$$
$$\{\dot{U}\}_{t+\Delta t} = \{\dot{U}\}_t + a_6\{\ddot{U}\}_t + a_7\{\ddot{U}\}_{t+\Delta t} , \tag{B.3}$$

where the coefficients $a_0 \sim a_7$ are given as follows:

$$a_0 = \frac{1}{\beta \Delta t^2}, \quad a_1 = \frac{\gamma}{\beta \Delta t}, \quad a_2 = \frac{1}{\beta \Delta t},$$
$$a_3 = \frac{1}{2\beta} - 1, \quad a_4 = \frac{\gamma}{\beta} - 1, \quad a_5 = \frac{\Delta t}{2}\left(\frac{\gamma}{\beta} - 2\right), \tag{B.4}$$
$$a_6 = \Delta t(1-\gamma), \quad a_7 = \gamma \Delta t .$$

Substituting the preceding expression (B.3) into Eq. (B.1) yields the *equivalent stiffness equations*

$$[K_{\text{eff}}]\{U\}_{t+\Delta t} = \{P_{\text{eff}}\}_{t+\Delta t}\,, \tag{B.5}$$

where the effective stiffness matrix $[K_{\text{eff}}]$ and the effective load vector $\{P_{\text{eff}}\}_{t+\Delta t}$ are defined as follows:

$$[K_{\text{eff}}] = a_0[M] + a_1[C] + [K]\,,$$

$$\begin{aligned}\{P_{\text{eff}}\}_{t+\Delta t} &= \{P\}_{t+\Delta t} + [M](a_0\{U\}_t + a_2\{\dot{U}\}_t + a_3\{\ddot{U}\}_t) \\ &\quad + [C](a_1\{U\}_t + a_4\{\dot{U}\}_t + a_5\{\ddot{U}\}_t)\,.\end{aligned} \tag{B.6}$$

From Eq. (B.5), the structural displacements $\{U\}$ at time $t+\Delta t$ can be solved as

$$\{U\}_{t+\Delta t} = [K_{\text{eff}}]^{-1}\{P_{\text{eff}}\}_{t+\Delta t}\,. \tag{B.7}$$

It follows that the velocities and accelerations at time $t+\Delta t$ can be obtained from Eq. (B.3). Since all the structural responses at time $t+\Delta t$ have been made available, one can proceed to the next time step, treating all the responses solved for the structure at this step as the initial conditions, updating the structural and loading configurations, including the mass, damping, stiffness matrices, $[M]$, $[C]$, $[K]$ and applied loads $\{P\}$ when necessary. By repeating the above procedure for a certain number of time steps, one can compute the time-history response of the structure throughout the duration in which the structural behavior is of interest.

## Appendix C

# Vertical Frequency of Vibration of Curved Beam

By letting the determinant in Eq. (5.15) equal to zero, one obtains

$$(a_1 - \omega_{v1}^2)(b_1 - \omega_{v1}^2) - a_2 b_2 = 0\,, \tag{C.1}$$

from which the solution can be written as

$$\omega_{v1} = \frac{(a_1 + b_1) \pm \sqrt{(a_1 + b_1)^2 - 4(a_1 b_1 - a_2 b_2)}}{2}\,. \tag{C.2}$$

Let us discuss whether the term $a_1 b_1 - a_2 b_2$ is less than zero. From Eq. (5.11), it can be shown that

$$\begin{aligned} a_1 b_1 &= c_1[\pi^2 R^2 L^2 (E^2 I_z^2 + G^2 J^2) + E I_z G J (L^4 + \pi^4 R^4)]\,, \\ a_2 b_2 &= c_1[\pi^2 R^2 L^2 (E^2 I_z^2 + G^2 J^2) + E I_z G J (2\pi^2 R^2 L^2)]\,, \end{aligned} \tag{C.3}$$

where the constant $c_1$ is

$$c_1 = -\frac{1}{\rho J}\frac{1}{m R^4 L^4}\left(\frac{\pi}{L}\right)^2 < 0\,. \tag{C.4}$$

It follows that

$$a_1 b_1 - a_2 b_2 = c_1 E I_z G J (L^2 - \pi^2 R^2)^2\,. \tag{C.5}$$

Since $c_1$ is less than zero, it can be ascertained that $a_1 b_1 - a_2 b_2$ is also less than zero. On the other hand, because $\omega_{v1}$ is not an imaginary number, the positive sign in Eq. (C.2) should be selected. Thus, the frequency of vibration, $\omega_{v1}$, should be computed as

$$\omega_{v1} = \sqrt{\frac{(a_1 + b_1) + \sqrt{(a_1 - b_1)^2 + 4 a_2 b_2}}{2}}\,. \tag{C.6}$$

# Appendix D

# Horizontal Frequency of Vibration of Curved Beam

Letting the determinant in Eq. (5.35) equal to zero yields

$$(\bar{a}_1 - \omega_{h1}^2)(\bar{b}_1 - \omega_{h1}^2) - \bar{a}_2\bar{b}_2 = 0\,, \tag{D.1}$$

from which the solution can be written as

$$\omega_{h1} = \frac{(\bar{a}_1 + \bar{b}_1) \pm \sqrt{(\bar{a}_1 + \bar{b}_1)^2 - 4(\bar{a}_1\bar{b}_1 - \bar{a}_2\bar{b}_2)}}{2}\,. \tag{D.2}$$

Let us determine whether the term $\bar{a}_1\bar{b}_1 - \bar{a}_2\bar{b}_2$ is less than zero. From Eq. (5.32), it can be shown that

$$\begin{aligned} \bar{a}_1\bar{b}_1 &= c_2\left\{\pi R I_y\left[\left(\frac{\pi}{L}\right)^2 + \frac{1}{R^2}\right]^2 + \pi\frac{\bar{a}_1}{R^2}\right\}, \\ \bar{a}_2\bar{b}_2 &= c_2\left(\pi\frac{\bar{a}_1}{R^2} - \frac{8\bar{a}_1}{\pi R^2}\right), \end{aligned} \tag{D.3}$$

where the constant $c_2$ is

$$c_2 = \frac{E^2\bar{a}_1\pi\left(\dfrac{4}{\pi^2} - \dfrac{1}{2}\right)}{m^2L^2\left(\dfrac{8}{\pi^2} - \dfrac{5}{6}\right)} > 0\,. \tag{D.4}$$

It follows that

$$\bar{a}_1\bar{b}_1 - \bar{a}_2\bar{b}_2 = c_2\left\{\left[\left(\frac{\pi}{L}\right)^2 + \frac{1}{R^2}\right]^2 + \frac{8\bar{a}_1}{\pi R}\right\}. \tag{D.5}$$

Since $c_2$ is greater than zero, it is ascertained that $\bar{a}_1\bar{b}_1 - \bar{a}_2\bar{b}_2$ is also greater than zero. For the frequency of vibration, $\omega_{h1}$, to be a real number, only the negative sign in Eq. (D.2) should be selected. Thus,

$$\omega_{h1} = \sqrt{\frac{(\bar{a}_1 + \bar{b}_1) - \sqrt{(\bar{a}_1 - \bar{b}_1)^2 + 4\bar{a}_2\bar{b}_2}}{2}}. \tag{D.6}$$

## Appendix E

# Derivation of Residual Vibration for Curved Beam in Eq. (5.53)

The derivation to be presented below follows basically the procedure of Appendix A. By noting that the time lag for the $j$th moving load is $t_j = (j-1)d/v$, the residual vibration response of the bridge caused by $N-1$ moving loads that passed the beam, as given in Eq. (5.52), can be rewritten as

$$\begin{aligned} U_{N,2}\left(\frac{L}{2}, t\right) &= -2PS_1 \cos\frac{\omega_1 L}{2v} \\ &\quad \times \left\{ \sum_{j=1}^{N-1} \sin\omega_1 \left[\left(t - \frac{L}{2v}\right) - (j-1)\frac{d}{v}\right] \right\} \\ &\quad \times H\left(t - t_{N-1} - \frac{L}{v}\right), \quad \text{for} \quad t - t_{N-1} \geq \frac{L}{v}. \end{aligned} \tag{E.1}$$

The series term in Eq. (E.1) can be separated into two terms as

$$\begin{aligned} &\sum_{j=1}^{N-1} \sin\omega_1 \left[\left(t - \frac{L}{2v}\right) - (j-1)\frac{d}{v}\right] \\ &\quad = \sin\omega_1 \left(t - \frac{L}{2v}\right) + \sum_{j=1}^{N-2} \sin\omega_1 \left[\left(t - \frac{L}{2v}\right) - j\frac{d}{v}\right]. \end{aligned} \tag{E.2}$$

From Manulis' (1965) handbook (Eq. (9), p. 104), the following equality is shown to be valid:

$$\sum_{n=1}^{m} \sin(\alpha + n\theta)$$
$$= \frac{-\sin\alpha + \sin(\alpha+\theta) - \sin[\alpha + (m+1)\theta] + \sin(\alpha + m\theta)}{2(1-\cos\theta)}. \tag{E.3}$$

By letting $n = j$, $m = N-2$, $\alpha = \omega_1(t - L/2v)$, and $\theta = -\omega_1 d/v$ in Eq. (E.3), one can rearrange the last term in Eq. (E.2) as

$$\sum_{j=1}^{N-2} \sin\omega_1 \left[\left(t - \frac{L}{2v}\right) - j\frac{d}{v}\right] = \frac{C}{2(1-\cos(\omega_1 d/v))}, \tag{E.4}$$

where

$$\begin{aligned} C = &-\sin\omega_1\left(t - \frac{L}{2v}\right) + \sin\omega_1\left(t - \frac{L}{2v} - \frac{d}{v}\right) \\ &- \sin\omega_1\left[t - \frac{L}{2v} - (N-1)\frac{d}{v}\right] \\ &+ \sin\omega_1\left[t - \frac{L}{2v} - (N-2)\frac{d}{v}\right]. \end{aligned} \tag{E.5}$$

By the relation

$$\sin a - \sin b = 2\cos\frac{a+b}{2}\sin\frac{a-b}{2}, \tag{E.6}$$

one can rewrite the expression in Eq. (E.5) as

$$C = 4\sin\frac{\omega_1 d}{2v}\sin\omega_1\left(t - \frac{L}{2v} - \frac{N-1}{2}\frac{d}{v}\right)\sin\omega_1\left(\frac{N-2}{2}\frac{d}{v}\right). \tag{E.7}$$

On the other hand, it is known that

$$1 - \cos\frac{\omega_1 d}{v} = 2\left(\sin\frac{\omega_1 d}{2v}\right)^2. \tag{E.8}$$

Using the relations given in Eqs. (E.2), (E.4), (E.7) and (E.8), one can derive from Eq. (E.1) the following

$$U_{N,2}\left(\frac{L}{2},t\right) = -2PS_1 \cos\frac{\omega_1 L}{2v}\left[\sin\omega_1\left(t-\frac{L}{2v}\right) + \sin\omega_1\left(\frac{N-2}{2}\frac{d}{v}\right)\right.$$
$$\left.\times\ \sin\omega_1\left(t-\frac{L}{2v}-\frac{N-1}{2}\frac{d}{v}\right)\sin^{-1}\frac{\omega_1 d}{2v}\right]$$
$$\times H\left(t-t_{N-1}-\frac{L}{v}\right), \quad \text{for} \quad t-t_{N-1} > \frac{L}{v}. \tag{E.9}$$

# Appendix F

# Beam Element and Structural Damping Matrix

In the first part of this appendix, the mass matrix and stiffness matrix of the three-dimensional beam element used throughout the book will be summarized. In the second part, the procedure for determining the damping matrix of the structure based on the hypothesis of *Rayleigh damping* will be presented.

## F.1. Equation of Motion for Beam Element

The beam element considered is a prismatic, three-dimensional beam of solid cross sections, with the effect of warping deformations excluded. The longitudinal axis of the beam is denoted by $x$, and the two transverse principal axes of the cross section of the beam by $y$ and $z$. There are six degrees of freedom (DOFs), i.e., three translations and three rotations, associated with each of the two ends $A$ and $B$ of the element, as shown in Fig. F.1. For the case with no external loads, the equation of motion for the beam element can be written as follows:

$$[m_b]\{\ddot{u}_b\} + [c_b]\{\dot{u}_b\} + [k_b]\{u_b\} = \{0\}, \tag{F.1}$$

where $\{u_b\}$ is the displacement vector of the element, which consists of 12 DOFs, an overdot denotes differentiation with respect to time, $[m_b]$, $[c_b]$ and $[k_b]$ denote the mass, damping and stiffness matrices of the element. Both the mass matrix $[m_b]$ and stiffness matrix $[k_b]$ are available in most textbooks on structural and dynamic analyses, e.g., see Paz (1991) and McGuire *et al.* (2000). The element displacement

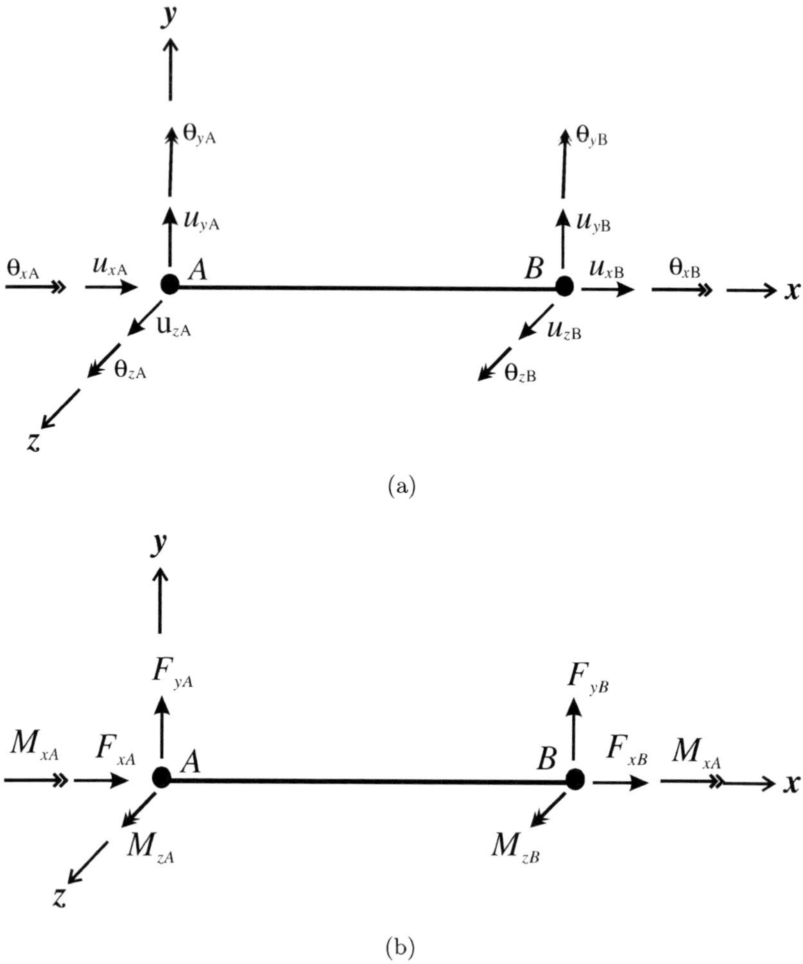


Fig. F.1. Three-dimensional beam element: (a) nodal DOFs and (b) nodal forces.

vector $\{u_b\}$ is

$$\{u_b\} = \langle\, u_{xA} \quad u_{yA} \quad u_{zA} \quad \theta_{xA} \quad \theta_{yA} \quad \theta_{zA} \quad u_{xB} \quad u_{yB} \quad u_{zB} \quad \theta_{xB} \quad \theta_{yB} \quad \theta_{zB} \,\rangle^T . \tag{F.2}$$

Correspondingly, there are 12 nodal forces for the element, as shown in Fig. F.1(b). The mass matrix $[m_b]$ is

$$[m_b] = \frac{mL}{420}\begin{bmatrix} [m_1] & [m_2]^T \\ [m_2] & [m_3] \end{bmatrix}, \tag{F.3}$$

where $L$ is the length of the element and $m$ is the mass per unit length. By letting $A$ denote the cross area and $I_p$ the polar moment of inertia of the beam, the submatrices in Eq. (F.3) can be given as follows:

$$[m_1] = \begin{bmatrix} 140 & & & & Symm. & \\ 0 & 156 & & & & \\ 0 & 0 & 156 & & & \\ 0 & 0 & 0 & \dfrac{140I_p}{A} & & \\ 0 & 0 & -22L & 0 & 4L^2 & \\ 0 & 22L & 0 & 0 & 0 & 4L^2 \end{bmatrix}, \tag{F.4}$$

$$[m_2] = \begin{bmatrix} 70 & 0 & 0 & 0 & 0 & 0 \\ 0 & 54 & 0 & 0 & 0 & 13L \\ 0 & 0 & 54 & 0 & -13L & 0 \\ 0 & 0 & 0 & \dfrac{70I_p}{A} & 0 & 0 \\ 0 & 0 & 13L & 0 & -3L^2 & 0 \\ 0 & -13L & 0 & 0 & 0 & -3L^2 \end{bmatrix}, \tag{F.5}$$

$$[m_3] = \begin{bmatrix} 140 & & & & Symm. & \\ 0 & 156 & & & & \\ 0 & 0 & 156 & & & \\ 0 & 0 & 0 & \dfrac{140I_p}{A} & & \\ 0 & 0 & 22L & 0 & 4L^2 & \\ 0 & -22L & 0 & 0 & 0 & 4L^2 \end{bmatrix}. \tag{F.6}$$

The stiffness matrix $[k_b]$ can be given as follows:

$$[k_b] = \begin{bmatrix} [k_1] & [k_2]^T \\ [k_2] & [k_3] \end{bmatrix}. \tag{F.7}$$

By letting $E$ and $G$ denote the elastic and shear moduli, respectively, $J$ the torsional constant, and $I_y$ and $I_z$ respectively the moments of inertia about the $y$- and $z$-axes of the element, the submatrices in Eq. (F.7) can be given as follows:

$$[k_1] = \begin{bmatrix} \frac{EA}{L} & & & & & Symm. \\ 0 & \frac{12EI_z}{L^3} & & & & \\ 0 & 0 & \frac{12EI_y}{L^3} & & & \\ 0 & 0 & 0 & \frac{GJ}{L} & & \\ 0 & 0 & -\frac{6EI_y}{L^2} & 0 & \frac{4EI_y}{L} & \\ 0 & \frac{6EI_z}{L^2} & 0 & 0 & 0 & \frac{4EI_z}{L} \end{bmatrix}, \tag{F.8}$$

$$[k_2] = \begin{bmatrix} -\frac{EA}{L} & 0 & 0 & 0 & 0 & 0 \\ 0 & -\frac{12EI_z}{L^3} & 0 & 0 & 0 & -\frac{6EI_z}{L^2} \\ 0 & 0 & -\frac{12EI_y}{L^3} & 0 & \frac{6EI_y}{L^2} & 0 \\ 0 & 0 & 0 & -\frac{GJ}{L} & 0 & 0 \\ 0 & 0 & -\frac{6EI_y}{L^2} & 0 & \frac{2EI_y}{L} & 0 \\ 0 & \frac{6EI_z}{L^2} & 0 & 0 & 0 & \frac{2EI_z}{L} \end{bmatrix}, \tag{F.9}$$

$$[k_3] = \begin{bmatrix} \frac{EA}{L} & & & & & Symm. \\ 0 & \frac{12EI_z}{L^3} & & & & \\ 0 & 0 & \frac{12EI_y}{L^3} & & & \\ 0 & 0 & 0 & \frac{GJ}{L} & & \\ 0 & 0 & \frac{6EI_y}{L^2} & 0 & \frac{4EI_y}{L} & \\ 0 & -\frac{6EI_z}{L^2} & 0 & 0 & 0 & \frac{4EI_z}{L} \end{bmatrix}. \quad \text{(F.10)}$$

The damping matrix $[c_b]$ is usually not computed on the element level, but implicitly implied as a part of the structural damping matrix to be shown in the following section.

## F.2. Structural Damping Matrix

The damping of structures can appear in various forms. By *classical damping*, it means that the damping matrix of the structure can be expressed as some linear combinations of the mass and stiffness matrices of the structure. With such a property, the natural modes of vibration are orthogonal not only with respect to the mass and stiffness matrices, but also to the damping matrix. One benefit from this is that the equations of motion of the structure become decoupled when transformed to the generalized coordinates.

Let us consider the equations of motion for a structure that is free of external loads:

$$[M_b]\{\ddot{U}_b\} + [C_b]\{\dot{U}_b\} + [K_b]\{U_b\} = \{0\}, \quad \text{(F.11)}$$

where $\{U_b\}$ denotes the nodal displacements, and $[M_b]$ and $[K_b]$ the mass and stiffness matrices of the structure. Based on the assumption of Rayleigh or classical damping, the damping matrix $[C_b]$ of the structure can be expressed in a general form as a combination of the mass matrix $[M_b]$ and stiffness matrix $[K_b]$ (Caughey, 1960), that is,

$$[C_b] = [M_b] \sum_{i=0}^{N-1} a_i([M_b]^{-1}[K_b])^i \,, \tag{F.12}$$

where $N$ is the total number of DOFs of the structure and $a_i$ are the coefficients to be determined. By the orthogonality properties of the vibration modes with respect to the mass and stiffness matrices, the $n$th modal damping coefficient $C_n$ can be expressed as follows:

$$C_n = \{\phi\}_n^T[C_b]\{\phi\}_n = \{\phi\}_n^T[M_b] \sum_{i=0}^{N-1} a_i([M_b]^{-1}[K_b])^i\{\phi\}_n \,, \tag{F.13}$$

where $\{\phi\}_n$ denotes the $n$th vibration mode of the structure with no damping. The following is the modal equation that must be satisfied by the $n$th vibration mode $\{\phi\}_n$:

$$[K_b]\{\phi\}_n = \omega_n^2[M_b]\{\phi\}_n \,, \tag{F.14}$$

where $\omega_n$ is the $n$th vibration frequency of the structure. The following is the definition for the $n$th modal mass:

$$M_n = \{\phi\}_n^T[M_b]\{\phi\}_n \,. \tag{F.15}$$

By using Eqs. (F.14) and (F.15), the $n$th modal damping coefficient $C_n$ in Eq. (F.13) can be written as

$$C_n = \sum_{i=0}^{N-1} a_i\omega_n^{2i} M_n \,. \tag{F.16}$$

On the other hand, the damping coefficient for the $n$th modal equation in terms of the generalized coordinates is

$$C_n = 2\xi_n\omega_n M_n \,, \tag{F.17}$$

where $\xi_n$ is the $n$th damping ratio. For the case with the modal mass normalized as $M_n = 1$, the $n$th damping ratio can be solved from Eqs. (F.16) and (F.17) as

$$\xi_n = \frac{1}{2\omega_n} \sum_{i=0}^{N-1} a_i\omega_n^{2i} \,. \tag{F.18}$$

For most engineering structures, it is impractical to consider all the $N$ modes of damping ratios. If only the first two modes are considered, which is the case known as *Rayleigh damping*, the damping matrix $[C_b]$ in Eq. (F.12) reduces to

$$[C_b] = a_0[M_b] + a_1[K_b]\,. \tag{F.19}$$

The Rayleigh coefficients $a_0$ and $a_1$ can be determined only if the damping ratios $\xi_i$, $\xi_j$ and frequencies $\omega_i$, $\omega_j$ are given for any two vibration modes, i.e., from Eq. (F.18),

$$\begin{Bmatrix} a_0 \\ a_1 \end{Bmatrix} = 2\begin{bmatrix} \omega_i^{-1} & \omega_i \\ \omega_j^{-1} & \omega_j \end{bmatrix}^{-1} \begin{Bmatrix} \xi_i \\ \xi_j \end{Bmatrix}. \tag{F.20}$$

For the case when the frequencies of vibration of the first two modes, i.e., $i = 1$ and $j = 2$, are given and the damping ratios for the two mode are assumed to be the same, i.e., $\xi_1 = \xi_2 = \xi$, the preceding equation reduces to

$$\begin{Bmatrix} a_0 \\ a_1 \end{Bmatrix} = \frac{2\xi}{\omega_1 + \omega_2} \begin{Bmatrix} \omega_1\omega_2 \\ 1 \end{Bmatrix}. \tag{F.21}$$

The damping matrix $[C_b]$ can then be computed from Eq. (F.19).

# Appendix G

# Partitioned Matrices and Vector for Vehicle, Eq. (9.4)

The partitioned matrices and vectors for the vehicle equation in Eq. (9.4) can be given as follows:

$$[m_{uu}] = \begin{bmatrix} M_c & 0 & 0 & 0 & 0 & 0 \\ & I_c & 0 & 0 & 0 & 0 \\ & & M_t & 0 & 0 & 0 \\ & & & I_t & 0 & 0 \\ & Symm. & & & M_t & 0 \\ & & & & & I_t \end{bmatrix}, \tag{G.1}$$

$$[m_{uw}] = [m_{wu}]^T = [0]\,, \tag{G.2}$$

$$[m_{ww}] = \begin{bmatrix} M_w & 0 & 0 & 0 \\ 0 & M_w & 0 & 0 \\ 0 & 0 & M_w & 0 \\ 0 & 0 & 0 & M_w \end{bmatrix}, \tag{G.3}$$

$$[l_w] = \begin{bmatrix} 1 & 0 & 0 & 0 \\ 0 & 1 & 0 & 0 \\ 0 & 0 & 1 & 0 \\ 0 & 0 & 0 & 1 \end{bmatrix}, \tag{G.4}$$

$$[c_{uu}] = \begin{bmatrix} 2c_s & 0 & -c_s & 0 & -c_s & 0 \\ & 2l_c^2 c_s & -l_c c_s & 0 & l_c c_s & 0 \\ & & c_s + 2c_p & 0 & 0 & 0 \\ & & & 2l_t^2 c_p & 0 & 0 \\ & Symm. & & & c_s + 2c_p & 0 \\ & & & & & 2l_t^2 c_p \end{bmatrix}, \quad \text{(G.5)}$$

$$[c_{uw}] = [c_{wu}]^T = \begin{bmatrix} 0 & 0 & 0 & 0 \\ 0 & 0 & 0 & 0 \\ -c_p & -c_p & 0 & 0 \\ -l_t c_p & l_t c_p & 0 & 0 \\ 0 & 0 & -c_p & -c_p \\ 0 & 0 & -l_t c_p & l_t c_p \end{bmatrix}, \quad \text{(G.6)}$$

$$[c_{ww}] = \begin{bmatrix} c_p & 0 & 0 & 0 \\ 0 & c_p & 0 & 0 \\ 0 & 0 & c_p & 0 \\ 0 & 0 & 0 & c_p \end{bmatrix}, \quad \text{(G.7)}$$

$$[k_{uu}] = \begin{bmatrix} 2k_s & 0 & -k_s & 0 & -k_s & 0 \\ & 2l_c^2 k_s & -l_c k_s & 0 & l_c k_s & 0 \\ & & k_s + 2k_p & 0 & 0 & 0 \\ & & & 2l_t^2 k_p & 0 & 0 \\ & Symm. & & & k_s + 2k_p & 0 \\ & & & & & 2l_t^2 k_p \end{bmatrix}, \quad \text{(G.8)}$$

$$[k_{uw}] = [k_{wu}]^T = \begin{bmatrix} 0 & 0 & 0 & 0 \\ 0 & 0 & 0 & 0 \\ -k_p & -k_p & 0 & 0 \\ -l_t k_p & l_t k_p & 0 & 0 \\ 0 & 0 & -k_p & -k_p \\ 0 & 0 & -l_t k_p & l_t k_p \end{bmatrix}, \quad \text{(G.9)}$$

$$[k_{ww}] = \begin{bmatrix} k_p & 0 & 0 & 0 \\ 0 & k_p & 0 & 0 \\ 0 & 0 & k_p & 0 \\ 0 & 0 & 0 & k_p \end{bmatrix}, \tag{G.10}$$

$$\{f_{ue}\}_{t+\Delta t} = \{0\}, \tag{G.11}$$

$$\{f_{we}\}_{t+\Delta t} = \langle -W \quad -W \quad -W \quad -W \rangle^T. \tag{G.12}$$

Here, we use $M_w$, $M_t$, and $M_c$ to denote the mass of the wheel, bogie, and car body, respectively. It follows that the vehicle weight $W$ in Eq. (G.12) can be expressed as

$$W = (M_w + 0.5M_t + 0.25M_c)g, \tag{G.13}$$

with $g$ denoting the acceleration of gravity.

## Appendix H

# Related Matrices and Vectors for CFR Element

The following are the related matrices and vectors for the system matrices of the CFR element as given in Eq. (9.10):

$$[\psi_u^0] = \begin{bmatrix} \frac{l}{3} & \frac{l}{6} \\ \frac{l}{6} & \frac{l}{3} \end{bmatrix} \quad (\, u_1 \quad u_2 \,), \tag{H.1}$$

$$[\psi_u^1] = \begin{bmatrix} \frac{1}{l} & -\frac{1}{l} \\ -\frac{1}{l} & \frac{1}{l} \end{bmatrix} \quad (\, u_1 \quad u_2 \,), \tag{H.2}$$

$$[\psi_v^0] = \frac{l}{420} \begin{bmatrix} 156 & 22l & 54 & -13l \\ & 4l^2 & 13l & -3l^2 \\ & & 156 & -22l \\ Symm. & & & 4l^2 \end{bmatrix} \quad (\, v_1 \quad \theta_1 \quad v_2 \quad \theta_2 \,), \tag{H.3}$$

$$[\psi_v^2] = \begin{bmatrix} 12/l^3 & 6/l^2 & -12/l^3 & 6/l^2 \\ & 4/l & -6/l^2 & 2/l \\ & & 12/l^3 & -6/l^2 \\ Symm. & & & 4/l \end{bmatrix} \quad (\, v_1 \quad \theta_1 \quad v_2 \quad \theta_2 \,), \tag{H.4}$$

$$[\psi_u^0]_l = \frac{1}{2\lambda_u} \quad (u_2), \tag{H.5}$$

$$[\psi_u^1]_l = \frac{\lambda_u}{2} \quad (u_2)\,, \tag{H.6}$$

$$[\psi_v^0]_l = \begin{bmatrix} \dfrac{3}{4\lambda_v} & -\dfrac{1}{4\lambda_v^2} \\ -\dfrac{1}{4\lambda_v^2} & \dfrac{1}{8\lambda_v^3} \end{bmatrix} \quad (\,v_2 \quad \theta_2\,)\,, \tag{H.7}$$

$$[\psi_v^2]_l = \begin{bmatrix} \lambda_v^3 & -\lambda_v^2 \\ -\lambda_v^2 & \dfrac{3\lambda_v}{2} \end{bmatrix} \quad (\,v_2 \quad \theta_2\,)\,, \tag{H.8}$$

$$[\psi_u^0]_r = \frac{1}{2\lambda_u} \quad (u_1)\,, \tag{H.9}$$

$$[\psi_u^1]_r = \frac{\lambda_u}{2} \quad (u_1)\,, \tag{H.10}$$

$$[\psi_v^0]_r = \begin{bmatrix} \dfrac{3}{4\lambda_v} & \dfrac{1}{4\lambda_v^2} \\ \dfrac{1}{4\lambda_v^2} & \dfrac{1}{8\lambda_v^3} \end{bmatrix} \quad (\,v_1 \quad \theta_1\,)\,, \tag{H.11}$$

$$[\psi_v^2]_r = \begin{bmatrix} \lambda_v^3 & \lambda_v^2 \\ \lambda_v^2 & \dfrac{3\lambda_v}{2} \end{bmatrix} \quad (\,v_1 \quad \theta_1\,)\,, \tag{H.12}$$

$$\lambda_u = \sqrt{\frac{k_{bh}}{2E_rA_r}}\,, \tag{H.13}$$

$$\lambda_v = \sqrt[4]{\frac{k_{bv}}{8E_rI_r}}\,. \tag{H.14}$$

The parenthesized symbols in the above expressions indicate the degrees of freedom of the element with which the column and row entries in each matrix are associated.

# Appendix I

# Related Matrices and Vectors for 3D Vehicle Model

The following are the partitioned mass matrices involved in the equation of motion for the 3D vehicle model in Eq. (10.4):

$$[m_{uu}] = diag[\, M_c \quad M_c \quad I^*_{cx} \quad I^*_{cy} \quad I^*_{cz} \quad M_t \quad M_t \quad I^*_{tx} \quad I^*_{ty} \quad I^*_{tz} \quad M_t \quad M_t \quad I^*_{tx} \quad I^*_{ty} \quad I^*_{tz}\,], \tag{I.1}$$

$$[m_{ww}] = diag[\, M_w \quad M_w \quad I^*_w \quad M_w \quad M_w \quad I^*_w \quad M_w \quad M_w \quad I^*_w \quad M_w \quad M_w \quad I^*_w\,], \tag{I.2}$$

$$[m_{uw}] = [m_{wu}]^T = [0]\,, \tag{I.3}$$

where *diag* denotes that the entries following represent the diagonal elements of the matrix and that all the nondiagonal elements are zero. The following are the partitioned damping matrices:

$$[c_{uu}] = \begin{bmatrix}
a & 0 & 0 & 0 & 0 & -b & 0 & 0 & 0 & 0 & -b & 0 & 0 & 0 & 0 \\
 & c & -d & 0 & 0 & 0 & -e & -f & 0 & 0 & 0 & -e & -f & 0 & 0 \\
 & & g & 0 & 0 & 0 & h & i & 0 & 0 & 0 & h & i & 0 & 0 \\
 & & & j & 0 & 0 & k & l & 0 & 0 & 0 & -k & -l & 0 & 0 \\
 & & & & m & -n & 0 & 0 & 0 & 0 & n & 0 & 0 & 0 & 0 \\
 & & & & & o & 0 & 0 & 0 & 0 & 0 & 0 & 0 & 0 & 0 \\
 & & & & & & p & q & 0 & 0 & 0 & 0 & 0 & 0 & 0 \\
 & & & & & & & r & 0 & 0 & 0 & 0 & 0 & 0 & 0 \\
 & & & & & & & & s & 0 & 0 & 0 & 0 & 0 & 0 \\
 & & & & & & & & & t & 0 & 0 & 0 & 0 & 0 \\
 & & & & Symm. & & & & & & o & 0 & 0 & 0 & 0 \\
 & & & & & & & & & & & p & q & 0 & 0 \\
 & & & & & & & & & & & & r & 0 & 0 \\
 & & & & & & & & & & & & & s & 0 \\
 & & & & & & & & & & & & & & t
\end{bmatrix}, \tag{I.4}$$

where

$$a = 4c_{sy}\,, \quad b = 2c_{sy}\,, \quad c = 4c_{sz}\,, \quad d = 4h_{cs}c_{sz}\,, \quad e = 2c_{sz}\,, \quad f = 2h_{ts}c_{sz}\,, \quad g = 4d_s^2 c_{sy} + 4h_{cs}^2 c_{sz}\,,$$
$$h = 2h_{cs}c_{sz}\,, \quad i = -2d_s^2 c_{sy} + 2h_{cs}h_{ts}c_{sz}\,, \quad j = 4l_c^2 c_{sz}\,, \quad k = 2l_c c_{sz}\,, \quad l = 2h_{ts}l_c c_{sz}\,, \quad m = 4l_c^2 c_{sy}\,,$$
$$n = 2l_c c_{sy}\,, \quad o = 2c_{sy} + 4c_{py}\,, \quad p = 2c_{sz} + 4c_{pz}\,, \quad q = 2h_{ts}c_{sz} - 4h_{tp}c_{pz}\,,$$
$$r = 2d_s^2 c_{sy} + 2h_{ts}^2 c_{sz} + 4d_p^2 c_{py} + 4h_{tp}^2 c_{pz}\,, \quad s = 4l_t^2 c_{pz}\,, \quad t = 4l_t^2 c_{py}\,, \tag{L5a–t}$$

$$[c_{uw}] = [c_{wu}]^T$$
$$= \begin{bmatrix} 0 & 0 & 0 & 0 & 0 & 0 & 0 & 0 & 0 & 0 & 0 & 0 \\ 0 & 0 & 0 & 0 & 0 & 0 & 0 & 0 & 0 & 0 & 0 & 0 \\ 0 & 0 & 0 & 0 & 0 & 0 & 0 & 0 & 0 & 0 & 0 & 0 \\ 0 & 0 & 0 & 0 & 0 & 0 & 0 & 0 & 0 & 0 & 0 & 0 \\ 0 & 0 & 0 & 0 & 0 & 0 & 0 & 0 & 0 & 0 & 0 & 0 \\ -2c_{py} & 0 & 0 & -2c_{py} & 0 & 0 & 0 & 0 & 0 & 0 & 0 & 0 \\ 0 & -2c_{pz} & 0 & 0 & -2c_{pz} & 0 & 0 & 0 & 0 & 0 & 0 & 0 \\ 0 & 2h_{tp}c_{pz} & -2d_p^2 c_{py} & 0 & 2h_{tp}c_{pz} & -2d_p^2 c_{py} & 0 & 0 & 0 & 0 & 0 & 0 \\ 0 & 2l_t c_{pz} & 0 & 0 & -2l_t c_{pz} & 0 & 0 & 0 & 0 & 0 & 0 & 0 \\ -2l_t c_{py} & 0 & 0 & 2l_t c_{py} & 0 & 0 & 0 & 0 & 0 & 0 & 0 & 0 \\ 0 & 0 & 0 & 0 & 0 & 0 & -2c_{py} & 0 & 0 & -2c_{py} & 0 & 0 \\ 0 & 0 & 0 & 0 & 0 & 0 & 0 & -2c_{pz} & 0 & 0 & -2c_{pz} & 0 \\ 0 & 0 & 0 & 0 & 0 & 0 & 0 & 2h_{tp}c_{pz} & -2d_p^2 c_{py} & 0 & 2h_{tp}c_{pz} & -2d_p^2 c_{py} \\ 0 & 0 & 0 & 0 & 0 & 0 & 0 & 2l_t c_{pz} & 0 & 0 & -2l_t c_{pz} & 0 \\ 0 & 0 & 0 & 0 & 0 & 0 & -2l_t c_{py} & 0 & 0 & 2l_t c_{py} & 0 & 0 \end{bmatrix}, \tag{L.6}$$

$$[c_{ww}] = diag[\,2c_{py} \quad 2c_{pz} \quad 2d_p^2 c_{py} \quad 2c_{py} \quad 2c_{pz} \quad 2d_p^2 c_{py} \quad 2c_{py} \quad 2c_{pz} \quad 2d_p^2 c_{py} \quad 2c_{py} \quad 2c_{pz} \quad 2d_p^2 c_{py}\,]\,. \tag{I.7}$$

The matrix $[k_{uw}] = [k_{wu}]^T$ is identical in form to $[c_{uw}]$ given in Eq. (I.6), except that the terms $c_{py}$, $c_{pz}$ should be replaced by $k_{py}$, $k_{pz}$, respectively. The matrix $[k_{uu}]$ is identical in form to the matrix $[c_{uu}]$ given in Eq. (I.4) except that the nonzero entries are given as follows:

$$\begin{gathered}
a = 4k_{sy}\,, \quad b = 2k_{sy}\,, \quad c = 4k_{sz}\,, \quad d = 4h_{cs}k_{sz}\,, \\
e = 2k_{sz}\,, \quad f = 2h_{ts}k_{sz}\,, \quad g = 4d_s^2 k_{sy} + 4h_{cs}^2 k_{sz}\,, \\
h = 2h_{cs}k_{sz}\,, \quad i = -2d_s^2 k_{sy} + 2h_{cs}h_{ts}k_{sz}\,, \quad j = 4l_c^2 k_{sz}\,, \\
k = 2l_c k_{sz}\,, \quad l = 2h_{ts}l_c k_{sz}\,, \quad m = 4l_c^2 k_{sy}\,, \quad n = 2l_c k_{sy}\,, \\
o = 2k_{sy} + 4k_{py}\,, \quad p = 2k_{sz} + 4k_{pz}\,, \quad q = 2h_{ts}k_{sz} - 4h_{tp}k_{pz}\,, \\
r = 2d_s^2 k_{sy} + 2h_{ts}^2 k_{sz} + 4d_p^2 k_{py} + 4h_{tp}^2 k_{pz}\,, \\
s = 4l_t^2 k_{pz}\,, \quad t = 4l_t^2 k_{py}\,.
\end{gathered} \tag{I.8}$$

The following are the related force vectors:

$$\begin{aligned}
\{f_{ue}\}_{t+\Delta t} &= \{0\}\,, \\
\{f_{we}\}_{t+\Delta t} &= \langle -W \quad -W \quad -W \quad -W\,\rangle^T\,,
\end{aligned} \tag{I.9}$$

where $W$ denotes the weight lumped from the vehicle, $W = (M_c + 2M_t + 4M_w)g/4$.

# Appendix J

# Mass and Stiffness Matrices for Rail and Bridge Elements

## J.1. Mass and Stiffness Matrices of the CFR Element for Both Tracks

$$
[m_{A0}] = [m_{B0}] = \frac{m_t l}{420}
$$

$$
\times \begin{bmatrix}
140 & 0 & 0 & 0 & 0 & 0 & 70 & 0 & 0 & 0 & 0 & 0 \\
 & 156 & 0 & 0 & 0 & 22l & 0 & 54 & 0 & 0 & 0 & -13l \\
 & & 156 & 0 & -22l & 0 & 0 & 0 & 54 & 0 & 13l & 0 \\
 & & & \dfrac{140I_t^*}{m_t} & 0 & 0 & 0 & 0 & 0 & \dfrac{70I_t^*}{m_t} & 0 & 0 \\
 & & & & 4l^2 & 0 & 0 & 0 & -13l & 0 & -3l^2 & 0 \\
 & & & & & 4l^2 & 0 & 13l & 0 & 0 & 0 & -3l^2 \\
 & & & & & & 140 & 0 & 0 & 0 & 0 & 0 \\
 & & & & & & & 156 & 0 & 0 & 0 & -22l \\
 & & & Symm. & & & & & 156 & 0 & 22l & 0 \\
 & & & & & & & & & \dfrac{140I_t^*}{m_t} & 0 & 0 \\
 & & & & & & & & & & 4l^2 & 0 \\
 & & & & & & & & & & & 4l^2
\end{bmatrix}, \tag{J.1}
$$

$$[k_{A0}] = [k_{B0}]$$

$$= \begin{bmatrix}
a & 0 & 0 & 0 & 0 & 0 & -a & 0 & 0 & 0 & 0 & 0 \\
 & b & 0 & 0 & 0 & g & 0 & -b & 0 & 0 & 0 & g \\
 & & c & 0 & -e & 0 & 0 & 0 & -c & 0 & -e & 0 \\
 & & & d & 0 & 0 & 0 & 0 & 0 & 0 & 0 & 0 \\
 & & & & f & 0 & 0 & 0 & e & 0 & i & 0 \\
 & & & & & h & 0 & -g & 0 & 0 & 0 & j \\
 & & & & & & a & 0 & 0 & 0 & 0 & 0 \\
 & & & & & & & b & 0 & 0 & 0 & -g \\
 & & & Symm. & & & & & c & 0 & e & 0 \\
 & & & & & & & & & d & 0 & 0 \\
 & & & & & & & & & & f & 0 \\
 & & & & & & & & & & & h
\end{bmatrix}, \tag{J.2}$$

where

$$a = \frac{E_t A_t}{l}, \quad b = \frac{12 E_t I_{tz}}{l^3}, \quad c = \frac{12 E_t I_{ty}}{l^3}, \quad d = 0,$$

$$e = \frac{6 E_t I_{ty}}{l^2}, \quad f = \frac{4 E_t I_{ty}}{l}, \quad g = \frac{6 E_t I_{tz}}{l^2}, \quad h = \frac{4 E_t I_{tz}}{l}, \tag{J.3}$$

$$i = \frac{2 E_t I_{ty}}{l}, \quad j = \frac{2 E_t I_{tz}}{l}.$$

## J.2. Mass and Stiffness Matrices of the Bridge Element

The mass matrix $[m_{b0}]$ of the bridge element is identical in form to that given in Eq. (J.1) except that the two parameters $m_t$ and $I_t^*$ should be replaced by $m_b$ and $I_b^*$, respectively. In addition, the stiffness matrix $[k_{b0}]$ of the bridge element is identical in form to that of Eq. (J.2) but with the following substitutions:

$$
\begin{aligned}
a &= \frac{E_t A_t}{l}, \quad b = \frac{12E_b I_{bz}}{l^3}, \quad c = \frac{12E_b I_{by}}{l^3}, \quad e = \frac{6E_b I_{by}}{l^2}, \\
d &= \frac{E_b I_{bx}}{2(1+\nu_b)l}, \quad f = \frac{4E_b I_{by}}{l}, \quad g = \frac{6E_b I_{bz}}{l^2}, \\
h &= \frac{4E_b I_{bz}}{l}, \quad i = \frac{2E_b I_{by}}{l}, \quad j = \frac{2E_b I_{bz}}{l}.
\end{aligned} \tag{J.4}
$$

## J.3. Mass and Stiffness Matrices for the LSR Element

$$
[m_{l0}] = \begin{bmatrix}
\frac{m_t}{2\lambda_u} & 0 & 0 & 0 & 0 & 0 \\
 & \frac{3m_t}{4\lambda_v} & 0 & 0 & 0 & -\mathrm{sgn}\frac{m_t}{4\lambda_v^2} \\
 & & \frac{3m_t}{4\lambda_{wv}} & 0 & \mathrm{sgn}\frac{m_t}{4\lambda_w^2} & 0 \\
 & & & \frac{I_t^*}{2\lambda_\theta} & 0 & 0 \\
 & Symm. & & & \frac{m_t}{8\lambda_w^3} & 0 \\
 & & & & & \frac{m_t}{8\lambda_v^3}
\end{bmatrix}, \tag{J.5}
$$

$$
[k_{l0}] = \begin{bmatrix}
\frac{E_t A_t \lambda_u}{2} & 0 & 0 & 0 & 0 & 0 \\
 & E_t I_{tz}\lambda_v^3 & 0 & 0 & 0 & -\mathrm{sgn} E_t I_{tz}\lambda_v^2 \\
 & & E_t I_{ty}\lambda_w^3 & 0 & \mathrm{sgn} E_t I_{ty}\lambda_w^2 & 0 \\
 & & & 0 & 0 & 0 \\
 & Symm. & & & \frac{3}{2}E_t I_{ty}\lambda_w & 0 \\
 & & & & & \frac{3}{2}E_t I_{tz}\lambda_v
\end{bmatrix}, \tag{J.6}
$$

where the sign "sgn" is taken as positive.

## J.4. Mass and Stiffness Matrices of the RSR Element

The mass and stiffness matrices of the RSR element, i.e., $[m_{r0}]$ and $[k_{r0}]$, are identical to those of the LSR element except that the sign "sgn" is taken as negative.

## J.5. Related Matrices and Vectors for the Rail Elements

$$[\psi_u] = \begin{bmatrix} \frac{l}{3} & \frac{l}{6} \\ \frac{l}{6} & \frac{l}{3} \end{bmatrix} (u_1, u_2)\,, \tag{J.7}$$

$$[\psi_v] = \begin{bmatrix} \frac{13l}{35} & \frac{11l^2}{210} & \frac{9l}{70} & \frac{-13l^2}{420} \\ 0 & \frac{l^3}{105} & \frac{13l^2}{420} & \frac{-l^3}{140} \\ 0 & 0 & \frac{13l}{35} & \frac{-11l^2}{210} \\ 0 & 0 & 0 & \frac{l^3}{105} \end{bmatrix} (v_1, \psi_1, v_2, \psi_2)\,, \tag{J.8}$$

$$[\psi_\theta] = \begin{bmatrix} \frac{l}{3} & \frac{l}{6} \\ \frac{l}{6} & \frac{l}{3} \end{bmatrix} (\theta_1, \theta_2)\,, \tag{J.9}$$

$$[\psi_w] = \begin{bmatrix} \frac{13l}{35} & \frac{-11l^2}{210} & \frac{9l}{70} & \frac{13l^2}{420} \\ 0 & \frac{l^3}{105} & \frac{-13l^2}{420} & \frac{-l^3}{140} \\ 0 & 0 & \frac{13l}{35} & \frac{11l^2}{210} \\ 0 & 0 & 0 & \frac{l^3}{105} \end{bmatrix} (w_1, \varphi_1, w_2, \varphi_2)\,, \tag{J.10}$$

$$[\psi_{v\theta}] = \begin{bmatrix} \frac{7l}{20} & \frac{3l}{20} \\ \frac{l^2}{20} & \frac{l^2}{30} \\ \frac{3l}{20} & \frac{7l}{20} \\ \frac{-l^2}{30} & \frac{-l^2}{20} \end{bmatrix} (v_1, \psi_1, v_2, \psi_2) \text{ and } (\theta_1, \theta_2)\,, \tag{J.11}$$

$$[\psi_{w\theta}] = \begin{bmatrix} \frac{7l}{20} & \frac{3l}{20} \\ \frac{-l^2}{20} & \frac{-l^2}{30} \\ \frac{3l}{20} & \frac{7l}{20} \\ \frac{l^2}{30} & \frac{l^2}{20} \end{bmatrix} (w_1, \varphi_1, w_2, \varphi) \text{ and } (\theta_1, \theta_2)\,, \tag{J.12}$$

$$[\psi_u]_l = \left[\frac{1}{2\lambda_u}\right] (u_2)\,, \tag{J.13}$$

$$[\psi_\theta]_l = \left[\frac{1}{2\lambda_\theta}\right] (\theta_2)\,, \tag{J.14}$$

$$[\psi_v]_l = \begin{bmatrix} \frac{3}{4\lambda_v} & -\frac{1}{4\lambda_v^2} \\ -\frac{1}{4\lambda_v^2} & \frac{1}{8\lambda_v^3} \end{bmatrix} (v_2, \psi_2)\,, \tag{J.15}$$

$$[\psi_w]_l = \begin{bmatrix} \frac{3}{4\lambda_w} & \frac{1}{4\lambda_w^2} \\ \frac{1}{4\lambda_w^2} & \frac{1}{8\lambda_w^3} \end{bmatrix} (w_2, \varphi_2)\,, \tag{J.16}$$

$$[\psi_u]_r = \left[\frac{1}{2\lambda_u}\right] (u_1)\,, \tag{J.17}$$

$$[\psi_\theta]_r = \left[ \frac{1}{2\lambda_\theta} \right] (\theta_1) \,, \tag{J.18}$$

$$[\psi_v]_r = \begin{bmatrix} \dfrac{3}{4\lambda_v} & \dfrac{1}{4\lambda_v^2} \\ \dfrac{1}{4\lambda_v^2} & \dfrac{1}{8\lambda_v^3} \end{bmatrix} (v_1, \psi_1) \,, \tag{J.19}$$

$$[\psi_w]_r = \begin{bmatrix} \dfrac{3}{4\lambda_w} & -\dfrac{1}{4\lambda_w^2} \\ -\dfrac{1}{4\lambda_w^2} & \dfrac{1}{8\lambda_w^3} \end{bmatrix} (w_1, \varphi_1) \,. \tag{J.20}$$

The parenthesized symbols in the above expressions indicate the DOFs associated with the row and column entries of the matrix.

# References

Aida, T., Green, R., and Hosogi, Y. (1990). "Dynamic behavior of railway bridges under unsprung masses of a multi-vehicle train," *J. Sound & Vibr.*, **142**(2), 245–260.

Akin, J. E. and Mofid, M. (1989). "Numerical solution for response of beams with moving mass," *J. Struct. Eng.*, ASCE, **115**(1), 120–131.

Au, F. T. K., Wang, J. J., and Cheung, Y. K. (2001a). "Impact study of cable-stayed bridge under railway traffic using various models," *J. Sound & Vibr.*, **240**(3), 447–465.

Au, F. T. K., Cheng, Y. S., and Cheung, Y. K. (2001b). "Effects of random road surface roughness and long-term deflection of prestressed concrete girder and cable-stayed bridges on impact due to moving vehicles," *Comp. & Struct.*, **79**, 853–872.

Ayre, R. S., Ford, G., and Jacobsen, L. S. (1950). "Transverse vibration of a two-span beam under action of a moving constant force," *J. Appl. Mech.*, **17**(1), 1–12.

Ayre, R. S. and Jacobsen, L. S. (1950). "Transverse vibration of a two-span beam under the action of a moving alternating force," *J. Appl. Mech.*, **17**(3), 283–290.

Bhatti, M. H., Garg, V. K., and Chu, K. H. (1985). "Dynamic interaction between freight train and steel bridge," *J. Dyn. Syst., Measurement & Control*, **107**(1), 60–66.

Biggs, J. M. (1964). *Introduction to Structural Dynamics*, McGraw-Hill, New York, N.Y.

Blejwas, T. E., Feng, C. C., and Ayre, R. S. (1979). "Dynamic interaction of moving vehicles and structures," *J. Sound & Vibr.*, **67**(4), 513–521.

Bolotin, V. V. (1964). *The Dynamic Stability of Elastic Systems*, Holden-Day, Inc., San Francisco, California, U.S.A.

Cai, Y., Chen, S. S., Rote, D. M., and Coffey, H. T. (1994). "Vehicle/guideway interaction for high speed vehicles on a flexible guideway," *J. Sound & Vibr.*, **175**(5), 625–646.

Caughey, T. K. (1960). "Classical normal modes in damped linear dynamic systems," *J. Appl. Mech.*, ASME, **27**, 269–271.

Chang, C. H. (1997). *Rigid vehicle–bridge interaction response analysis and impact responses of horizontally curved beams*, Master's thesis, National Taiwan University, Taipei, Taiwan, R.O.C. (in Chinese).

Chang, D. and Lee, H. (1994). "Impact factors for simple-span highway girder bridges," *J. Struct. Eng.*, ASCE, **120**(3), 704–715.

Chatterjee, P. K. and Datta, T. K. (1995). "Dynamic analysis of arch bridges under travelling loads," *Int. J. Solids Struct.*, **32**(11), 1585–1594.

Chatterjee, P. K., Datta, T. K., and Surana, C. S. (1994). "Vibration of suspension bridges under vehicular movement," *J. Struct. Eng.*, ASCE, **120**(3), 681–703.

Chen, G. and Zhai, W. M. (1999). "Numerical simulation of the stochastic process of railway track irregularities," *J. Southwest Jiaotong Univ.*, **34**(3), 138–142 (in Chinese).

Chen, Y. H. (1978). "Dynamic analysis of continuous beams subjected to moving loads," *Chinese J. Civil & Hydraulic Eng.*, **5**(2), 1–7 (in Chinese).

Chen, Y. H. and Li, C. Y. (2000). "Dynamic response of elevated high-speed railway," *J. Bridge Eng.*, ASCE, **5**(2), 124–130.

Cherchas, D. B. (1979). "A dynamic simulation for a high speed magnetically levitated guided ground vehicle," *J. Dyn. Syst. Measurement & Control*, **101**(3), 223–229.

Cheung, Y. K., Au, F. T. K., Zheng, D. Y., and Cheng, Y. S. (1999). "Vibration of multi-span non-uniform bridges under moving vehicles and trains by using modified beam vibration functions," *J. Sound & Vibr.*, **228**(3), 611–628.

Chu, K. H., Garg, V. K., and Dhar, C. L. (1979). "Railway–bridge impact: Simplified train and bridge model," *J. Struct. Div.*, ASCE, **105**(9), 1823–1844.

Chu, K. H., Garg, V. K., and Wang, T. L. (1986). "Impact in railway prestressed concrete bridges," *J. Struct. Eng.*, ASCE, **112**(5), 1036–1051.

Claus, H. and Schiehlen, W. (1998). "Modelling and simulation of railway bogie structural vibrations," *Vehicle Syst. Dyn. Supplement*, **28**, 538–552.

Coussy, O., Said, M., and van Hoove, J. P. (1989). "The influence of random surface irregularities on the dynamic response of bridges under suspended moving loads," *J. Sound & Vibr.*, **130**(2), 313–320.

Delfosse, P. (1991). "Very-high-speed tests on the French railway track," *Vehicle Syst. Dyn.*, **20**, 195–206.

Delgado, R. M. and dos Santos, S. M. (1997). "Modelling of railway bridge–vehicle interaction on high speed tracks," *Comp. & Struct.*, **63**(3), 511–523.

Deutsche, B. (1993). DS804 — Vorschrift fur Eisenbahnbrucken und sonstige Ingenieurbauwerke (VEI).

Diana, G. and Cheli, F. (1989). "Dynamic interaction of railway systems with large bridges," *Vehicle Syst. Dyn.*, **18**, 71–106.

Diana, G., Cheli, F., and Bruni, S. (2000). "Railway runnability and train–track interaction in long span cable supported bridges," *Advances in Struct. Dyn.*, Vol. 1, eds. Ko, J. M. and Xu, Y. L., Elsevier, Amsterdam, 43–54.

Dodds, C. J. and Robson, J. D. (1973). "The description of road surface roughness," *J. Sound & Vibr.*, **31**(2), 175–183.

Dukkipati, R. V. and Dong, R. (1999). "Idealized steady state interaction between railway vehicle and track," *Proc. Inst. Mech. Engrs. Part F, J. Rail & Rapid Transit*, **213**(2), 15–29.

Dugush, Y. A. and Eisenberger, M. (2002). "Vibrations of non-uniform continuous beams under moving loads," *J. Sound & Vibr.*, **254**(5), 911–926.

Elkins, J. A. and Carter, A. (1993). "Testing and analysis techniques for safety assessment of rail vehicles: The sate-of-the-art," *Vehicle Syst. Dyn.*, **22**, 185–208.

Esveld, G. (1989). *Modern Railway Track*, MRT Productions, Duisburg, Germany.

European Committee for Standardization (1995). EUROCODE 1: Basis of design and actions on structures, Part 3: Traffic loads on bridges, ENV 1991-3.

Fertis, D. G. (1973). *Dynamics and Vibration of Structures*, John Wiley & Sons, New York, N.Y.

Fries, R. H. and Coffey, B. M. (1990). "A state-space approach to the synthesis of random vertical and crosslevel rail irregularities," *J. Dyn. Syst., Measurement, and Control*, **112**(1), 83–87.

Frýba, L. (1972). *Vibration of Solids and Structures under Moving Loads*, Noordhoff International Publishing, Groningen, The Netherlands.

Frýba, L. (1996). *Dynamics of Railway Bridges*, Thomas Telford, London, England.

Frýba, L. (2001). "A rough assessment of railway bridges for high speed trains," *Eng. Struct.*, **23**(5), 548–556.

Galdos, N. H., Schelling, D. R., and Sahin, M. A. (1993). "Methodology for impact factor of horizontally curved box bridges," *J. Struct. Eng.*, ASCE, **119**(6), 1917–1934.

Garg, V. K. and Dukkipati, R. V. (1984). *Dynamics of Railway Vehicle Systems*, Academic Press, New York, N.Y.

Gbadeyan, J. A. and Oni, S. T. (1995). "Dynamic behaviour of beams and rectangular plates under moving loads," *J. Sound & Vibr.*, **182**(5), 677–695.

Genin, J. and Chung, Y. I. (1979). "Response of a continuous guideway on equally spaced supports traversed by a moving vehicle," *J. Sound & Vibr.*, **67**, 245–251.

Genin, J., Ginsberg, J. H., and Ting, E. C. (1975). "A complete formulation of inertial effects in the guideway–vehicle interaction problem," *J. Sound & Vibr.*, **38**(1), 15–26.

Genin, J. and Ting, E. C. (1979). "Vehicle–guideway interaction problems," *The Shock & Vibr. Dig.*, **11**(12), 3–9.

Genin, J., Ting, E. C., and Vafa, Z. (1982). "Curved beam response due to a moving vehicle," *J. Sound & Vibr.*, **65**, 569–575.

Gere, J. M. and Timoshenko, S. P. (1990). *Mechanics of Materials*, PWS-KENT, Boston.

Gorman, D. J. (1975). *Free Vibration Analysis of Beams and Shafts*, John Wiley & Sons, N.Y.

Grandil, J. and Ramondenc, P. (1990). *The Dynamic Behavior of Railways on High Speed Lines*, SNCF, France.

Green, M. F. and Cebon, D. (1994). "Dynamic response of highway bridges to heavy vehicle loads: Theory and experimental validation," *J. Sound & Vibr.*, **170**(1), 51–78.

*Guide Specifications for Horizontally Curved Highway Bridges* (1980). American Association for State Highway and Transportation Officials (AASHTO), Washington, D.C.

Guo, W. H. and Xu, Y. L. (2001). "Fully computerized approach to study cable-stayed bridge–vehicle interaction," *J. Sound & Vibr.*, **248**(4), 745–761.

Gupta, R. K. (1980). "Dynamic loading of highway bridges," *J. Eng. Mech. Div.*, ASCE, **106**(2), 337–393.

Guyan, R. J. (1965). "Reduction of stiffness and mass matrices," *AIAA J.*, **3**(2), 380.

Henchi, K., Fafard, M., Talbot, M., and Dhatt, G. (1998). "An efficient algorithm for dynamic analysis of bridges under moving vehicles using a coupled modal and physical components approach," *J. Sound & Vibr.*, **212**(4), 663–683.

Hetényi, M. (1979). *Beams on Elastic Foundation*, The University of Michigan Press, Ann Arbor.

Hsu, L. C. (1996). *Impact responses of bridges traveled by high speed trains*, Master's thesis, National Taiwan University, Taipei, Taiwan, R.O.C. (in Chinese).

Huang, D., Wang, T. L., and Shahawy, M. (1993). "Impact studies on multi-girder concrete bridges," *J. Struct. Eng.*, ASCE, **119**(8), 2387–2402.

Humar, J. L. and Kashif, A. M. (1993). "Dynamic response of bridges under travelling loads," *Can. J. Civ. Eng.*, **20**, 287–298.

Hwang, E. S. and Nowak, A. S. (1991). "Simulation of dynamic load for bridges." *J. Struct. Eng.*, ASCE, **117**(5), 1413–1434.

Ichikawa, M., Miyakawa, Y., and Matsuda, A. (2000). "Vibration analysis of the continuous beam subjected to a moving mass," *J. Sound & Vibr.*, **230**, 492–506.

Inbanathan, M. J. and Wieland, M. (1987). "Bridge vibrations due to vehicle moving over rough surface," *J. Struct. Eng.*, ASCE, **113**(9), 1994–2008.

Inglis, C. E. (1934). *A Mathematical Treatise on Vibration in Railway Bridges*, The University Press, Cambridge, England.

Iwnick, S. (1998). "Manchester benchmarks for rail vehicle simulation," *Vehicle Syst. Dyn.*, **30**, 295–313.

Jeffcott, H. H. (1929). "On the vibrations of beams under the action of moving loads," *Phil. Magazine*, Ser. 7, **8**(48), 66–97.

Ju, S. H. and Lin, H. T. (2003). "Numerical investigation of a steel arch bridge and interaction with high speed trains," *Eng. Struct.*, **25**, 241–250.

Kawatani, M. and Kim, C. W. (2001). "Computer simulation for dynamic wheel loads of heavy vehicles," *Struct. Eng. & Mech.*, **12**(4), 409–428.

Kim, S. and Nowak, A. S. (1997). "Load distribution and impact factors for I-girders bridges," *J. Bridge Eng.*, ASCE, **2**(3), 97–104.

Kou, J. W. and DeWolf, J. T. (1997). "Vibrational behavior of continuous span highway bridge — Influencing variables," *J. Struct. Eng.*, ASCE, **123**(3), 333–344.

Kurihara, M. and Shimogo, T. (1978a). "Stability of a simply-supported beam subjected to random spaced moving loads," *J. Mech. Design*, ASME, **100**(7), 507–513.

Kurihara, M. and Shimogo, T. (1978b). "Vibration of an elastic beam subjected to discrete moving loads," *J. Mech. Design*, ASME, **100**(7), 514–519.

Lin, C. L. (2001). *Reduction in vibration of isolated railway bridges*, Master's thesis, National Taiwan University, Taipei, Taiwan, R.O.C. (in Chinese).

Li, J. Z. and Su, M. B. (1999). "The resonant vibration for a simply supported girder bridge under high-speed trains," *J. Sound & Vibr.*, **224**(5), 897–915.

Lin, B. H., Yau, J. D., and Yang, Y. B. (2000). "Impact factors for simple beams subjected to moving loads," *Struct. Eng.*, **15**(2), 3–20 (in Chinese).

Lowan, A. N. (1935). "On transverse oscillations of beams under the action of moving variable loads," *Phil. Magazine*, Ser. 7, **19**(127), 708–715.

Ma, K. Q. and Zhu, J. L. (1998). "Seismic responses analysis for a system of high-speed train and continuous rigid-frame bridge," *J. Shanghai Tiedao Univ.*, **19**(10), 29–36 (in Chinese).

Mangulis, V. (1965). *Handbook of Series for Scientists and Engineers*, Academic Press, N.Y.

Marcondes, J., Burgess, G. J., Harichandran, R., and Snyder, M. B. (1991). "Spectral analysis of highway pavement roughness," *J. Transp. Eng.*, ASCE, **117**(5), 540–549.

Marchesiello, S., Fasana, A., Garibaldi, L., and Piombo, B. A. D. (1999). "Dynamics of multi-span continuous straight bridges subject to multi-degrees of freedom moving vehicle excitation," *J. Sound & Vibr.*, **224**(3), 541–561.

Matsuura, A. (1976). "A study of dynamic behavior of bridge girder for high speed railway," *Proc. JSCE*, **256**, 35–47 (in Japanese).

McGuire, W., Gallagher, R. H., and Ziemian, R. D. (2000). *Matrix Structural Analysis*, 2nd edn., John Wiley & Sons, New York, N.Y.

Meisenholder, S. G. and Weidlinger, P. (1974). "Dynamic interaction aspects of cable-stayed guideways for high speed ground transportation," *J. Dyn. Syst., Measurement, & Control*, ASME, **74**-Aut-R, 180–192.

Miura, S. (1996). "Deformation of track and safety of train in earthquake," *Quarterly of RTRI*, **37**(3), 139–147.

Miyamoto, T., Ishida, H., and Matsuo, M. (1997). "Running safety of railway vehicle as earthquake occurs," *Quarterly of RTRI*, **38**(3), 117–122.

Museros, P., Romero, M. L., Poy, A., and Alarcón, E. (2002). "Advances in the analysis of short span railway bridges for high-speed lines," *Comp. & Struct.*, **80**, 2121–2132.

Nassif, H. H. and Nowak, A. S. (1995). "Dynamic load spectra for girder bridges," *Transp. Res. Rec. 1476*, Transp. Res. Board, Nat. Res. Council, Washington, D.C., 69–83.

Newmark, N. M. (1959). "A method of computation for structural dynamics," *J. Eng. Mech. Div.*, ASCE, **85**(3), 67–94.

Nielsen, J. C. O. and Abrahamsson, T. J. S. (1992). "Coupling of physical and modal components for analysis of moving non-linear dynamic systems on general beam structures," *Int. J. Num. Meth. Eng.*, **33**, 1843–1859.

O'Connor, C. and Pritchard, R. W. (1984). "Impact studies on small composite girder bridge," *J. Struct. Eng.*, ASCE, **111**(3), 641–653.

*Ontario Highway Bridge Design Code.* (1983). Ministry of Transportation and Communication, Downsview, Ontario, Canada.

Pan, T. C. and Li, J. (2002). "Dynamic vehicle element method for transient response of coupled vehicle–structure systems," *J. Struct. Eng.*, ASCE, **128**(2), 214–223.

Paz, M. (1991). *Structural Dynamics — Theory and Computation*, Van Nostrand Reinhold, New York, N.Y.

Paultre, P., Chaallal, O., and Proulx, J. (1992). "Bridge dynamics and dynamic amplification factors — A review of analytical experimental findings," *Can. J. Civ. Eng.*, **19**, 260–278.

Pesterev, A. V., Yang, B., Bergman, L. A., and Tan, C. A. (2001). "Response of elastic continuum carrying multiple moving oscillators," *J. Eng. Mech.*, ASCE, **127**(3), 260–265.

Pesterev, A. V., Bergman, L. A., Tan, C. A., Tsao, T. C., and Yang, B. (2003). "On asymptotics of the solution of the moving oscillator problem," *J. Sound & Vibr.*, **260**, 519–536.

Rao, G. V. (2000). "Linear dynamics of an elastic beam under moving loads," *J. Vibr. & Acous.*, ASME, **122**(7), 281–289.

Richardson, H. H. and Wormley, D. N. (1974). "Transportation vehicle/beam-elevated guideway dynamic interactions: A state-of-the-art review," *J. Dyn. Syst. Measurement & Control*, **96**(2), 169–179.

Sadiku, S. and Leipholz, H. H. E. (1987). "On the dynamics of elastic systems with moving concentrated masses," *Ingenieur-Archiv*, **57**, 223–242.

Savin, E. (2001). "Dynamic amplification factor and response spectrum for the evaluation of vibrations of beams under successive moving loads," *J. Sound & Vibr.*, **248**, 267–288.

Schupp, G. and Jaschinski, A. (1996). "Investigation of running behaviour and derailment criteria for freight cars by dynamic simulation," *COMPRAIL 96, Computers in Railways V*, **2**, Computational Mechanics Publications, Southampton, UK, 13–22.

Shen, R. (1996). "Study of analytic methods for vertical vibration of high-speed railway girder bridge," *Bridge Construction*, **2**, 46–49 (in Chinese).

Senthilvasan, J., Thambiratnam, D. P., and Brameld, G. H. (2002). "Dynamic response of a curved bridge under moving truck load," *Eng. Struct.*, **24**, 1283–1293.

Smith, C. C., Gilchrist, A. J., and Wormley, D. N. (1975). "Multiple and continuous span elevated guideway–vehicle dynamic performance," *J. Dyn. Syst., Measurement & Control*, **97**(1), 30–39.

Sridharan, N. and Mallik, A. K. (1979). "Numerical analysis of vibration of beams subjected to moving loads," *J. Sound & Vibr.*, **65**, 147–150.

*Standard Specifications for Highway Bridges.* (1989). 14th edn., American Association of State Highway and Transportation Officials (AASHTO), Washington, D.C.

Stanišič, M. M. (1985). "On a new theory of the dynamic behavior of the structures carrying moving masses," *Ingenieur-Archiv*, **55**, 176–185.

Stanišić, M. M. and Hardin, J. C. (1969). "On the response of beams to an arbitrary number of concentrated moving masses," *J. Franklin Inst.*, **287**, 115–123.

Stokes, G. G. (1849). "Discussion of a differential equation relating to the breaking of railway bridges," *Trans. Cambridge Phil. Soc.*, **8**(5), 707–735.

Taheri, M. R., Ting, E. C., and Kukreti, A. R. (1990). "Vehicle–guideway interactions: A literature review," *The Shock & Vibr. Dig.*, **22**(6), 3–9.

Tan, C. P. and Shore, S. (1968a). "Dynamic response of a horizontally curved bridge," *J. Struct. Div.*, ASCE, **94**(3), 761–781.

Tan, C. P. and Shore, S. (1968b). "Response of horizontally curved bridges to moving load." *J. Struct. Div.*, ASCE, **94**(9), 2135–2151.

Tan, G. H., Brameld, G. H., and Thambiratnam, D. P. (1998). "Development of analytical model for treating bridge–vehicle interaction," *Eng. Struct.*, **20**(1), 54–61.

Timoshenko, S. P. (1922). "On the forced vibrations of bridges," *Phil. Magazine*, Ser. 6, **43**, 1018–1019.

Timoshenko, S. P. and Young, D. H. (1955). *Vibration Problems in Engineering*, 3rd edn., D. Van Nostrand, New York, N.Y.

Ting, E. C. and Genin, J. (1980). "Dynamics of bridge structures," *SM Archives*, **5**(3), 217–252.

Ting, E. C., Genin, J., and Ginsberg, J. H. (1974). "A general algorithm for the moving mass problem," *J. Sound & Vibr.*, **33**(1), 49–58.

Ting, E. C., Genin, J., and Ginsberg, J. H. (1975). "Dynamic interaction of bridge structures and vehicles," *The Shock & Vibr. Dig.*, **7**(11), 21–29.

Ting, E. C. and Yener, M. (1983). "Vehicle–structure interactions in bridge dynamics," *The Shock & Vibr. Dig.*, **15**(12), 3–9.

*T. Y. Lin Taiwan* (1993). "Vibrations of elevated bridge structures caused by train loadings of the west Taiwan High Speed Rail Project — Supplementary final report," *Provisional Eng. Office of High Speed Rail*, Ministry of Transportation and Communications, Taiwan, R.O.C.

Veletsos, A. S. and Huang, T. (1970). "Analysis of dynamic response of highway bridges," *J. Eng. Mech. Div.*, ASCE, **96**(5), 593–620.

Vellozzi, J. (1967). "Vibration of suspension bridges under moving loads," *J. Struct. Div.*, ASCE, **93**(4), 123–138.

Wakui, H., Matsumoto, N., Matsuura, A., and Tanabe, M. (1995). "Dynamic interaction analysis for railway vehicles and structures," *J. Struct. Mech. & Earthquake Eng.*, JSCE, **513**, 129–138 (in Japanese).

Wang, R. T. (1997). "Vibration of multi-span Timoshenko beams to a moving force," *J. Sound & Vibr.*, **207**(5), 731–742.

Wang, T. L., Garg, V. K., and Chu, K. H. (1991). "Railway bridge/vehicle interaction studies with new vehicle model," *J. Struct. Eng.*, ASCE, **117**(7), 2099–2116.

Wang, T. L. and Huang, D. (1992). "Cable-stayed bridge vibration due to road surface roughness," *J. Struct. Eng.*, ASCE, **118**(5), 1354–1374.

Wang, T. L., Huang, D., and Shahawy, M. (1992). "Dynamic response of multigirder bridges," *J. Struct. Eng.*, ASCE, **118**(8), 2222–2238.

Warburton, G. B. (1976). *The Dynamical Behaviour of Structures*, Pergamon, Oxford, England.

Weaver, W. and Johnston, P. R. (1987). *Structural Dynamics by Finite Elements*, Prentice-Hall, Englewood Cliffs, N.J.

Weaver, W., Timoshenko, S. P., and Young, D. H. (1990). *Vibration Problems in Engineering*, 5th edn., John Wiley & Sons, New York, N.Y.

Wen, R. K. (1960). "Dynamic response of beams traversed by two-axle loads," *J. Eng. Mech. Div.*, ASCE, **86**(EM5), 91–111.

White, R. C., Limbert, D. A., Hedrick, J. K., and Cooperrider, N. K. (1978) "Guideway-suspension tradeoffs in rail vehicle systems," Arizona State University Engineering Research Center Report ERC-R-78035.

Willis, R. (1849). "Appendix to the report of the commissioners appointed to inquire into the application of iron to railway structures," *H. M. Stationary Office*, London, England.

Wiriyachai, A., Chu, K. H., and Garg, V. K. (1982). "Bridge impact due to wheel and track irregularities," *J. Eng. Mech.*, ASCE, **108**(4), 648–666.

Wu, C. M. (1998). *Dynamic response of curved bridges due to high speed trains and vibration of continuous bridges*, Master's thesis, National Taiwan University, Taipei, Taiwan, R.O.C.

Wu, J. S. and Dai, C. W. (1987). "Dynamic responses of multispan nonuniform beam due to moving loads," *J. Struct. Eng.*, ASCE, **113**(3), 458–474.

Wu, J. S., Lee, M. L., and Lai, T. S. (1987). "The dynamic analysis of a flat plate under a moving load by the finite element method," *Int. J. Num. Meth. Eng.*, **24**, 743–762.

Wu, Y. S. (2000). *Dynamic interaction of train–rail–bridge system under normal and seismic conditions*, Ph.D. thesis, National Taiwan University, Taipei, Taiwan, R.O.C.

Wu, Y. S. and Yang, Y. B. (2003). "Steady-state response and riding comfort of trains moving over a series of simply supported bridges," *Eng. Struct.*, **25**(2), 251–265.

Wu, Y. S., Yang, Y. B., and Yau, J. D. (2001). "Three-dimensional analysis of train–rail–bridge interaction problems," *Vehicle Syst. Dyn.*, **36**(1), 1–35.

Xia, H., Xu, Y. L., and Chan, T. H. T. (2000). "Dynamic interaction of long suspension bridges with running trains," *J. Sound & Vibr.*, **237**(2), 263–280.

Xu, Y. L., Xia, H., and Yan, Q. S. (2003). "Dynamic response of suspension bridge to high wind and running train," *J. Bridge Eng.*, ASCE, **8**(1), 46–55.

Yang, F. and Fonder, G. A. (1996). "An iterative solution method for dynamic response of bridge–vehicles systems," *Earthquake Eng. Struct. Dyn.*, **25**, 195–215.

Yang, F. and Fonder, G. A. (1998). "Dynamic response of cable-stayed bridges under moving loads," *J. Eng. Mech.*, ASCE, **214**(7), 741–747.

Yang, Y. B., Chang, C. H., and Yau, J. D. (1999). "An element for analysing vehicle–bridge systems considering vehicle's pitching effect," *Int. J. Num. Meth. Eng.*, **46**, 1031–1047.

Yang, Y. B. and Kuo, S. R. (1987). "Effect of curvature on stability of curved beams," *J. Struct. Eng.*, ASCE, **113**(6), 1185–1202.

Yang, Y. B. and Kuo, S. R. (1994). *Theory and Analysis of Nonlinear Framed Structures*, Prentice-Hall, Englewood Cliffs, New Jersey.

Yang, Y. B., Liao, S. S., and Lin B. H. (1995). "Impact formulas for vehicles moving over simple and continuous beams," *J. Struct. Eng.*, ASCE, **121**(11), 1644–1650.

Yang, Y. B., Liao, S. S., and Lin, B. H. (1997a). Closure to "impact formulas for vehicles moving over simple and continuous beams," Proc. No. 08917, *J. Struct. Eng.*, ASCE, **123**(4), 533–534.

Yang, Y. B. and Lin, B. H. (1995). "Vehicle–bridge interaction analysis by dynamic condensation method," *J. Struct. Eng.*, ASCE, **121**(11), 1636–1643.

Yang, Y. B., Lin, C. L., Yau, J. D., and Chang, D. W. (2004). "Mechanism of resonance and cancellation for train-induced vibrations on bridges with elastic bearings," *J. Sound & Vibr.*, **269**(1–2), 345–360.

Yang, Y. B., Lin, C. W., and Hsu, T. P. (1996). "Impact responses for bridges subject to random traffic flows," *Proc. 1996 Asia-Pacific Regional Meeting*, eds. Chou, C. P. and Chang, S. K., Taipei, R.O.C., **4**, 43–49.

Yang, Y. B., Wu, C. M., and Yau, J. D. (2001). "Dynamic response of a horizontally curved beam subjected to vertical and horizontal moving loads," *J. Sound & Vibr.*, **242**(3), 519–537.

Yang, Y. B. and Wu, Y. S. (2001). "A versatile element for analysing vehicle–bridge interaction response," *Eng. Struct.*, **23**, 452–469.

Yang, Y. B. and Wu, Y. S. (2002). "Dynamic stability of trains moving over bridges shaken by earthquakes," *J. Sound & Vibr.*, **258**(1), 65–94.

Yang, Y. B. and Yau, J. D. (1997). "Vehicle–bridge interaction element for dynamic analysis," *J. Struct. Eng.*, ASCE, **123**(11), 1512–1518 (Errata: **124**(4), 479).

Yang, Y. B. and Yau, J. D. (1998). "Vehicle–bridge interactions and applications to high speed rail bridges," *Proc. Natl. Science Coun. R.O.C. (A)*, **22**(2), 214–224.

Yang, Y. B., Yau, J. D., and Hsu, L. C. (1997b). "Vibration of simple beams due to trains moving at high speeds," *Eng. Struct.*, **19**(11), 936–944.

Yau, J. D. (1996). *Dynamic response of bridges travelled by trains — Analytical and numerical approaches*, Ph.D. thesis, National Taiwan University, Taipei, Taiwan, R.O.C.

Yau, J. D., Wu, Y. S., and Yang, Y. B. (2001). "Impact response of bridges with elastic bearings to moving loads," *J. Sound & Vibr.*, **248**(1), 9–30.

Yau, J. D. and Yang, Y. B. (1998). "Effect of axle interactions on the dynamic response of the simple beam and vehicles moving over it," *Proc. 6th East Asia-Pacific Conf. Struct. Eng. & Construction*, **2**, National Taiwan University, Taipei, Taiwan, R.O.C., January, 1437–1442.

Yau, J. D. and Yang, Y. B. (2004). "Vibration reduction for cable-stayed bridges traveled by high speed trains," *Fin. Elem. Analysis & Design*, **40**, 341–359.

Yau, J. D., Yang, Y. B., and Kuo, S. R. (1999). "Impact response of high speed rail bridges and riding comfort of rail cars," *Eng. Struct.*, **21**(9), 836–844.

Zhang, Q. L., Vrouwenvelder, A., and Wardenier, J. (2001a). "Numerical simulation of train-bridge interactive dynamics," *Comp. & Struct.*, **79**, 1059–1075.

Zhang, Q. L., Vrouwenvelder, A., and Wardenier, J. (2001b). "Dynamic amplification factors and EUDL of bridges under random traffic flows," *Eng. Struct.*, **23**, 663–672.

Zheng, D. Y., Cheung, Y. K., Au, F. T. K., and Cheng, Y. S. (1998). "Vibration of multi-span non-uniform bridges under moving loads by using modified beam vibration functions," *J. Sound & Vibr.*, **212**, 455–467.

# Subject Index